ENCYCLOPEDIA OF SOUTH AMERICAN AQUATIC INSECTS:
EPHEMEROPTERA

Encyclopedia of South American Aquatic Insects: Ephemeroptera

Illustrated Keys to Known Families, Genera, and Species in South America

by

Charles W. Heckman

Forestry Sciences Laboratory

KLUWER ACADEMIC PUBLISHERS

DORDRECHT / BOSTON / LONDON

A C.I.P. Catalogue record for this book is available from the Library of Congress.

ISBN 1-4020-0775-2

Published by Kluwer Academic Publishers,
P.O. Box 17, 3300 AA Dordrecht, The Netherlands.

Sold and distributed in North, Central and South America
by Kluwer Academic Publishers,
101 Philip Drive, Norwell, MA 02061, U.S.A.

In all other countries, sold and distributed
by Kluwer Academic Publishers,
P.O. Box 322, 3300 AH Dordrecht, The Netherlands.

Cover: One of the earliest color illustrations of a South American mayfly:
Siphlonella guttta, as illustrated by Pictet (1843-45)

Printed on acid-free paper

Printed in the Netherlands

This book is dedicated to my mother,

Ruth A. Heckman

for her support that permitted the completion of this volume.

Contents

Introduction

This work was begun to provide keys to the aquatic insect species known from Brazil. The original goal was to include all genera known from South America and all species from Brazil, but for most groups, the scope was expanded to encompass all species in South America, and, in some cases, to include terrestrial species of orders comprising both terrestrial and aquatic taxa. In no case is a taxonomic revision of any group undertaken, although recommendations for such revisions are included where appropriate, and probable synonymy of nominal species still treated as valid in the literature is noted.

Two different approaches will be employed according to the taxon being treated. For phylogenetic groups encompassing overwhelmingly or exclusively aquatic species, such as the orders Plecoptera and Ephemeroptera or the families Dytiscidae and Culicidae, keys are provided to distinguish all genera and species known to occur in South America. An effort has been made to include every identifiable species so that the user of the key can determine with reasonable certainty whether or not his specimen belongs to a species that has already been described or whether it is one that is not yet known to science. Where feasible, complete keys will be prepared for groups containing both aquatic and terrestrial species that do not encompass an extraordinarily large number of species. This has already been done for the order Collembola.

The second approach will be used for aquatic species belonging to predominantly terrestrial taxa, such as the order Lepidoptera or the families Curculionidae and Muscidae. In such cases, the number of terrestrial species involved is too great to deal with conveniently. For example, a work confined to aquatic insects cannot include a key to all of the approximately 45,000 weevil species (O'Brien and Wibmer, 1978) or even the portion of its enormous fauna inhabiting South America because only a small percentage of them live in or on aquatic plants. However, without such a comprehensive key, it becomes difficult for a person not specializing in the taxonomy of this group to be sure in which genus or subfamily his specimen belongs. For aquatic species in such groups, a descriptive approach is used. An attempt has been made to describe the morphology completely enough for the non-specialist to recognize his aquatic species. At the same time, descriptions or keys to higher taxa within these groups are provided as necessary so that the aquatic species can be distinguished from the terrestrial ones and phylogenetic affinities of the specimens can be recognized.

The geographical limitations of this work are not strictly maintained throughout. For some genera, keys are provided that include mention or descriptions of species still known only from Central America or Caribbean islands, while for others, the keys are strictly limited to the South American fauna. The choice of comprehensiveness was dictated mainly by convenience. For genera encompassing few species and those that have been revised by competent taxonomists, keys to most or all Neotropical species could easily be provided. In the cases of other genera encompassing poorly described species, those which have not been reported south of Panama and the Antilles were not included in the keys.

In general, the ranges of the aquatic insect species are very poorly known. The reported occurrences indicate more the locations at which entomologists have worked than the actual distribution of the species. The type specimens of a great many South American species were collected during expeditions to individual regions of the continent, and systematic surveys of the fauna have only been undertaken for a few groups. Among the prominent early collectors was Charles Darwin. Larger expeditions were undertaken during the late 19th and early 20th century. Considerable numbers of species have been described after examination of the specimens collected in Patagonia and South Chile during the early part of the 20th century. In Brazil, Rio de Janeiro and Santa Catarina appear very frequently in the distribution reports due to the extensive collection in those states by F. Plaumann, while many other species were described from Amazonas because of the presence of a large research station, the Instituto Nacional de Pesquisas da Amazônia, at Manaus. Naturally, a great many species were described from specimens collected at São Paulo and Rio de Janeiro, where most of the Brazilian institutes of research are located. Reports from other Brazilian states are generally much rarer, so the ranges of Brazilian species can generally be expected to be much more extensive than indicated by the published reports. Notable research efforts in other parts of South America have resulted in large numbers of species being recorded for Surinam and the Amazon region of Peru. Knowledge of the fauna in other regions varies considerably from taxon to taxon, depending upon the efforts of talented individuals specializing in individual orders or families. Generally, the insect taxa with the greatest impact on public health and agriculture are best known.

For some rarer species in neglected taxa, the ranges are scarcely defined at all. Specimens described during the last century were sometimes reported from "Brazil" or even "South America" without any more exact collecting data. The fact that the descriptions of such species are generally sketchy, at best, makes it doubtful whether they can be recognized again. There is mention in many of the review papers cited in this work about the confusion caused by falsely labelled specimens, incomplete collecting data, and doubt among the collectors about the names of the places from which their specimens originated. Names such as "Chapada" in Brazil frequently appear without mention of which Chapada is meant. This problem will continue to cause confusion in the future as it has in the past.

The problems encountered in systematic biology worldwide are especially pronounced in the Neotropical Region. This is due to the dearth of competent taxonomic studies on the regional fauna. Theoretically, the scientist who needs to know the correct name of a specimen can accomplish this by a procedure that assures a high probability of success, although a good deal of time and expense may be required. The first step in this procedure is the use of a comprehensive key treating the higher taxon to which the specimen belongs. This process is much easier for a specialist familiar with the group than for a person who needs to know the identity of species being used in physiological experiments or surveyed in synecological studies. In most cases, after the specimens have been identified using the keys, original or revised descriptions listed by the author of the key should be consulted to confirm the identifications. If the available descriptions are so poor that the identity of the species remains uncertain, a comprehensive taxonomic review of the taxon it belongs to must be consulted. If such a review has not yet been published, the specimen to be identified must be compared with type specimens of each species belonging to the higher taxon to which the specimen belongs. These should be found in museum collections.

According to taxonomic convention, the author of a published original description of a new species should designate one of his specimens as a holotype. This type then becomes the standard on which the identifications of all specimens collected in the future are based. The species to which the holotype belongs is then referred to by the name proposed by the author, at least when the same name has not previously been applied to another species.

In addition to the holotype, other specimens believed to be of the same species by the author are designated as paratypes. If a holotype is not available, a paratype can be examined to confirm identifications. If the holotype has been lost or extensively damaged, a specialist revising the taxonomy of the group may designate a lectotype apparently belonging to the species in question. This then replaces the holotype as a standard for recognizing the species, at least until the lost holotype or paratype is found.

Anyone wishing to take the trouble to identify a specimen beyond the shadow of a reasonable doubt should be able to follow this procedure. If his specimen is not the same species as any of the holotypes preserved in museums, he should prepare a description of his specimen and have it published with his proposed name for the new species or give it to a specialist who is interested in doing this.

Unfortunately, the procedures described above often fail to work, especially when South American insects are involved. First of all, keys to identify the species are seldom available, and when they are, they are generally out of date or incomplete for the region being studied. This leaves the researcher with the chore of collecting a large number of original descriptions to match with his specimens. These descriptions are sometimes very sketchy, leaving the reader with no reliable way of identifying his specimen from the available literature. Comparisons with type specimens may not only be difficult in many cases because the types were deposited in museums on other continents, they are often impossible because some authors have failed to mention the name of the museum in which they intended to place their type specimens or because the specimen cannot be found again due to war damage or sloppiness by the curators of the museum in question. Even when the type can be found, it often proves to be in very poor condition, and relatively few paratypes of South American species have been designated that can be examined in the absence of a usable holotype.

Using established procedures for identifying South American insects is made even more problematic by the practice of some authors of describing new species on the basis of only one life stage. Not only have many species been described by taxonomists who examined only adults, some are known only from larvae. Furthermore, countless species are known only from adults of one sex. While the description of an adult may provide a legitimate basis for establishing a new species, it is difficult to find justification for naming a species based only on larvae. Some authors have designated larvae as the types of species congeneric with others known only as adults. These individuals seemed to be in such a hurry to publish that they deliberately left the arduous task of matching larvae with adults to other researchers.

In practice, convention should dictate which gender and life stage should be chosen as the holotype. For example, adult male chironomids are presently those on which species descriptions should be based. Earlier descriptions of female adults are generally useless for determining a species unless someone has taken the trouble to match the female to a described male. In an ideal system of nomenclature, descriptions of adults of both sexes and of the larval stages should be provided. However, the state of the art still requires specimens to be identified according to partial descriptions of one stage and often one sex. Therefore, before fully workable systems for identifying South American species will be possible, an enormous number of revisions and supplemental descriptions will be necessary. This work is meant as a first step in the process: providing keys as reliable as the available publications permit.

Finally, it is necessary to note that no keys to taxa higher than family are provided in this work. Keys to the insect orders are difficult to prepare and use because of the many exceptions to the general morphological characteristics, such as wingless dipterans and heteropterans as well as ephemeropterans that lack legs. Descriptions of the orders can be found in any number of textbooks on entomology or invertebrate zoology, and most specimens can be quickly assigned to the correct higher taxon by browsing the volumes of this series and comparing them to the illustrations. With a little experience, anyone can learn to recognize the order of an insect almost at a glance.

An appeal for quality in taxonomic work

This appeal is addressed to two groups, the first consisting of those responsible for deciding who obtains what portion of the available research funds and the second being the taxonomists themselves.

It has long been recognized by experienced ecologists that identification of the species present in a community is an absolute necessity for thorough ecological research. The biota of any water body, for example, is more than a quantity of "biomass" or a "pathway for energy". Many ecologists, especially those beginning their research careers, have the greatest difficulty in identifying the species present, and the quality of their work is limited by this difficulty. The ecologist may indeed investigate systems using methods very different from those engaged in descriptive biology, but he is nevertheless dependent on a sound basis of taxonomic information for the proper reporting of his results. Briefly stated, ten may be equal to ten, but ten oranges are not equal to ten cows, and ten of one dragonfly species are not equal to ten of another. The "emergency measure" of identifying a species only as far as the genus or family is not satisfactory because two congeneric species may have completely different habitat preferences, feeding habits, and seasonal activity cycles. In short, for one ecologist to properly compare his work with that of another, he must know the name of the species he has encountered and of those that have influenced the biotic communities he has been studying.

In the past, ecologists usually relied on taxonomic specialists to identify their specimens. Since the specimens invariably belonged to a wide variety of taxa, they were distributed among a number of taxonomists, each of whom was an expert only for his own small phylogenetic group. This practice has ceased to function in recent years because of the failure of those in charge of distributing research funds to support scientists who wish to engage themselves in taxonomic work professionally. As the taxonomists reached retirement age, they ceased active research without being able to pass their knowledge on to successors. As time went on, more and more personal knowledge and skill that is difficult or impossible to get from books was permanently lost.

The disappearance of the best taxonomists was accompanied by a decline in the quality of ecological work. This may not be considered totally undesirable by persons in government and business who are constantly confronted by the demands of environmentalists. As ecology moves increasingly out of the laboratory and into the courtroom, the ground rules change, and where principles remain unclear, the lawyers have more room to maneuver, using their skills of persuasion rather than being forced to confront research results already proven beyond the shadow of a reasonable doubt. Furthermore, laws passed to protect endangered species can be better circumvented if no people can be found who are capable of identifying such species.

Although there may well be incentives to suppress taxonomic research, the virtual demise of taxonomy as a field of biology has been instigated mainly by the scientific community itself. Ambition and the desire for quick success have motivated the staffs of many institutions to seek persons promising to achieve great breakthroughs, and no place was left for routine systematic work, which is by nature slow and methodical. Often, it was the number publications rather than their quality or length that determined whether or not a person could succeed in finding gainful employment in science. Recently, science citation indices have become the criterion for judging success, prompting young researchers to investigate whatever everyone else is investigating in order to assure themselves of a maximum number of other scientists who would have reason to cite their work. Pressure was placed on the young scientist to come up with something clever very quickly, leaving no time to learn in depth about the organisms he was encountering. It is especially unfortunate that the worldwide elimination of taxonomist positions came just prior to the development of computer equipment that makes it very easy to establish data banks. Thus, science was robbed of the personnel competent to systematize and disseminate the great body of information that had accumulated during two centuries of intensive work by systematic biologists in all parts of the world.

A long time went by without any crisis becoming evident because the taxonomists and systematic biologists already working in the museums and universities continued donating their time to help identify specimens collected or used in the laboratory by ecologists, physiologists, geneticists, and biochemists. Even after their retirements, many continued contributing time to help younger colleagues in other fields of biology. Moreover, a number of excellent specialists were employed in other fields of work rather than in the biological sciences and made their contributions to the knowledge of the world's flora and fauna as a hobby.

With the gradual elimination of the competent taxonomists and systematic biologists through death or disability, however, the problems of defining community structure, determining species diversity, and identifying the organisms causing some problem in the field have been intensifying. For many taxa, there are no longer experts who can be consulted, or those experts that are still active are hopelessly overworked.

The result is the loss of a vast body of information about the flora and fauna in all parts of the world. Without knowledge of the species that have been encountered, the ecologist is faced with a dilemma. He must learn to identify a vast array of species belonging to the widest variety of phylogenetic groups, or he must limit his research to superficial phenomena. Hence there has been an increasing tendency to use numbers instead of names and to try to equate quantities of unlike and undefined substances or objects, although this cannot yield any reliable results, as mentioned above.

The second aspect of the problem must be dealt with by the taxonomists themselves, who have in the past often been guilty of laxity in quality control of their work. The ambition to gain status by naming new species has certainly been a motivating factor for many of these scientists, and for this purpose, it is easier to write short papers with sketchy descriptions of aberrant specimens than to produce comprehensive reviews compiling the available information on genera or families together with thorough, well illustrated descriptions of new taxa. As mentioned above, it is evident that the examination of type specimens for positive identification of South American insects is often impossible because none were designated, or if they were, they have been lost or greatly damaged. For better or worse, the published literature frequently provides the only criteria for identifying species.

While compiling this key, the enormously broad spectrum of quality in the taxonomic publications became evident. Some papers are of very high quality, such as the works of Belle on the Odonata. These publications provide descriptions of all essential morphological characters, the exact locations of type specimens, and precise collecting data. Any morphological details lacking in the text can be seen in the detailed illustrations. It is clear from these publications that the species described are distinct from all other known species.

In general, more recent publications tend to include better descriptions than older ones, but this is not the case with all taxa. Many works more than a century old are still useful for identifying species, and some of the information they provide cannot be found in any more recent publications. On the other hand, many recent publications are extremely poor, providing either very sketchy descriptions or showing a very poor comprehension of the concept of species as a closed phylogenetic grouping of individuals, which may possess a considerable degree of individual variability (Mayr, 1963).

Many of the better taxonomic publications are rather long, but length is no guarantee of quality. There are some excellent concise works, which can be used with great reliability for the identification of species. Some very long works, including incredible amounts of detail, on the other hand, are very poorly organized and omit much essential information while including vast amounts of trivia. Authors who are more collectors than biologists have used very keen, practiced observation abilities to find the most minor differences among single populations to produce massive numbers of nominal species that are indistinguishable to less skilled observers. Among the European fauna, long lists of synonyms attest to the vast amount of printer's ink that has been squandered on the description and suppression of spurious taxa. Unfortunately, revisions of many South American taxa are still lacking, and many of the nominal species included in the keys will certainly prove to be junior synonyms when more is known about the variability of natural populations.

In the following sections, minimum criteria will be suggested for describing new taxa. All of the features an author should describe to define a new species or subspecies within a particular taxonomic group will be mentioned. It is further suggested, as a general rule, to base descriptions on a holotype and several paratypes. New species described from only one or two specimens should be regarded as questionable, particularly when the morphological differences between them and specimens of other nominal species are very slight. The practice of choosing type specimens that are missing various structures, such as legs, abdomen, or antennae, should be strongly discouraged. With modern means of mass transportation available, there should be no serious obstacle to collecting more material from the locations at which the type specimen of the alleged new species was encountered, and funds should be provided for such collecting for the reasons outlined above.

It would probably be justified to declare many names of species poorly described and without known type specimens as *nomena inquerendae* and omit them from the key. However, whenever distinctive features were found in the original description that can be used to distinguish a species from all other known species in the genus, the name was included in the keys. Names from the keys must therefore be used with caution because one or more undescribed species may also have the characteristics in question. Consulting the original description cited in the bibliography is recommended, but many of these descriptions are too sketchy to be of much help. Where such situations are apparent, it is hoped that specialists familiar with the genus or family, if any are presently active, will designate lectotypes for the species most likely to have been referred to by the earlier authors. Holotypes, paratypes, and original descriptions should be provided for any similar, congeneric species that were hitherto undescribed.

Scope of the work

The families recognized are mainly those found in the key of O'Brien and Wibmer (1978), and some of the new families that have been described by various authors after the appearance of that key are omitted because they have not been generally accepted, are too poorly defined, are too similar to other families to recognize any general definitive characteristics for use in the key, or seem for another reason to be of doubtful validity.

Fig. 1: The geographical scope of this work is limited to continental South America and the offshore islands shown on this map.

Fig. 2: The Brazilian states where the species were found are generally reported as shown on this map. Older reports, however, may not reflect the modern political boundaries, so it often requires knowledge of the travels of entomologists to determine the precise locations of known occurrence. D.F. designates the Federal District of Brasilia.

Geographically, the work is limited to continental South America and islands very close to the coast, such as Tierra del Fuego (**Fig. 1**). Islands far enough offshore to be zoogeographically distinctive, such as the Galapagos Islands, are not included, even though politically, they are parts of South American countries. Except in the case of Brazil, the distribution reported in the keys usually refers only to the country in which the species have been found. When they have been reported in the literature, the Brazilian states are shown, as depicted in **Figure 2**. However, some inaccuracies may have been inadvertently introduced in the cases of states which have changed their boundaries since the insects were described. For example, Mato Grosso formerly encompassed the present states of Mato Grosso, Mato Grosso do Sul, and Rondônia, so species reported in older literature from Mato Grosso may actually be confined to any one of these three present states.

Acknowledgements

The preparation of these keys required the search of a large amount of literature. A great deal of help was provided in obtaining many obscure or archaic papers by Frau G. Lechner of the Max-Planck-Institut für Limnologie Library in Plön and the Timberland Public Library System in the State of Washington, U.S.A. Other works were provided courtesy of Senhora Marilia Junqueira of the Fundação Centro Tecnológico de Minas Gerais in Belo Horizonte. Some of the most recent publications were found in data banks through the generous help of Sean Cash, doctoral candidate at the Agriculture and Resource Economics Department, University of California at Berkeley. I am also grateful to Dr. Alain Thomas of the Université Paul Sabatier, Toulouse, for rushing me three recent reprints by airmail so that I could complete the manuscript without excessive delay.

The Ephemeroptera Galactica website, maintained by the Museum of Aquatic Entomology of Florida A & M University, was of considerable help. It provided a checklist of South American species that was nearly complete and made some of the most recent publications available over the Internet.

Literature

Mayr, E., 1963: Animal Species and Evolution. - Harvard Univ. Press, Cambridge, Mass. 797 pp.

O'Brien, C. W., and G. J. Wibmer, 1978: Genera and species of Curculionidae (Coleoptera). - Ent. News 89(2-3):89-92.

Section 2

Ephemeroptera

The mayflies are unique among the insects in having two instars bearing wings. The first, called the subadult or subimago, emerges from the last aquatic instar and, after a short period of time, molts again. This must be kept in mind when examining winged specimens, because the features used to identify the species are usually, but not always, those of the imago, or adult, and not the subimago, or subadult.

Unfortunately, the majority of the South American species have been described from specimens of only one sex, subadults, or even larvae. Because the adults of both sexes or the immature stages of many of their congeneric species have still not been described, it is likely that some species have been given more than one name, with the male and female or the adult and larva being regarded as different species. This problem has been aggravated by the practice of designating only one or two type specimens. In preparing and using keys to the species, the ensuing confusion frequently becomes evident. There is no remedy at the present time except to undertake basic research, preferably cultivating the species through their entire life cycles in the laboratory and matching the larvae, subadults, and adults. Determining which larvae develop into which adults is also possible in the field, but it is more time consuming to eliminate errors by such methods as collecting the exuvia of emerging subadults, finding habitats supporting only one species in a given genus or family, or dissecting wing buds.

It should be noted that the larvae of mayflies and other insects with incomplete metamorphosis were formerly called nymphs or sometimes naiads, the name given to mythological water nymphs. This practice has been discontinued in recent years, and the term larva has been applied the immature stages of all insects belonging to the Pterygota, that is, insects with wings or secondarily wingless. Therefore, in the case of mayflies, the word larva used herein is equivalent to "nymph" and "naiad" in older works.

While there have been some exemplary publications on individual genera, many taxonomic works on South American mayflies have been of poorer quality than those on insects belonging to most other orders. As a result, the keys include some unavoidable omissions that can be frustrating to researchers conducting faunistic or ecological studies. The most annoying of these is the lack of descriptions of adults of species in some genera and of the larvae of species in others. A second is the uncertain taxonomy, which will most certainly be changed in the future as more comprehensive studies of individual genera and families are completed. Many names that appeared in earlier literature are already recognized to be junior synonyms, and from the descriptions of others, it appears likely that the lists of synonyms will grow considerably as knowledge of the fauna increases. Finally, a number of mistakes were noted in various publications, including apparent mismatches of legends and figures and some discrepancies between the verbal descriptions and illustrations. Due to these problems, many identifications will have to be regarded as tentative until the specimens are compared with types or even until descriptions of all stages of each species become available.

While the great gaps in the knowledge of South American mayflies make this insect order especially difficult for ecologists, zoogeographers, and other scientists to work with, the amount of information still to be gained makes the group especially attractive for taxonomists who would like to bring some order out of the chaos. In the complete absence of knowledge about the larvae of some genera and the adults of others, any descriptive work at all on these taxa is sure to yield useful findings. Hence, it is hoped that entomologists will give some needed attention to the South American Ephemeroptera, investing enough time and energy to provide the kind of thorough observations needed to fill in the gaps in the knowledge revealed in the keys.

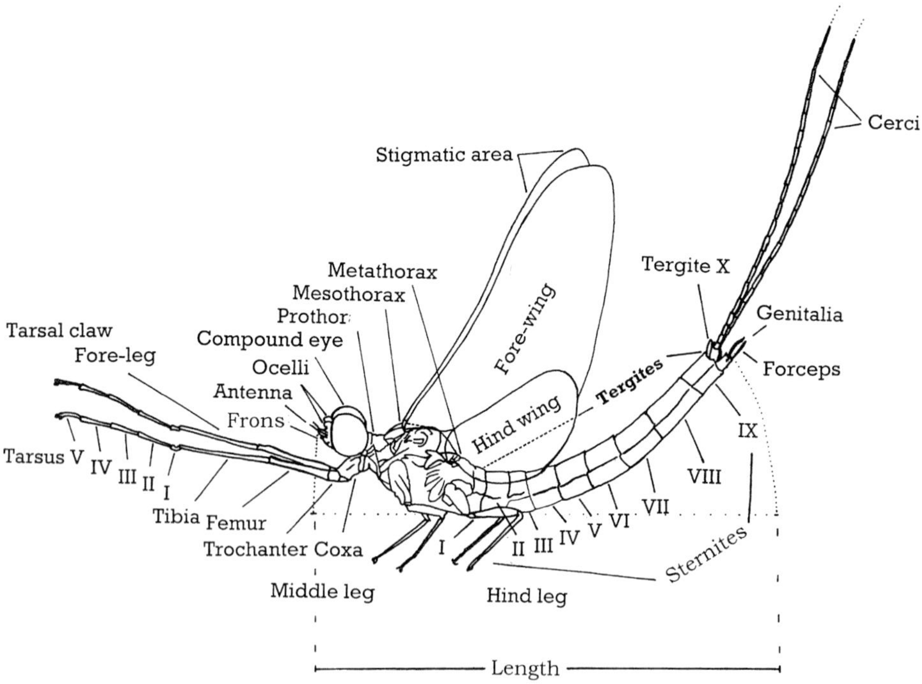

Fig. 2.1 The general terminology for the prominent external morphological structures of an adult *Hexagenia* sp., illustrated semi-diagrammatically. The structures of the subadult are similar in shape, but the genitalia usually do not appear to be freely functional. Alternative terminology for individual structures is discussed in the text.

2.1. Morphology

Adult and subadult insects belonging to the order Ephemeroptera can usually be recognized at a glance by their fragile-looking bodies, large fore-wings, and long, thin filaments arising from the apex of the abdomen. On closer inspection, these filaments include a pair of lateral caudal filaments, often called cerci, and a median caudal filament (**Fig. 2.1**). Three such filaments are present in most mayfly species, but the median filament of a few may be greatly shortened or even absent.

Characteristics of larval mayflies include tracheal gills on the abdomen, unpaired tarsal claws, and an enlargement of the mesothorax (Edmunds, 1959). Like the adults, most larvae bear three filaments at the apex of the abdomen, which permit them to be distinguished immediately from all dragonfly and stonefly larvae. Only a few have shortened median filaments, sometimes leading the casual observer to confuse them with stonefly larvae.

For the purposes of specific identification, a number of morphological features are especially important. For adults, the features used most often in the keys include the shape and relative size of the front and rear wings, their venation, the characteristics of the antennae and caudal filaments, the structure and relative proportions of the leg segments and claws, the relative lengths and widths of the palp segments, and color patterns on the body surface and legs. The genitalia, however, are of primary importance for the taxonomy, which makes the descriptions of species according to the morphology of the larvae alone highly problematic.

For the larvae, important morphological characteristics include generally the same characteristics used for the adults with the exception of the genitalia. Even the general wing vein pattern has been determined from final instar larvae, which contain the incipient wings in the dorsal wing buds, or pods, sometimes making the features of the wing important even for identification of the larvae. The larval morphology often reflects ecological requirements. Species inhabiting streams with fast flowing water will often be flattened and armed with claws suitable for clinging to rocks. Some of these species show modification for burrowing in the sediment or in wood, while other have gills capable of rapid movement for creating water circulation in lentic water with oxygen concentrations less than optimal for the stream-dwellers.

The general morphological features of an adult are shown in **Figure 2.1**. A commonly used set of abbreviations for the wing veins is illustrated in **Figure 2.2.** Unfortunately, this is not the only set of abbreviations in common use, and an alternative set is illustrated in **Figure 2.3**. It is important to be familiar with both because both are encountered in recent literature. The former set appears in the instructive key to the adult Ephemeroptera in the British Isles (Elliott and Humpesch, 1983) and in many publications by Demoulin (1952a), while the latter is preferred by Edmunds *et al.* (1976) and appeared in a publication of Ulmer (1943), who had used a different system for wing vein nomenclature in his earlier works. Also keep in mind that the specimens depicted, or any other mayflies for that matter, cannot be considered typical of the entire group. As relatively primitive insects that first appeared on earth as early as the Permian (Meglitsch, 1967), mayflies have certainly been around long enough for the individual groups to show considerable morphological divergence. Unfortunately, all specialists have not agreed upon the homology of the individual veins in various mayfly families. Therefore, some discrepancies among the designations of individual veins by different authors can be encountered in the literature. For example, when Ulmer (1943) described the origin of vein M, he was referring to the same vein that Needham and Murphy (1924) called Cu_1 and Demoulin (1952a) called CuA. This occurred even though Ulmer had revised his nomenclature from his earlier works (Ulmer, 1920a). Note that abbreviations such as M_1 and Cu_2 correspond roughly with MA and CuP, the A and P referring to anterior and posterior, respectively.

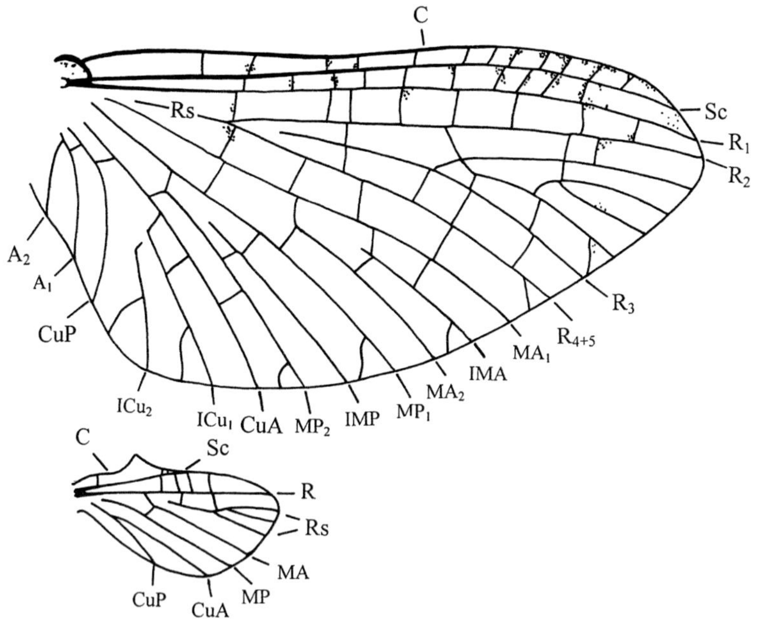

Fig. 2.2 The fore and hind wings of *Miroculis fittkaui* showing the nomenclature of the long veins according to Savage (1983): C = costa, Sc = subcosta, R = radial, Rs = radial sector, MA = anterior media, MP = posterior media, IMA = anterior intermedia, IMP = posterior intermedia, CuA = anterior cubital, CuP = posterior cubital, ICu = intercubital, A = anal. This nomenclature corresponds to one of those in general use for insect wings. The branches of the veins are numbered from the anterior distal root around the circumference to the posterior distal root of the wing. The anterior edge is delimited by the costa. Because most mayflies have a relatively large number of long veins, the numbering becomes complex. Some numbered branches of the radial vein, for example, are considered to be two running together, such as R_{4+5}. In some species, this vein may fork before reaching the wing margin forming R_4 and R_5.

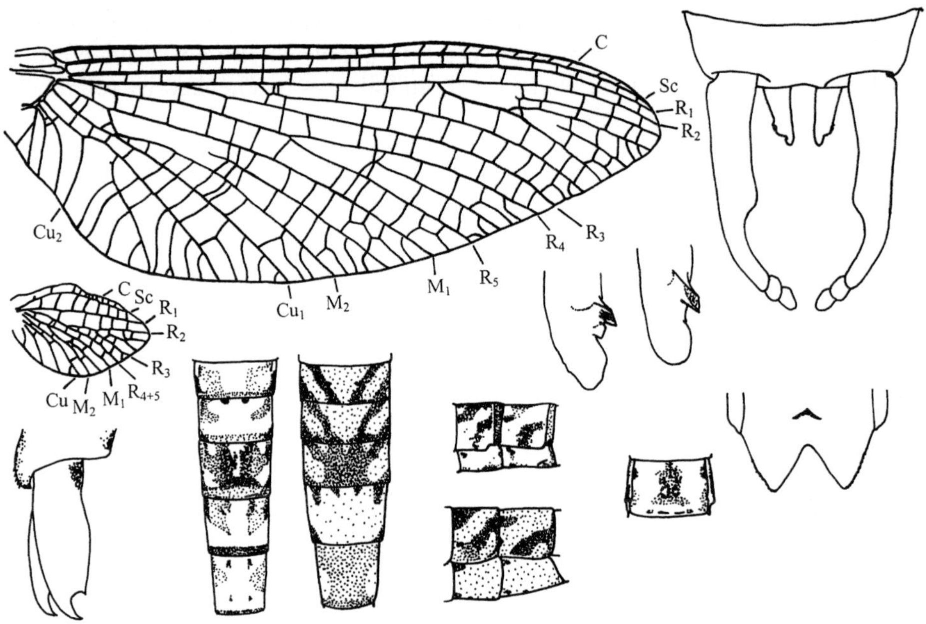

Fig. 2.3 Morphological features of an adult *Massartella brieni:* the fore and hind wings (upper left) with the veins labelled according to the abbreviations of Ulmer (1943), which conform roughly with those of Needham and Murphy (1924) and Edmunds *et al.* (1976); tarsal claws on the fore-leg (lower left), patterns on the fourth through eighth abdominal tergites of a male and a female (lower right center, left and right, respectively), the sixth and seventh abdominal segments of a male and a female in lateral view (lower center, above and below, respectively), the sixth sternite of a male (lower right center), the male genitalia in ventral view with enlargements of the penis lobes in lateral view (upper and middle right), and the ninth sternite of a female abdomen in ventral view (lower right).

The confusion is exacerbated by the fact that the descriptions of the South American mayflies have been written in several languages, each with peculiarities in the abbreviations for insect wing veins. Fortunately, the use of Latin names for these veins has minimized such differences in the abbreviations, but where they occur, they can cause confusion. The peculiarities of the individual taxa are shown in the keys, which usually rely on the original descriptions by the authors who named them. It is therefore recommended to identify specimens using the illustrations of the wings provided with the keys rather than trying to identify the individual veins by name. Since the specialists have not been able to agree among themselves about which vein is which, the verbal descriptions of the wing venation should always be checked by reference to illustrations, if at all possible.

Another feature of the wing veins is their variability. It is not uncommon to find individuals with differences in the presence of absence of veins between the left and right wings. It should be remembered that differences in the location or even the presence of individual cross veins is not a reason to conclude that two specimens belong to different species. While intraspecific differences in the long veins are generally not very great, anomalies do occur, and it is always desirable to check other anatomical features, especially the genitalia, to confirm identification made from the wing venation. The problems caused by individual variability is exacerbated in South America because many species have been described from relatively few specimens, making it impossible to reliably determine the extent of the intraspecific differences in size, coloration, and relative dimensions of the appendages, as well as the wing venation. In spite of the individual differences, the wing venation is probably one of the more reliable general characteristics of adult mayflies, and it is certainly one of the easiest to observe.

Like most other aquatic insects, a primary characteristic of each mayfly species is the genitalia. The reason the copulatory organs of each species are distinct, according to the widely held belief of entomologists, is that the survival of any species in greatly enhanced if accidental cross breeding with other species is prevented. Although the synchronized emergence adult males and females at the same time of year tends to produce large reproductive swarms of individual species for relatively short periods each year, there is nearly always some overlap in the spawning season of various species that may develop in the same body of water. Hence, some females might mate with males of the wrong species in mixed breeding swarms and produce infertile eggs, reducing the reproductive success of the whole population, if a physical impediment did not exist to prevent interspecific copulation. The great differences between the genitalia of each species are therefore seen as hindrances to interspecific mating. The male organ matches the female organ only of members of the same species, and it is mechanically impossible for members of different species to mate. A minor difference in size of shape will not be sufficient for this purpose, so the differences between species, especially those considered to be phylogenetically related species, must be considerable, making it relatively easy to distinguish one species from another by the copulatory organs, especially those of the male.

In addition to the genitalia, which are usually the definitive features for recognizing species, other characteristics of a given species also may be relatively easy to observe. As already mentioned, the wing veins are convenient features to examine under low magnification if the specimen has been satisfactorily preserved, either by drying or in ethanol. While there is a natural variability in the venation from individual to individual, there are many differences that seem to be species specific. Good illustrations of the wings are far better than written descriptions, but such descriptions are very often encountered in the literature, so anyone attempting to identify a specimen should be familiar with the common terminology, as presented in **Figures 2.2** and **2.3.**

Other important structures (**Fig. 2.3**), some of which are used to distinguish genera and families, include the number of wings present, the relative lengths of the caudal filament and cerci, the shape of the labrum and outline of its anterior margin, the morphology of the maxilla and labium and their palps, the absence or degeneration of leg pairs, the number and relative lengths of the tarsal segments, the structure of the tarsal claws, the morphology of the eyes, and the structure of the thorax. The color pattern is frequently the feature that permits the fastest identification of individual species, although variability must be taken into consideration. In the discussion of the larvae, the nomenclature of the mouthparts is provided. The mouths of adults are not functional. The chitinous plates forming the exoskeleton are particularly important for identification because they are rigid and retain their form, although some crumbling and folding is often apparent in dried specimens. The upper plates covering the thorax and abdomen are called tergites in the keys, although some authors call them terga. The corresponding plates on the underside are called sternites, or sterna. The lateral regions between the tergites and sternites are known as pleura. The thoracic plates are often greatly modified, and the tergal or dorsal plates of the thorax are called the pronotum, mesonotum, and metanotum, covering the prothorax, mesothorax, and metathorax, respectively. Each of these three sections of the thorax bears one pair of legs. The plates on the underside of these sections are called the prosternum, mesosternum, and metasternum, respectively. Various pleural plates cover the lateral part of the thorax, but, for some reason, they have seldom been mentioned by taxonomists or used as features to distinguish mayfly species.

The appendages are also important features for the identification of specimens. The legs are attached to their respective sections of the thorax by a segment called the coxa. The short trochanter is located between the coxa and femur, which is usually the longest and most robust segment. The tibia is usually a long segment between the femur and the tarsus. There may be one or more tarsal segments, the most apical of which bears the tarsal claws, which are often distinctive enough to be an important feature for the identification of genera and sometimes species.

The antenna consists of a flagellum mounted on two basal segments called the scape and pedicel, which are usually more robust than the remaining segments which form the flagellum. The length of the antenna and number of its segments can vary considerably, making this appendage valuable for recognizing some species. Unfortunately, the long antennae of some species may break off, limiting the use of this feature for identification. Generally, alternative characteristics should also be provided in keys.

The filaments at the posterior end of the body are even more fragile than the antennae, and they are seldom all intact in preserved specimens. They can, however, provide some help in identifying specimens. If the middle filament, which is generally referred to as the caudal filament, is much shorter than the two lateral cerci, this may provide an important clue to the taxon to which the specimen belongs. It can generally be determined right away whether or not the apical section of the filament has broken off. Breaks make the filaments appear to end abruptly rather than taper evenly to a point. The relative lengths of the cerci and caudal filaments and number of segments, or cercomeres, each unbroken cercus consists of would also be valuable features to identify the species if it were not for the fragility of these appendages. In practice, their most useful morphological features are their color patterns. Some are light in color with darkened joints between the segments, while others have distinctive alternations of color. The length of the setae on the cerci is a more important feature for identifying larvae than adults.

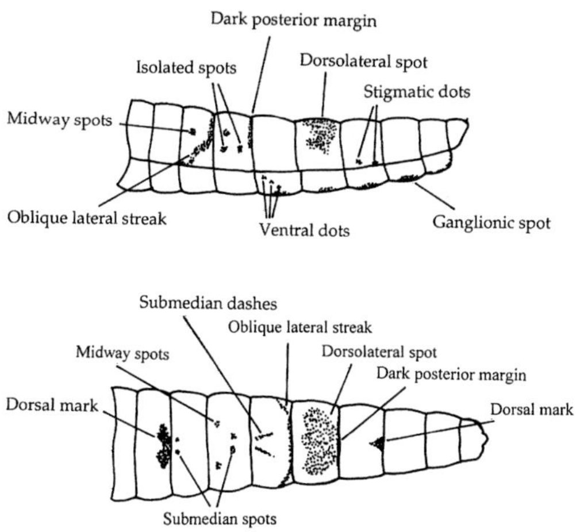

Fig. 2.4 The nomenclature for abdominal markings proposed by Traver and Edmunds (1967) and applied to mayflies in the genus *Thraulodes*.

An attempt was made by Traver and Edmunds (1967) to give standardized names to the markings on the abdomen (**Fig. 2.4**), but they have not come into general usage, and there is no satisfactory substitute to an illustration of the color pattern. However, these names appear in the key to the species of *Thraulodes*, for which they were introduced.

Of primary importance for the taxonomy of adult mayflies are the structures of the male genitalia (**Fig. 2.5**). There are considerable differences among the mayfly families in these structures. Some have a simple, undivided penis, while in others, the penis is divided into two lobes. Some are divided so deeply that the insects effectively have two penes, one beside the other. The female genitalia are simpler, but most species can be distinguished by the size and shape of the subgenital plate, which is located on the ventral surface, where it appears as a patch or plate on the eighth sternite.

Females are frequently captured with eggs attached to the posterior portion of the abdomen. Typically, large numbers of eggs are scattered in the water, and their release is almost continuous. Some form balls or clumps that cling to the abdomen until the female washes them off by dipping the tip of the abdomen in the water to touching it to water plants. The structure of the eggs can therefore also be a useful feature for identifying the adults that are carrying them.

The subadult or subimago is similar in appearance to the adult. Even though this instar is usually passed rapidly, it is particularly vulnerable to capture because much of its time is taken up by molting from the aquatic larval stage or preparing for its final molt. It can be distinguished from the adult by its duller appearance and its translucent rather than transparent wings. When it molts, a thin layer is shed from the body surfaces, including the wings (Pennak, 1978).

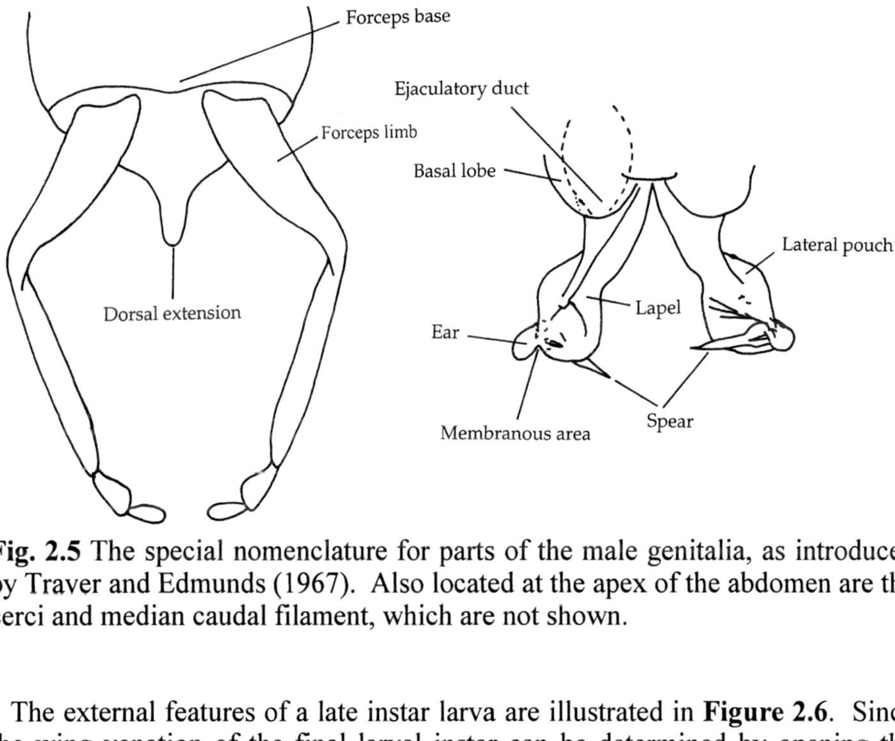

Fig. 2.5 The special nomenclature for parts of the male genitalia, as introduced by Traver and Edmunds (1967). Also located at the apex of the abdomen are the cerci and median caudal filament, which are not shown.

The external features of a late instar larva are illustrated in **Figure 2.6**. Since the wing venation of the final larval instar can be determined by opening the wing cases, some authors have used this as a morphological feature for distinguishing species. However, there are better and more reliable ways of recognizing the individual species. The family of most larvae can be recognized at a glance by the trained eye from the general habitus. However, there are a few families that may sometimes be confused with one another. The most reliable feature for distinguishing the families is the presence, position, and shape of the external gills. Gills continue to be important for distinguishing some genera, but other features usually become more important when species must be identified.

There are some less obvious features that have frequently been used to distinguish the species of larval specimens. Some of them must be dissected away for examination under a microscope to confirm the identification of the larva to species. Some of these structures are illustrated in **Figure 2.7**. These include the individual mouthparts with their palps and other structures, the legs with their tarsal claws, the margins of the tergites as well as other structures on them, and the gills.

A few words on basic conventions in observing and measuring specimens are necessary. The length of an insect is measured from the anteriormost point on the margin of the head to the posteriormost point of the genitalia. The antennae and cerci should never be included in the total length, which is also sometimes referred to as the body length. However, confusion may occur if the head is being compared to specific parts of the body, in which case body length may include the body without the head. It is unusual to encounter this kind of description, so generally body length should be taken as the total length of the head and body but without appendages, as described above. Larvae with "tusks" present a problem, and sometimes authors are not clear about whether these structures are included in the total length. Fortunately, this problem does not arise often.

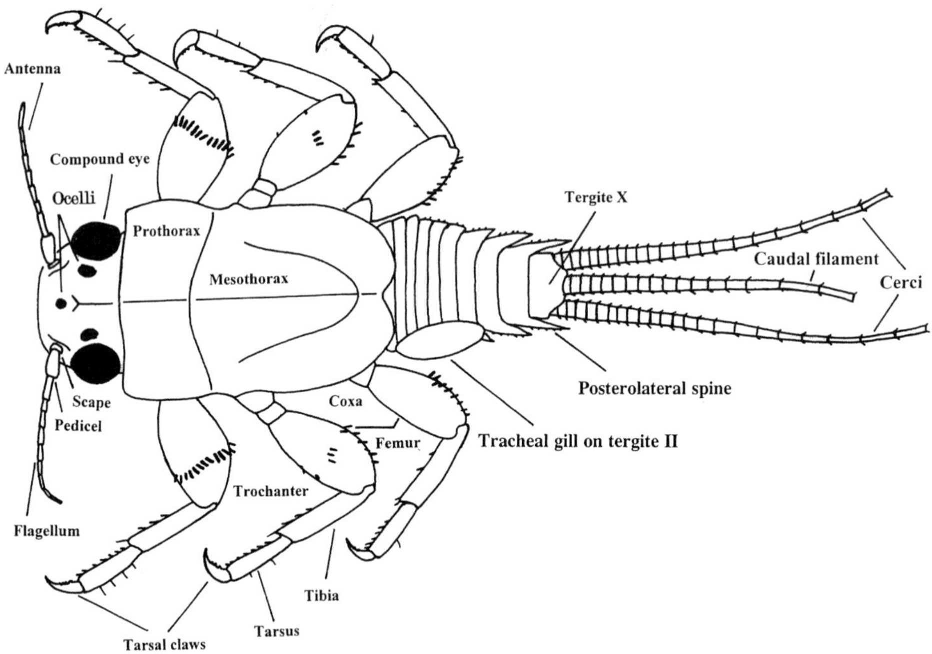

Fig. 2.6 The main features of the external morphology of an unidentified larva of *Leptohyphes* sp. The apical portions of the cerci are not shown. The illustration was redrawn from one by Demoulin (1966a), who had called it *Tricorythodes* sp.

Finally, a word of caution is needed concerning teneral specimens. After insects molt, they require time for the chitinous parts to harden and assume their normal coloration. Although mayflies tend to undergo this process faster than many other insects, a specimen captured immediately after molting is likely to be more subject to distortion of the structures used for identification. Because the color pattern is also an important feature for identifying some species, it must be remembered that the colors of teneral specimens tend to be paler. For this reason, it is a good policy to collect insects alive whenever possible and maintain them under near natural conditions for at least several days. It should be remembered, however, that many species adapted to oxygen-rich streams succumb quickly in the standing water of a collecting jar. For that reason, ideal methods for collection and preservation cannot always be employed in the field, and some teneral specimens will be encountered in most collections.

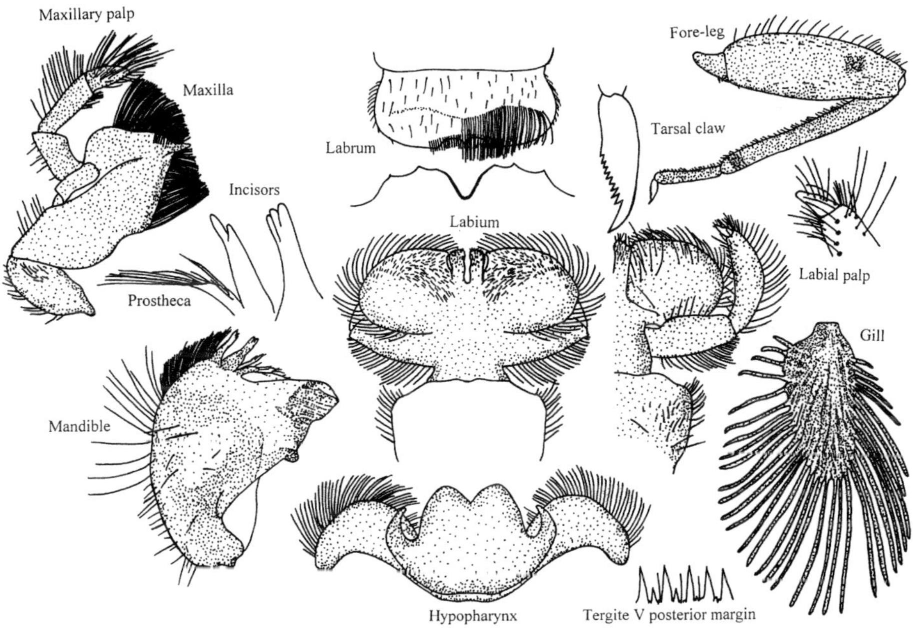

Fig. 2.7 Structures commonly used by taxonomists to characterize the species of mayfly larvae they describe. The larva is an unidentified species of *Ulmeritus* in the subgenus *Ulmeritoides*, which was recently proposed for the status of genus by Domínguez (1991). Because the differences between these two nominal genera are very minor, however, they were retained in one genus in the key. The incisors and prostheca are appendages of the mandible. The labrum has the hair-like setae on the right side removed, and the outline just below it is that along its median indentation. The labium in the center is shown in dorsal view, but just to its right is its left half in ventral view. The apex of its palp is enlarged to its right.

2.2. Ecology

The larvae of most mayfly species develop in flowing water rather rich in dissolved oxygen. However, some survive in standing or slowly flowing waters, even those in which oxygen shortages may develop from time to time. In the Neotropical lowlands, the number of species that can survive at high temperatures and rather low oxygen concentrations is, as expected, greater than in cooler parts of the world. Although no known South American mayfly larva can be classified as terrestrial, at least one species, *Cloeodes hydation*, has been found to be able to survive for as long as nine hours out of water (Nolte *et al.*, 1996), facilitating its development in temporary water bodies.

The mayfly larvae display a wide variety of habits, from clinging to the undersides stones in rapidly flowing mountain streams to burrowing in the soft sediments of tropical lakes. Some species, such as those of *Asthenopus*, are capable of constructing burrows in wood (Sattler, 1967).

Mayfly larvae are generally considered to subsist either on plants or on detritus. The diet, however, probably varies considerably from species to species. There have been contradictory reports on the preferred foods of common European species, although they have been studied for well over a century. The nutritional requirements of South American species are still virtually unknown.

Adult mayflies characteristically have very short life spans. Those species that have been studied are not known to take any nourishment during the subadult and adult stages. Some have only vestigial legs, because the adults do not usually need to rest on objects during their short lives. They mate in flight and spawn immediately, dying rapidly thereafter. While others may survive somewhat longer, none are robust enough to permit them a relatively long life after they emerge from the water. This feature also makes it hard to obtain preserved specimens of adults with all of the anatomical features intact. The caudal filaments are especially prone to breakage, but antennae, legs, and palps are also frequently lost.

2.3. Preservation and examination

While adult mayflies have traditionally been killed in jars containing cyanide or ether, pinned, and kept in insect cases (Needham, 1899), many specimens in collections are now maintained in alcohol or mounted on slides. Preservation in ethanol with 5% to 10% glycerin added maintains the specimens in a rather lifelike condition; however, the colors may slowly fade. The color of dried specimens, however, tends to fade much faster. Furthermore, because of the fragility of their bodies, dried specimens tend to become greatly shrunken and distorted. Legs and cerci are easily broken off and lost, especially if the specimen is dry.

While some taxonomists suggest preserving nymphs in 70% or 80% ethanol, others recommend using 95% ethanol (Edmunds, 1959). My personal recommendation is to use the commonly marketed 96% solution because after the specimens with small amounts of water are introduced into the alcohol, the final concentration will usually be diluted to roughly 80%. However, specimens transferred to 96% ethanol show no notable distortions or unusual discoloration. By adding 5% glycerine to the preservative, the specimens are kept somewhat less brittle and are protected from inadvertent short-term desiccation.

Preservatives other than ethanol are not recommended because of the damage they do to the chitin of the exoskeleton. Aqueous solutions, such as formalin, also have a tendency to cause swelling of the specimens, although some authors recommend using 2% formalin. However, insects so preserved must be first fully moistened with ethanol to keep them from floating on the surface of the formalin and failing to be effectively preserved. I have often observed the wings or legs of mayflies becoming distorted during the preservation process in formalin, making them difficult to examine. For this reason, as well as the health hazards of working with it, I cannot recommend formalin for long-term preservation. Specimens may be fixed in ether, a mixture of formalin and alcohol, or acetic acid mixtures and then transferred to ethanol as soon as possible after fixation to avoid serious distortion.

For over a century, the great value of maintaining collections of exuviae together with the newly emerged adults for taxonomic study has been recognized (Needham, 1899). Not only for basic taxonomic studies but also for general ecological surveys of aquatic species, exuviae should be collected and preserved in alcohol, whenever possible, together with the adults. The cast larval exoskeletons generally preserve the basic body structures of the immature stages. If the subadult has just emerged, it should be maintained alive until its exoskeleton has hardened before it is preserved. If conditions permit, it is desirable to maintain the subadult until it has molted again, and the adult should be maintained alive at least until its exoskeleton has hardened. A series from larva to adult greatly facilitates the correct identification of the species and also provides specimens of great value for taxonomic study. Complete series of exuviae from all instars of individuals cultured in the laboratory are of even greater value for morphological studies, permitting the external anatomy of the larval stages, subadults, and adults of each species to be fully described without problems of confusing the larvae of one species with the adult of another, as sometimes has occurred during field studies. Although the ideal way of eliminating all chance of error is to raise larvae in the laboratory, this is not always feasible. Generally, those species inhabiting standing water are relatively easy to raise, but those from streams usually require an aquarium with a circulation system and considerable aeration to duplicate the conditions in oxygen-rich lotic water. Once the subadult has emerged, keeping it alive until metamorphosis to the final adult instar is relatively easy. The greatest danger to the winged stages is falling into the water or becoming entrapped in condensed water droplets.

A less satisfactory way of obtaining conspecific specimens of larvae, subadults, and adults is to collect them in the field and assume that common congeneric specimens probably belong to the same species. For stream dwelling mayflies that require cold, fast-flowing water rich in oxygen, this method may be the only practical one because laboratory facilities for their successful rearing may be too elaborate to construct. Just as in the laboratory, it is necessary to allow each newly emerged instar obtained in the field some time to stay alive after metamorphosis to allow the chitinous structures to develop and harden and for the color pattern to appear. Preserved specimens appearing unusually etiolated and fragile had probably just molted before fixation and will therefore be less valuable for study than those with normally hardened exoskeletons.

2.4. Taxonomic problems

The biggest taxonomic problem facing anyone working with South American mayflies is the lack of descriptions of all life stages and both sexes of each species. The most vexing aspect of this problem has resulted from the habit of some taxonomists to describe larvae only, making it impossible to determine whether or not the species has already been described as an adult. In several cases, only descriptions of the larvae of one or more congeneric species are available, while the rest of the congeneric species are known only as adults. Moreover, some of the adults have been described only from male or only from female specimens. Given the sexual dimorphism shown by some species, it cannot be ruled out that each sex has been described as a separate species.

While these problems could be easily be eliminated by detailed studies of each species, little effort is being made to facilitate such studies. The quality of the taxonomic work has varied greatly, making it possible to easily recognize members of one genus and almost impossible to positively identify members of another. Many taxonomic revisions must be completed before the necessary ecological studies of the South American species can be effectively undertaken. This is particularly regrettable in light of the value of mayflies for characterizing various aspects of the water quality of the streams and lakes they inhabit.

It would be Utopian at the present time to plead for extensive surveys of South American water bodies and taxonomic revisions of particularly poorly known families and genera. A general deterioration in the handling of scientific funding has made it difficult for competent entomologists to find financial support for basic scientific research on a particular taxa for an extended period of time. This situation favors researchers who are satisfied with superficial surveys and imprecise descriptions of one or two specimens and prevents the formation of groups of scientists working cooperatively on a single taxon at a level lower than order. For example, in discussions of various insect orders, I have often heard that the expert on a particular family works at a particular institute somewhere in the world, while the expert for another family works at another institute on another continent. It does not seem to occur to funding agencies that it might be useful to have at least three or four persons, perhaps in different countries, working on the same family and thereby forcing each other to maintain a minimum standard of quality. At the present time, a few groups of experts are actively working on a few genera or families and describing many new South American species, while other genera or families have not been investigated at all for many years.

It is probably moot to stress the importance of mayflies as food for economically important fish species or to mention how their disappearance from a given stream can provide a warning of deteriorating water quality. These and similar facts have been repeated endlessly by students of mayflies, but this has not motivated scientific research institutes or international funding agencies to support sufficient studies of the group, particularly in South America.

As a result of these problems, students of the order Ephemeroptera in South America will have to cope with an inordinate degree of uncertainty in the identification of many groups. It is safe to say that in most cases, valuable ecological studies on mayfly habitats cannot be undertaken by ecologists because of the uncertainty in the identification of their specimens. This key can provide relief only for certain taxa in limited geographical areas, while at other locations, the ecologist or limnologist will probably encounter mainly species that have yet to be described. In other words, a basic knowledge of the fauna is necessary before the ecology of the individual member species in that fauna can be elucidated. Unless a name or other generally understood designation can be attached to a given species, the most carefully gathered body of data on that species cannot be made useful to the general scientific community. In the case of the South American mayflies, much work is still required to provide minimally satisfactory taxonomic baseline information.

2.5. Suggestions for improvement

Field investigations should be encouraged throughout South America to provide descriptions of the larval instars, subadults, and male and female adults of each species. Only after these descriptions are available will it be possible to meaningfully incorporate Ephemeroptera into routine ecological studies of sufficient depth to provide reliable characterization of specific kinds of habitat. To improve the quality of taxonomic work, sufficient care will have to be taken to match the larval specimens with the adult of each species. This may be done in several ways. Artificial habitats can be created in the laboratory for rearing larvae to adulthood. In this way, all stages can be described as they develop. Because natural stream habitats may be difficult to duplicate, the method of collecting emerging mayflies and preserving the larval and pre-adult exuvia in the field can also prove useful. In this case, only descriptions of the early instars would be lacking. Finally, parallel collections of larvae and adults from individual water bodies can often permit taxonomists to correctly match individual larval forms to their corresponding adults. In a few cases, however, congeneric species might be confused, so such efforts must be made with great caution.

Descriptions of new species of mayfly or previously unknown life stages of known species should be prepared in a way that can be readily understood by other scientists, especially those who have not specialized in the taxonomy of the group. The importance of good illustrations cannot be overemphasized, and several authors cited in the keys below have provided exemplary illustrated descriptions that make identification by non-specialists relatively easy.

It would be encouraging if minimum standards were set demanding descriptions of the final instar larvae, the subadults, and adults of both sexes. However, the rules of scientific nomenclature do not demand this, and it cannot be expected that authors in the future will provide such complete sets of descriptions in every case. Therefore, the points that should be provided in descriptions of each life stage will be outlined below. One practice that should be strongly discouraged, however, is the description of a new species based on a life stage that has not been described for all other members of the genus. For example, if five congeneric species have been described only from adult female specimens, the description of a sixth based only on larval specimens should not be permitted. In such a case, it is obviously not possible to rule out the possibility that the sixth "species" is already among those previously described. An advantage of having more than one "expert" working on any particular taxon is that conventions can be established by mutual agreement to designate the definitive life stage and sex and to define the minimum satisfactory elements of the description of a new species within a given genus or family.

The following elements should be included in any description of an adult:

1. The genitalia of both sexes should be illustrated in dorsal or ventral and lateral view. If specimens of both sexes are not available, the description should await the collection of additional specimens. If this is deemed impossible, then the adult male should be described. Larvae and females should not be described as new species unless they are so remarkable that they are likely to be members of a hitherto undescribed family or unless it is clear that the species is parthenogenetic. Male subadults should not be described instead of the adult;
2. The vein pattern of both the fore and hind wing should be clearly illustrated, and if sexual dimorphism is strong, the wings of both males and females should be illustrated. Markings on the wings should be described, and any tint or clouding of the wing membrane should be recorded;
3. The lengths of the head and body, fore and hind wings, antenna, caudal filament, and cerci should be reported;
4. The color patterns of both sexes and the larvae should be described in detail;
5. Detailed descriptions of the head should include the location of the ocelli and the presence of turbinate eyes in either of the sexes as well as the relative height and width of the eyestalks and the space between them relative to the maximum diameter of an eye;

6. The mouthparts, legs, tarsal claws, and antennae of both the adults and larvae should be described and illustrated. The description should include the color pattern, prominent setae, relative lengths of the segments, and length to width ratios of each.
7. Thorough descriptions of the eggs carried by the females.
8. Every anatomical character introduced by earlier authors as key features for recognizing species in the respective genus or family.

Recently, a tendency to record average dimensions has been noted. This should be strongly discouraged. For a person trying to identify a specimen 4 mm long, knowing that the average length of a species is 5 mm is of no help at all. For one species with an average length of 5 mm, the length may range from 4.5 to 5.5 mm, while for another, the range may be 3 to 7 mm. Obviously, the range rather than the average should be recorded.

A note on the figures

Most illustrations were redrawn from the figures appearing with the original descriptions. In a few cases, they are based on figures prepared to justify taxonomic revisions or with descriptions of taxa later found to be synonyms of species already described. When using illustrations of specimens other than the holotype, there is a risk of perpetuating earlier published errors in identification. For future taxonomic research, the original figures on which the illustrations in this book are based should be consulted as primary sources. Most of them can be found in the publications cited in the introductions to the appropriate keys.

2.6. Key to the families in South America

Information for the keys was taken from Ulmer (1920a, b), Needham and Murphy (1924), Traver (1944), Demoulin (1955a, b, c, d), Edmunds *et al.* (1963), Day (1963), Edmunds (1984), and Domínguez *et al.* (1995). The assignment of species to families generally follows the scheme of McCafferty (1991). Larvae are identified reliably only from features of the final instar.

Adults and subadults

1. In the fore-wing, veins MP_2 and CuA are strongly divergent from MP_1, and CuP, and the A diverge strongly from each other at their bases. Vein MA in the hind wing is always unforked (**Fig. 2.8**). 2
- Veins CuP and A_1 in the fore-wing run parallel from the base or diverge only very slightly; the veins may be greatly reduced in number to only 3 or 4 longitudinal veins (**Fig. 2.2**). 6
2. The subcosta in the fore-wing is concealed in a fold in the membrane under the radius and is visible only basally and not apically. The branches of R and M

run parallel in pairs. Both wings are clouded and only translucent. The legs of the female are short and weak, and the tibiae and tarsi of the males are transversely frosted. There are only two posterior filaments, which are shorter than the body.
....Palingeniidae

Although species assigned to *Palingenia* have been reported from South America, all but one have been transferred to genera in other families. The only South American species remaining that could be assigned to this family is *Palingenia atrostoma* (Weber, 1801), reported from Brazil. It has not been found since the early 19th century, and Domínguez *et al.* (2001) list it as a *nomen dubium.*

- The subcosta in the fore-wing is fully formed and visible for its whole length **(Fig. 2.8)**. 3

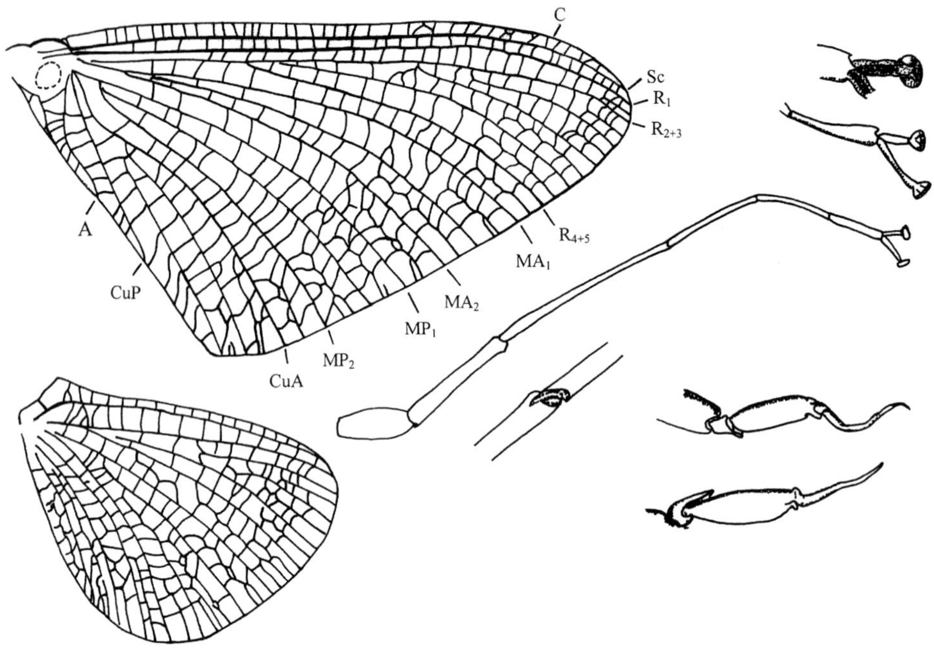

Fig. 2.8 *Asthenopus picteti:* the fore and hind wings (left); the fore-leg (middle to the right margin), showing two enlargements of the pad-like claws and one of the joint between the tibia and first tarsal segment; the vestigial middle (lower middle right) and hind (lower right) legs.

3. The middle and hind legs of the male and often all legs of the female are feeble and non-functional. The wings are not hyaline but translucent (**Fig. 2.8**).
....4

- All legs in both sexes are well developed and functional (**Fig. 2.1**).
....5

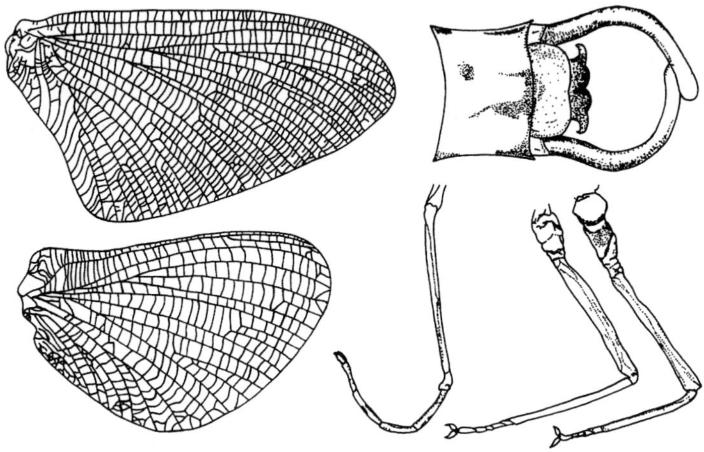

Fig. 2.9 *Campylocia bocainensis:* the fore and hind wings (left), male genitalia (upper right), and the fore, middle, and hind legs (lower right, left to right).

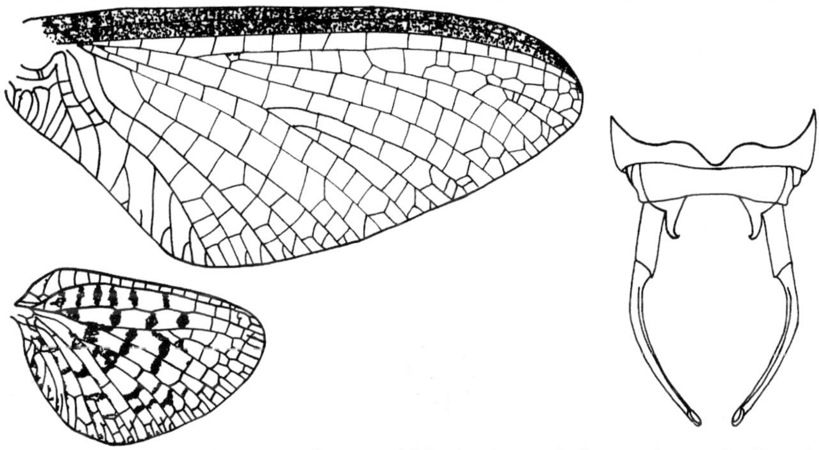

Fig. 2.10 *Hexagenia albivitta:* fore and hind wings (left); male genitalia (right).

4. Vein M in the fore-wing forks beyond the first 1/4 of the vein and at least at the same level as R. If it forks closer to the base, then from 0 to 3 veins in the anal cell have sigmoid cross veins leading from the last of these veins to the edge of the wing. Both wings are translucent with a dull shine in the male or completely dull in the female. There are no short cross veins at the posterior edges of the wings. The legs are weak, although the forelegs of the male may be long. The hind legs are usually short, and they are always weak (**Fig. 2.9**).
....Euthyplociidae..p. 45

- Vein M in the fore-wing forks near the base and proximal to the fork in R. Two long, simple veins or two to nine veins joined by many cross veins are present in the first anal cell. No sigmoid cross veins join the last of the long veins with the edge of the wing. The body color is extremely pale. The wings are colorless and somewhat translucent, or they have a gray or purplish gray tinge (**Fig. 2.8**).

....Polymitarcyidae..p. 51

5. Vein A_1 in the fore-wing is forked near the wing margin. The abdomen is usually yellowish, sometimes with reddish lateral stripes or spots on the tergites. ....Potamanthidae

No South American species are assigned to this family.

- Vein A_1 in the fore-wing is unforked, but it is joined to the hind margin by three or more veinlets. Many short cross veins fill the space between the long veins along the posterior edge, especially in the hind wing. Both wings are hyaline and shiny, except for areas distinctly marked (**Fig. 2.10**). The abdomen of most species has a striking dark pattern.

....Ephemeridae..p. 87

6. The cubital intercalaries of the fore-wing do not consist of a series of veinlets that sometimes fork or curve and join vein CuA to the hind margin. The hind tarsi sometimes consist of five segments (**Fig. 2.11**). 7

- The cubital intercalaries of the fore-wing consist of a series of veinlets that often fork or curve sinuously and join vein CuA to the hind margin. The hind tarsi always consist of four segments. The subcosta in the fore-wing is fully developed and completely separate from the radius (**Fig. 2.2**). 11

7. Vein CuP in the fore-wing is strongly curved (**Fig. 2.11**). 8

- Vein CuP in the fore-wing is only slightly curved or straight (**Fig. 2.12**). 9

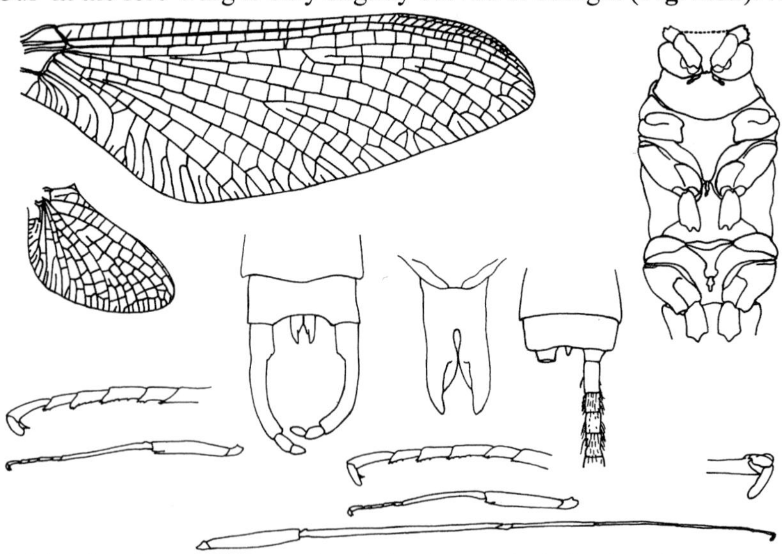

Fig. 2.11 *Murphyella needhami* male adult: fore and hind wing (upper left), the ventral side of the thorax with the legs removed (upper right), the apex of the abdomen in ventral view with an enlargement of the penis and in dorsal view showing the base of one of the cerci (middle, left to right), the fore-leg with an enlargement of the claw (below, center to right), and the middle (lower left) and hind (lower center) legs below enlargements of their tarsi.

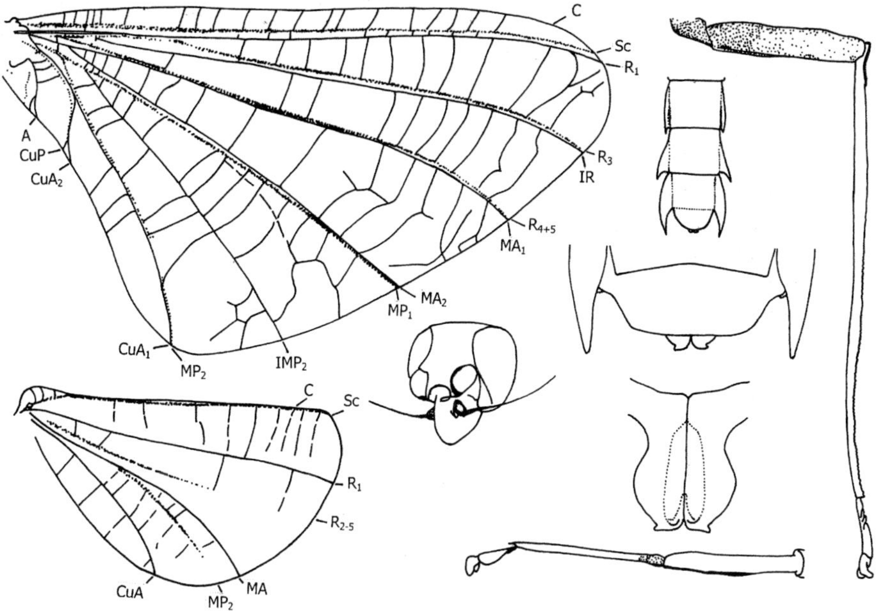

Fig. 2.12 *Fittkauneuria carina* adult male: fore and hind wing (left), head in oblique frontal view (center), fore-leg (lower right center), hind leg (right), and the apical segments of the abdomen, apex enlarged, and penis, all in ventral view (right center, top to bottom).

8. Hind wings are always present and relatively large; several of their veins are forked. The costal projection is shorter than the width of the wing (**Fig. 2.11**).
....Coloburiscidae

This family has been reduced by some taxonomists to a subfamily of Oligoneuriidae. The only known genus from South America in this family is *Murphyella* Lestage, 1930, and its only known South American species is *Murphyella needhami* Lestage, 1930, from Chile. Its head is reddish orange with brown ocelli. Its antennae are brownish red. The total body length is 13 mm; its fore-wing length is 16 mm; its hind wing reaches 5 mm. Synonyms include *Heptagenia? molinai* Navás, 1930; *Dictyosiphlon molinai* (Navás, 1930); *Metamonius needhami* Navás, 1935 (nec Lestage); and *Murphyella molinai* (Navás, 1930). Its larva was illustrated by Needham and Murphy (1924) as *Metamonius*? sp.

- If a hind wing is present, it is small and has only two or three unforked veins and a very large, curved costal projection. The fork in vein MA of the fore-wing is almost symmetrical. The genital forceps of the male consists of two or three segments (**Fig. 2.13**).
....Leptohyphidae..p. 88

9. The fore-legs are mainly or entirely dark, and the middle and hind legs are pale, or the fore-wing venation is greatly reduced, apparently to only three or four longitudinal veins, and the body is dark brown, gray, or black. Vein MP of the hind wing is forked near the wing margin. The subcosta in the fore-wing is not visible as a complete vein or visible only at the base; it is usually fused with the radius or completely absent. The wings are milky or gray clouded, often with a very simple network of veins, including only three to seven long veins in the fore-wing and cross veins only in the second through fifth anterior spaces. If many cross veins are present, they form an irregular anastomosing network in the apical part of the wing. The hind wing lacks complete cross veins or has only a few in the basal part (**Fig. 2.12**). Middle-sized to large mayflies.
....Oligoneuriidae..p. 143
- The leg coloration is other than dark fore-legs and pale middle and hind legs, and the fore-wing venation is not greatly reduced. Vein MP of the hind wing forks near its base or near its mid-length (**Fig. 2.14**). 10
10. The claws on each tarsus are similar, both ending in acute tips (**Fig. 2.14**).
....Siphlonuridae..p. 156
- The claws on each tarsus are dissimilar (**Fig. 2.15**).
....Ameletopsidae..p. 158
11. The two cerci are well developed, but the median terminal filament is absent or rudimentary. Typical intercalary veins are present (**Fig. 2.16**). 12
- There are three well-developed terminal filaments (**Fig. 2.17**). 13

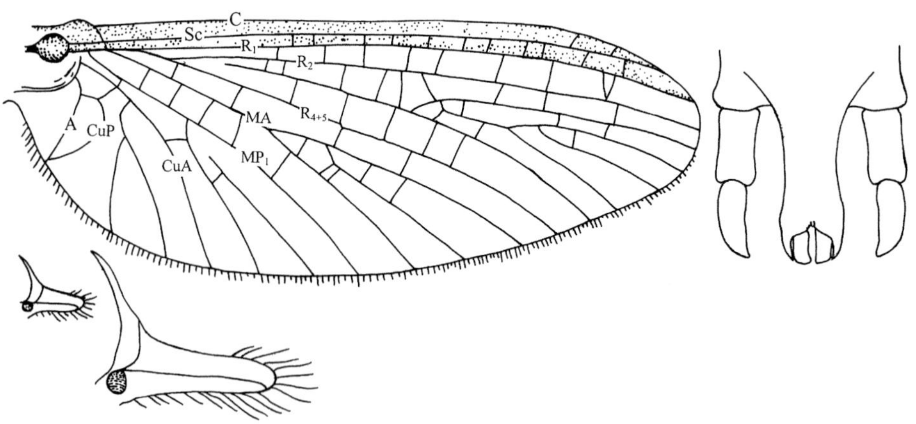

Fig. 2.13 *Haplohyphes huallaga*: the fore-wing with the wing vein abbreviations of Allen (1966) in his original description (upper left), the hind wing drawn to the scale of the fore-wing and enlarged to show details (lower left), and the male genitalia in dorsal view (right).

12. There are short, basally detached, single or double marginal intercalary veins present in each interspace of the fore-wings. M in the fore-wing is not forked. M_1 is followed by an unattached vein, which is followed by another corresponding to M_2 that is not attached to M_1. The fore-wing has few cross veins. The hind wing is reduced or absent; when present, it has only two or three long veins and few cross veins. The wings are hyaline. The penis lobes are membranous, and the upper portion of the eyes of the male are turbinate (**Fig. 2.16**).
....Baetidae..p. 160
- The marginal intercalary veins are attached basally to other veins. MA_2 and MP_2 are attached basally, making M clearly forked in the fore-wing. The hind wings are relatively large. The males have well developed penis lobes and lack turbinate eyes.
....Heptageniidae
No species of this family are known from South America.

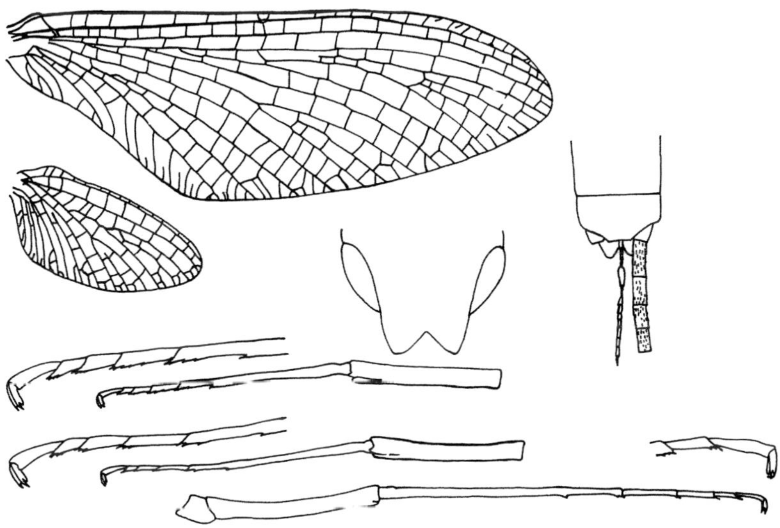

Fig. 2.14 *Metamonius anceps* female adult: fore and hind wings with the hind wing enlarged (upper left); the apex of the abdomen in dorsal view with one cercus removed (upper right) and a ventral view showing the process of the ninth sternite covering the tenth (center); the hind, middle, and fore-legs below enlargements of the apical sections of their tarsi (below, top to bottom).

13. The hind wing is present and usually fairly large, with one or more forked veins. The costal projection is shorter than the width of the wing (**Fig. 2.2, 2.17**). 14
- The hind wing is absent or very small and has only two or three unforked veins. The costal projection is 1.5 to 3 times the width of the wing and straight or recurved (**Fig. 2.18**). 15

14. A_1 and A_2 in the fore-wing are well separated at the base, but A_2 is close to A_3. There are no unattached veins between Cu_2 and A_1 or between the lower branch of Cu and the long vein posterior to it (**Figs. 2.2, 2.17**).
....Leptophlebiidae..p. 232

\- A_1 and A_2 in the fore-wing are close together near the base; A_2 is well separated from A_3. There are usually two unattached veins between Cu_2 and A_1 or between the lower branch of Cu and the long vein posterior to it (**Fig. 2.19**).
....Ephemerellidae

There is only one ephemerellid species known from South America: *Melanemerella brasiliana* Ulmer, 1920.

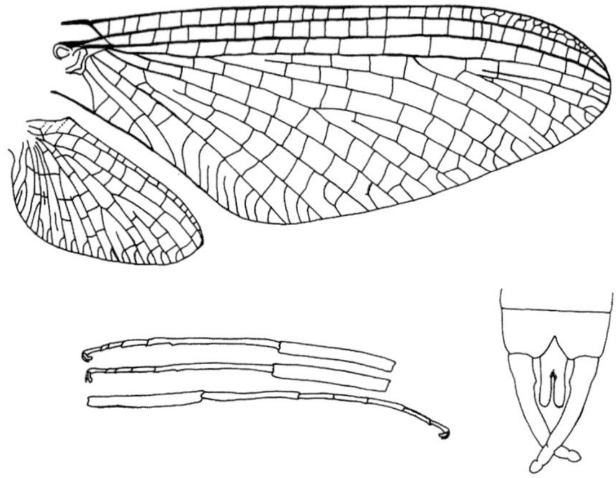

Fig. 2.15 *Chaquihua penai*: fore and hind wings (above), genitalia of a subadult male (lower right), and hind, middle, and fore-legs (lower left, top to bottom).

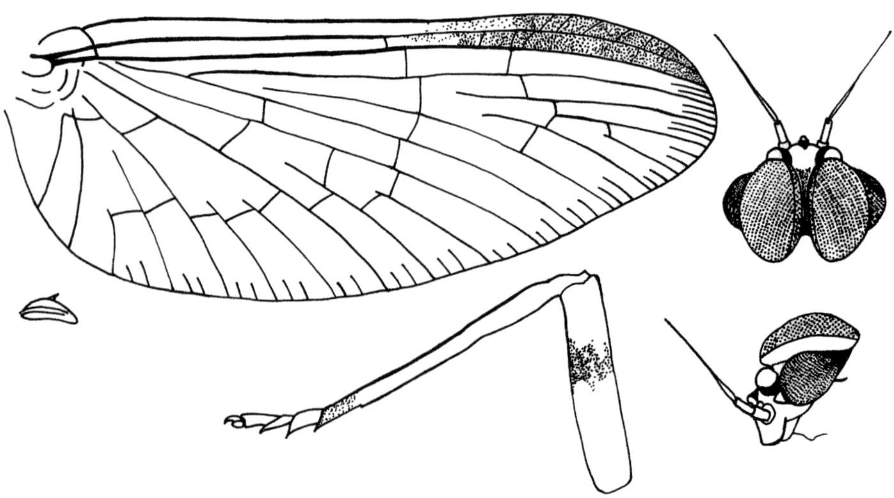

Fig. 2.16 *Cloeodes penai:* the fore and hind wings of a male drawn to scale (left), the head of a male in dorsal (upper right) and lateral (lower right) view, and the hind leg of a female (lower center).

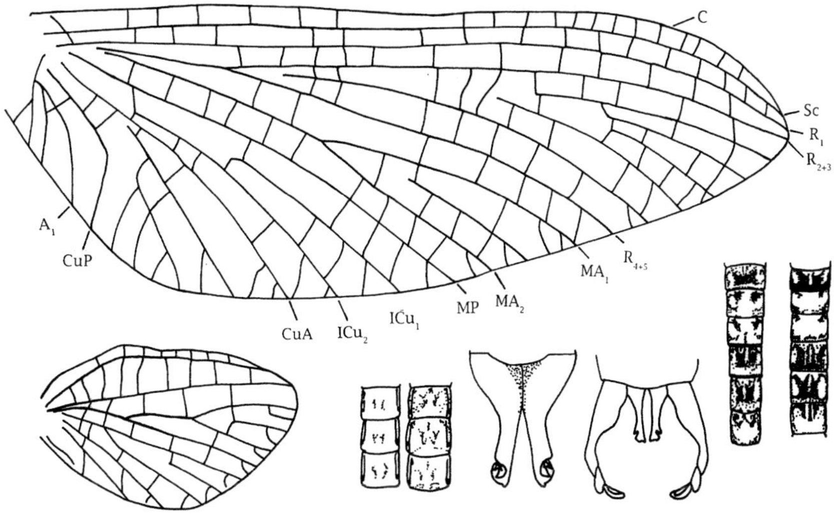

Fig. 2.17 *Penaphlebia fulvipes:* the fore and hind wings (above and left), two color patterns on the third through eighth tergites (lower right) and on the fifth through seventh sternites of the male (lower left center), and the male genitalia (lower right center) with an enlargement of the penis lobes (lower center).

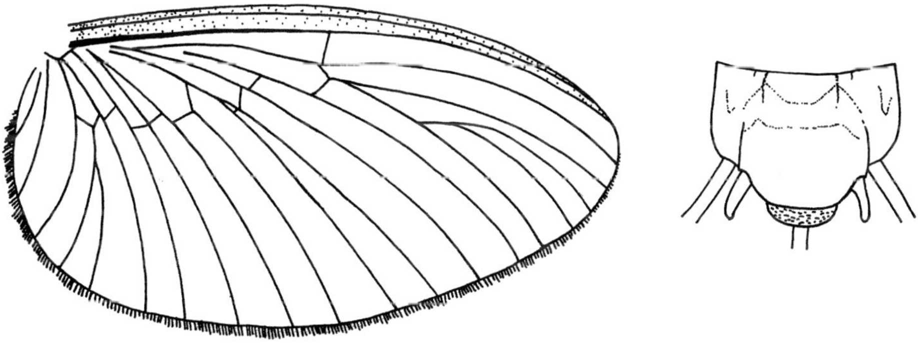

Fig. 2.18 *Caenis cigana:* the wing (left) and genitalia of the male with only the basis of the caudal filaments showing (right).

15. The lower branch of Cu and the interstitial vein just posterior to it are subequal in length to the anterior branch of Cu and reach the base of the wing. The two veins in the second anal cell form a very acute fork and leave only a narrow space between them. A_2 and A_3 form a similar angle. There is never more than one cross vein in a cell except for the radial cell, which contains two or three. The wings are wide and milky, and the anal cell broadens toward the body (**Fig. 2.18**).
....Caenidae..p. 367

- The lower branch of Cu and the interstitial vein just posterior to it are much shorter than the anterior branch of Cu and are interrupted before reaching the base of the wing. The two veins in the second anal cell form a divergent fork and run rather far apart. A_2 and A_3 are sharply bent and run parallel, usually without joining at a fork. There are usually several cross veins in each cell (**Fig. 2.20**). Each forceps arm of the male genitalia consists of four segments, and the subgenital plate has a deep, U-shaped incision.
....Oniscigastridae..p. 387

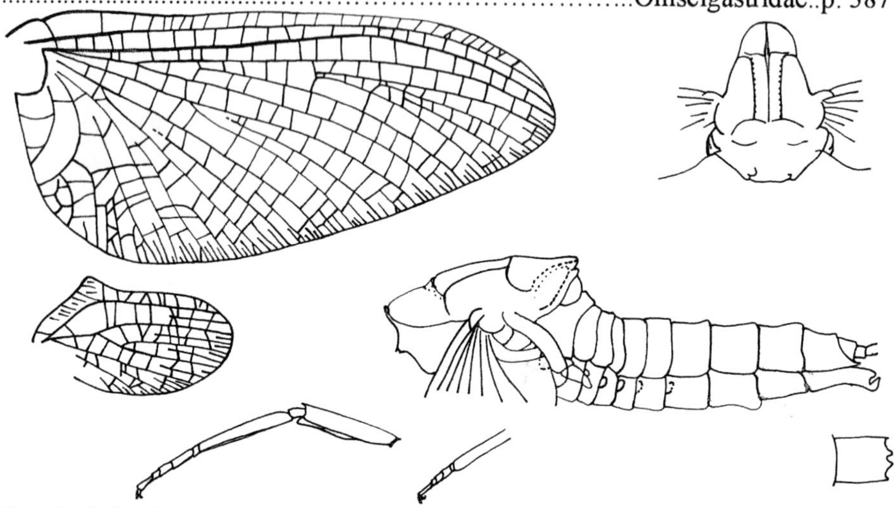

Fig. 2.19 *Melanemerella brasiliana:* fore and hind wings with the hind wing enlarged to about twice its size in relation to the fore-wing (upper left), thorax in anterior view (upper right), and the thorax and abdomen of a female in lateral view (lower right) above parts of the fore and middle legs and ninth sternite.

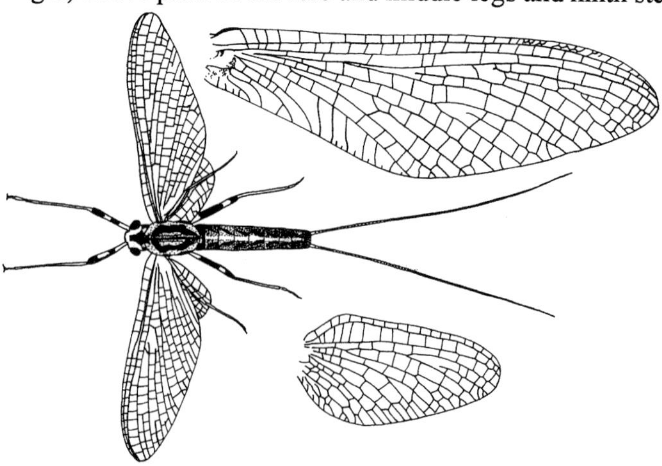

Fig. 2.20 The habitus of *Siphlonella guttata* (left) and the fore and hind wing of *Tasmanophlebia* sp. (right), a genus not found in South America.

Larvae

1. Abdominal gills are absent. Respiratory pits are present on the thoracic sternites. The femora and tibiae of the fore-legs are lined with a dense coat of setae on the anterior edge (**Fig. 2.21**).
....Coloburiscidae
This family has been reduced by some taxonomists to a subfamily of Oligoneuriidae. The only known genus from South America in this family is *Murphyella* Lestage, 1930, and its only known South American species is *Murphyella needhami* Lestage, 1930, from Chile. Synonyms include *Heptagenia? molinai* Navás, 1930; *Dictyosiphlon molinai* (Navás, 1930); *Metamonius needhami* Navás, 1935 (nec Lestage); and *Murphyella molinai* (Navás, 1930). Its larva was illustrated by Needham and Murphy (1924) as *Metamonius*? sp.
- Abdominal gills are present (**Fig. 2.22**). 2

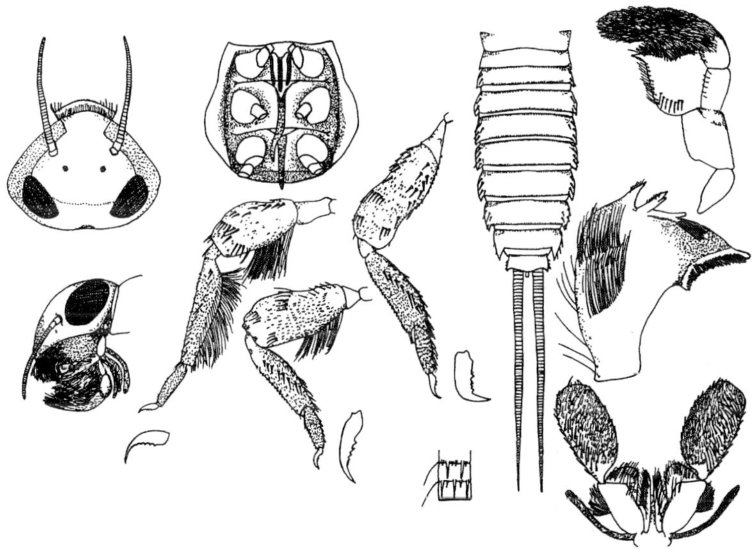

Fig. 2.21 *Murphyella needhami* larva: head in dorsal (upper left) and lateral (lower left) view; the thorax in ventral view with the legs removed to expose the gills (upper left center); the abdomen, cerci, and vestigial median filament in dorsal view with an enlargement of cercal segments, or cercomeres, to the left of the cerci (right center); the fore, middle, and hind legs with enlargements of each tarsal claw (lower left center, left to right); and the maxilla, mandible, and labium (right, top to bottom).

2. The gills on abdominal segments 2 through 7 are forked and have fringed margins. The mandibles bear large, tusk-like processes that project forward and are usually visible from above the head (**Fig. 2.22**). 3

- If the abdominal gills are forked, then their margins are not fringed. Tusk-like processes are not present on the mandible, except in the case of some leptophlebiids (**Figs. 2.23**). 7

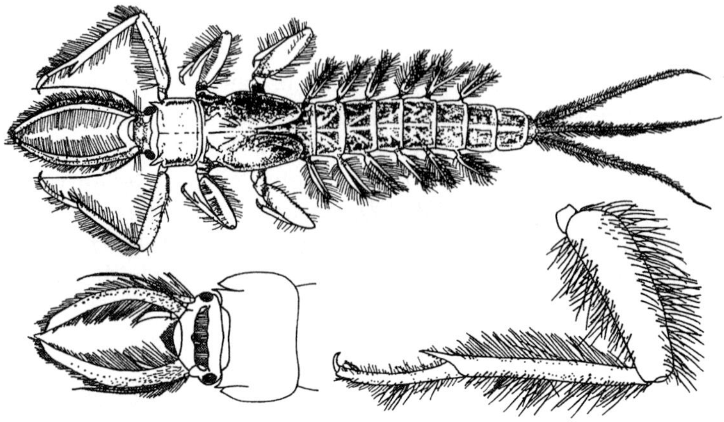

Fig. 2.22 The habitus of a larva of *Campylocia sikorai*, a species unknown in South America (above), and an enlarged head and pronotum of *Campylocia anceps* (lower left), and its fore-leg (lower right).

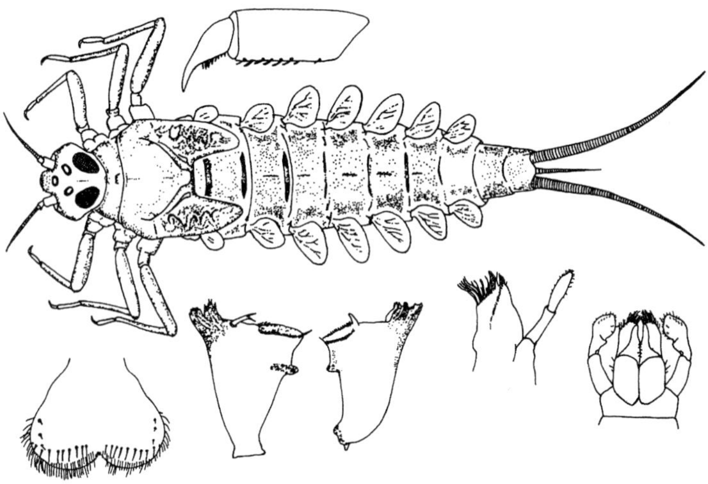

Fig. 2.23 *Mayobaetis ellenae* larva: habitus (middle), an enlarged tarsus with the tarsal claw (above), and (below, left to right): the labrum, mandibles, apex of a maxilla, and labium.

3. The fore tibiae are cylindrical and unmodified. The abdominal gills are held laterally (**Fig. 2.22**). 4

- The fore-tibiae are flattened or bear tubercles to adapt them for burrowing. The abdominal gills are folded dorsally to partially cover the abdominal tergites (**Fig. 2.24**). 5

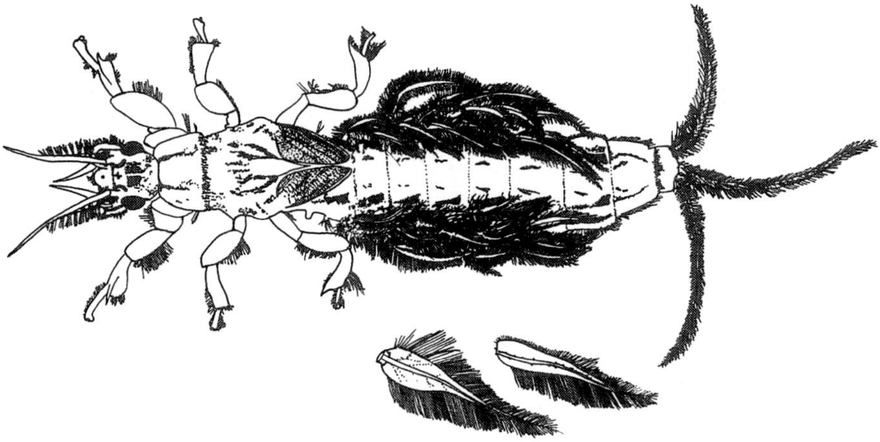

Fig. 2.24 Habitus of a *Hexagenia albivitta* larva (above) with two gills of an unidentified *Hexagenia* sp. (lower right).

4. The anterior margin of the head projects over the labrum. The mandibular tusks are pubescent.
....Potamanthidae
No South American species are assigned to this family.
- The anterior margin of the head is truncate or concave, exposing the labrum. The mandibular tusks are densely hairy (**Fig. 2.22**).
....Euthyplociidae..p. 45
5. The front of the head is rounded with no conspicuous process, or if such a process is present, mandibular tusks viewed laterally curve downward apically and possess 20 or more tubercles on the upper surface (**Fig. 2.25**).
....Polymitarcyidae..p. 51
- Viewed laterally, the mandibular tusks curve upward apically; their upper surface usually bears hairs or spines but no tubercles. The ventral apex of the hind tibia is produced into an acute point, and there is a conspicuous frontal process on the head between the antenna bases (**Fig. 2.24**). 6
6. The mandibular tusks are generally circular in cross section and lack a pronounced keel (**Fig. 2.24**).
....Ephemeridae..p. 87
- The mandibular tusks bear a distinct dorsolateral keel that is toothed as far as the apical half and armed with a row of spurs along the toothed edge.
....Palingeniidae
The only South American species in this family is *Palingenia atrostoma* (Weber, 1801), reported from Brazil and last found in the early 19th century.

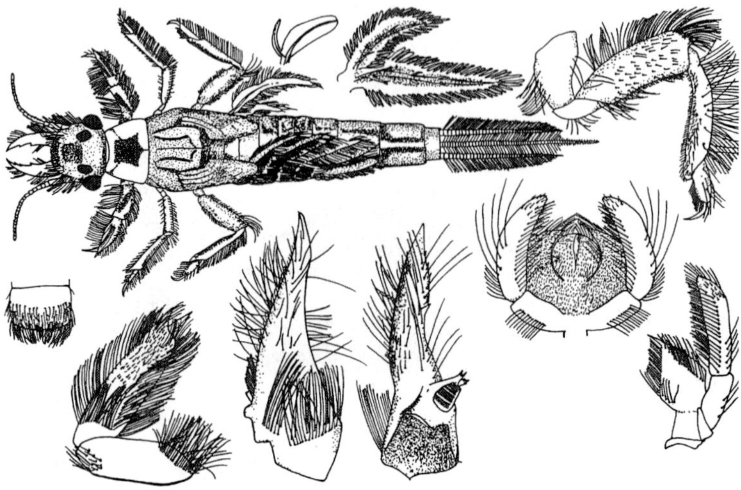

Fig. 2.25 *Campsurus* sp. larva: habitus (upper left) with the first and fourth gills enlarged above it and its hind leg to the right, and (lower row, left to right); the labrum, fore-leg, left mandible in dorsal view, right mandible in ventral view, labium, and maxilla.

7. There is a dense double row of long setae on the inner surface of the fore-legs, a tuft of gills at the base of each maxilla, and sometimes gills at the bases of the fore-coxae. The first pair of abdominal gills is inserted ventrally; the other pairs are inserted dorsally. The outer caudal filaments bear a heavy fringe of hairs on the inner side only, and the posterolateral angles of the apical abdominal segments are prolonged into thin lateral spines (**Fig. 2.26**). ....Oligoneuriidae..p. 143

- There is no dense double row of long setae on the inner surface of the fore-legs and no gill tufts on the fore-coxae or maxilla (**Figs. 2.23**). 8

8. There is an operculate gill cover formed instead of gills on either the first or second abdominal segment (**Fig. 2.27**). 9

- Gills may be absent from the first and second segments of the abdomen, but they are never replaced by covers for successive pairs (**Figs. 2.23**). 11

9. Gills, which consist of double lamellae, are present only on the first four segments of the abdomen. The upper lamella of each gill on the first abdominal segment is modified into an operculate cover for the gills on all four segments. There are sharp lateral spines on the first nine abdominal segments; those on the posterior segments are long and flattened. The tibiae and tarsi appear to form a single segment (**Fig. 2.28**). ....Oniscigastridae..p. 387

- There are no gills on the first abdominal segment; those on the second are strongly modified into operculate or semi-operculate structures that cover the more posterior gills (**Figs. 2.29**). 10

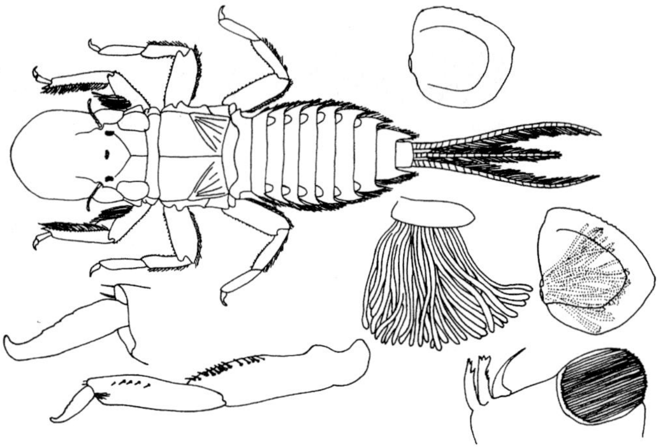

Fig. 2.26 The habitus of *Spaniophlebia* sp. (upper left) with enlargements of the tibia and tarsus of the middle leg and its tarsal claw (lower left), the chewing portion of the right mandible (lower right), and the gill on the first (right center), fourth (middle right), and seventh (upper right) abdominal segments.

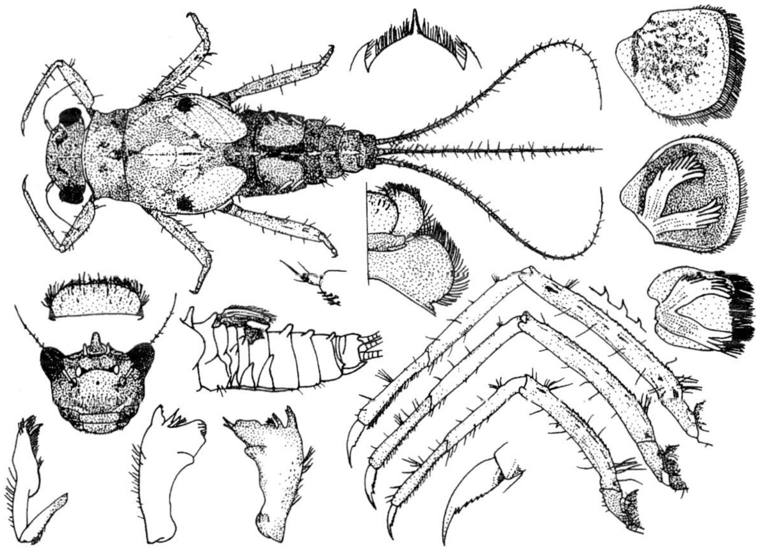

Fig. 2.27 *Coryphorus aquilus* larva: habitus (upper left) below an enlargement of the structure that closes the gill chamber on the sixth abdominal segment (upper left); outline of the abdomen in lateral view (center, below habitus); frontal view of the head beneath an enlargement of the labrum (middle left); the maxilla and two mandibles (lower left, left to right); half of the labium in dorsal view with the apex of a palp enlarged (center below posterior part of habitus); an enlargement of the spines on the edge of the femur, the hind, middle, and foreleg with a tarsal claw (lower right, top to bottom); dorsal and ventral views of the second gill and a ventral view of the third gill (upper right, top to bottom).

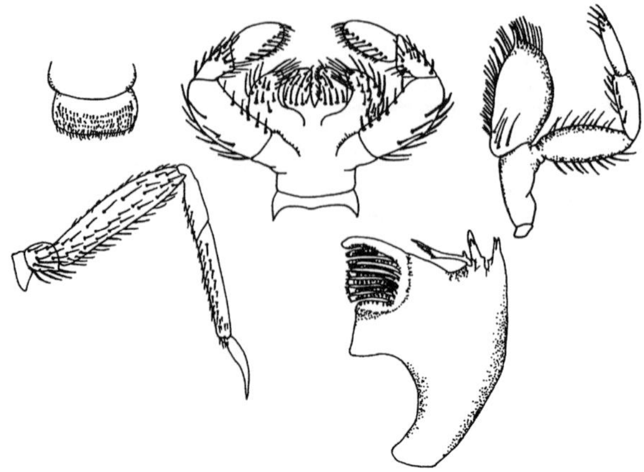

Fig. 2.28 *Siphlonella ventilans* larva (upper row, left to right): labrum, labium, and maxilla; hind leg (lower left), and a mandible (lower right).

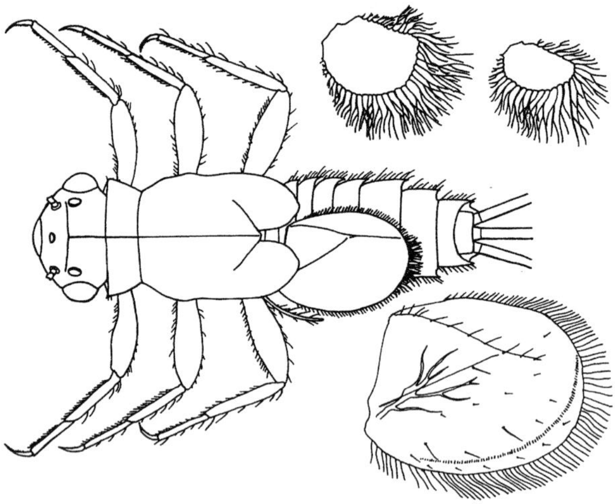

Fig. 2.29 *Caenis cuniana* larva: habitus (left) with an enlarged gill cover from the second abdominal segment (lower right) and the gills from the third (upper center) and sixth abdominal segments (upper right).

10. The gill covers on the second abdominal segment are roughly rectangular with broad, rounded edges; the gills on the third through sixth segments have fringed margins (**Fig. 2.29**).
....Caenidae..p. 367
- Gill covers on the second abdominal segment are triangular or oval; gills on succeeding segments never have fringed margins (**Fig. 2.27**).
....Leptohyphidae..p. 88

11. Gills are absent on the second abdominal segment and rudimentary or absent on the first segment. The third or fourth through seventh segments bear anterior or dorsal gills in the form of oval lamellae and posterior or ventral gills in the form of lamellae with numerous lobes. The abdominal tergites often bear paired tubercles **(Fig. 2.30)**. ....Ephemerellidae

There is only one ephemerellid species known from South America: *Melanemerella brasiliana* Ulmer, 1920.

- Gills are always present on the first through fifth abdominal segments and usually on the first through seventh **(Fig. 2.23)**.
....12

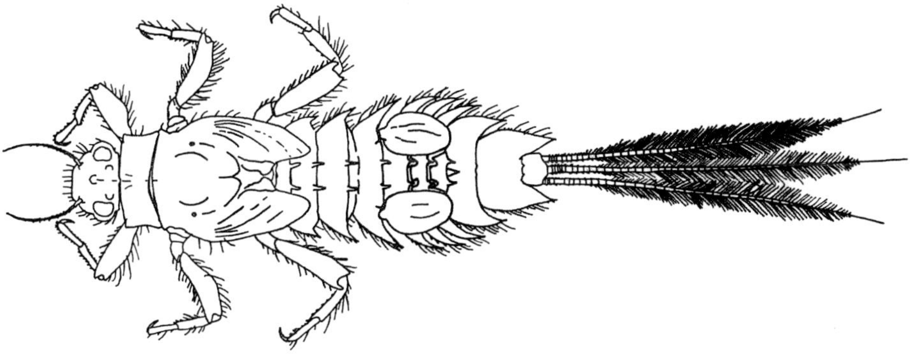

Fig. 2.30 *Ephemerella* sp., an unidentified member of a genus not reported from South America. Dorsal gills are evident only on the fourth abdominal segment.

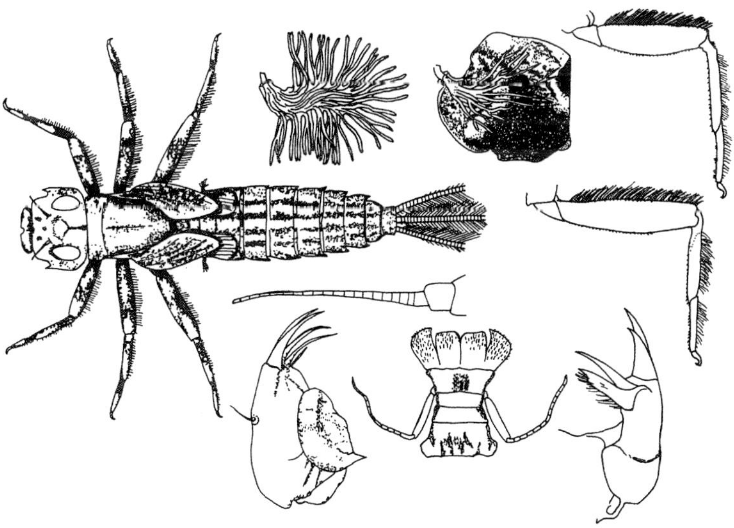

Fig. 2.31 *Chiloporter eatoni* larva: habitus (left), antenna (below abdomen of habitus), maxilla (lower left center), labium with palps (lower right center), mandible (lower right), gills from the first (upper center) and second (upper right center) pairs, and the fore (upper right) and hind legs (middle right).

12. The larva, including the head, is distinctly flattened. The eyes and antennae are borne dorsally, but the mandibles are not visible in dorsal view.
....Heptageniidae
No species of this family are known from South America.
- The larva is not flattened, or if it is, then the mandibles are visible and form part of the flattened dorsal surface of the head (**Fig. 2.23**). 13
13. The maxillary and labial palps are filiform, consisting of many segments. The gills on the first abdominal segment are plumose (**Fig. 2.31**).
....Ameletopsidae..p. 158
- The maxillary and labial palps do not consist of more than three segments. The gills on the first abdominal segment are not plumose (**Fig. 2.23**). 14
14. The gills are usually forked, or if not, they are fringed. They are arranged in clusters of filaments or bilamellate with margins fringed or terminating in a filament or point (**Fig. 2.32**). The apicolateral margin of each maxilla bears a dense tuft of hair-like setae.
....Leptophlebiidae..p. 232
- The gills are unfringed, or rarely fringed only on the inner margin, and consist of single, double, or triple laminae never terminating in points or filaments. The apicolateral margin of the maxilla never bears a dense tuft of hair-like setae (**Fig. 2.23**).
....15

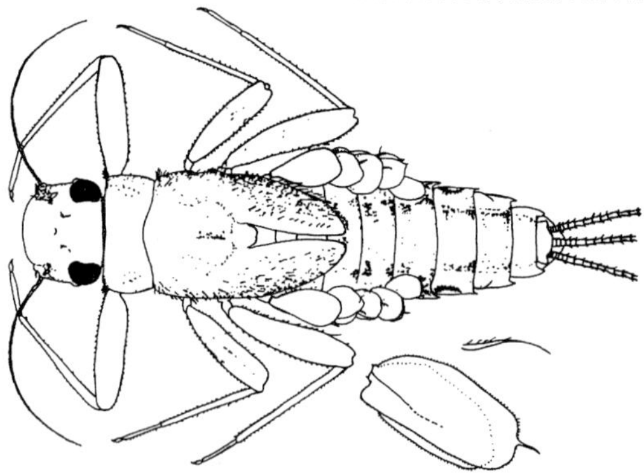

Fig. 2.32 The habitus of the last larval instar of *Hapsiphlebia anastomosis* and enlarged gills from the third (below) and seventh (above) abdominal segments.

15. The median terminal filament is shorter than the tenth tergite, or, if it is well developed and longer, then the antenna is longer than twice the width of the head (**Fig. 2.23**).
....Baetidae..p. 160
- The median terminal filament is subequal to the cerci, and the antennae are shorter than twice the diameter of the head. Gills are present only on the first four abdominal segments; they are elytroid on the first segment. There are mid-dorsal hooks on the first abdominal segment (**Fig. 2.33**).
....Siphlonuridae..p. 156

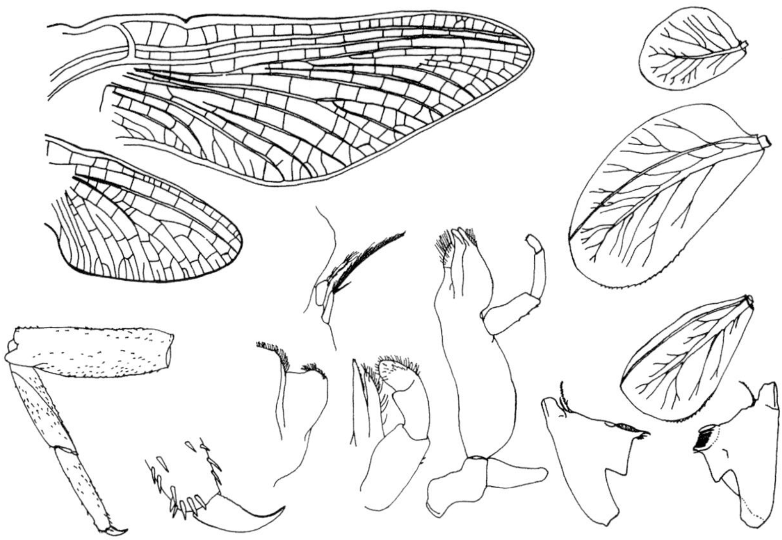

Fig. 2.33 A larva of *Metamonius anceps* with the wings of the subadult dissected from the wing pods (upper left); fore-leg with an enlargement of the claw (lower left); mandibles (lower right); the prostheca of the right mandible enlarged (middle); the left sides of the hypopharynx (lower left center); half of the labium (lower center); maxilla (lower right center); and the gills of the first, fourth, and seventh segments of the abdomen (upper and middle right, top to bottom).

Euthyplociidae

Key to the genera of adults in South America

The information for the key was provided by Demoulin (1952a, 1966b).

1. There is usually at least one intercalary vein in Cu. The middle and hind tarsi each consist of four segments. The gonostylus consists of only one segment (**Fig. 2.9**).
....*Campylocia* Needham and Murphy, 1924..p. 48
- Vein Rs in the fore-wing forks proximal to the fork in MA, and there is no intercalary vein in Cu. There are usually 12 to 15 sigmoidal cross veins in the cubital cell. The middle and hind tarsi each consist of five segments. There are two segments in the gonostylus (**Fig. 2.34**). 2
2. The hind wing is generally triangular in shape and more than half as long as the fore-wing. Vein MP forks near its base and is connected to CuA by sigmoidal cross veins. Vein CuP is somewhat recurved (**Fig. 2.34**).
....*Euthyplocia* Eaton, 1871..p. 47

- The hind wing is lanceolate and somewhat longer than 1/4 the length of the fore-wing. Vein MP forks near its midlength. There are few sigmoidal cross veins joining CuA. The middle and hind tarsi bear triangular claws with hooked tips (**Fig. 2.35**).

....*Mesoplocia* Demoulin, 1952

The only species in this genus is *Mesoplocia intermedia* Demoulin, 1952, which inhabits Ecuador.

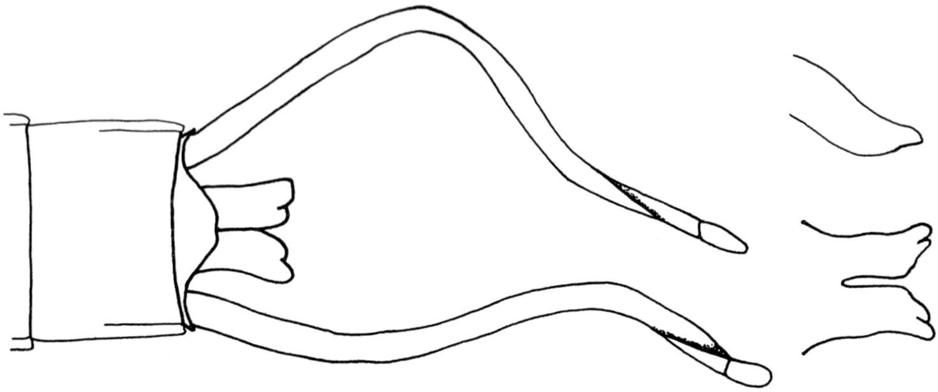

Fig. 2.34 The male genitalia of *Euthyplocia hecuba* in ventral view (left) with enlargements of the penis in lateral (upper right) and ventral view (lower right).

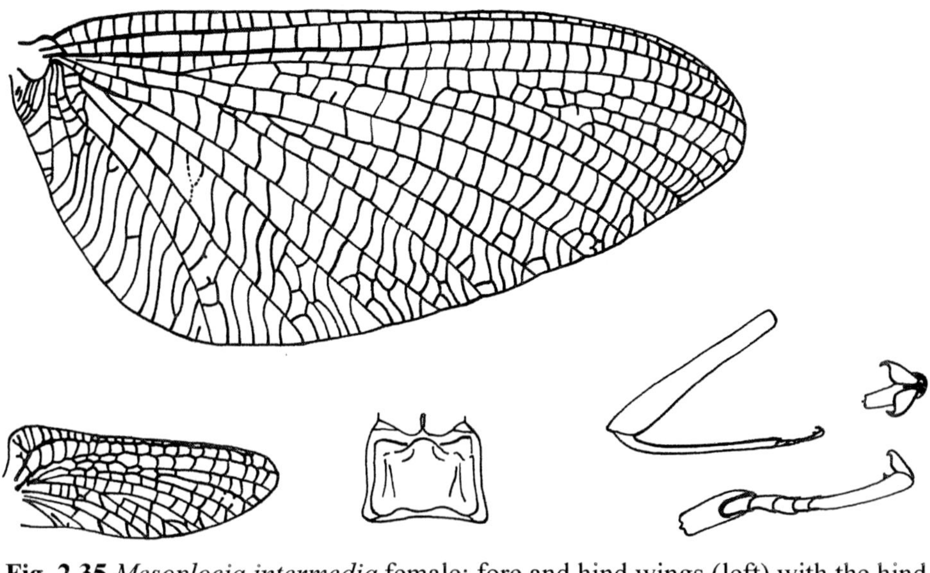

Fig. 2.35 *Mesoplocia intermedia* female: fore and hind wings (left) with the hind wing enlarged relative to the fore-wing, the prothorax (lower center), and the hind leg with enlargements of the tarsi and the claws (lower right).

Key to the genera of known larvae in South America

Information for the key was provided by Ulmer (1920c, 1942). The larva of *Mesoplocia* has not yet been described.

1. The antenna is about 3.5 times as long as the mandibles. There is an apical process on the tarsus of the fore-leg. The apical spine on the tibia of the fore-leg is at least half as long as the tarsus. The mandibles are shorter than the combined length of the head and pronotum, and their long processes are strongly and evenly curved, smooth, and end in a narrow point (**Fig. 2.36**).
....*Euthyplocia* Eaton, 1871
- The antenna is shorter or only slightly longer than the mandibles. There is no apical process on the tarsus of the fore-leg. The apical spine on the tibia of the fore-leg is only 1/6 to 1/4 as long as the tarsus. The mandibles are longer than the combined length of the head and pronotum, and their long processes are only moderately curved, armed with a row of stout spines, and end in an apex that is delimited from the rest of the process on its outer surface (**Fig. 2.22**).
....*Campylocia* Needham and Murphy, 1924..p. 48

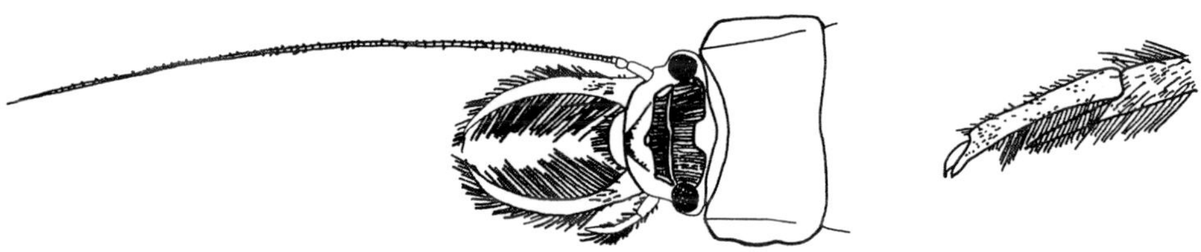

Fig. 2.36 The head and pronotum with the left antenna and right palp removed (left) and details of the anterior tarsus (right) of a *Euthyplocia hecuba* larva.

Key to the species of adult *Euthyplocia* in South America

Information for the key was provided by Demoulin (1952a). There is no key to the larvae because that of *Euthyplocia haenschi* has not been described.

1. The color is rather light. The last segment of the gonostylus is only twice as long as wide (**Fig. 2.34**).
....*Euthyplocia hecuba* (Hagen, 1861)
(Mexico, Central America, Colombia, Ecuador, Peru, Venezuela, Argentina, Brazil).
- The color is dark. The last segment of the gonostylus is four times as long as wide (**Fig. 2.37**).
....*Euthyplocia haenschi* Ulmer, 1942
(Ecuador).

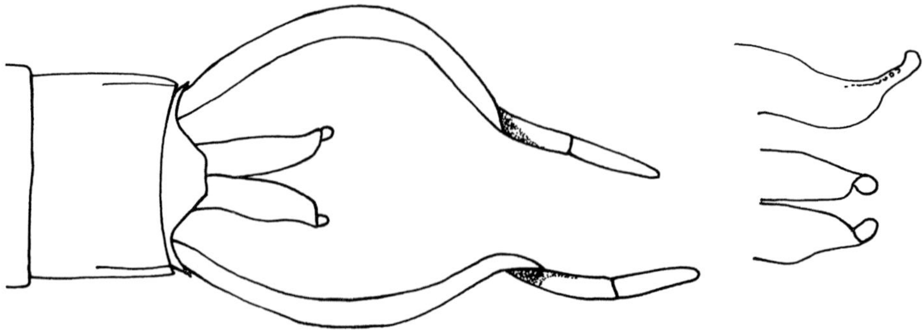

Fig. 2.37 The male genitalia of *Euthyplocia haenschi* in ventral view (left) with enlargements of the penis in lateral (upper right) and ventral view (lower right).

Key to the species of adult *Campylocia* in South America

Information for the key was provided by Navás (1920a), Berner and Thew (1961), and Pereira and da Silva (1990a). Wing venation may prove unreliable for identification, so the male genitalia should be considered definitive.

1. The caudal filaments of the male are 68 to 70 mm long, and those of the female are 80 to 90 mm. There are three intercalary veins in the cubital cell of the fore-wing. In the hind wing, there are two intercalary veins between MA and MP_1, two between IMP and MP_2, and one between IR_3 and R_2. MP_1, IMP, and MP_2 originate at the same point in the hind wing (**Fig. 2.9**). Length of male: 18 to 26 mm; of female: 30 to 33 mm. Fore-wing length of male: 19 to 23 mm; of female: 30 to 39 mm. Color: brown with lighter and darker markings.
....*Campylocia bocainensis* Pereira and da Silva, 1990 (São Paulo).
- The caudal filaments are not more than about 64 mm long. The wings show different characteristics (**Fig. 2.38**). 2
2. Veins CuP and A_1 in the fore-wing arise from a common fork and are connected by about four cross veins in the proximal half. The apical tergite partially covering the male genitalia is evenly convex medially (**Fig. 2.38**). Length of male: 11 to 20 mm; of male fore-wing: 10 to 22 mm. Length of female: 22 to 26 mm; of female fore-wing: 15 to 34 mm.
....*Campylocia anceps* (Eaton, 1889) (Colombia, Ecuador, Venezuela, Guyana, Surinam, French Guiana, Peru, Brazil). Syn: *Euthyplocia anceps* Eaton, 1883; *Euthyplocia burmeisteri* (Hagen, 1888); *Euthyplocia intercalata* Banks, 1918; *Euthyplocia guntheri* Navás, 1920; *Campylocia ampla* Needham and Murphy, 1924; *Longinella guntheri* (Navás, 1920).
- Veins CuP and A_1 in the fore-wing arise independently and are connected by only one or two veins in the proximal half. The apical tergite partially covering

the male genitalia is emarginate and convex in the middle **(Fig. 2.39)**. Length of male body: 15 to 22 mm; of fore-wing: 19 to 22.5 mm. Length of female body: 22 mm; of fore-wing: 31 to 35 mm.
....*Campylocia dochmia* Berner and Thew, 1961 (Minas Gerais).

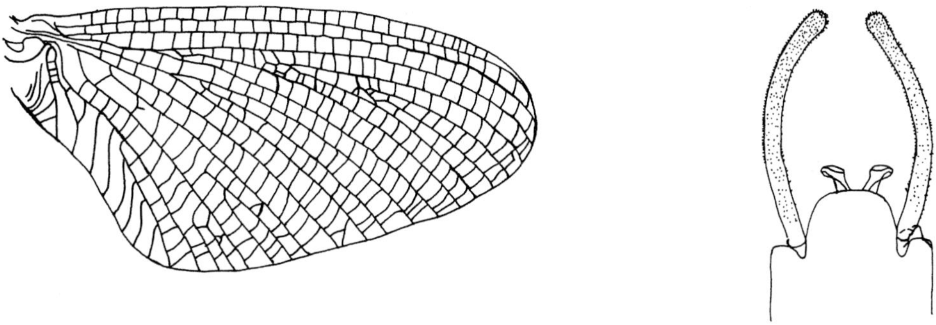

Fig. 2.38 The fore-wing of *Campylocia anceps* (left) and the male genitalia in dorsal view (right).

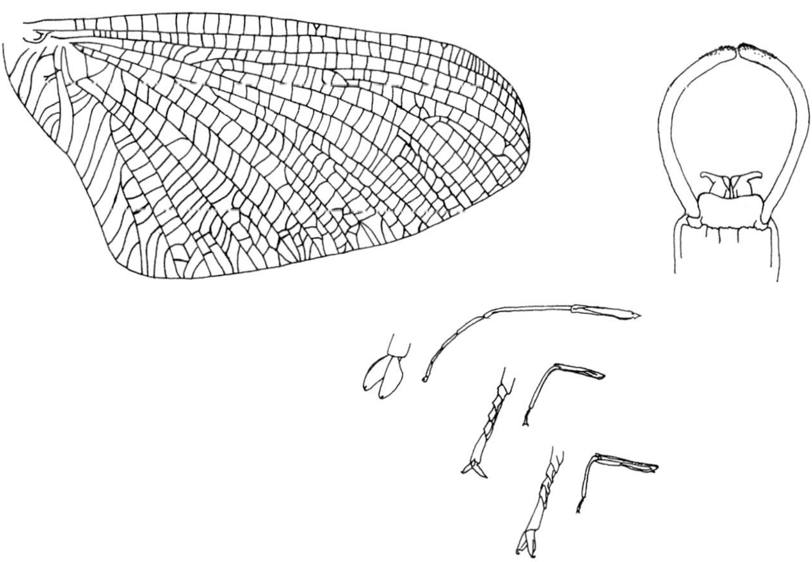

Fig. 2.39 *Campylocia dochmia:* the fore-wing (upper left), genitalia of a male (upper right), and the fore, middle, and hind legs with enlargements of the claws (below, top to bottom).

Key to the species of *Campylocia* larvae in South America

Information for the key was taken from Needham and Murphy (1924) and Pereira and da Silva (1990a). The larva of *Campylocia dochmia* has not yet been described.

1. The branches on the first gill differ greatly in size. The antennae are approximately 2/3 the length of a mandibular tusk. There is a slight excavation on the inner surface of the femur containing a dense group of setae (**Fig. 2.40**). Length: about 35 mm. Length of cerci: 18 mm. Length of caudal filament: 21 mm. Coloration: mainly brown with a few darker markings.
....*Campylocia bocainensis* Pereira and da Silva, 1990 (São Paulo).

- The branches on the first gill do not differ greatly in size. The antennae are more than 2/3 the length of a mandibular tusk. There is no excavation filled with a dense group of setae on the hind femur.
....*Campylocia anceps* (Eaton, 1889) (Colombia, Ecuador, Venezuela, Guyana, Surinam, French Guiana, Peru, Brazil). Syn: *Euthyplocia anceps* Eaton, 1883; *Euthyplocia burmeisteri* (Hagen, 1888); *Euthyplocia intercalata* Banks, 1918; *Euthyplocia guntheri* Navás, 1920; *Campylocia ampla* Needham and Murphy, 1924; *Longinella guntheri* (Navás, 1920).

Fig. 2.40 *Campylocia bocainensis* female larva (left to right): the head and pronotum showing the relative lengths of the mandibular tusks, antennae, and palps; the tusks (above) and labrum (below); the apex of the right maxilla with the palp (above) and the labium (below); the fore, middle, and hind legs; and the gills on the first (above) and second (below) abdominal segment.

Polymitarcyidae

Key to the genera of adults in South America

Information for the key was taken from Spieth (1943), Traver (1947a), Day (1963), and Hubbard and Domínguez (1988).

1. The pronotum is very short and ring-like, not broader posteriorly than anteriorly. The fore legs of the male are almost as long or longer than the body. The forceps is stout (**Fig. 2.8**).
.... *Asthenopus* Eaton, 1871..p. 53

A revision of the subfamily Asthenopodinae has placed *Asthenopodes* Ulmer, 1924, in synonymy with this genus (Hubbard and Domínguez, 1988).

- The pronotum is about as long as wide and much broader posteriorly than anteriorly. The front leg of the male is about half as long as the body. The forceps is long and slender (**Fig. 2.41**).
....2

2. Marginal veinlets are present.
.... *Ephoron* Williamson, 1802

The only species reported from South America is *Ephoron umbrata* (Hagen, 1888), which inhabits Brazil. However, it is not known whether this is really a species of *Ephoron*, and Domínguez *et al.* (2001) consider it a *nomen dubium*.

- Marginal veinlets absent (**Fig. 2.41**). 3

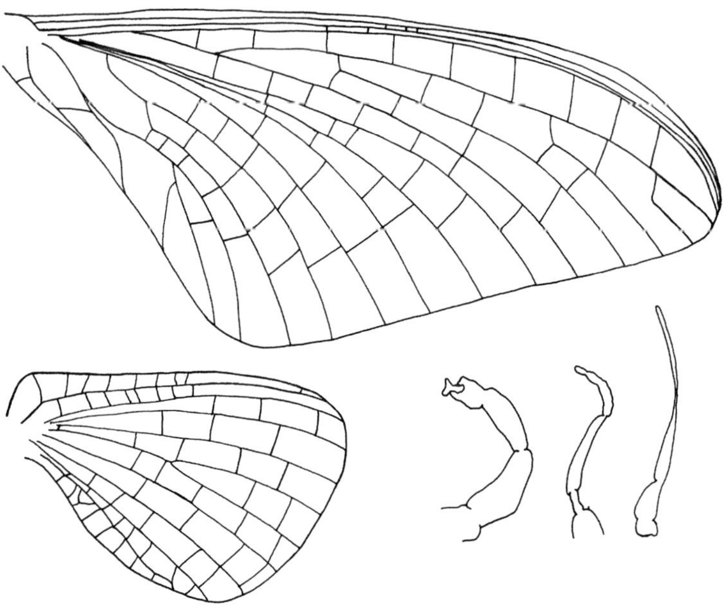

Fig. 2.41 *Tortopus igaranus:* fore and hind wings (left) and the fore, middle, and hind legs (lower right, left to right).

3. The middle and hind legs are considerably reduced with all parts beyond the trochanters missing; they are broadened and fan-like. No claw-like appendage extends from the base of the forceps distal to the forceps processes (**Fig. 2.42**).
....*Campsurus* Eaton, 1868..p. 55
- The middle and hind legs are reduced to non-functional and membranous appendages, but all parts are recognizable, and they are not broadened or fan-like (**Fig. 2.41**). A claw-like appendage extends from the base of the forceps distal to the forceps processes.
....*Tortopus* Needham and Murphy, 1924..p. 82

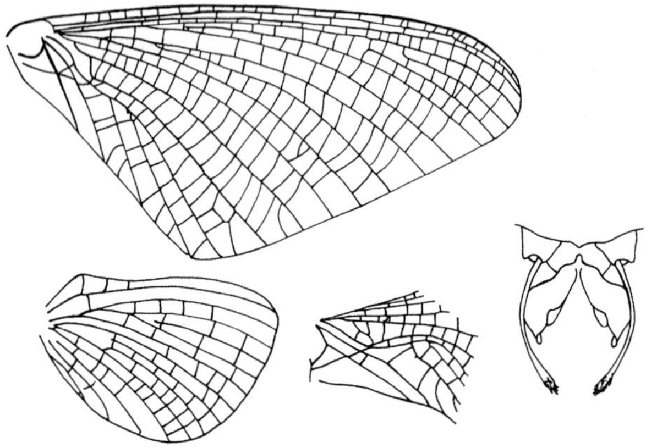

Fig. 2.42 *Campsurus evanidus:* fore and hind wings (left), a section of the fore-wing showing a variant vein pattern (lower center), and the male genitalia (lower right).

Key to the genera of known larvae in South America

Information for the key was provided by Eaton (1883-1888), Needham and Murphy, 1924, and, for general generic differences, Edmunds (1959) and Day (1963). Larvae in this family have been subject only to cursory study in South America.

1. The short, wide, robust mandibular tooth is modified for burrowing.
....*Asthenopus* Eaton, 1871..p. 53
- The mandibular tooth is elongate and narrow. (**Fig. 2.43**). 2
2. The dorsal and ventral surfaces of the mandibular tooth bear numerous tubercles.
....*Ephoron* Williamson, 1802
The only South American species reported is *Ephoron umbrata* (Hagen,1888), which inhabits Brazil. Because its original description is sketchy, this species is of uncertain status and requires reexamination (Domíguez *et al*., 2001). Its assignment to the genus *Ephoron* is therefore doubtful.

- The mandibular tooth lacks tubercles on the dorsal and external surfaces, while the internal surface is serrated and bears one or more distinct tubercles. 3

3. There are several teeth on the inner margin of the mandibular tusk, which is crenulated apically. The lateral margin of the mandible bears numerous setae. The larvae construct U-shaped tunnels. The front of the head is rounded and lacks a frontal process (**Fig. 2.43**).
....*Campsurus* Eaton, 1868..p. 55

- The mandibular tusk bears only one prominent tubercle on the inner margin. There are only a few long setae on the lateral margin of the mandible. The larvae burrow in the sediment.
....*Tortopus* Needham and Murphy, 1924..p. 82

Fig. 2.43 *Campsurus melanocephalus* larva: labrum (upper left); maxilla (upper left center); apical part of mandible and tusk (upper right center); labium (upper right); fore, middle, and hind legs (lower left, left to right); and the first and third gills (lower right, top and bottom, respectively).

Key to the adults in the genus *Asthenopus* in South America

Information for the key was taken from Berner (1978), Domínguez (1988a), and Hubbard and Domínguez (1988). Descriptions of the larvae are lacking.

1. No short intercalary veins are present at the apical margin of the fore-wing, but some cross veins, shorter than the distance between the longitudinal veins, may reach the margin. Caution: the veins of the fore-wings show sexual

dimorphism, which resulted in descriptions of males and females as different species. The fore-legs of the male are not nearly as long as the body (**Fig. 2.44**). ....*Asthenopus curtus* (Hagen, 1861) (Ecuador, Venezuela, Guyana, Surinam, Paraguay, Pará, Amazonas). Syn: *Palingenia albifilum* var. Walker, 1853; *Palingenia curta* Hagen, 1861; *Campsurus curtus* Hagen, 1861; *Campsurus amazonicus* Hagen, 1888; *Asthenopus amazonicus* (Hagen, 1888).

- Short intercalary veins produced as extensions of cross veins are present at the apical margin of the fore-wing (**Fig. 2.8**). 2

2. The prothorax of the male is twice as wide as long. The fore-legs of the male are almost as long as the fore-wings. The penis lobes are slender from the base to the apex (**Fig. 2.8**).
....*Asthenopus picteti* (Hubbard, 1975) (Guyana, Uruguay, Brazil). Syn: *Palingenia albicans* Pictet, 1943; *Campsurus albicans* (Pictet, 1843); *Asthenopodes albicans* (Pictet, 1843) = mistaken identifications; *Asthenopodes picteti* Hubbard, 1975.

- The prothorax of the male is three times as wide as long. The fore-legs of the male are about half as long as the fore-wings. The penis lobes are broad at the base and narrow toward the apex (**Fig. 2.45**). Length: 7.6 to 8.0 mm. Fore-wing length: 7.8 to 8.4 mm. Hind wing length: 3.2 to 3.8 mm. Color: mainly brownish to yellow with a blackish wash.
....*Asthenopus gilliesi* Domínguez, 1988 (Uruguay).

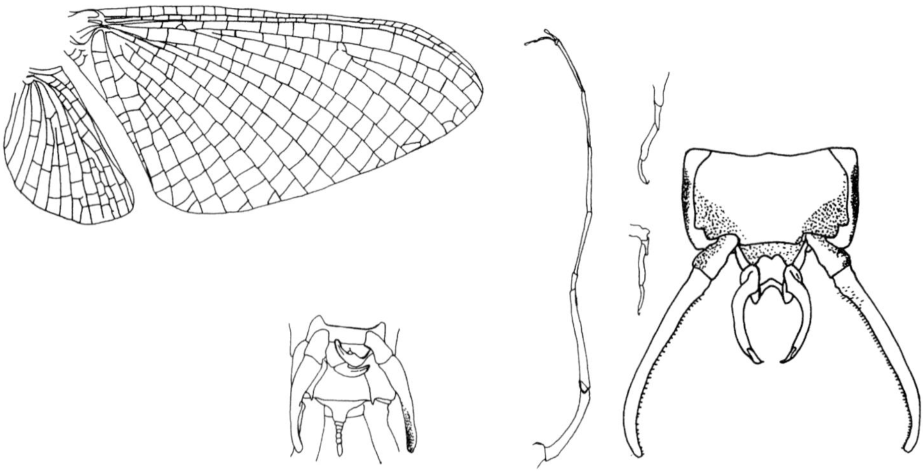

Fig. 2.44 *Asthenopus curtus:* fore and hind wing of a male (upper left); fore (center), middle (lower right center), and hind leg (upper right center); genitalia of a male in ventral view (lower left center) as illustrated by Demoulin (1955a), whose identification of the species as *Asthenopus amazonicus* was questionable; genitalia of a male in ventral view (right) as illustrated by Domínguez (1988a).

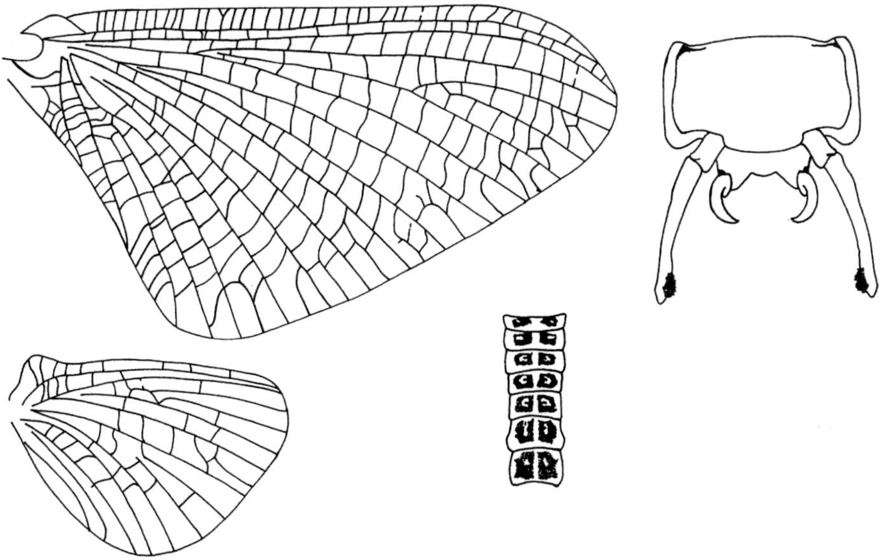

Fig. 2.45 *Asthenopus gilliesi* male: fore and hind wing (left), color pattern on the second through eighth abdominal tergites (lower center), and the genitalia in ventral view (right).

Key to the adults in the genus *Campsurus* in South America

Information for the key was taken from Percheron (1838), Eaton (1883-85), Weyenbergh (1883), Esben-Petersen (1912), Needham and Murphy (1924), Navás (1920b, 1927, 1930a, 1931a, b, 1932, 1934a), Ulmer (1920a, 1942), Spieth (1943), Traver (1944, 1947a, 1950), Puthz (1973), Pereira and da Silva (1991), and Heckman (1998a, b). Because the larvae have not been described, only a key to the adults can be provided

1. Lateral extensions of the apical sternite of the male abdomen extend well beyond the bases of the forceps and form double triangular processes on each side (**Fig. 2.46**). Length of the male: 12 mm; fore-wing length of the male: 13 mm. Color: light brown with an ochre yellow abdomen, which has a dark longitudinal stripe crossing the midline of the anterior six tergites. The wings are clouded, and the subcostal and radial veins are blackened. The female has not been described, and the description of the male is very sketchy. A redescription, particularly of the wing veins, is urgently needed.
....*Campsurus quadridentatus* Eaton, 1871 (Peru, Argentina, Pará).

- The lateral extensions of the apical sternite of the male abdomen do not extend well beyond the bases of the forceps, or they do not form double triangular processes on each side (**Fig. 2.47**). In case of doubt, the abdomen is not ochre yellow with a midline stripe on the six anterior tergites of the abdomen. 2

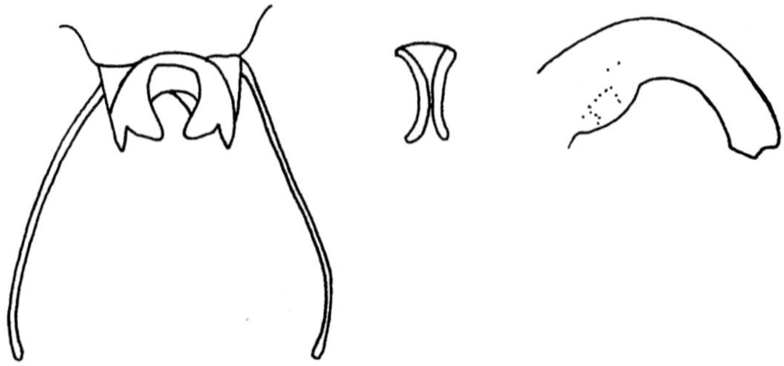

Fig. 2.46 *Campsurus quadridentatus* (left to right): the male genitalia in ventral view, the penis in posterior view with the apex pointed upward, and the penis in lateral view with the apex pointed downward.

2. The apical sternite on the abdomen of the male forms a median process that exceeds the length of the rest of the sternite at its narrowest point (**Fig. 2.47**). The cerci of the male are entirely pale and extend more than three times the length of the body. Length of the male: 9 mm; of the female: 7 mm. Fore-wing length of male: 9 mm; of female: 9 to 10 mm. Hind wing length of male: 5 mm; of female: 4 mm. Length of cerci of the male: 34 mm. Color: mainly piceus with lighter markings. The basal antenna segment and much of the abdomen is translucent; the abdomen is piceus dorsally and yellow ventrally. The female is somewhat darker than the male. This species also requires a redescription, especially with regard to the arrangement of the wing veins.
....*Campsurus litaninensis* Spieth, 1943 (Surinam).

- The median part of the apical abdominal sternite of the male does not exceed the minimum length of the sternite (**Fig. 2.48**), or the cerci of the male are not entirely pale and longer than three times the body length. 3

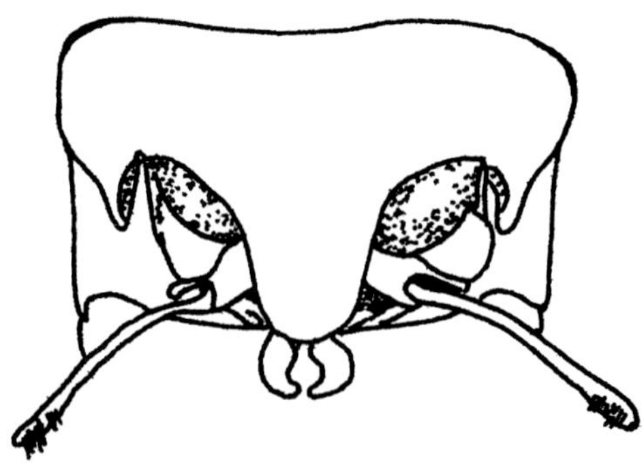

Fig. 2.47 A ventral view of the male genitalia of *Campsurus litaninensis*.

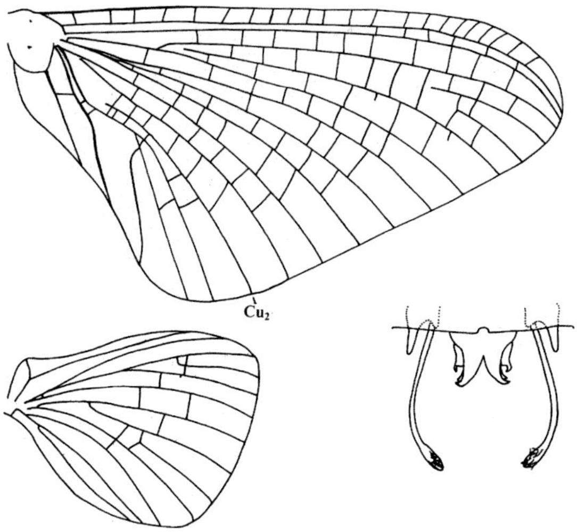

Fig. 2.48 The fore and hind wing of *Campsurus lucidus* (left) and the male genitalia (lower right).

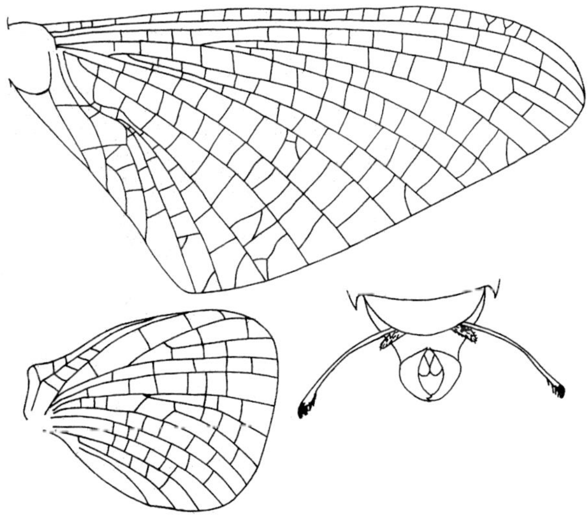

Fig. 2.49 *Campsurus violaceus:* fore and hind wing (left) and the male genitalia (lower right).

3. The proximal end of vein Cu_2 in the fore-wing runs midway between the first anal vein and the bisector of the cubital fork and usually appears to be joined to the bisector or, exceptionally, to Cu_1 itself. *C. violaceus* is placed in this group because, although Cu_2 runs closer to the first anal vein for most of its length, it forms a wide curve to join the bisector of the cubital fork at its base (**Fig. 2.49**). 4

- The proximal end of vein Cu_2 in the fore-wing is closer to the first anal vein and usually appears to arise from it (**Fig. 2.50**). 18

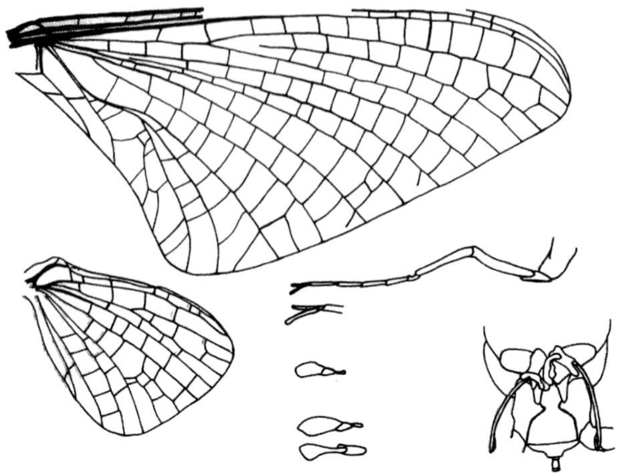

Fig. 2.50 *Campsurus notatus:* the wings showing damage to the specimen's fore-wing (left); the fore-leg with an enlargement of the claw, the middle leg, and two views of the hind leg (center, top to bottom); the male genitalia in ventral view (lower right).

4. Vein Cu_2 ends proximally in a disconnected spur (**Fig. 2.48**). Small species, generally no longer than about 7 mm. 5

- Vein Cu_2 joins the bisector, or in the females of some species, Cu_1 itself, at the end of a rounded curve (**Fig. 2.42**). 9

5. There are fewer than 10 cross veins in the hind wing (**Fig. 2.48**). Body length of male: 6 mm. Fore-wing length of male: 7 mm. Color: pale with a black head. ....*Campsurus lucidus* Needham and Murphy, 1924 (Colombia, Peru, Guyana, Argentina, Brazil).

- There are much more than 10 cross veins in the hind wing. The fore-legs are dark with paler tarsi (**Fig. 2.50**). 6

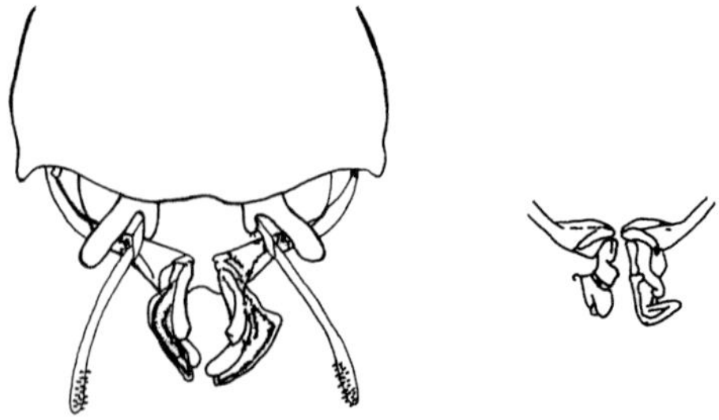

Fig. 2.51 The male genitalia and details of the penis of *Campsurus cuyuniensis.*

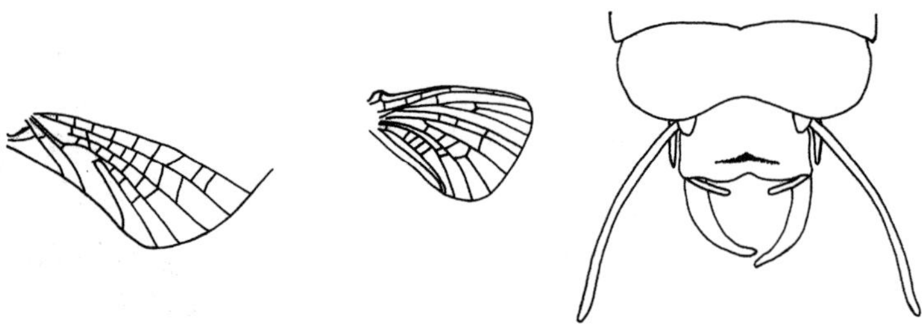

Fig. 2.52 *Campsurus indivisus* (left to right): the veins in the inner, posterior part of the fore-wing, encompassing the cubital and anal regions; the hind wing; the male genitalia in ventral view.

6. The first through eighth abdominal segments are yellowish, paler than the thorax, and pale between the segments with only a slight darkening of their posterior margins. The ninth and tenth segments are brownish black. There is a black line between the eyes. The apices of the penis lobes are blunt and finger-like (**Fig. 2.51**). Length: 5 to 7 mm; fore-wing length: 5.5 to 7 mm. Color: the head and thorax are mainly reddish brown.
....*Campsurus cuyuniensis* Traver, 1947 (Guyana). This and the species in the following couplet belong to the *Campsurus latipennis* group, as defined by Traver (1947a), who expressed doubt about the true identity of *C. latipennis* and the validity of the species he described. However, he insisted that more than one species belongs to this group. The similarity of the genitalia and other common morphological features leaves a good deal of doubt about the validity of all three species.
- The anterior portion of the abdomen is not yellowish and paler than the thorax. The apices of the penis lobes are not blunt and finger-like (**Fig. 2.52**). 7
7. The head is brownish yellow with wide black borders around the ocelli and black lines along the midline. The pronotum is light grayish violet dorsally and yellowish laterally. The mesonotum is brownish yellow with five dark brown longitudinal stripes, and the metanotum is yellow. The abdomen is light gray with blackish gray markings dorsally that are not largely confined to prominent bands along the posterior margins and dorsal midline. Ventrally, the abdomen is yellow. The apical sternites of the male are excavated medially (**Fig. 2.52**). Body length: 7 mm; length of the fore-wing: 7.5 mm; length of the cerci: about 23 mm in the male and 3 mm in the female.
....*Campsurus indivisus* Ulmer, 1942 (Brazil).
- The abdominal segments are prominently banded along the posterior margins and may also have markings along the dorsal midline. The underside of the abdomen is not mainly yellow. 8
8. The thorax is dark yellow, and the abdomen is "umber grey, more distinctly so" on the fourth to tenth segments. They are darker where the segments join and on the "dorsal vessel". The ventral part of the abdomen is testaceus. The

penis lobes are rounded at their apices. A description of the male genitalia has been provided, but it is not certain that it pertains to this species (**Fig. 2.53**).
....*Campsurus latipennis* (Walker, 1853) (Guyana, Brazil). Syn: *Palingenia latipennis* Walker, 1853.
- The head and thorax are dark reddish brown, and the pronotum is yellowish brown. There are blackish markings along the mid-ventral line of the thorax, on the sternal borders of the prothorax, and at the wing roots. The abdomen is yellow with wide black posterior borders of the segments and a black dorsal midline. The sternites have a blackish brown shading. The penis lobes end in acute processes (**Fig. 2.54**). Length: 6 mm; fore-wing length: 6 mm.
....*Campsurus essequibo* Traver, 1947 (Guyana).

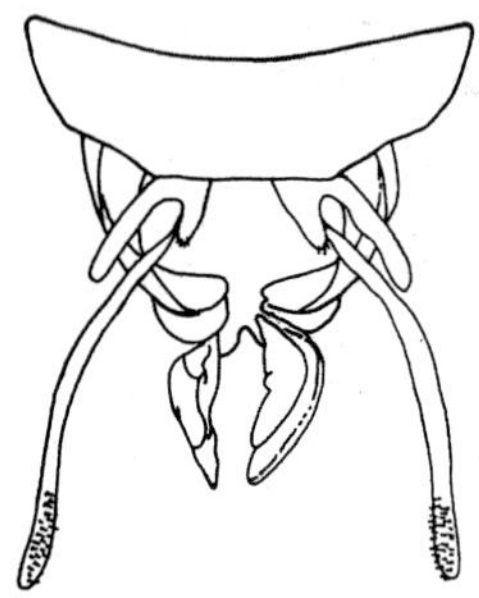

Fig. 2.53 The male genitalia of a specimen believed by Traver (1947a) to be a true *Campsurus latipennis*.

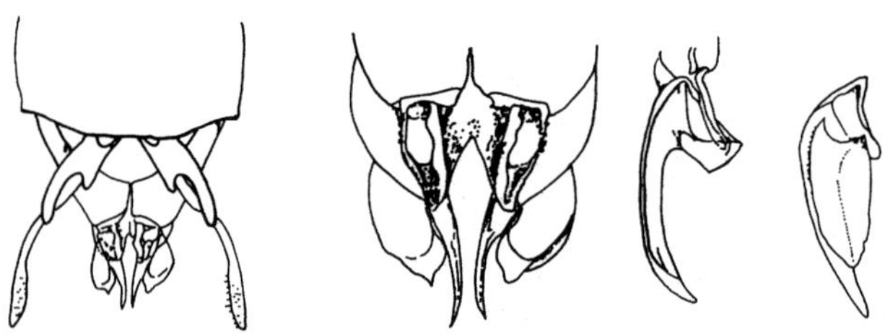

Fig. 2.54 The male genitalia (left), penis (center), and two views of penis lobes of *Campsurus essequibo*.

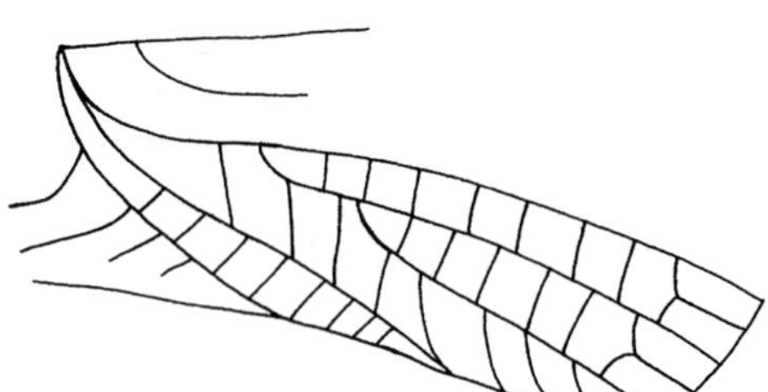

Fig. 2.55 The veins near the base in the posterior part of the fore-wing of a male *Campsurus paraguarius*.

9. The length of the male is about 6 mm, and that of the female is about 4.6 mm. The fore-wing of both is about 9.0 to 9.3 mm long (**Fig. 2.55**). The hind wing is 3.8 to 4.4 mm long. The proximal halves of the costal and subcostal spaces are lightly tinted with a dark violet color. The wing veins are mainly white. The head is black dorsally and yellowish white ventrally, and the pronotum is whitish. The abdomen is mainly whitish with dark violet pigment near the apex and a pair of violet spots on the eighth tergite. The species is not well described. ....*Campsurus paraguarius* Navás, 1920 (Paraguay). Spelling error: *Campsurus paraquarius* Navás, 1920.

- The length of the body is at least 8 mm. 10

10. There is a combined total of more than 10 cross veins in the costal and subcostal spaces of the hind wing, and the cross veins are distributed as far as the distal ends of the spaces (**Fig. 2.56**). 11

- There are fewer than six cross veins in the costal and subcostal spaces of the hind wing, usually confined to the proximal half of these spaces. There may be no cross veins except for the humeral. The total length does not exceed about 12 mm (**Fig. 2.42**). 13

11. The head and body are almost entirely light yellow with only a slight darkening on the fore-tibiae and black compound eyes and ocelli. The posterior margin of the tenth tergite is wide and nearly straight (**Fig. 2.56**). Length of male: 13 mm. Fore-wing length of male: 20 mm. ....*Campsurus major* Needham and Murphy, 1924 (Argentina).

- The head is blackish dorsally, and there are extensive dark markings on the thorax and abdomen or at least distinctly darkened bands along the posterior margins of the abdominal tergites. 12

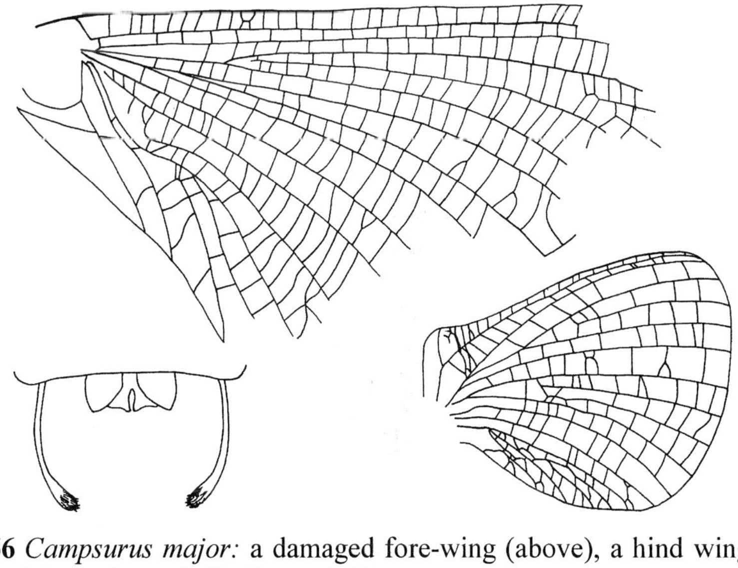

Fig. 2.56 *Campsurus major:* a damaged fore-wing (above), a hind wing (lower right), and the male genitalia (lower left).

12. Vein M_2 in the fore-wing forms an arc as it originates from the medial intercalary vein; it is joined by cross veins to both the medial intercalary and Cu_1 as it runs an equal distance from each (**Fig. 2.57**). The description by Ulmer (1942) is imprecise but suggests that its size and approximate coloration is the same as that of *C. dorsalis*.
....*Campsurus burmeisteri* Ulmer, 1942 (Brazil).
- Vein M_2 in the fore-wing seems to arise from a fork in Cu_2. The penis lobes end in claw-like processes (**Fig. 2.58**). Body length of the female: 11 to 16 mm. Fore-wing length of female: 15 to 22 mm. Length of the cerci: about 7 mm.
....*Campsurus dorsalis* (Burmeister, 1839) (Paraguay, Espirito Santo). The species description was provided by Ulmer (1942) based on specimens collected and identified by Pictet because the type specimens of Burmeister could not be found. Syn: *Palingenia dorsalis* Burmeister, 1939; *Asthenopus dorsalis* (Burmeister, 1839); *Tortopus dorsalis* (Burmeister, 1839).

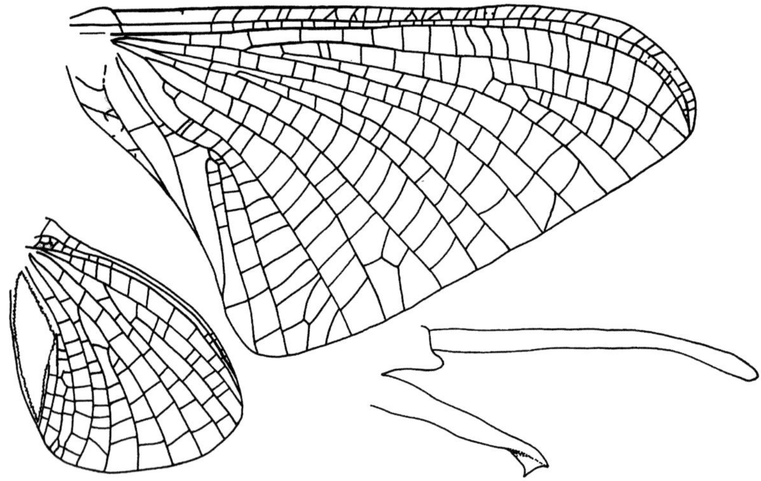

Fig. 2.57 *Campsurus burmeisteri:* fore and hind wings with a section torn out of the hind wing (left) and a lateral view of the male genital appendages (lower right).

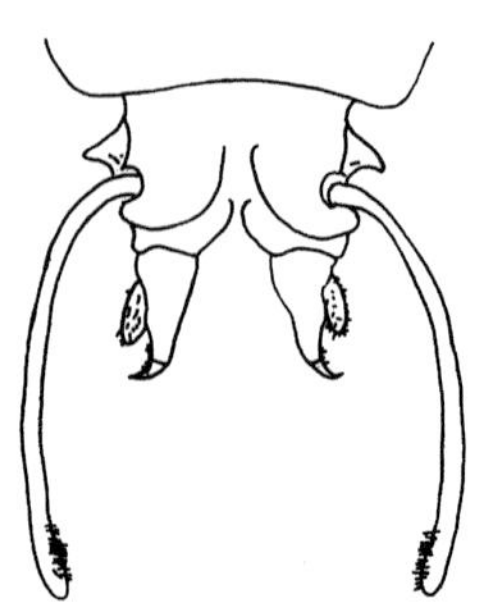

Fig. 2.58 The male genitalia of *Campsurus dorsalis.*

13. The length is 12 mm. There is a large, square, dark spot on the pronotum and two dark stripes on the mesonotum, which continue onto the metanotum, where they run together. The pattern on each abdominal segment is distinct. There are no cross veins in the costal and subcostal areas of the hind wing, except for one running from vein Sc to the peak of the costal prominence (**Fig. 2.59**).
....*Campsurus holmbergii* (Weyenbergh, 1883) (Argentina). Syn: *Ephemera holmbergii* Weyenbergh, 1883. The identity of this species is doubtful, and the wings in **Fig. 2.59** are from a specimen only presumed by Needham and Murphy (1924) to belong to this species. However, the features they introduced into their key do not conform exactly to the original description of Weyenbergh (1883). A lectotype should be designated if it is impossible to find the holotype for re-examination.
- The length is no greater than 11.5 mm. 14

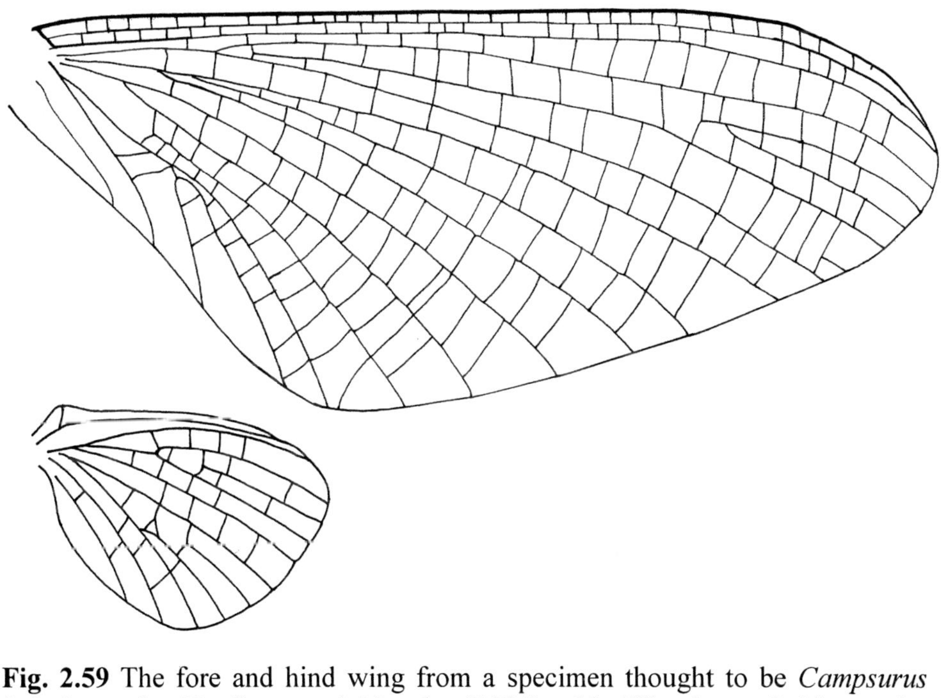

Fig. 2.59 The fore and hind wing from a specimen thought to be *Campsurus holmbergii* by Needham and Murphy (1924). Identifications of this species must remain tentative until a holotype is examined or a lectotype designated and a better description is published.

14. The two long intercalary veins are usually attached separately to the first anal vein in the fore-wing. There are two pairs of dark dorsal spots on the mesothorax, one anterior to and one posterior to the pale convex middle portion. The fore-legs are yellow tinged with purple. The basal curve at the origin of

vein Cu_2 is at the level of the fork between veins M_3 and M_4 (**Fig. 2.42**). Body length: 8 to 10.5 mm. Wing length: 8 to 12 mm. Caudal filament length of male: more than 26 mm.
....*Campsurus evanidus* Needham and Murphy, 1924 (Minas Gerais).
- The two long intercalaries join each other distal to their junction with the first anal vein of the fore-wing (**Fig. 2.60**). There is no tetrad of dark spots on the mesothorax, although there may be one pair of dark longitudinal stripes. The fore-legs are purplish or sepia, or they are yellowish white with a dark stripe on each tibia. 15

15. The fore-legs are yellowish white with a dark stripe on each tibia. The head is yellow with black on the vertex and occiput. The pronotum has a broad, dark median stripe narrowly divided in the middle, which fades out toward the posterior part. The mesonotum and metanotum are brown with a pair of dark longitudinal stripes on the mesonotum and a black marking on the metanotum that divides anteriorly to form a V. The abdomen is light yellow ventrally, and dorsally, it is orange yellow with a dark median stripe that is interrupted along the posterior margins of the tergites. The fore-wing is hyaline with dark veins. Vein Cu_2 in the fore-wing runs nearly equidistant from the bisector and first anal vein or slightly closer to the bisector (**Fig. 2.60**). Length of the male: 9.2 mm. Fore-wing length: 10 mm. Hind wing length: 4.7 mm.
....*Campsurus juradinus* Navás, 1930 (Argentina).
- The fore-legs are purplish or sepia. 16

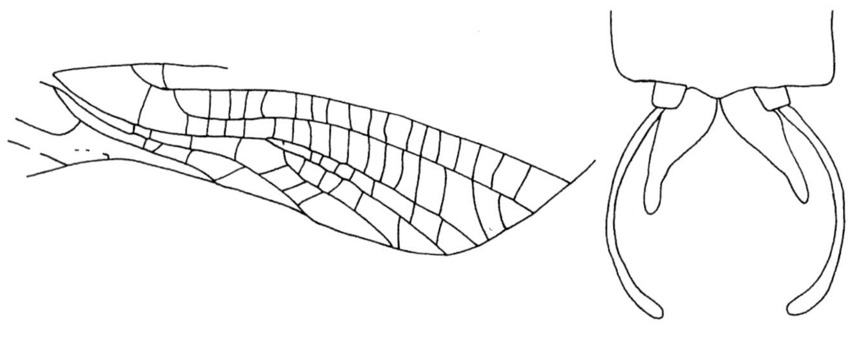

Fig. 2.60 *Campsurus juradinus:* the veins of the inner posterior section of the fore-wing (left) and the male genitalia (right).

16. The dorsal part of the prothorax has a large, round, dark median spot on a pale background behind the transverse suture. This spot is connected to the black marking on the occiput by a broad trapezoidal band. There is a pair of elongated pale spots on each abdominal segment aligned to point posteriad; they are located within the dark brown mid-dorsal stripe. Vein Cu_2 in the fore-wing runs closer to the first anal vein than to the bisector but curves widely to join the

bisector at its base. The paraprocts of the male extend nearly laterad (**Fig. 2.49**). Body length: 10 to 11.5 mm. Wing length: 10 to 14 mm. Caudal filaments of male: more than 32 mm.
....*Campsurus violaceus* Needham and Murphy, 1924 (Argentina).
- The dorsal part of the prothorax is not marked with a large, round, dark median spot connected to the black marking on the occiput. Pale spots in the brown mid-dorsal stripe on the abdomen, if present, are roundish (**Fig. 2.61**).
....17

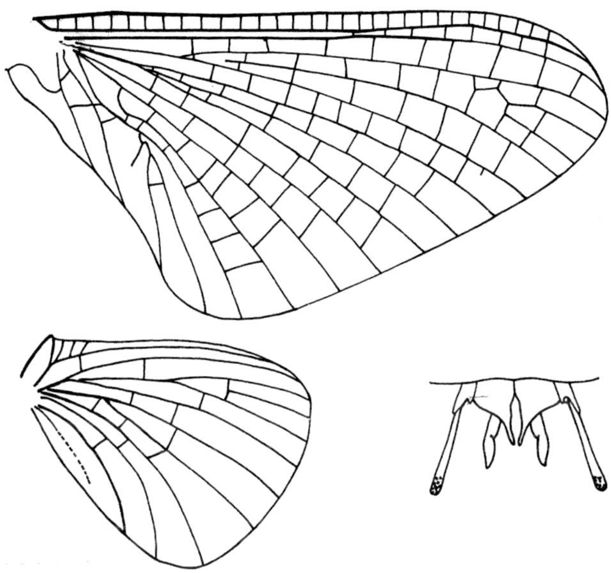

Fig. 2.61 *Campsurus scutellaris:* fore and hind wings (left) and the male genitalia (lower right).

17. Each abdominal segment has one pair of pale spots on the area of the brown mid-dorsal stripe (**Fig. 2.61**). Length of male: 8.5 mm. Length of wing: 7 mm.
....*Campsurus scutellaris* Needham and Murphy, 1924 (Argentina).
- Each abdominal segment has two pairs of pale spots on the area of the brown mid-dorsal stripe. The penis lobes are very narrow (**Fig. 2.62**). Length of male: 9 mm; female: 9.5 mm. Fore-wing length of male:10 mm; female: 11 mm. Color: gray or brown with violet and yellow markings. The genitalia depicted in **Fig. 2.62** conform with the illustrations of Esben-Petersen (1912), which Ulmer (1942) found to be inaccurate in some details. However, the discrepancies seem to have arisen due to the perspective of the drawings.
....*Campsurus jorgenseni* Esben-Petersen, 1912 (Argentina).

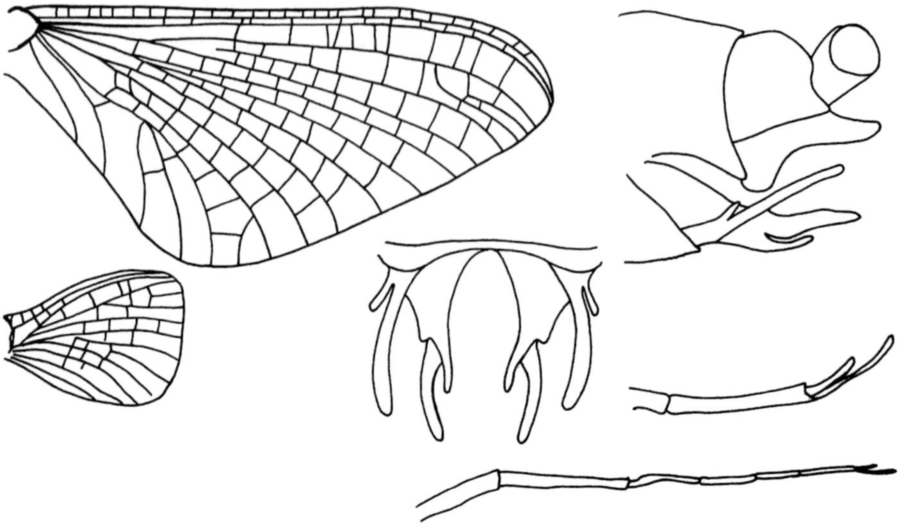

Fig. 2.62 *Campsurus jorgenseni:* fore and hind wings (left), the fore-leg of a male with an enlargement of the apical tarsal segment and claw (lower right), and the male genitalia in lateral (upper right) and ventral view (center).

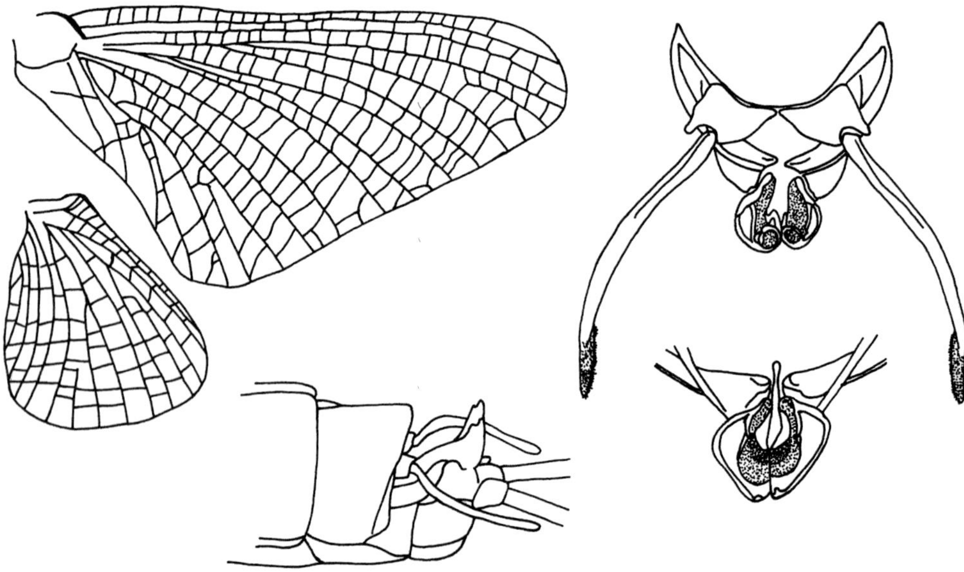

Fig. 2.63 *Campsurus brasiliensis:* fore and hind wing slightly restored from the original illustration (left), the male genitalia (upper right) and enlargement of the penis (lower right), and the male genitalia in lateral view with only the bases of the cerci (lower center). Traver (1944) reported that the left and right wing of his specimen differ in the position of vein M_2. In the wing not shown here, the vein appears to curve toward and join vein IM_1.

18. Vein Cu_2 forms a small curve where it joins A_2 at the proximal end of the enclosed space; the enclosed space is rather blunt. The anal intercalary veins arise from a common stalk (**Fig. 2.63**). Coloration: mainly pale. 19
- Vein Cu_2 runs very close to vein A_2; the enclosed space is acutely lanceolate, but may be slightly curved (**Fig. 2.50**). 25

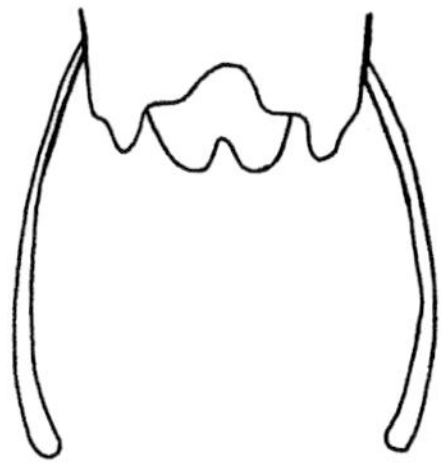

Fig. 2.64 The apex of the abdomen of a male *Campsurus dallasi.*

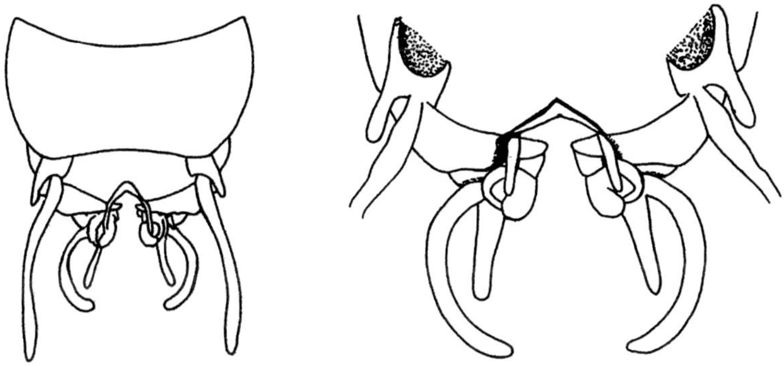

Fig. 2.65 The male genitalia (left) of *Campsurus ulmeri* and the structures of its penis enlarged (right).

19. The male is about 12 mm in length or longer. The fore-wing length of the male is greater than 12 mm. 20
- The body and wing length of the male is no longer than about 11 mm. 21
20. The body length of the male is 13 to 17.5 mm; its fore-wing length is 15 to 16 mm. The pale areas of the body are whitish. The posterior borders of the tenth tergite and ninth sternite at the apex of the abdomen of the male are convex (**Fig. 2.63**). Caudal filament length in male: 40 to 45 mm. The compound eyes, a ring around their bases, and a V-shaped mark between the eyes are black. There is a purplish red spot on the upper part of each ocellus. There is a grayish brown triangle pointed anteriad on the pronotum with purplish spots at each end and at the posterolateral angles of the pronotum. There are deep anterior and lateral grooves on the mesonotum that are amber in color. Areas C, Sc, and R on the fore-wing are purplish. There are pale, smoky lateral patches on a white background on the tergites. Traver (1944) reported that this species is very similar to *Campsurus major* and should be in the same group in

the key provided by Needham and Murphy (1924). However, the position of vein Cu_2 in his diagram (**Fig. 2.63**) shows that this is not the case.
....*Campsurus brasiliensis* Traver, 1944 (Rio Grande do Sul).

- The male is about 12 mm long, and its fore-wing is about 12.3 mm. The pale areas of the body are yellow, yellowish brown, or yellowish gray. The posterior borders of the tenth tergite and ninth sternite at the apex of the abdomen of the male are emarginate (**Fig. 2.64**). The head is yellow. There is blackish pigment around the eyes. The thorax is yellowish brown with narrow dark lines on the mesonotum. The abdomen is yellowish gray with dark longitudinal lines on the tergites which broaden in some segments into transverse markings.
....*Campsurus dallasi* Navás, 1927 (Argentina).

21. The body and wing length of the male is 11 mm. The wings have few cross veins. The penis lobes are widely and deeply separated at the base (**Fig. 2.65**). Color of specimens in alcohol: head, whitish below with brownish and blackish markings dorsally; thorax smoky yellowish with black markings; abdomen yellowish white with black markings; wing membrane and veins hyaline, except for the violet grey longitudinal veins.
....*Campsurus ulmeri* Traver, 1950 (Santa Catarina).

- The body of the male does not exceed about 9 mm. If the fore-wing exceeds 9 mm, then it is much longer than the body, and the wings have many cross veins in the basal portions of the subcostal and costal spaces (**Fig. 2.66**). 22

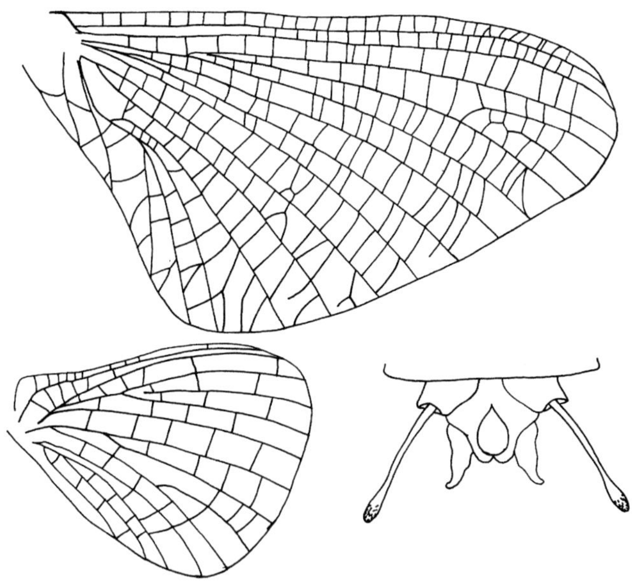

Fig. 2.66 *Campsurus pallidus:* the fore and hind wing (left) and the male genitalia (lower right).

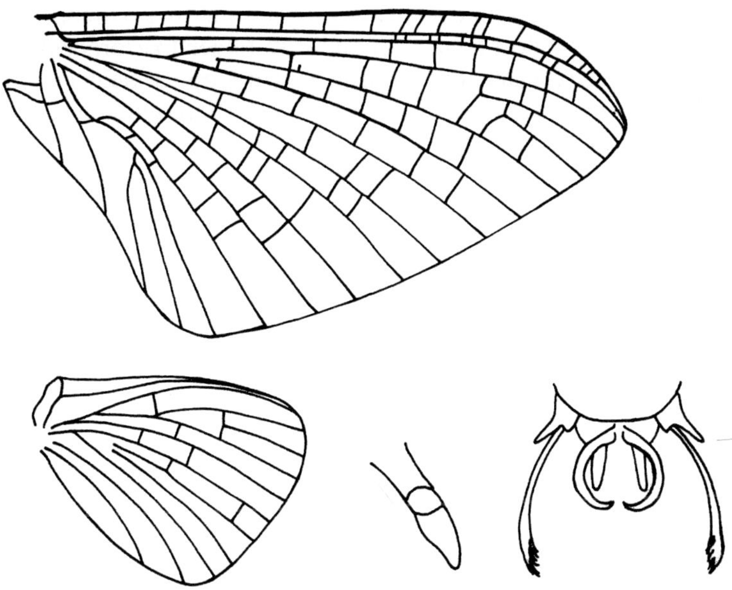

Fig. 2.67: *Campsurus segnis:* fore and hind wings (left), the vestige of the middle leg (lower center), and the genitalia of the male.

22. Cross veins are abundant in the hind wing, and numerous such veins are present in the basal portions of the subcostal and costal spaces (**Fig. 2.66**). Length of male: 9 mm. Fore-wing of male: 12 mm.
....*Campsurus pallidus* Needham and Murphy, 1924 (Argentina, Paraguay)
- There are few cross veins in the hind wing and none are in the costal and subcostal spaces except for the humeral (**Fig. 2.67**). 23
23. The thorax is rather uniformly pale brown, slightly darker along the sutures of the plates (**Fig. 2.67**). Body length: 7 to 7.5 mm. Wing length: 7 to 8 mm. Length of caudal filament of male: 21 mm.
....*Campsurus segnis* Needham and Murphy, 1924 (Guiana, Surinam, Pará).
- There are prominent dark markings on a light background on the pronotum. 24
24. The thorax is pale with a prominent dark, T-shaped area on the pronotum with the top of the T occupying the entire anterior third of the pronotum. The mesonotum bears dark longitudinal areas that join in a U form posteriorly. There are two relatively short apical segments at the end of each forceps arm (**Fig. 2.68**). Body length: 8 to 10 mm; fore-wing length: 7 to 16 mm.
....*Campsurus pedicellarius* Spieth, 1943 (Guyana, Surinam). There is a great similarity between this species and *Campsurus segnis*, making their status somewhat questionable.

- The pronotum has two dark lines close to the midline that extend from the anterior margin to the middle of the "notum." A finger-like process extends from the basal segment of each forceps arm, and the apical segment of each forceps arm is by far the longest segment (**Fig. 2.69**). Fore-wing length: 7 to 8 mm. The female has not been described.
....*Campsurus duplicatus* Spieth, 1943 (Guyana, Amazonas).

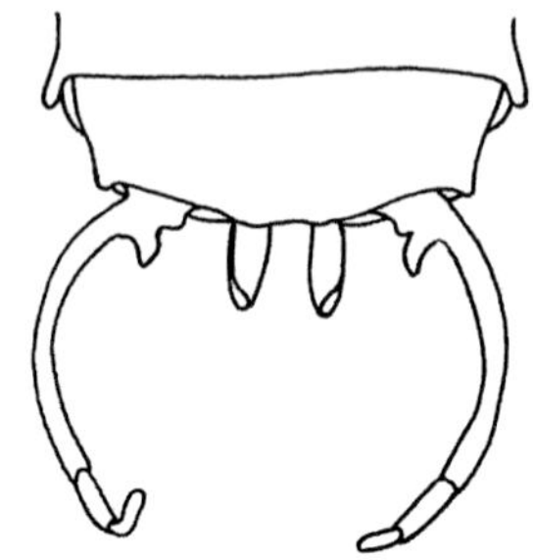

Fig. 2.68 A ventral view of the male genitalia of *Campsurus pedicellarius.*

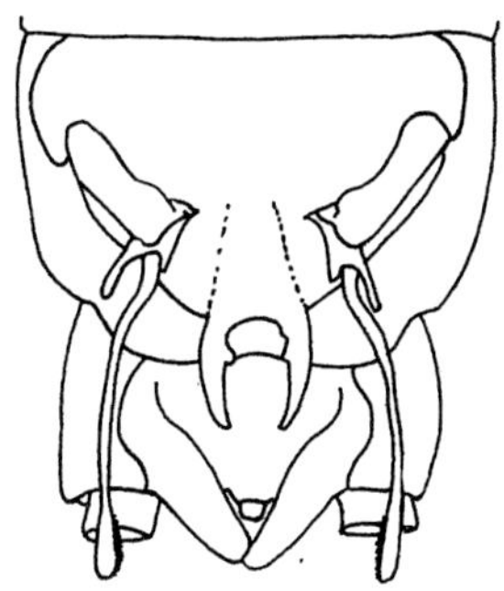

Fig. 2.69 A ventral view of the male genitalia of *Campsurus duplicatus.*

25. The abdominal tergites of the male are yellowish or grayish yellow with a blackish or dark gray midline and variable, paired lateral markings, which become larger on each successive tergite toward the apex of the abdomen. The lateral markings sometimes merge with the midline. The posterior margins of the tergites are not darkened. The posterior margin of the subgenital plate is lightly emarginated in the middle (**Fig. 2.70**). Length: 10 to 12 mm. Fore-wing length: 13 to 14 mm. Length of the cerci: about 35 mm. Color: the head is brown or reddish brown with black around the ocelli; the pronotum has a fine, dark median line surrounded by brownish violet and yellow laterally. The mesonotum and metanotum are yellowish brown with blackish lines on the mesonotumm.
....*Campsurus truncatus* Ulmer, 1920 (Peru, Bolivia, Espirito Santo).

- If two dark paired markings are present on each of the abdominal tergites, then there is no dark midline, or the posterior margins of the tergites are darkened. 26

26. There is a pair of pale, bracket-shaped marks placed back to back on the dorsum of each of the middle abdominal segments (**Fig. 2.71**). 27

- There are no bracket-shaped marks placed back to back on the dorsum of the abdominal segments. 28

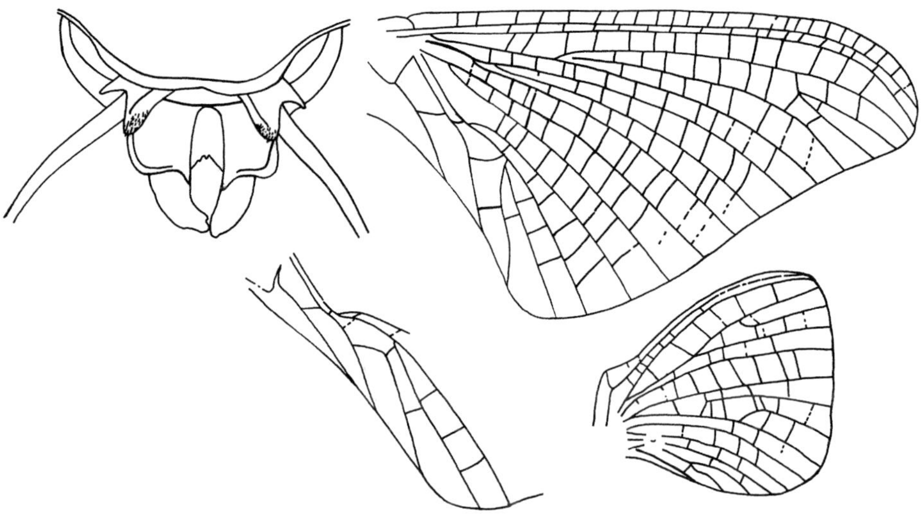

Fig. 2.70 *Campsurus truncatus:* the male genitalia in ventral view with the apices of the appendages removed (upper left), the fore and hind wings (right), and the anal section of the fore-wing of another specimen.

27. There are short setae on the inner surface of the forceps limbs. The head and pronotum are light with a dark pattern (**Fig. 2.72**). ....*Campsurus albifilum* (Walker, 1853) (French Guiana, Argentina, Brazil).

- The inner surface of the forceps limbs lack obvious short setae (**Fig. 2.71**). The vertex of the head is entirely covered with black pigment. Length: 14 to 16 mm. Fore-wing length: 13 to 15 mm. The thorax is mainly brown with dark brown markings, and the abdomen is light yellow with brown markings. .... *Campsurus melanocephalus* Pereira and da Silva, 1991 (Rio de Janeiro). A comparison of the male genitalia will show that this name is almost certainly a junior synonym of *Campsurus truncatus* (see couplet 25).

28. The apical margins of the abdominal segments are pale. 29
- The apical margins of the abdominal segments are fuscous or blackish. 36

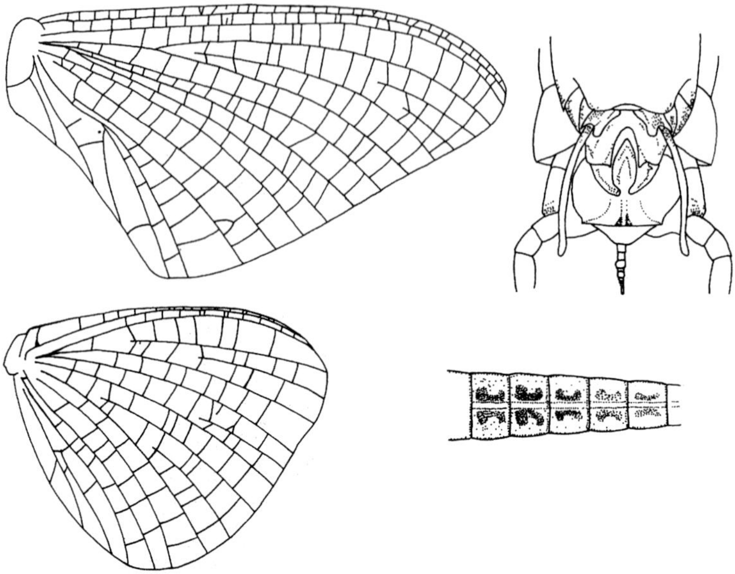

Fig. 2.71 *Campsurus melanocephalus:* fore and hind wing (left), pattern on the fourth through eighth abdominal tergites (lower right), and the apex of the abdomen of a male in ventral view (upper right).

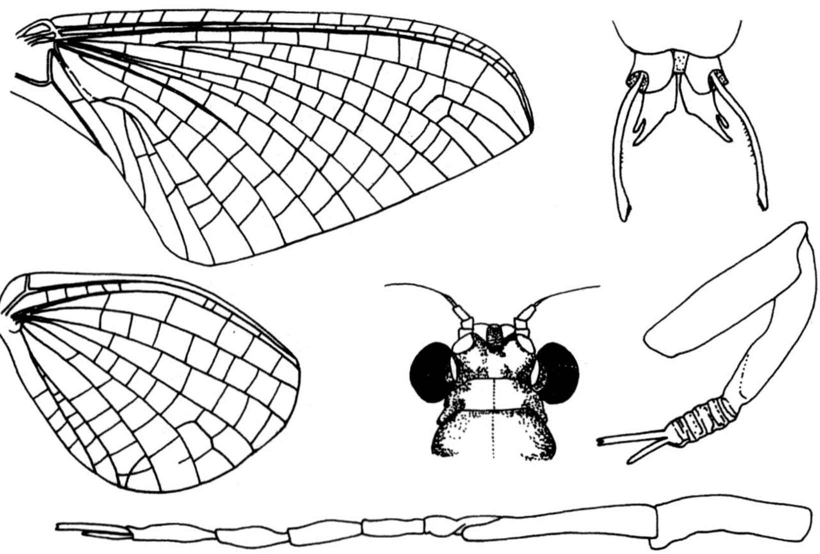

Fig. 2.72 *Campsurus albifilum:* fore and hind wings (left), the head of an adult male (center), male genitalia (upper right), the fore-leg of a subadult (middle right) and an adult (below).

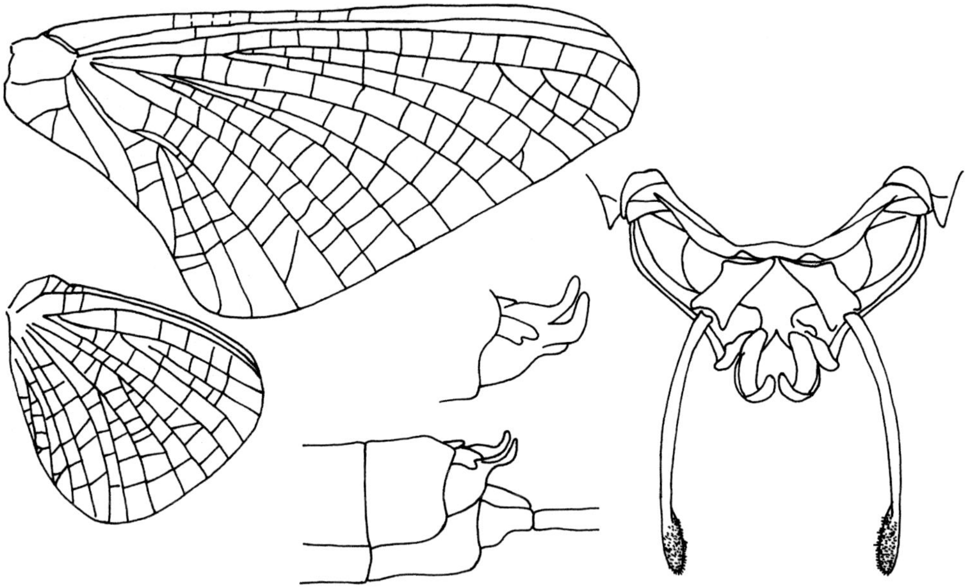

Fig. 2.73 *Campsurus assimilis:* fore and hind wing (left), male genitalia in ventral (right) and lateral view (lower center), and an enlargement of the penis in lateral view (center).

29. The pale markings on the body are whitish. There are no cross veins in the apical parts of areas C and Sc in the hind wing (**Fig. 2.73**). Length of male: 17 mm. Length of male fore-wing: 15 mm..
....*Campsurus assimilis* Traver, 1944 (Rio Grande do Sul).
- The pale markings on the body are yellowish. The total length of the male without caudal filaments is less than 15 mm. 30
30. The head, except for the blackish eyes, is brownish yellow, light brown, or rust color dorsally and yellow ventrally. The body is light yellow with few markings. 31
- The head is mainly blackish dorsally, and at least some thoracic or abdominal segments have prominent dark middorsal stripes, bands, or saddle marks. 33
31. In the fore-wing, the cubital veins are strongly curved in one direction to form arcs. More than two cross veins connect each pair of anal veins (**Fig. 2.74**). The head is light yellowish brown with blackish brown eyes. Length: 13 mm. Length of fore-wing: 18 mm; hind wing: 7 mm. Length of the cerci: 40 to 43 mm.
....*Campsurus wappaei* (Weyenbergh, 1883) (Argentina). Syn: *Ephemera wappaei* Weyenbergh, 1883.
- The cubital veins are either fairly straight or are somewhat sigmoid. In any case, they do not form regular arcs (**Fig. 2.75**). 32

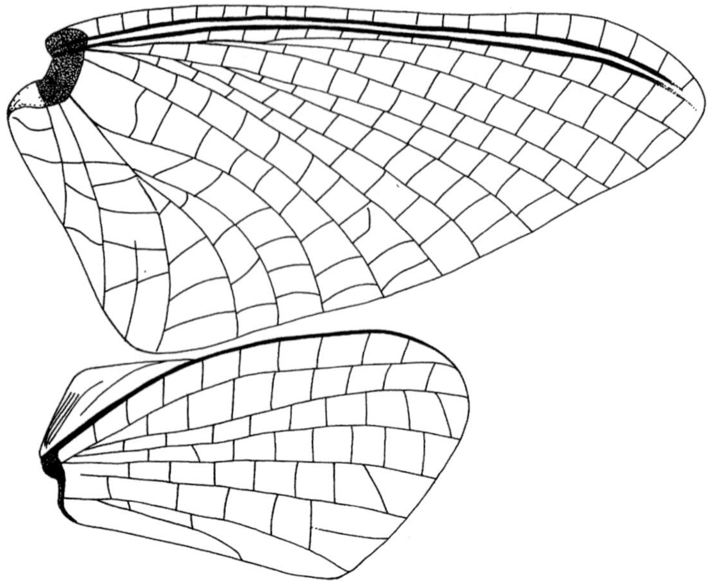

Fig. 2.74 The fore and hind wing of *Campsurus wappaei*.

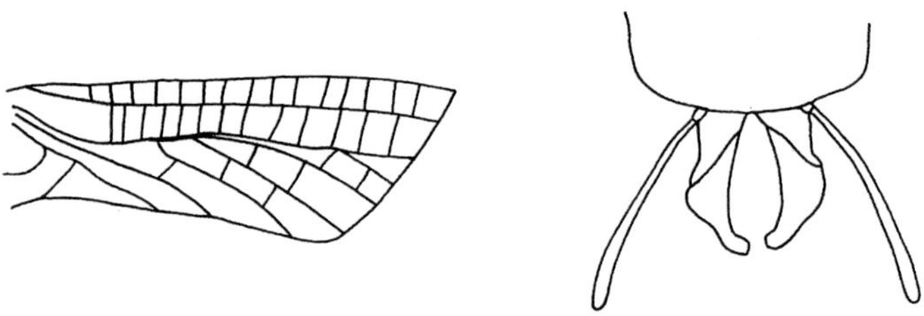

Fig. 2.75 *Campsurus pfeifferi*: the veins of the inner, posterior part of the fore-wing, called the anal and cubital sections (left), and the male genitalia (right).

32. The cubital veins run fairly straight and parallel to the edge of the wing. No more than two cross veins connect any pair of anal veins. The head is rust color above and yellow ventrally with black eyes. The basal segment of the male forceps is very small. The apical segment of the abdomen is smoothly rounded and nearly truncate without any sign of posterolateral processes (**Fig. 2.75**). Length of male: 12.5 mm. Length of male fore-wing: 12.5 mm. ....*Campsurus pfeifferi* Navás, 1931 (Argentina). Due to the poor descriptions of the species in couplets 31 and 32, the differences seem minimal. Although these species are all apparently valid, it will be difficult to distinguish them with certainty and to include them in any future taxonomic revisions unless well preserved and undamaged lectotypes are designated.

- The cubital veins are fairly straight, or they curve abruptly near the forks and then run straight or curve slightly in the other direction. There are more than two cross veins between some pairs of anal veins. The head is bright yellow with black eyes. The wing is hyaline with brownish veins and a blackened area between veins Sc and R. The apical segment of the abdomen of the male bears distinct triangular posterolateral processes (**Fig. 2.76**). Length: 12.7 mm. ....*Campsurus albicans* (Percheron, 1838) (Brazil). Syn: *Ephemera albicans* Percheron, 1839; *Palingenia albicans* (Percheron, 1838). The identity of this species must be regarded as dubious, and the designation of lectotypes would be extremely helpful to overcome confusion caused by authors who redescribed this species using misidentified specimens. Pictet (1841-43) provided a detailed description, supposedly of this species, together with an excellent color illustration of *Asthenopus picteti,* which he had apparently mistaken for the species of Percheron (1838). Eaton (1883-88) provided another description, supposedly of *Campsurus albicans.* He described an insect about 10 mm long with a wing span of about 30 mm, a black head, and a pronotum that is pale with a violet coloration laterally. The rest of the insect was mainly yellow, and the fore-wing and its veins had a violet tint. The identity of Eaton's specimen is unknown.

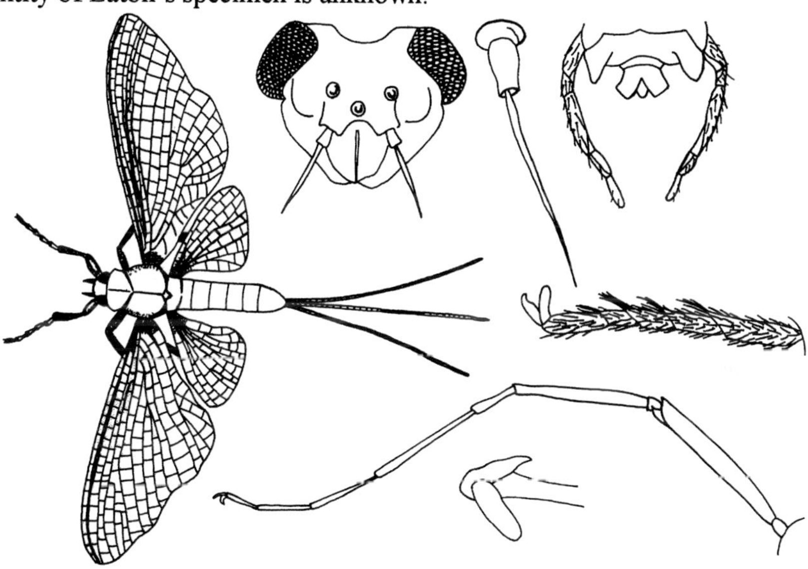

Fig. 2.76 *Campsurus albicans:* habitus (left), head in frontal view (upper left center), antenna (upper right center), male genitalia (upper right), tarsal segments of middle or hind leg, fore-leg, and an enlargement of the tarsal claws (lower right, top to bottom). This illustration, based on the drawing with the original description (Percheron, 1838), is apparently not morphologically accurate, considering the number of abdominal segments shown.

33. Each of the abdominal segments bears a purplish brown saddle mark. The head is blackish with a narrow brown line across the occiput. The basal segment

of the male forceps is stout, cylindrical, and elongated into an obtuse triangular lobe (**Fig. 2.50**). Length of male: 13.5 mm. Length of male fore-wing: 11.5 mm. Length of male caudal filaments: more than 30 mm.
....*Campsurus notatus* Needham and Murphy, 1924 (Argentina, Paraguay, Mato Grosso do Sul).
- The thorax and abdomen are yellowish with one to three dark longitudinal stripes. 34

34. The mesonotum is reddish brown with two dark longitudinal stripes. The abdomen is marked with three dark longitudinal stripes on a yellow background. The apical sternite of the male is broadly excavated (**Fig. 2.77**). Length of male: 9.5 mm. Fore-wing length of male: 12 mm; hind wing: 5.2 mm. Color: head blackish above and reddish brown below. The thorax and abdomen are mainly yellowish and brown with darker markings.
....*Campsurus paranensis* Navás, 1932 (Argentina).
- The thorax and abdomen are mainly yellowish or whitish yellow with a dark mid-dorsal stripe that is interrupted at the posterior margins of the segments. 35

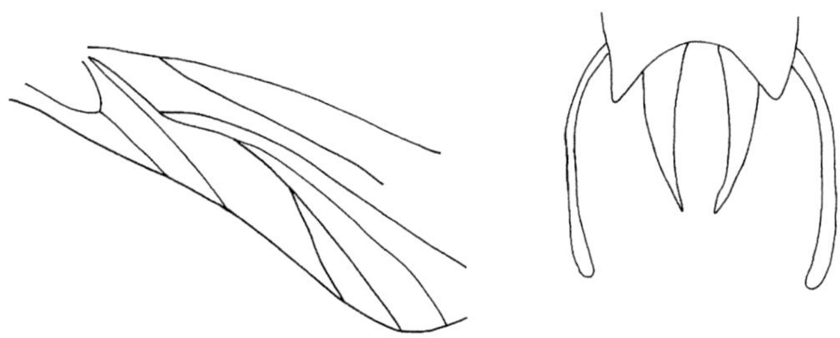

Fig. 2.77 The veins of the inner, posterior region of the fore-wing (left) and the male genitalia in ventral view (right) of *Campsurus paranensis.*

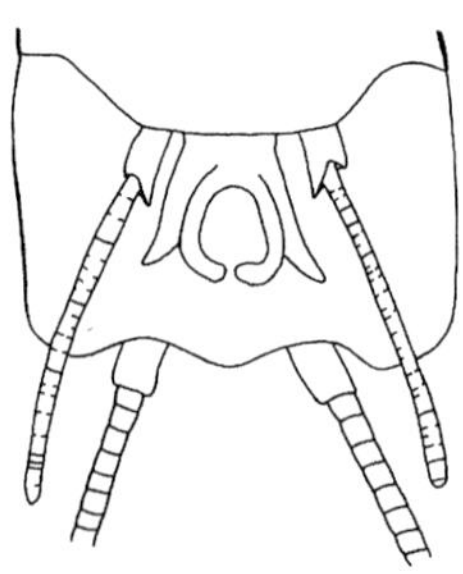

Fig. 2.78 The male genitalia of *Campsurus meyersi* in ventral view.

35. The basal segment of the male forceps is cylindrical with the inner margin somewhat produced at the apex. Each penis lobe is divided into a basal process directed posterolaterad and a narrow apical section curving strongly toward the

midline (**Fig. 2.78**). The apical segment of the abdomen is dark brown. The head is almost entirely dark brown dorsally with blackish eyes and yellowish white ventrally. The fore-wings have a slight pinkish tinge. Length of the male: 12 mm. Length of the fore-wing of the male: 12 mm. Length of the hind wing of the male: 5.6 mm.

....*Campsurus meyersi* Navás, 1934 (Argentina).

- The basal segment of the male forceps is short and nearly conical (**Fig. 2.79**). The apical segment of the abdomen is reddish brown. The head is pitch black above and yellowish ventrally. Length of male: 14.5 mm. Length of male fore-wing: 14.5 mm. Length of male hind wing: 6.7 mm.

....*Campsurus longicauda* Navás, 1931 (São Paulo).

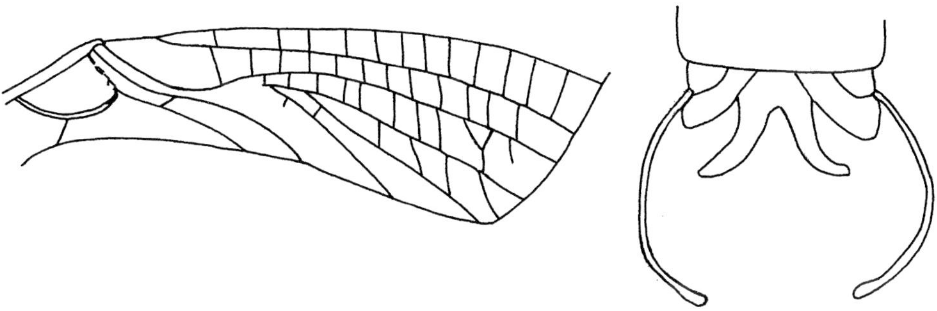

Fig. 2.79 *Campsurus longicauda:* veins of the inner, posterior part of the fore-wing, called the anal and cubital sections (left), and the male genitalia (right).

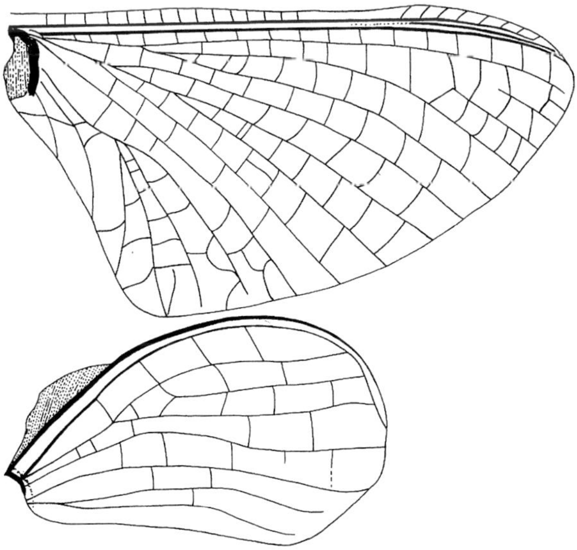

Fig. 2.80 The fore and hind wing of *Campsurus nappii.*

36. The pronotum is grayish yellow with a dark parallel band on each side of the midline, a darkened posterior margin, and sometimes a darker anterior margin, as well. The membrane between the costal vein and first long vein along the anterior margin of the hind wing is long and free of cross veins (**Fig. 2.80**). ....*Campsurus nappii* (Weyenbergh, 1883) (Argentina). Syn: *Palingenia nappii* (Weyenbergh, 1883
- The pronotum lacks a dark band on either side of the midline. 37

37. The dorsum of the pronotum is marked with a large yellow spot completely encircled with dark purple or brown. 38
- The pronotum is diffusely marked with yellow laterally with the yellowish markings sometimes confluent posteriorly. 42

38. The head posterior to the ocelli is pale. The pale spots in the longitudinal mid-dorsal stripe on the abdomen are well defined. Body length of female: 13.5 to 17 mm. Female fore-wing length: 14 to 19 mm (**Fig. 2.81**). ....*Campsurus corumbanus* Needham and Murphy, 1924 (Rio Grande do Sul).
- The head posterior to the ocelli is blackish. The pale spots in the longitudinal mid-dorsal stripe are diffuse or barely evident, or there is a translucent whitish midline running through most or the tergites. 39

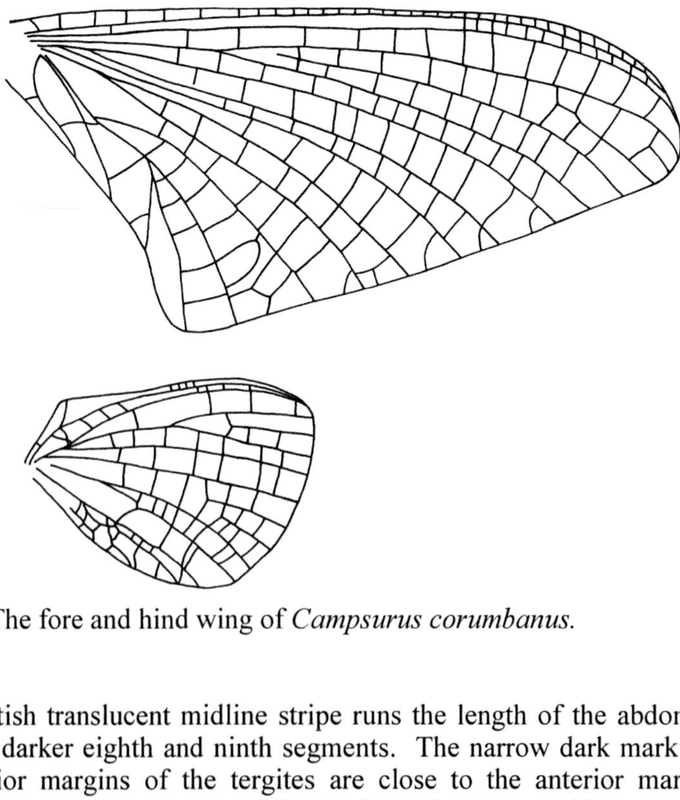

Fig. 2.81 The fore and hind wing of *Campsurus corumbanus.*

39. A whitish translucent midline stripe runs the length of the abdomen to the somewhat darker eighth and ninth segments. The narrow dark markings along the posterior margins of the tergites are close to the anterior markings and limited to about half the width of the tergite. The dark markings are brownish.

The penis lobes are curved evenly toward each other (**Fig. 2.82**). Length: 10 to 12 mm; fore-wing length: 10 to 11 mm; hind wing length: 4.9 mm.
....*Campsurus mahunkai* Puthz, 1973 (Bolivia). That this species is distinct from *Campsurus truncatus* is very doubtful. It was described without an adequate description of the wing, the author relying on its similarity in all respects with *C. truncatus*. The position in the key relies on the validity of this allegation. The sketches provided of the male genitalia of both species (Puthz, 1973) reveal differences that could be due to minor distortions of the specimens and none sufficient to prevent cross-mating.

- The mid-dorsal stripe encloses diffuse pale spots that are hard to distinguish. The dark markings are usually purplish. 40

40. The depth of the upper median fork of the hind wing is greater than half the length of the wing (**Fig. 2.83**). Length of female: 13 mm. Female fore-wing length: 15 to 16 mm.
....*Campsurus mutilus* Needham and Murphy, 1924 (Amazonas).

- The depth of the upper median fork of the hind wing is less than half the length of the wing (**Fig. 2.84**). 41

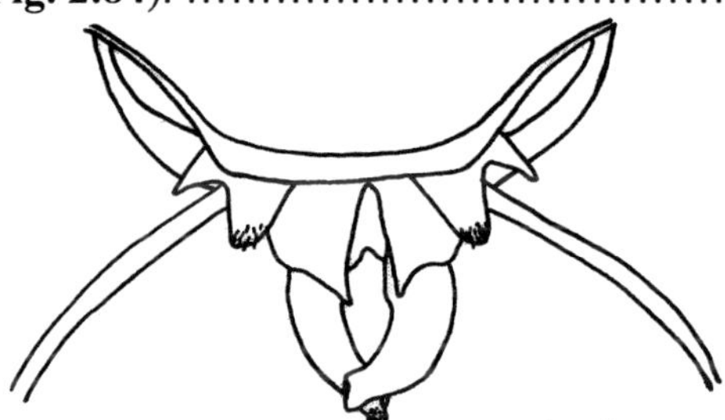

Fig. 2.82 The male genitalia of *Campsurus mahunkai* in ventral view.

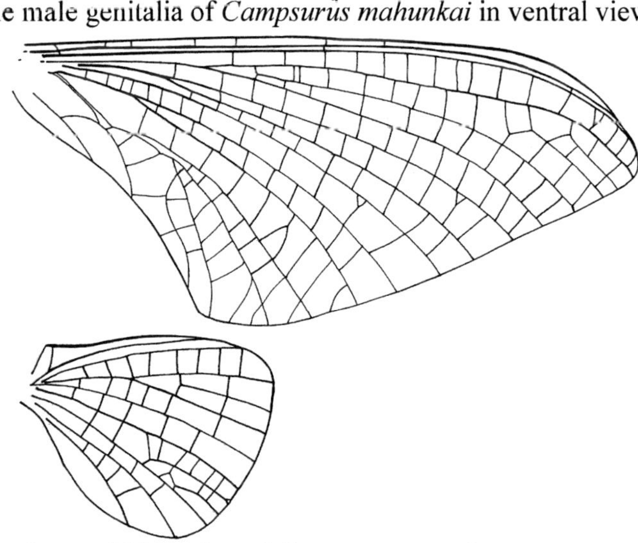

Fig. 2.83 The fore and hind wing of *Campsurus mutilus*.

41. Length of female: 9 mm. Female fore-wing length: 13 mm (**Fig. 2.84**). The male was not described.

....*Campsurus striatus* Needham and Murphy, 1924 (Mato Grosso do Sul, Mato Grosso).

- Length of female: 17 mm; of male: 12.5 mm. Fore-wing of female: 20.5 mm; of male: 14 mm. Hind wing of female: 10 mm; of male: 6 mm. The ninth tergite of the male has a broad, median, bilobed process coverning the bases of the forceps (**Fig. 2.85**)

....*Campsurus zikani* Navás, 1934 (Rio de Janeiro).

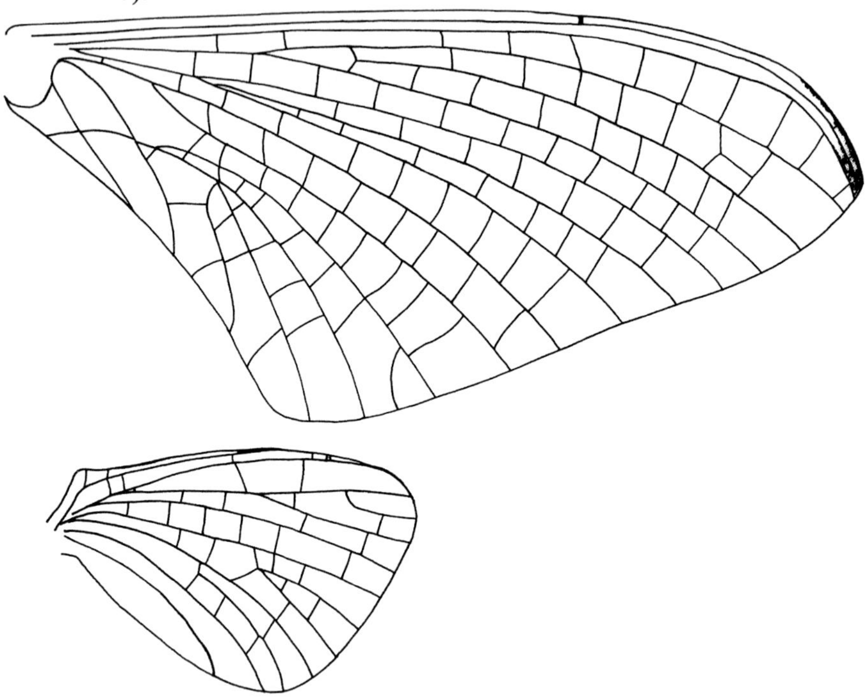

Fig. 2.84 The fore and hind wing of *Campsurus striatus.*

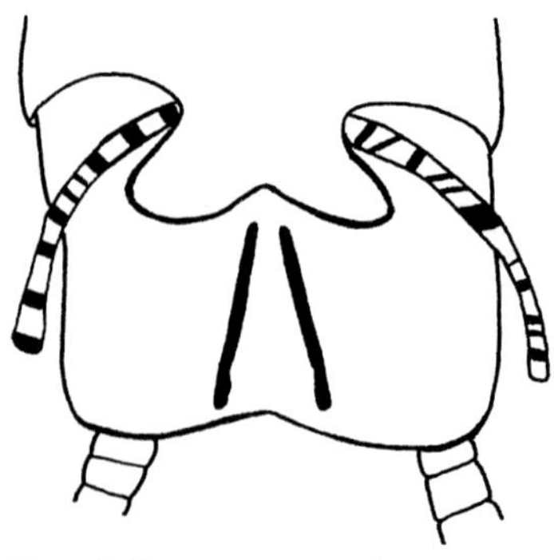

Fig. 2.85 The male genitalia of *Campsurus zikani* in ventral view.

42. The depth of the upper median fork of the hind wing is greater than half the length of the wing. A short apical portion of the forceps limbs is obviously narrowed (**Fig. 2.86**). Length: 13 mm; length of fore-wing: 13 mm. Color: head dark; prothorax grayish violet; mesonotum paler with three longitudinal stripes; abdomen grayish violet dorsally with a white median line and an indistinct pale spot on each side of the second through eighth tergites.
....*Campsurus argentinus* Esben-Petersen, 1912 (Argentina).
- The depth of the upper median fork of the hind wing is less than half the length of the wing (**Fig. 2.87**).Length of female: 12 mm. Female fore-wing length: 16 mm.
....*Campsurus claudus* Needham and Murphy, 1924 (Minas Gerais).

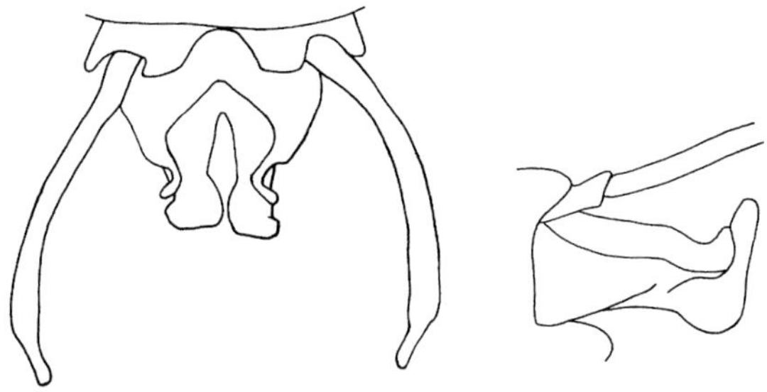

Fig. 2.86 The male genitalia of *Campsurus argentinus* in ventral (left) and lateral view (right).

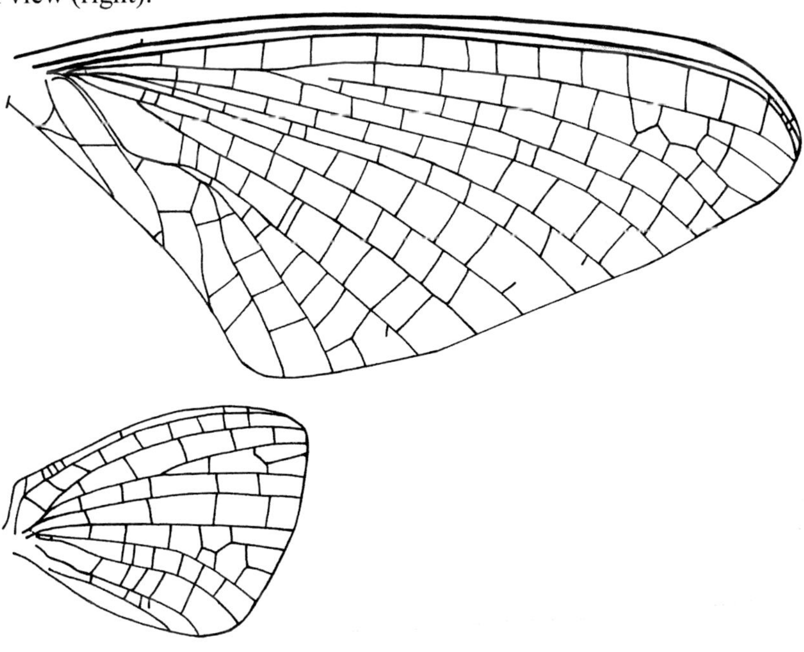

Fig. 2.87 The fore and hind wing of *Campsurus claudus*.

Key to the adults in the genus *Tortopus* in South America

Information for the key taken from Banks (1918), Navás (1920c, 1926a), Needham and Murphy (1924), Ulmer (1920a, 1942), Traver (1950), and Domínguez (1985). Some species were described after examination of only one sex, and some descriptions are sketchy, raising the possibility that some of the names are synonyms. Descriptions of the larvae of the South American species are not available.

1. The fore-wing of the male is about 13.4 mm, and that of the female is about 15.5 mm. The respective lengths of the males and females are 11.7 mm and 14.5 mm, and their hind wings are 6.5 mm and 7.4 mm. The cerci are about 38 mm long. The head is yellowish with blackish eyes and white antennae with the two basal segments darkened. The pronotum is yellow with a slight darkening in the middle, while the rest of the thorax is dark and shiny dorsally. The abdomen is uniformly yellow. The cerci are white. The legs are yellow with darker markings. The fore-tarsus of the male is much longer than the tibia. The wings are hyaline with white veins, except for the light violet C, Sc, and R of the fore-wing and the darkened basal fourth of Sc in the hind wing. There is a light violet area near the apex of the fore-wing. There are about 23 cross veins in the stigmatic area of the costal space. The subgenital plate of the male has an elongate triangular shape with a small apical emargination at the midline (**Fig. 2.88**).
....*Tortopus zottai* (Navás, 1920)
(Argentina). Syn: *Campsurus zottai* Navás, 1920.
- The fore-wing is either longer or shorter, its length not falling between about 12.5 and 17 mm. Other parts of the description will not match, as well. 2
2. The fore-wing is longer than 17 mm. The dorsal markings on the thorax and abdomen of the female are not in the form of solid dark longitudinal stripes (**Fig. 2.89**). Unfortunately, the males of the species in this group have not been described. 3
- The fore-wing is not longer than about 12.5 mm. 5

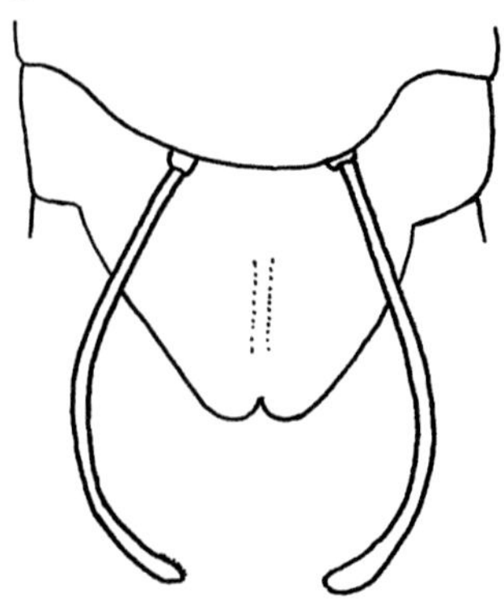

Fig. 2.88 Apex of the abdomen and genital forceps of a male *Tortopus zottai.*

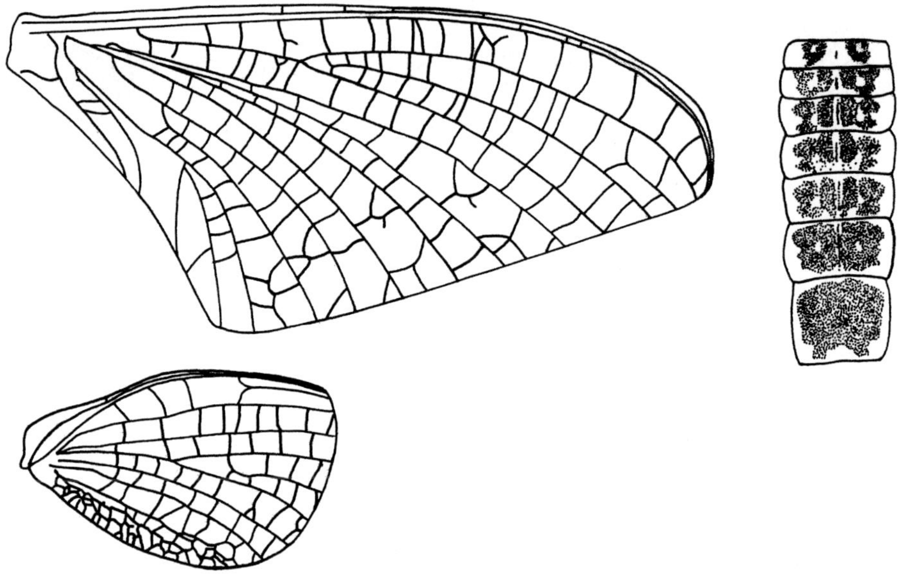

Fig. 2.89 The fore and hind wings (left) and the second through eighth abdominal segments in dorsal view (right) of *Tortopus obscuripennis.*

3. All of the wing veins are brown. Vein Sc in the hind wing runs close to C, and R almost reaches the wing tip (**Fig. 2.89**). The middle and hind legs are yellow. Total length: 19 to 20 mm; fore-wing length: 19.5 to 20.5 mm; hind wing length: 9 to 9.3 mm.
....*Tortopus obscuripennis* Domínguez, 1985 (Argentina).
- The long veins in the fore-wing, except sometimes C, Sc, and R, are yellowish or white (**Fig. 2.90**). 4
4. The length of the female is about 14.5 mm, and its fore-wing is about 17.5 mm long. Veins C, Sc, and R in the fore-wing are dark, while the other long veins are yellow (**Fig. 2.91**). The hind wing is about 8.4 mm long. There are many cross veins in the basal half of the costal and subcostal spaces in the hind wing. The legs are yellow, and the fore-leg is slightly darkened.
....*Tortopus bruchianus* (Navás, 1926) (Argentina). Syn: *Campsurus bruchianus* Navás, 1926.
- The length of the female is 18.5 to 19 mm, and its fore-wing is 17.5 to 18 mm long. The long veins in the wings are yellow. Vein Sc in the hind wing ends in C before the middle of the wing (**Fig. 2.90**). The middle and hind legs are whitish. Total length: 18.5 to 19 mm; fore-wing length: 17.5 to 18 mm; hind wing length: 7 to 7.5 mm.
....*Tortopus sarae* Domínguez, 1985 (Argentina).

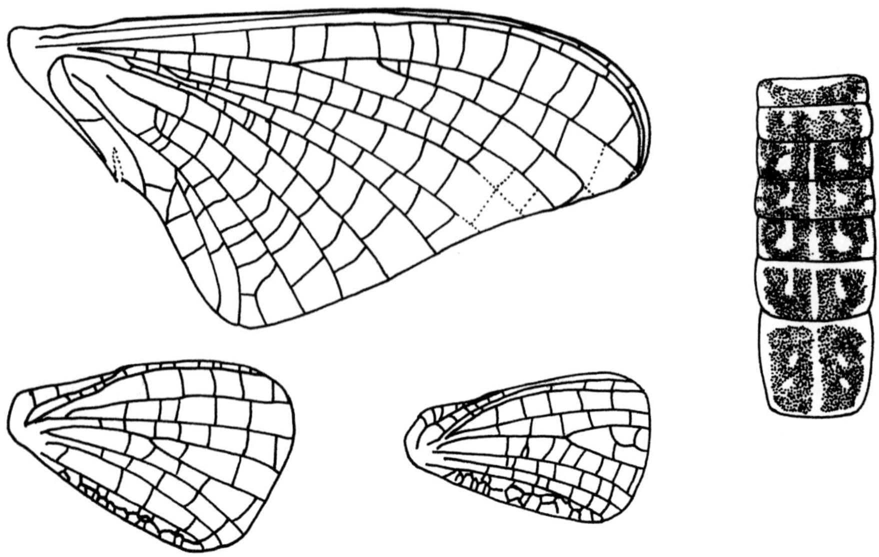

Fig. 2.90 *Tortopus sarae:* the fore-wing showing the long veins that are actually yellow in the living insects (upper left), varying hind wings from two different individuals (lower left and center), and the color pattern on the second through eighth abdominal tergites (right).

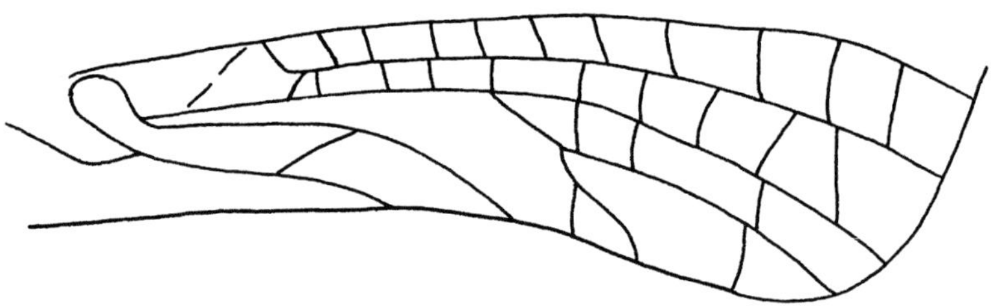

Fig. 2.91 The inner posterior part of the fore-wing of *Tortopus bruchianus,* showing the structure of the veins, which are yellow in life.

5. The male genitalia include two pairs of long, narrow appendages; the inner pair is somewhat more than half the length of the outer pair. There are no basal processes at the base of the inner pair (**Fig. 2.92**). The costal vein in the fore-wing is somewhat grayish, making the costal space appear darker than the rest of the wing, which is hyaline with whitish veins. The fore-legs are blackish with white claws that are about as long as the apical tarsal segment. The "expanse" of the wings is about 24 mm, meaning that the fore-wing should be approximately 10 mm. The female has not been described, and the wing veins have never been depicted or adequately described.
....*Tortopus parishi* (Banks, 1918) (Ecuador). Syn: *Campsurus parishi* Banks, 1918.

- The inner pair of long appendages on the male genitalia have a spine-like appendage at the base (**Fig. 2.93**), or most of the wing veins are darkened, or the fore-legs are pale. 6

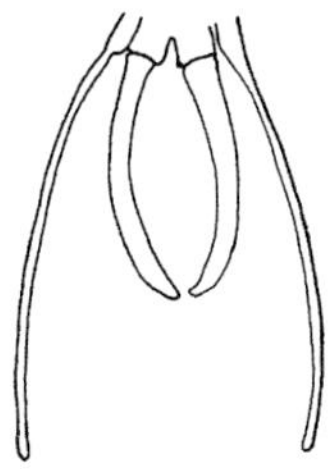

Fig. 2.92 Apex of the abdomen and genital forceps of a male *Tortopus parishi.*

6. Vein Sc in the hind wing reaches nearly to the wing tip, and there are at least about eight cross veins in the basal half of the costal space in both the fore and hind wing. These cross veins in the fore-wing may be very weak. About 13 longitudinal veins and their branches reach the costal vein in the broad apical region of the wing. There are two long, narrow appendages on the male genitalia; each of the inner pair has a small, spine-like process at the base (**Fig. 2.93**). Length: c. 10 mm. Fore-wing length: 10 to 12 mm. Color: yellow with darker markings. Blackish markings, sometimes with a violet tinge, are present around the compound eyes and between the ocelli. There are about five longitudinal stripes on the mesonotum. The abdomen of the female has a dark midline stripe, while that of the male has less extensive dark markings. The thickest longitudinal veins in the fore-wing are brownish violet or blackish with a violet tinge.
....*Tortopus unguiculatus* (Ulmer, 1920) (Central America, Colombia). The description of *T. parishi* by Banks (1918) is so sketchy that Ulmer (1942) speculated that this species might actually prove to be a synonym of *T. parishi* after its wings have been described and the male genitalia reexamined for the presence of a basal process.

- If vein Sc in the hind wing extends beyond the middle of the wing, then there are no discernible cross veins in the costal space of the fore-wing or there are much fewer than eight cross veins in the costal space of the hind wing (**Fig. 2.94**). 7

7. The dark markings on the head of the female are purplish and form a cross band between the eyes. The abdomen is obscure brownish, sometimes with traces of a median light stripe and parallel stripes. There are apparently no cross veins in the costal area of the fore-wing, but there are about nine such cross veins in the hind wing, all but one in the basal section (**Fig. 2.41**). The wing veins are purplish. Length: 6 to 10 mm. Fore-wing length: 9.5 to 12.5 mm. The male has not been described.
....*Tortopus igaranus* Needham and Murphy, 1924 (Peru).

- The dark markings on the head are not purplish. The abdomen is light yellow with extensive darker markings on the dorsum. Dark markings include blackish at the bases of the ocelli and gray longitudinal stripes beside whitish median stripes on the thorax and abdomen. Vein Sc in the hind wing terminates in C in the basal half of the wing, and there are two cross veins in its costal space (**Fig. 2.94**). Length: 8.5 mm. Fore-wing length: 9 mm. Female not described. ....*Tortopus harrisi* Traver, 1950 (Mato Grosso do Sul).

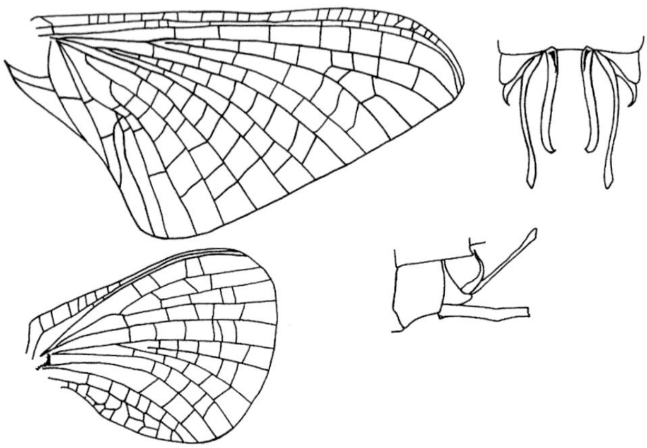

Fig. 2.93 *Tortopus unguiculatus:* the wings of a female (left), the male genitalia in lateral (lower right) and ventral view (upper right).

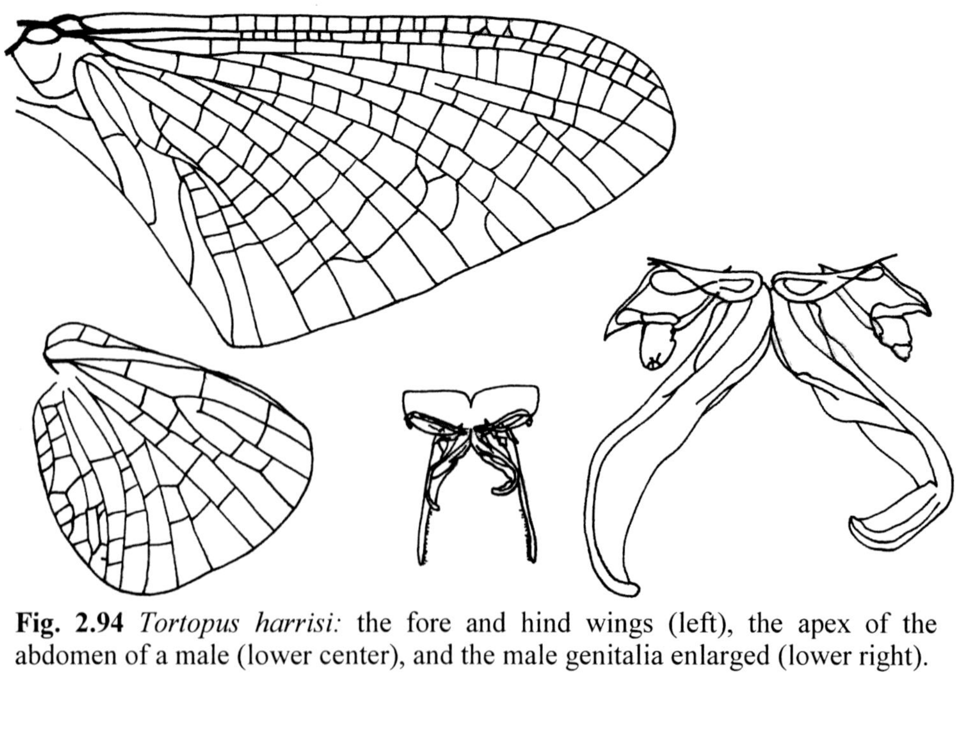

Fig. 2.94 *Tortopus harrisi:* the fore and hind wings (left), the apex of the abdomen of a male (lower center), and the male genitalia enlarged (lower right).

Ephemeridae

This family is represented in South America by a single genus, *Hexagenia* Walsh, 1863.

Key to the species of adult *Hexagenia* in South America

Information for the key was provided by Navás (1922a), Spieth (1941), and McCafferty (1968). Enough information about the larvae of the South American ephemerids is not available to produce a key. The information on larvae in this genus was reviewed by McCafferty (1970).

1. The wing membrane is a uniform deep brown.
......................................*Hexagenia (Pseudeatonica) mexicana* Eaton, 1883 (Mexico, Central America, Colombia, Ecuador, Peru).
- The wing membrane is mainly hyaline but may be darkened in the costal and subcostal spaces, and the hind wing may have clouded cross veins **(Fig. 2.95)**.
....2

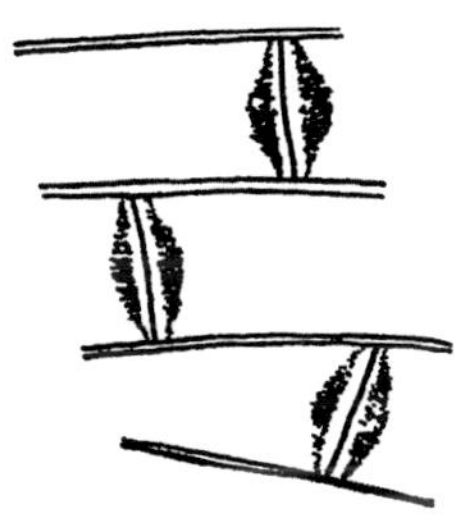

Fig. 2.95 Clouded cross veins of the otherwise hyaline hind wing of *Hexagenia callineura.*

2. The abdominal sternites are infuscated laterally (**Fig. 2.10**).
....*Hexagenia (Pseudeatonica) albivitta* (Walker, 1853) (Central America, Colombia, Guyana, Paraguay, Argentina, Uruguay, Brazil). Syn: *Baetis albivitta* Walker, 1853; *Palingenia continua* Walker, 1860; *Palingenia dorsigera* Hagen, 1861; *Hexagenia benedicta* Navás, 1922; *Hexagenia dominans* Navás, 1936
- The abdominal sternites are not infuscated laterally (**Fig. 2.95**). The wing span measured from wing tip to wing tip on an insect with fully extended wings is about 44 mm. Coloration: yellowish with black markings, including a band from one compound eye to the other crossing the ocelli, lateral stripes from the pronotum to the metanotum, and dark stripes on each side of the midline of the abdominal tergites, broadening from the base to the apex of each segment. The wings are strikingly colored with most longitudinal veins yellowish but Sc and R as well as the cross veins dark. The dark cross veins of the hind wings are

bordered by whitish on the membranes and lined with black fusiform marks. The fore-leg is mainly reddish with black on the apical tarsal segment, while the other legs are mainly pale.
....*Hexagenia (Pseudeatonica) callineura* Banks, 1914 (Colombia, Ecuador).

Leptohyphidae

This family is treated in older literature as a subfamily of Tricorythidae. One genus, *Coryphorus,* described from larvae, was tentatively placed in the subfamily Machadorythinae, otherwise known only from Africa, by Peters (1981). Recently, all South American species were placed in two subfamilies, Tricorythodinae and Leptohyphinae, by Wiersema and McCafferty (2000), and *Coryphorus* was included in Tricorythodinae. It is certain only that the systematic positions of the South American genera in this family will be revised again in the not too distant future.

Key to the genera of adults found in South America

The key is prepared according to the information provided by Traver (1958), Mayo (1968a), Domínguez (1984), Wiersema and McCafferty (2000), Molineri (2001a, b), and Molineri and Flowers (2001). Adults of *Coryphorus* Peters, 1981, and *Epiphrades* Wiersema and McCafferty, 2000, have not been described. Wiersema and McCafferty (2000) speculated that *Coryphorus* might eventually be found to be congeneric with *Cotopaxi* after its adults have been described. However, they placed the two in different subfamilies.

1. Both males and females lack hind wings and lack free blackish membranous processes joining the fore-wings and ending on or near the midline of the mesoscutellum. There are five tarsal segments and ventral extensions at the apices of the tibiae on all legs of the female and on the middle and hind legs of the male (**Fig. 2.96**).
....Subfamily Tricorythodinae.............2
- The males usually have hind wings, but they may sometimes be absent from females. In the case of *Cotopaxi*, hind wings are also lacking in males but the membranes of the fore-wings are joined by a membranous process passing across the posterior borders of the mesoscutellum and ending in well-developed, free blackish membranous processes near the midline. Except possibly in the case of *Cotopaxi,* there are four tarsal segments and no ventral extensions at the apices of the tibiae of females or the middle and hind tibiae of males (**Fig. 2.97**). The tibiae and tarsi were broken off the specimen of *Cotopaxi* described by Mayo (1968a).
....Subfamily Leptohyphinae.............5

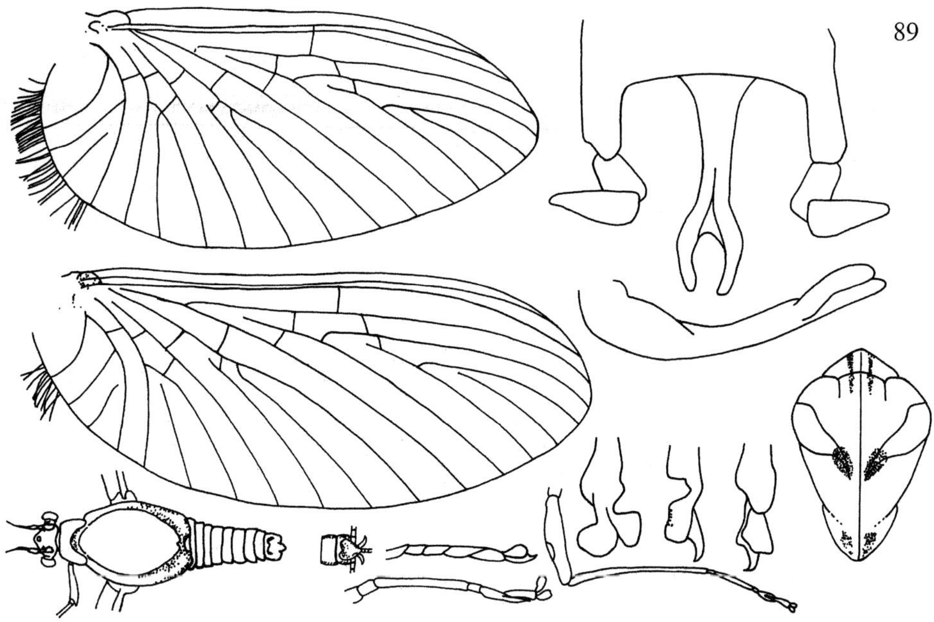

Fig. 2.96 *Tricorythopsis artigas:* the wing of an adult male (upper left) and female (middle left); the diagrammatic depiction of an adult male with an enlargement of the apex of its abdomen to its right (lower left); tarsal segments of the fore and hind-leg (lower center, below and above, respectively); the fore-leg (lower right center); the tarsal claws on the fore, middle, and hind legs (above fore-leg, left to right); the color pattern on the thorax (lower right); the male genitalia (upper right); and the penis in lateral view (middle right).

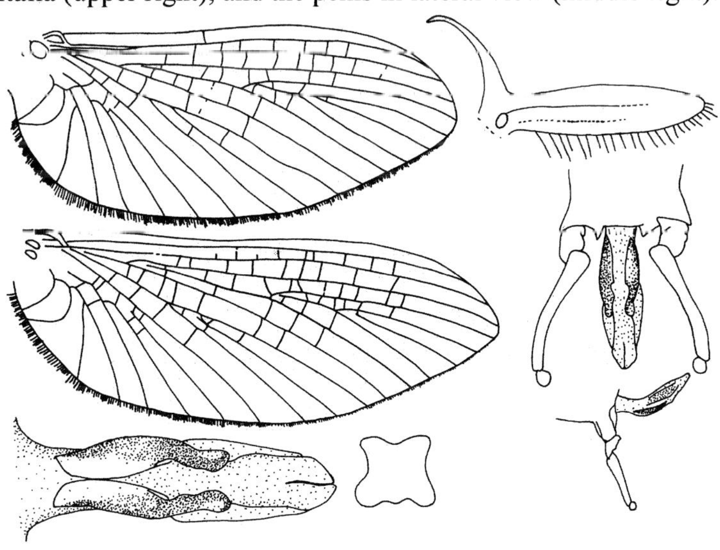

Fig. 2.97 *Yaurina mota:* the fore wing of a male (upper left) and a female (middle left), the hind wing of a male enlarged (upper right), the male genitalia in ventral (middle right) and lateral (lower right) view, the penis in ventral view (lower left), and a cross section showing its outline (lower center).

2. Vein IMP of fore-wing joins CuA basally forming a triangle; a curve is present in CuA at the point of union. MP_2 is shorter than IMP and appears to be an intercalary within this triangle. There are no paired occipital tubercles on the head. Each forceps arm of the male consists of two segments. The cerci of the female are never longer than the total length of the three apical segments of the abdomen. The female subadult has fore-legs, but only the fore-coxa and trochanter remain after the final molting. Both sides of the mesonotum have short, rounded processes (**Fig. 2.96**).
....*Tricorythopsis* Traver, 1958..p. 105
- Vein IMP is not so joined to CuA, which has no curve at the point of junction. IMP is often joined to MP_1 by a cross vein. MP_2 is at least as long as and usually longer than IMP (**Fig. 2.98**). There are paired occipital tubercles on the head just lateral to the ocelli. 3

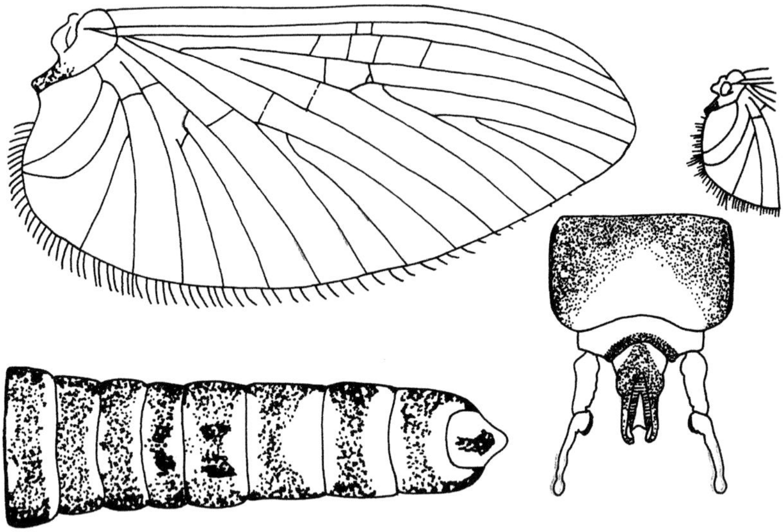

Fig. 2.98 *Tricoryhyphes popayanicus*: wing (upper left), the inner margin of the wing of another individual on a smaller scale (upper right), the markings on the dorsal side of the abdomen (lower left), and the male genitalia in dorsal view (lower right).

3. The tarsi on the fore-leg of the male are less than half the length of the fore-tibia. The combined length of the hind tibia and tarsi is equal to or shorter than the hind femur. The apical margin of the subgenital plate is produced beyond the bases of the forceps and is either truncate or shallowly emarginate (**Fig. 2.98**).
.............................*Tricoryhyphes* Wiersema and McCafferty, 2000..p. 115
- The tarsi on the fore-leg of the male are more than half the length of the fore-tibia and usually 2/3 to 3/4 of its length. The apical margin of the subgenital plate is not produced beyond the bases of the forceps, or, if it is, then the

combined length of the hind tibia and tarsi is at least 1/3 longer than the length of the hind femur. The combined length of the hind tibia and tarsi is equal to or shorter than the hind femur (**Fig. 2.99**). 4

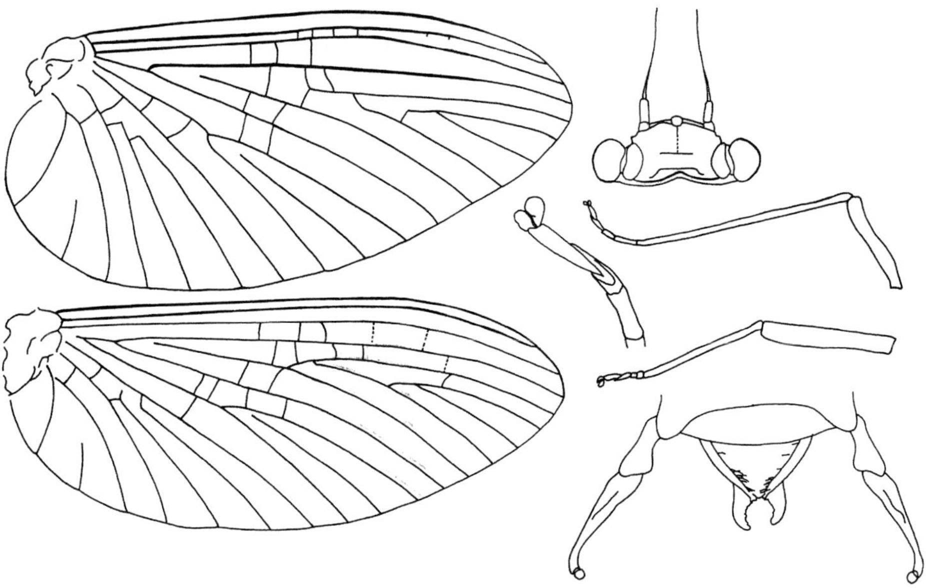

Fig. 2.99 *Tricorythodes australis:* the wing of a male (upper left) and a female (lower left) adult, the head of a male adult (upper right), the fore-leg of a male with an enlargement of the apical tarsal segments and claws (below head), the hind leg of a male (below fore-leg), and the male genitalia (lower right).

4. The length of the hind femur is greater than ¾ of the combined length of the hind tibia and tarsi. The length of the wing is about 2 to 4 mm.
....*Asioplax* Wiersema and McCafferty, 2000

The only South American species in this family is *Asioplax nicholsae* (Wang, Sites, and McCafferty, 1998), described as *Leptohyphes nicholsae* Wang, Sites, and McCafferty, 1998, from Ecuador. The characteristics used in the key are based on those of North American species because the adult of *Asioplax nicholsae* has not yet been described.

- The length of the hind femur is less than ¾ of the combined length of the hind tibia and tarsi (**Fig. 2.99**). The length of the wing is about 3 to 9 mm.
....*Tricorythodes* Ulmer, 1920..p. 116

5. The wing membranes are attached across the posterior borders of the mesoscutellum and end in well-developed, free blackish membranous processes

near the midline. Vein CuP is strongly arched. Hind wings are absent. Each forceps arm consists of a single segment (**Fig. 2.100**). Length: about 4 mm. Wing length: about 8 mm.

....*Cotopaxi* Mayo, 1968

The only known species in this genus is *Cotopaxi macuchae* Mayo, 1968, from Ecuador. It is 4 mm long with wings 8 mm long. Its head is black with amber circles around the eyes. The pronotum is black, and the remaining parts of the thorax are yellowish and honey brown with dark brown and black markings. The abdomen of the male is hyaline with narrow black markings on the first tergite and a blackened apical 1/3 of the other tergites. The sternites are pale.

- The wings lack membranous interconnections with black processes (**Fig. 2.101**). 6

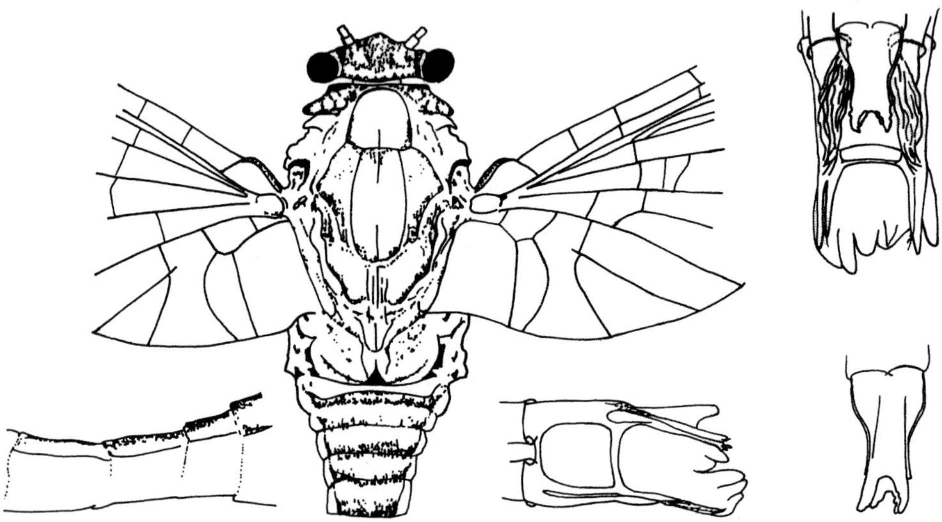

Fig. 2.100 *Cotopaxi macuchae*: habitus with the proximal parts of the wing (upper left), apical portion of the metanotum and first three abdominal segments in lateral view (lower left), the male genitalia in dorsal (upper right) and lateral (lower right center) view, and the penis in ventral view (lower right).

6. The eyes of the male are large and conceal almost the entire rest of the head, and those of the female are also relatively large (**Fig. 2.101**).

....*Leptohyphodes* Ulmer, 1920

According to Traver (1958), *Leptohyphodes inanis* (Pictet, 1843), from Brazil, should be the only species included in this genus. He treated the species *Tricorythus australis* Banks, 1913, which was placed in the genus *Leptohyphodes* by Ulmer (1920a), as a species of *Tricorythodes* (see below).

- The eyes of the male are not large enough to conceal most of the head. 7

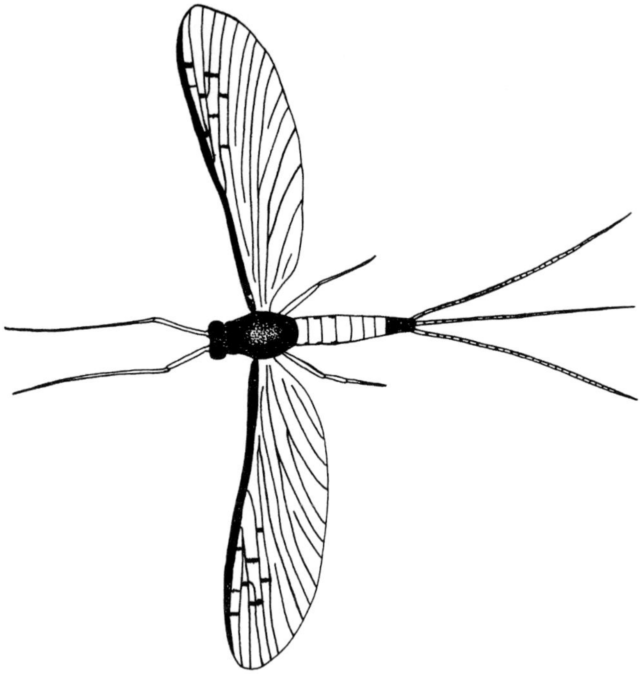

Fig. 2.101 The habitus of a female *Leptohyphodes inanis*.

7. Hind wings are present in both sexes. The forceps are short and robust and consist of two segments (**Fig. 2.102**). The mesoscutellum does not bear plumed setae, called plumidia.
....*Haplohyphes* Allen, 1966..p. 120
- Males have hind wings, but females of most species do not. When present, the hind wing is usually only a vestige possessing two longitudinal veins, no cross veins, and a prominent costal angulation. The forceps consist of three segments. The eyes of both sexes are relatively small and well separated (**Fig. 2.103**). The mesoscutellum usually bears well-developed plumidia. 8
8. The lateral margins of the styliger plate are extended posteriad beyond the bases of the second segment of the forceps, and there are acutely pointed projections along its posterior margins near the base of the forceps. The penis is dorsovertrally flattened and has a pair of conical spines arising from the dorsolateral or dorsal surface. It is not divided into distinct lobes but has a fissure along the midline at the apex. There is a dorsal projection arising at its base, and its basal 1/3 forms a sclerotized ring. It is also sclerotized laterally (**Fig. 2.103**). There are long, slender membranous processes on the mesonotum. The costal projection on the hind wing is long and curves toward the body.
....*Traverhyphes* Molineri 2001..p. 122

- The lateral margins of the styliger plate do not extend posteriad beyond the bases of the second segment of the forceps, or the penis bears different kinds of appendages (**Fig. 2.104**). 9

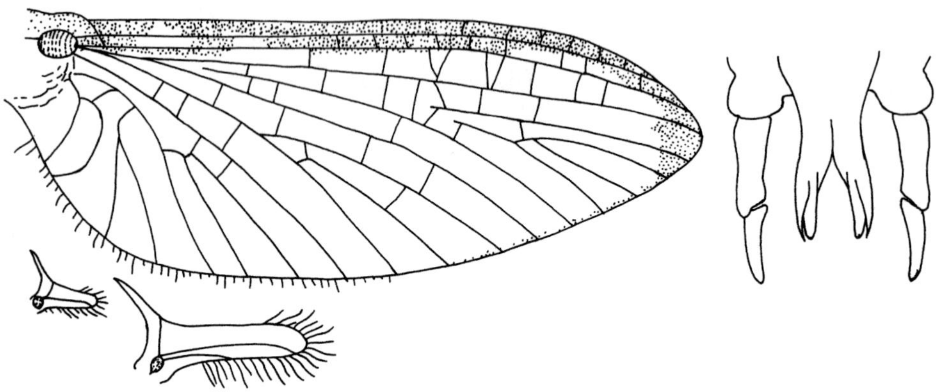

Fig. 2.102 *Haplohyphes mithras*: the fore-wing with the wing veins (upper left), the hind wing drawn to the scale of the fore-wing and enlarged to show details (lower left), and the male genitalia in dorsal view (right).

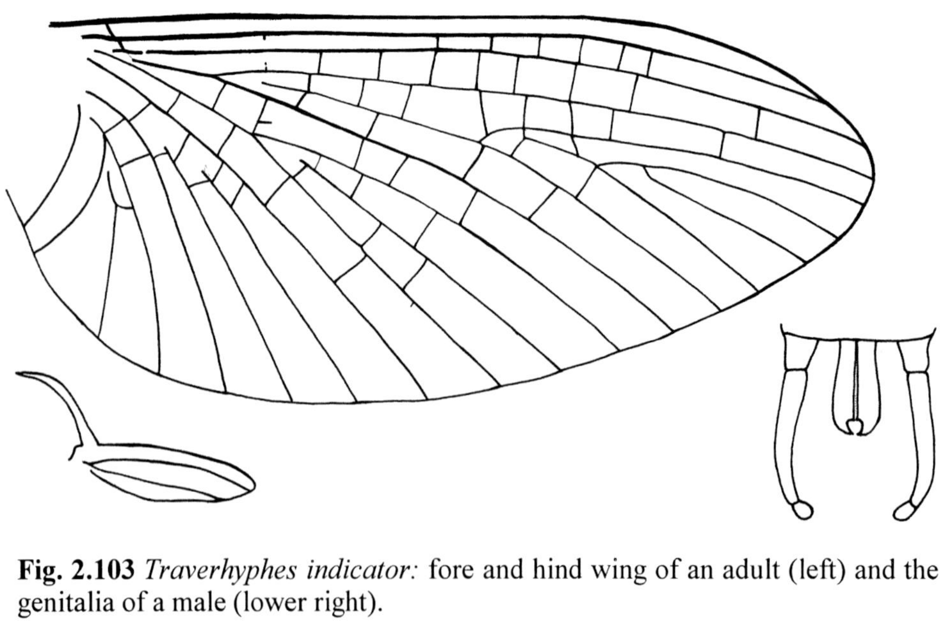

Fig. 2.103 *Traverhyphes indicator:* fore and hind wing of an adult (left) and the genitalia of a male (lower right).

9. The costal projection on the male hind wing is 0.50 to 0.55 times the hind wing length, and it is strongly curved. The styliger plate is extended posteriorly at the base of the forceps. The penis lobes are completely fused except along the small furrow at its apex, and two long, spine-like appendages arise from the base of the penis (**Fig. 2.97**).
....*Yaurina* Molineri, 2001..p. 123
- The costal projection on the male hind wing is shorter than 0.5 times the hind wing length, or the styliger plate is not extended posteriorly at the base of the forceps, or both. The penis lobes are not nearly completely fused, or they do not bear ventral, spine-like processes (**Fig. 2.104**). 10

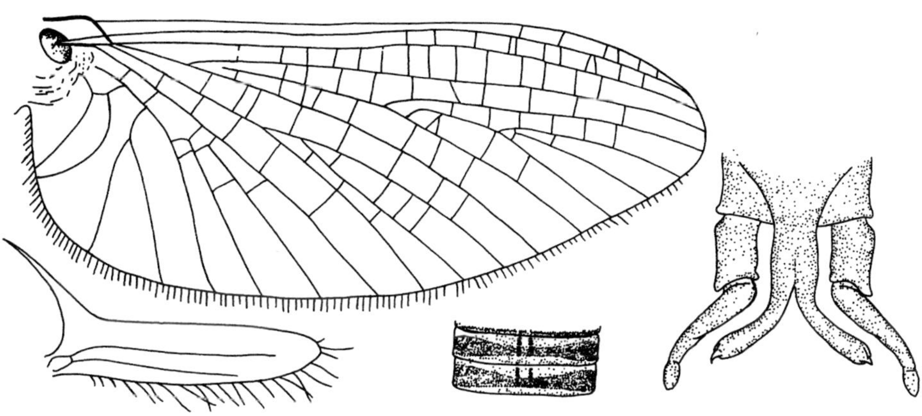

Fig. 2.104 *Leptohyphes petersi* male: the fore and hind wings with the hind wing drawn larger than scale (left), genitalia in dorsal view (right), and pattern on the fourth and fifth tergites.

10. Vein $ICuA_1$ in the fore-wing is not connected at its base to CuP. There are sharp, strong setae on the ventral side of the fore-tibiae of males. Females have a broadly convex posterior margin of the subgenital plate (**Fig. 2.104**). The penis is Y-shaped, each well separated lobe with an incision at the apex.
....*Leptohyphes* Eaton, 1882..p. 125
- Vein $ICuA_1$ in the fore-wing is connected at its base to CuP, often with a short cross vein. There are no sharp, strong setae on the ventral side of the fore-tibiae of males. Females have a concave posterior margin of the subgenital plate. The penis is divided by an incision at the apex, but the lobes run parallel to each other and do not diverge (**Fig. 2.105**).
..............................*Allenhyphes* Hofmann, Sartori, and Thomas, 1999..p. 141

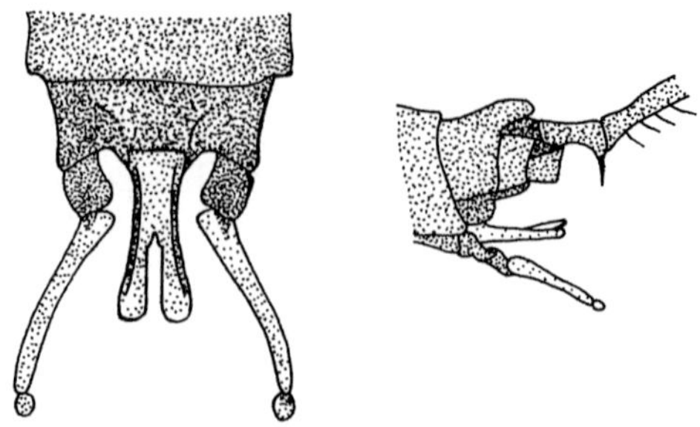

Fig. 2.105 The male genitalia of *Allenhyphes flinti* in ventral (left) and lateral view (right).

Key to the genera of larvae found in South America

The key is prepared according to the information provided by Edmunds *et al.* (1963), Peters (1981), Domínguez (1984), Wiersema and McCafferty (2000), Molineri (1999a, 2001a), and Molineri and Flowers (2001). The larvae of *Cotopaxi* have not been described.

1. The compound eyes are elevated above the vertex of the head. The gills on abdominal segments 2 through 5 are partially enclosed in a gill chamber covered by lateral flanges of the third through sixth tergites. The pedicel of the antenna is thick and at least 2.5 times as long as the scape. The anterior margin of the labrum is not deeply divided in the middle. Tubercles are present on the dorsal parts of the vertex, pronotum, and first to third and seventh to ninth abdominal segments (**Fig. 2.27**).
....*Coryphorus* Peters, 1981
The only known species in this genus is *Coryphorus aquilus* Peters, 1981, with a body length of 5.5 to 5.8 mm, from Colombia, Amazonas, and Pará.
- The compound eyes are widely separated and not elevated above the vertex of the head. The gills are not enclosed in a chamber but are operculate or semi-operculate (**Fig. 2.106**). 2
2. About five long spines are inserted on each side of the midline in the anterior half of the seventh abdominal tergite. There are small groups of shorter spines on the third through sixth abdominal segments near the gills. Gills are present on the second through sixth abdominal segments. The gill on the second segment is elongated egg-shaped and has two ventral lamellae. That on the third segment bears three ventral lamellae; those on the fourth and fifth segment each bear two ventral lamellae, and that on the sixth has only one ventral lamella. The maxillary palp consists of two segments. There is a transverse band of relatively long spines on the fore-femur. The submentum of the labium is large and expanded laterally (**Fig. 2.106**).
....*Traverhyphes* Molineri 2001..p. 122

- The seventh abdominal tergite does not have a row of five long spines inserted on either side of the midline in the anterior half, or small groups of relatively short spines are not present near the gills on the third through seventh abdominal segments (**Fig. 2.107**). 3

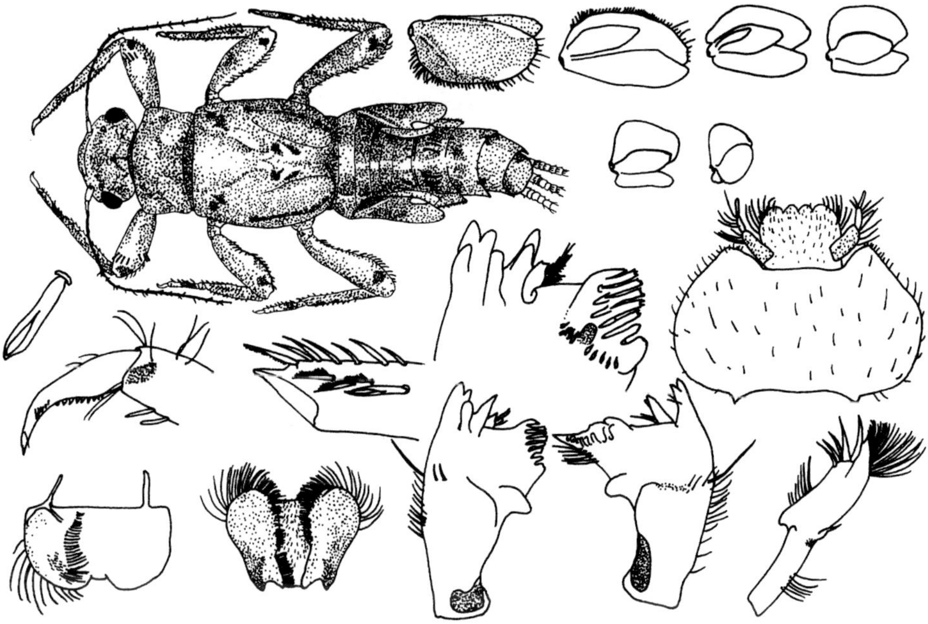

Fig. 2.106 *Traverhyphes indicator* larva: habitus (upper left); an enlarged spine from the transverse band on the fore-femur (upper middle left); tarsal claw (lower middle left); apex of the middle tibia (left center); labrum, hypopharynx, mandibles, and maxilla (lower row, left to right); enlarged apical area of the left mandible (above mandibles); labium (middle right); gill from the second abdominal segment in dorsal and ventral view and ventral views of the gills on the third and fourth segments (upper right, left to right); ventral views of the gills on the fifth and sixth segments (upper middle right, left and right).

3. There are no spines on the posterior margins of the six anterior abdominal tergites, but there may be tiny spines on the seventh through tenth tergites. Wing buds for hind wings are absent. The hind tibiae are less than twice as long as the hind tarsi (**Fig. 2.107**).
....Subfamily Tricorythodinae.............4
- The posterior margins of all, or at least the posterior nine tergites bear spines. The larvae of males and sometimes females have hind wing buds. The hind tibiae are at least about twice as long as the hind tarsi (**Fig. 2.108**).
....Subfamily Leptohyphinae.............8

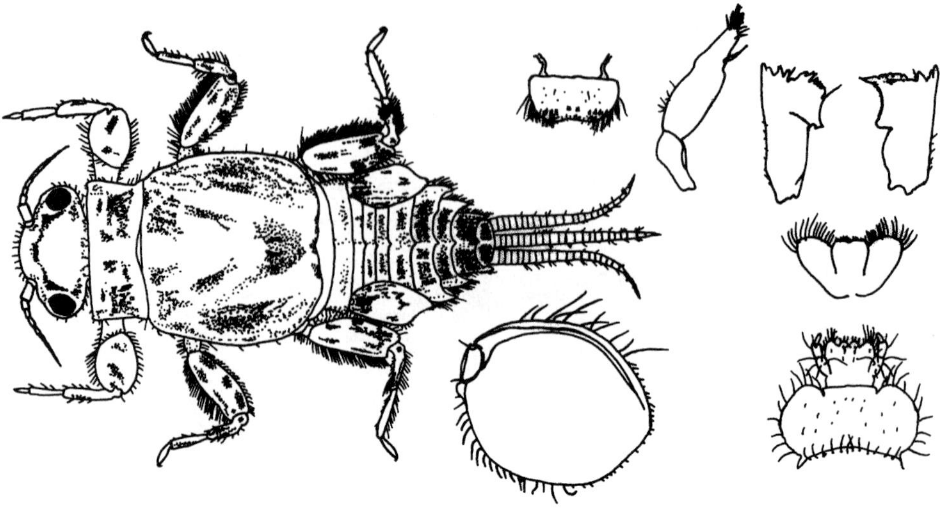

Fig. 2.107 *Asioplax nicholsae* larva: habitus (left); the labrum, maxilla, and mandibles (upper right, left to right); hypopharynx (middle right); labium (lower right); gill on the second abdominal segment (lower center).

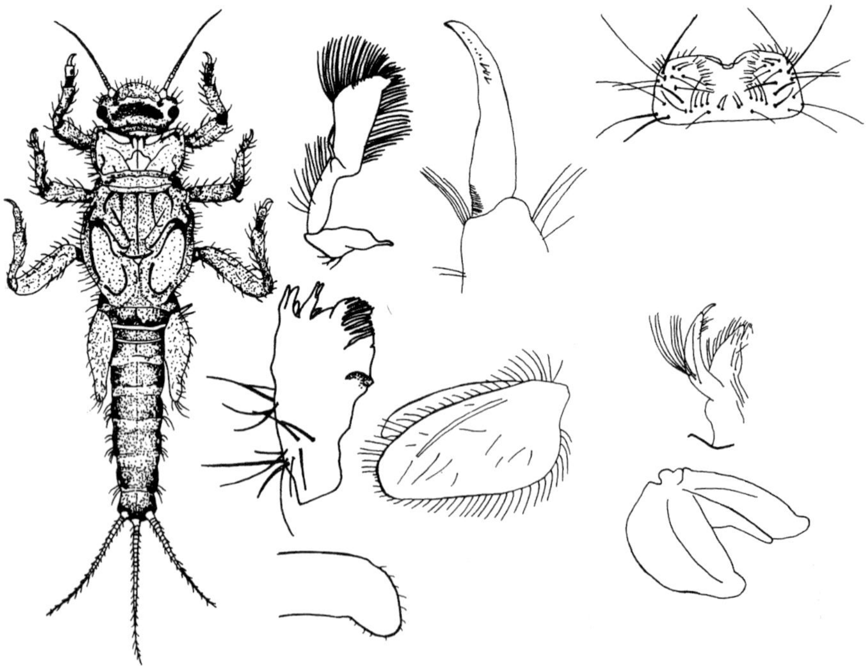

Fig. 2.108 *Haplohyphes baritu* larva: habitus (left), tarsal claw from the fore-leg (upper right center), labrum (upper right), maxilla (upper left center), mandible (left center), half of the labium (middle right), the second (right center) and third gill (lower right), and a posterolateral process from an abdominal sternite (lower left center).

4. There are tubercles present at the locations of the ocelli, the anterolateral part of the mesonotum, and sometimes on the midline of the occiput. The gills are generally rectangular in shape (**Fig. 2.109**).
.................................*Epiphrades* Wiersema and McCafferty, 2000..p.119
- There are no tubercles at the locations of the ocelli, the anterolateral part of the mesonotum, or the occiput. The gills tend to be roughly triangular or ovate (**Fig. 2.110**). 5

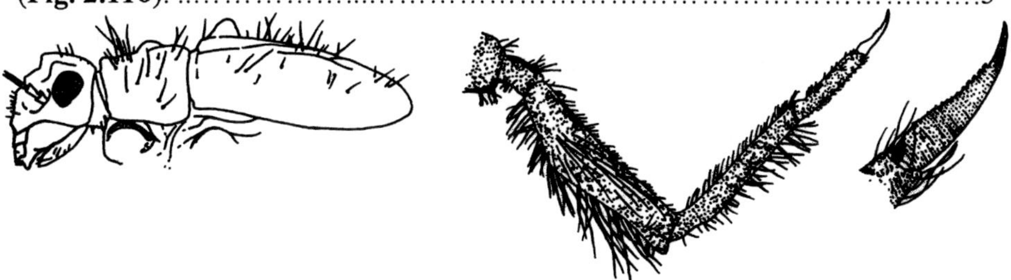

Fig. 2.109 A profile of the head and thorax and the fore-leg with an enlarged tarsal claw of a *Epiphrades cristatus* larva (left to right).

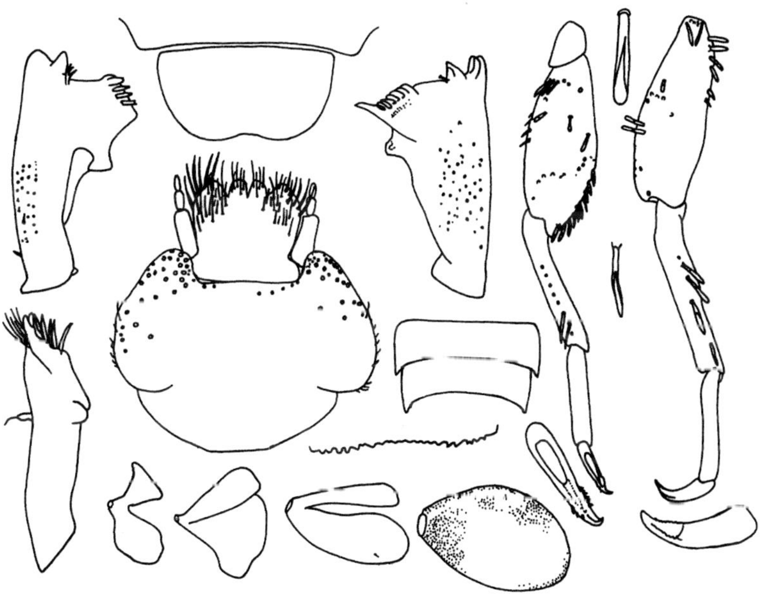

Fig. 2.110 *Tricorythopsis artigas* larva: the apex of the clypeus and labrum (upper left center), the mandibles (left and right of the labrum), a maxilla (lower left), a ventral view of the labium (below labrum), the middle and hind legs (right, left and right, respectively) with enlarged setae from the femur and tibia (between the respective segments of the two legs) and enlarged tarsal claws (lower right), the seventh and eighth abdominal tergites above an enlarged posterior margin of the seventh (middle right), and gills from the second, third, fourth, and fifth abdominal segments (below center, right to left).

5. There is a dorsoventral flattening of the body. The length along the midline of the seventh and the eighth abdominal tergite is shorter than its respective posterolateral spine. The width of the fore-femur is at least about 3/4 of its length. The fore-femur bears a fringe and a dorsal transverse band of setae crossing its basal third. If ventral lamellae are present at all on the inner side of the gills, they are less than half as long as the lamellae on the outer side (**Fig. 2.107**).
....*Asioplax* Wiersema and McCafferty, 2000
The only South American species in this family is *Asioplax nicholsae* (Wang, Sites, and McCafferty, 1998), described as *Leptohyphes nicholsae* Wang, Sites, and McCafferty, 1998, from Ecuador.
- The body shows no dorsoventral flattening. The length along the midline of the seventh and the eighth abdominal tergite is about equal to or longer than its respective posterolateral spine. The width of the fore-femur is no more than about 2/3 of its length. The fore-femur bears no fringe, and the dorsal transverse band of setae crosses it near the middle. Ventral lamellae are present on the inner side of the gills, and they are 1/2 to 3/4 as long as the lamellae on the outer side (**Fig. 2.110**). 6
6. The submentum, or basal part of the labium, abruptly narrows apically and lacks marginal setae along its apical third. Apical denticles are present on the tarsal claws, although they may be reduced to a single denticle on one side (**Fig. 2.110**). The maxilla lacks a crown of galealaciniae, and the apical denticles are oriented distally. Fully developed larvae do not exceed 3.4 mm in length. The gill on the second abdominal segment is ovoid, while those on the subsequent segments are triangular or irregular in shape.
....*Tricorythopsis* Traver, 1958..p. 105
- The submentum is evenly rounded laterally and is completely fringed with marginal setae. Rows of apical denticles are absent, although a pair of submarginal denticles may be present near the apex of the tarsal claws (**Fig. 2.111**). The maxilla bears a crown of galealaciniae, and the apical denticles are oriented distomedially. 7
7. The head bears large projections on the genae and frontoclypeus. On each hind leg, the femur is about as long as the tibia plus the tarsi. There is a cluster of one to four basal denticles and subapical denticles singly or in pairs on each tarsal claw (**Fig. 2.111**). The basal denticles may sometimes be lost.
....*Tricoryhyphes* Allen and Murvosh, 1987..p. 115
- The head lacks projections on the frontoclypeus, but such projections are sometimes present on the genae. On the hind leg, the femur is 1/2 to 3/4 as long as the tibia plus the tarsi. The claws bear many marginal denticles and sometimes also subapical ones. The gills bear two-lobed ventral lamellae (**Fig. 2.112**). The head is narrower than the pronotum.
....*Tricorythodes* Ulmer, 1920..p. 116
A genus unknown in South America, *Homoleptohyphes* Allen and Murvosh, 1987, is somewhat similar but can be distinguished by the considerably enlarged eyes of the males, which make the head wider than the pronotum.

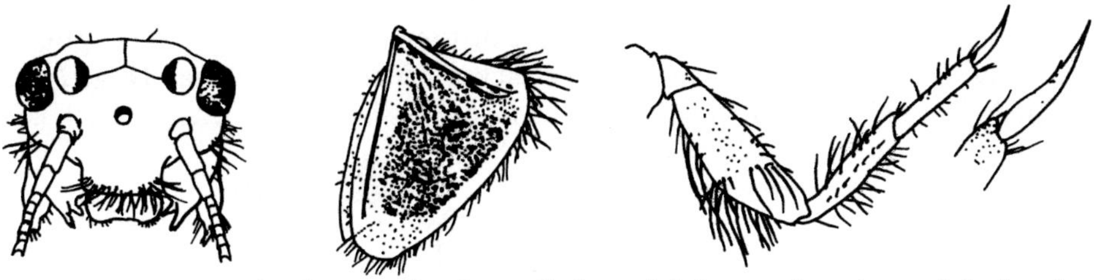

Fig. 2.111 *Tricoryhyphes ocellus* larva (left to right): anterior view of the head, gill, and the right fore-leg with an enlargement of the tarsal claw.

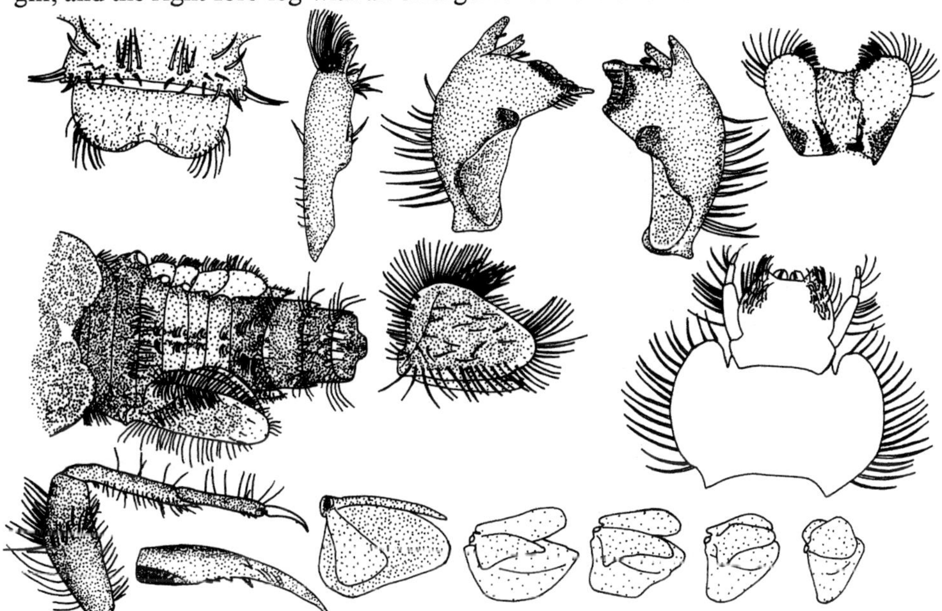

Fig. 2.112 *Tricorythodes hiemalis* larva (upper row, left to right): labrum in dorsal view, maxilla, mandibles, hypopharynx; (middle row, left to right): abdomen in dorsal view with the gills on the right side removed, the gill on the second abdominal segment in dorsal view, the labium in dorsal view; (lower row, left to right): fore-leg, tarsal claw, and gills on the second through sixth abdominal segments in ventral view.

8. The dorsal side of the fore-femur lacks a transverse band of spatulate setae. The submentum, which is the basal part of the labium, is long and constricted subapically. The gills are somewhat elongate and triangular, and they bear a fringe of long setae. Hind wing buds are present in both sexes (**Fig. 2.108**).
....*Haplohyphes* Allen, 1966..p. 120

- The dorsal side of the fore-femur bears a transverse band of spatulate setae. The submentum is not elongated and constricted subapically. The gills are roughly egg-shaped or rectangular and lack a fringe of long setae (**Fig. 2.113**). Female larvae lack hind wing buds. 9

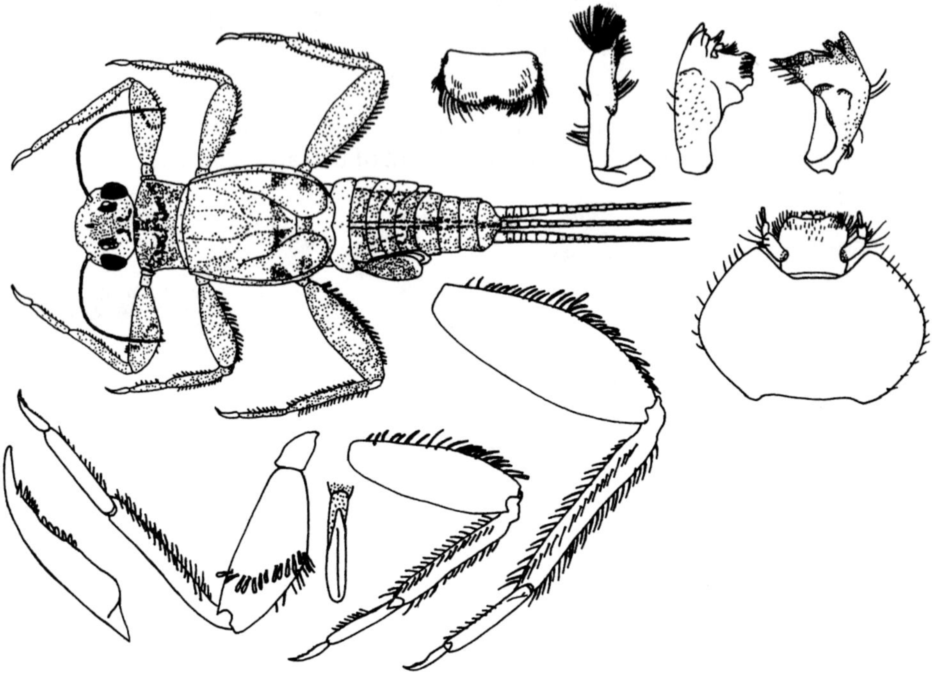

Fig. 2.113 *Yaurina yuta* larva: habitus (upper left); labrum, maxilla, and mandibles (upper right, left to right); labium showing the large submentum and relatively small mentum with the palps (middle right); the tarsal claw on the fore-leg enlarged, fore-leg, an enlarged seta from the fore-leg, middle leg, and hind leg (below, left to right).

9. The opercular gills are somewhat rectangular in shape and present only on the second through fifth abdominal tergites. The anterior margin of the labrum is deeply divided in the middle (**Fig. 2.114**). The compound eyes of the male are so large that they almost completely cover the head. There are no tubercles on the dorsum.
....*Leptohyphodes* Ulmer, 1920

According to Traver (1958), *Leptohyphodes inanis* (Pictet, 1843), from Brazil, should be the only species included in this genus. He treated the species *Tricorythus australis* Banks, 1913, which was placed in the genus *Leptohyphodes* by Ulmer (1920a), as a species of *Tricorythodes*.

- The opercular gills are roughly egg-shaped. Gills are present on the second through sixth abdominal tergites (**Fig. 2.113**). 10

Fig. 2.114 *Leptohyphodes* sp. larva (upper row, left to right): labrum, maxilla, and mandibles; (middle row, left to right): labium, hypopharynx, hind leg, and the ventral lamella from the gill on the second abdominal segment in dorsal view (below the hind leg); and (lower row, left to right): the gill on the second abdominal segment in ventral view, the gill on the third segment in lateral view, and the gill on the fifth segment in dorsal view. The larva from Brazil depicted here was presumed to belong to *Leptohyphodes* by Traver (1944, 1958) but not identified.

10. The gills on the second abdominal tergite almost completely cover the rest of the gills. Those on the second through fifth segments consist of three lamellae, while those on the sixth are formed by only one. The fore-femur bears spines that are long, thin, and have flattened distal parts. The maxillary palp is small and consists of two segments. The width of the widest part of the submentum, or basal part of the labium, is 3.2 times that of the mentum. There are five to seven blunt marginal denticles on the middle portion of the tarsal claw and a high apical row of four to seven slender submarginal denticles (**Fig. 2.113**).
....*Yaurina* Molineri, 2001..p. 123

- The anterior gills do not consist of three lamellae, or the gill on the sixth abdominal segment does not consist of a single lamella (**Fig. 2.115**). According to Molineri (2001a), all of the larvae of the South American species have not yet been examined to determine whether others belong to *Yaurina.* Any that key out at this point but are different from those already known should be carefully examined to determine whether their reassignment to this genus is necessary.
....11

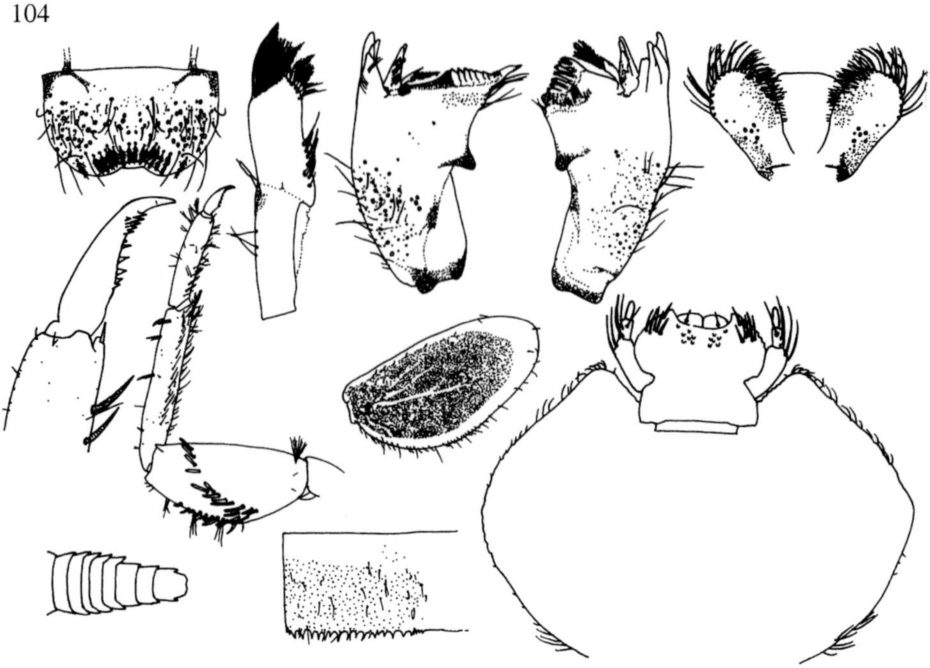

Fig. 2.115 *Allenhyphes flinti* larva (upper row, left to right): the labrum, maxilla, mandibles, and hypopharynx and the fore-leg to the right of its enlarged tarsal claw (middle left), the second gill (center), an outline of the abdomen in dorsal view (lower left), the microsculpture and apical margin of the fifth abdominal tergite (lower center), and the labium (lower right).

11. There is no elevated ridge running the length of any tibia. The gills bear a ventral lamella that is not constricted distally. The outer lamella is longer than the inner one, and its basal part lacks a beak-like process. There are basal and apical rows of denticles on the tarsal claws. The median filament of the male bears a large basal spine pointing ventrally. The fore-femur bears a transverse row of spatulate setae crossing its distal third (**Fig. 2.115**).
..........................*Allenhyphes* Hofmann, Sartori, and Thomas, 1999..p. 141
- A similar genus, *Vacupernius* Wiersema and McCafferty, 2000, which can be distinguished by its lack of the large, ventrally directed spine at the base of the median filament, has not yet been reported from South America. An elevated ridge runs the length of the middle and hind tibiae and usually the fore-tibia, as well. The gills bear a ventral lamella, which is nearly equal in length and width and has a constriction distally. The basal part of the outer lamella has a beak-like process. Marginal rows of denticles are present only near the base of the tarsal claws, but a single submarginal denticle may be present near the apex. The fore-femur bears a transverse row of spines (**Fig. 2.116**).
....*Leptohyphes* Eaton, 1882..p. 125

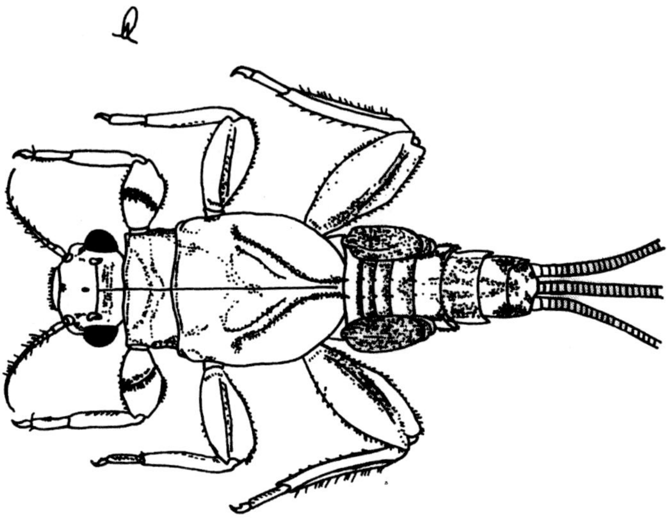

Fig. 2.116 The habitus of a *Leptohyphes plaumanni* larva with the apices of the cerci removed. The cerci are normally slightly longer than the body. Above left is an enlarged seta from the fore-femur.

Key to the adults in the genus *Tricorythopsis* reported from South America

Information for the key was provided by Traver (1958) and Molineri (1999a, 2002a). The descriptions are based mainly on the characteristics of the male adults, and some females cannot yet be identified with certainly.

1. There is a distinct, darkened sclerotization of the lateral margins of the penis lobes, which are brownish yellow. (**Fig. 2.117**). Vein CuP in the fore-wing is sometimes separated from ICuA at its base. 2
 - The lateral margins of the penis lobes are not sclerotized. Vein CuP in the fore-wing is always joined at the base to ICuA (**Fig. 2.118**). 3
2. The narrow black transverse band across the hind femur and sometimes across the middle femur is not interrupted. The penis is divided for more than 1/3 of its length (**Fig. 2.117**). Length: 2.4 to 2.8 mm. Fore-wing length: 2.5 to 2.8 mm. Color: mainly brownish yellow with blackish markings. The abdomen is generally yellowish, translucent, and marked with light gray.
....*Tricorythopsis undulatus* (Allen, 1967) (Argentina, Paraná). Syn: *Leptohyphes undulatus* Allen, 1967, and *Tricorythopsis petersorum* Molineri, 1999.
 - The single narrow black transverse bands across both the hind and middle femur is not interrupted by an unpigmented area. The penis is divided for only about 1/4 of its length (**Fig. 2.119**). Length of male: 1.8 to 1.9 mm; female: 2.3 to 3.0 mm. Fore-wing length of male: 2.4 to 2.5 mm; of female: 2.9 to 3.0 mm. Color: dark brown with yellow and blackish markings. The abdomen of the

male is brownish green washed with gray, darker on the first and the lateral margins of the second and third tergites. The sternites are somewhat lighter. The abdomen of the female is generally yellowish white with a light gray wash on the lateral parts of the tergites. The sternites are washed with gray.
....*Tricorythopsis gibbus* (Allen, 1967) (Minas Gerais, Rio de Janeiro). Syn: *Leptohyphes gibbus* Allen, 1967. The great similarity of the male genitalia of the two nominal species in this couplet suggests that they may well be conspecific.

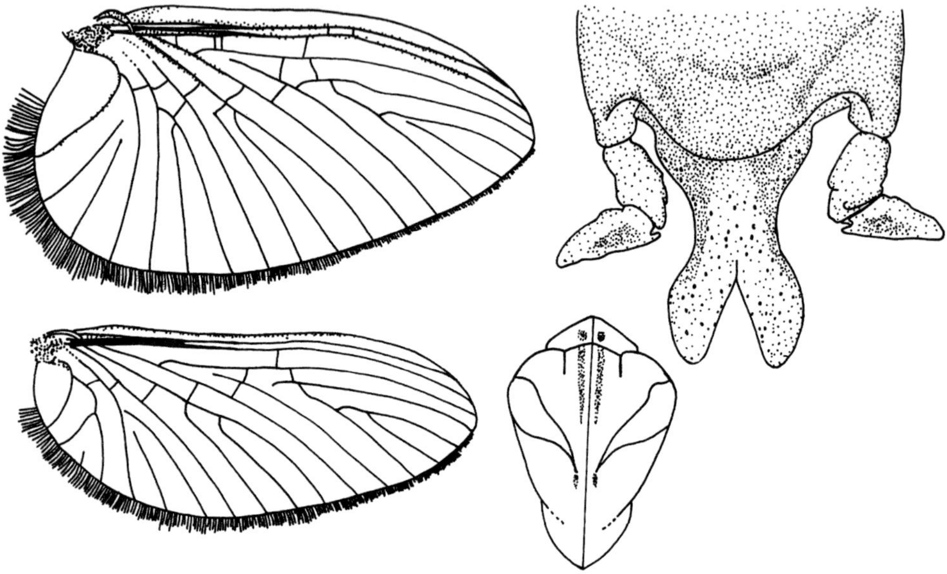

Fig. 2.117 *Tricorythopsis undulatus:* the fore-wing of a male (upper left) and a female (lower left), the male genitalia in ventral view (upper right), and the pattern of blackish markings on the mesonotum (lower right).

3. The base of the long, slender forceps is less than half the length of the first segment. The apices of the penis lobes are acutely pointed and curve toward the midline (**Fig. 2.118**). The cerci of the female do not reach beyond the tenth abdominal tergite. Length: 1.2 to 2.1 mm. Fore-wing length of the male: 1.6 to 1.8 mm; female: 2.3 to 2.5 mm. Color: the male is light orange yellow with blackish markings, while the female is generally whitish yellow. Both have a white, translucent abdomen.
....*Tricorythopsis volsellus* Molineri, 1999 (Venezuela).
- The base of the short, thick forceps is about as long as the first segment. The apices of the penis lobes are rounded (**Fig. 2.96**). 4

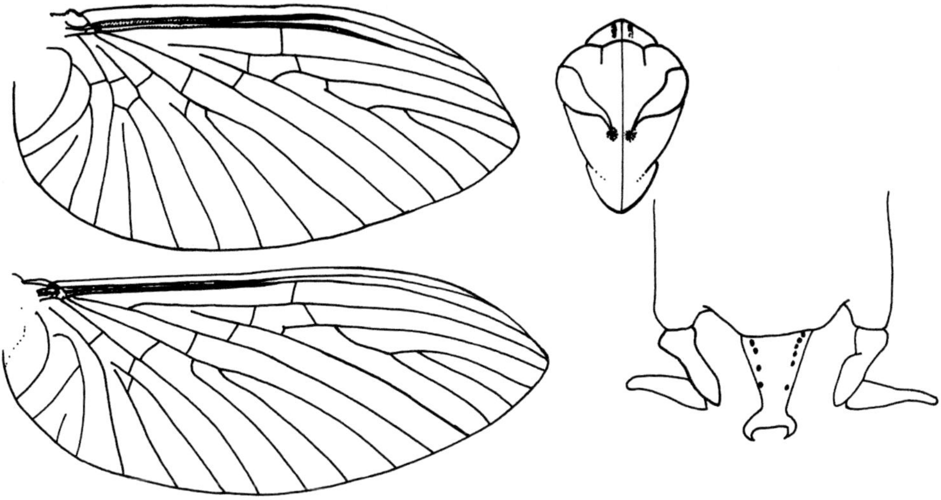

Fig. 2.118 *Tricorythopsis volsellus:* the fore-wing of a male (upper left) and a female (lower left), the pattern on the mesonotum (upper center), and the genitalia of a male in ventral view (lower right).

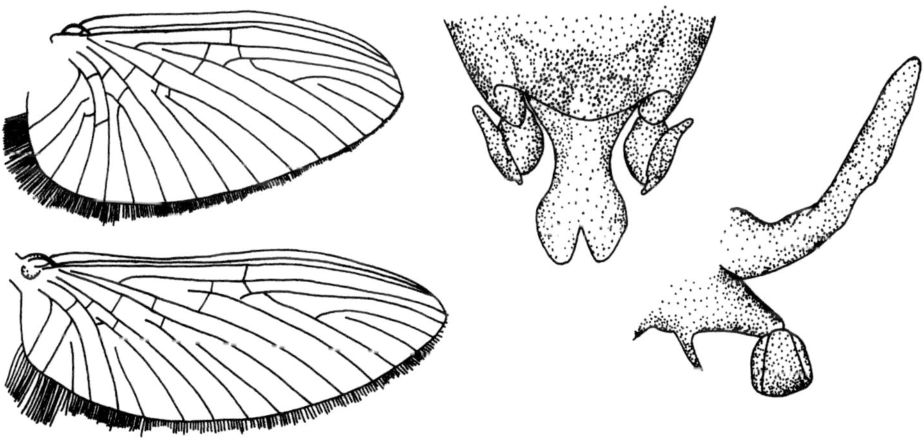

Fig. 2.119 *Tricorythopsis gibbus:* wing of a male (upper left) and a female (lower left), and the male genitalia in ventral (center) and lateral view (right).

4. The penis is divided for 1/3 to 1/4 of its length (**Fig. 2.96**). 5
 - Parts of the body are not covered with reddish markings. The penis is divided only for about 1/6 of its length (**Fig. 2.120**). 6

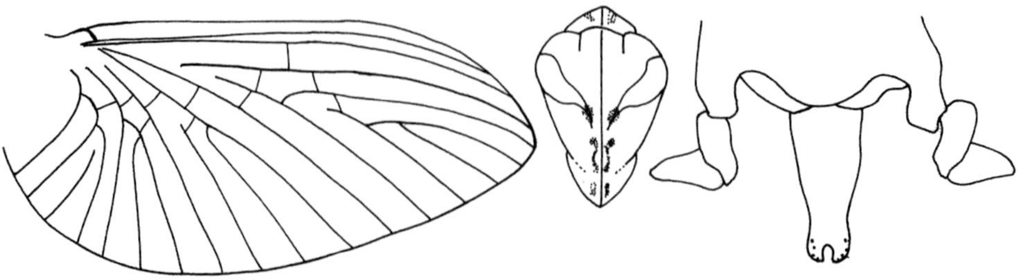

Fig. 2.120 *Tricorythopsis sigillatus* (left to right): the fore-wing of a male, the pattern on the mesonotum, and the male genitalia in ventral view.

5. There are reddish and blackish marks on the body, but they are confined to certain areas. The abdomen is whitish with some blackish and reddish markings. The penis is conical, tapering to about 1/5 of its basal thickness at its apex (**Fig. 2.96**).
.... *Tricorythopsis artigas* Traver, 1958 (Argentina, Uruguay, Rio Grande do Sul).). Syn: *Leptohyphes tinctus* Allen, 1973; *Allenhyphes tinctus* (Allen, 1973); *Tricorythopsis fictilis* Molineri, 1999.
- Reddish markings cover all parts of the body. The abdomen is heavily darkened with grayish red. The penis is relatively short and narrow (**Fig. 2.121**). Length: 1.8 to 2.3 mm. Fore-wing length of male: 1.8 to 2.0 mm. Color: brownish yellow with black and reddish markings.
....*Tricorythopsis yacutinga* Molineri, 2002 (Argentina).

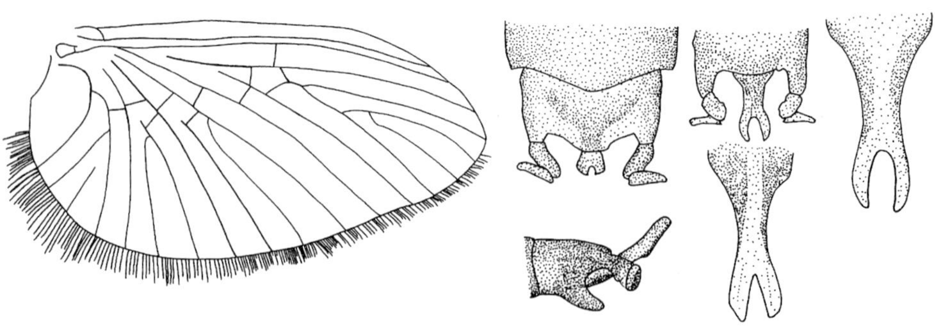

Fig. 2.121 *Tricorythopsis yacutinga* male adult: wing, genitalia extracted from a larva (upper center) and an adult in ventral (upper right center) and lateral view (lower center), and ventral views of two penes (right, above and below).

6. The penis is cylindrical and narrows only slightly toward its apex (**Fig. 2.120**). The underside of the thorax is whitish without a dark wash. Length of male: 1.5 to 2.2 mm. Fore-wing length of male: 1.8 to 2.0 mm. Color: very pale yellow with a translucent abdomen and blackish markings. The female has not been described.
....*Tricorythopsis sigillatus* Molineri, 1999 (Rio de Janeiro).
- The penis is conical with a narrow distal zone that is only about 1/5 as wide as the base. The underside of the thorax is darkened with a gray wash (**Fig. 2.122**).
....7

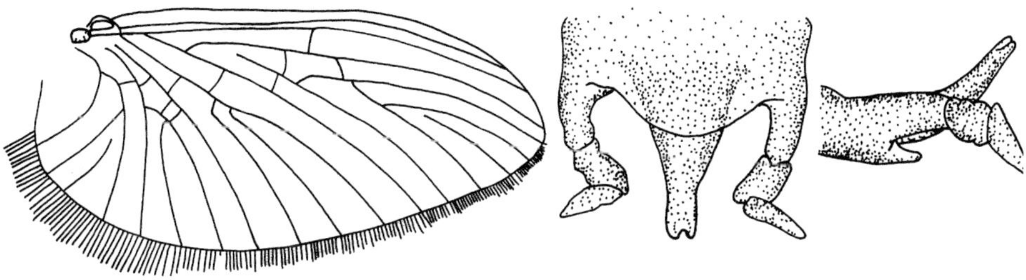

Fig. 2.122 *Tricorythopsis minimus* male adult (left to right): wing and genitalia in ventral and lateral view.

7. The abdominal sternites of the male are completely whitish except for the first and second, which are brownish yellow. Length: 1.5 to 2.25 mm. Fore-wing length of male: 1.75 to 2.0 mm; of female: 2.65 to 2.75 mm. Color: whitish yellow washed with grayish and blackish (**Fig. 2.122**).
....*Tricorythopsis minimus* (Allen, 1973) (Rio Grande do Sul). Syn: *Leptohyphes minimus* Allen, 1973, and *Leptohyphes viriosus* (Allen, 1973). Although Molineri (1999a) placed *Leptohyphes viriosus* in synonymy with this species, there are some discrepancies in the original description, including the total length and head markings.
- The first through ninth abdominal sternites of the male are translucent white with a black median mark and a pair of black lateral markings. Length: 1.4 to 1.85 mm. Fore-wing length of male: 1.5 to 1.75 mm; of female: 2.0 to 2.25 mm. Color: whitish yellow with a blackish wash and markings (**Fig. 2.123**).
....*Tricorythopsis chiriguano* Molineri, 2002 (Bolivia).

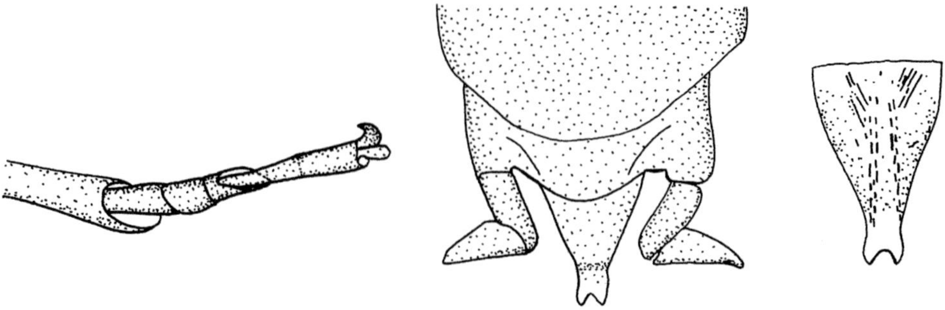

Fig. 2.123 *Tricorythopsis chiriguano* (left to right): the tarsal segments of the middle leg and the male genitalia and penis in ventral view.

Key to the larvae in the genus *Tricorythopsis* reported from South America

Information for the key was provided by Allen (1973) and Molineri (1999a, 2002a). The larva of *Tricorythopsis sigillatus* has not been described.

1. The body is short, broad, and flattened with disk-like proportions. All femora are greatly broadened to a nearly circular shape and fringed with long spines (**Fig. 2.124**). Length of male larva: 1.7 to 2.0 mm; female larva: 2.3 to 2.4 mm. Color: brownish yellow with whitish, reddish, and blackish markings. ....*Tricorythopsis yacutinga* Molineri, 2002 (Argentina).
- The body is not greatly flattened. The femora are clearly longer than wide. Long spines are not distributed all around the circumference of the femora (**Fig. 2.125**). 2
2. All coxae bear dorsal projections. The femoral spines are relatively short. The abdomen bears tubercles or undulating margins (**Fig. 2.125**). 3
- The coxae do not bear dorsal projections. The femoral spines are relatively long. The abdomen lacks tubercles and undulating margins (**Fig. 2.126**). 4
3. There is a bulbous projection at the apex of the mesonotum. Large tubercles are present in the middle of the fifth through eighth abdominal tergites, and there is a small tubercle present on the ninth tergite. There are no maxillary palps. The tarsal claws bear three to five marginal denticles and a row of two or three submarginal denticles near the apex (**Fig. 2.125**). The head has a median carina between the eyes. Length: 1.5 to 2.5 mm. Length of caudal filaments: 0.5 to 1.2 mm. Color: yellow with brown or black markings. The head is yellow, and thorax is yellow with variable black markings dorsally. The legs are light and dark brown. The abdominal tergites and gills are yellow with variable black markings, including blackened bases of the gills. The caudal filaments are yellow.
....*Tricorythopsis gibbus* (Allen, 1967) (Minas Gerais, Rio de Janeiro). Syn: *Leptohyphes gibbus* Allen, 1967.

- There is no bulbous projection at the apex of the mesonotum. No tubercles are present on the abdominal tergites, but the dorsal and lateral margins have undulations. The maxillary palp consists of only one segment and usually bears a seta at the apex. The general color is brown with dark brown or blackish markings. The legs are uniform brown, and the tarsal claws bear five to seven marginal denticles and often a single subapical denticle. The abdominal tergites are brown, sometimes with red or black markings, including undulating lateral margins of the tergites. The operculate gills are brown with a pale transverse line near the apex. The abdominal sternites are brown. The seventh through ninth abdominal segments bear posterolateral projections (**Fig. 2.127**). Length: 2.5 to 3.5 mm. Caudal filament length: 3 mm.
....*Tricorythopsis undulatus* (Allen, 1967) (Paraná). Syn: *Leptohyphes undulatus* Allen, 1967.

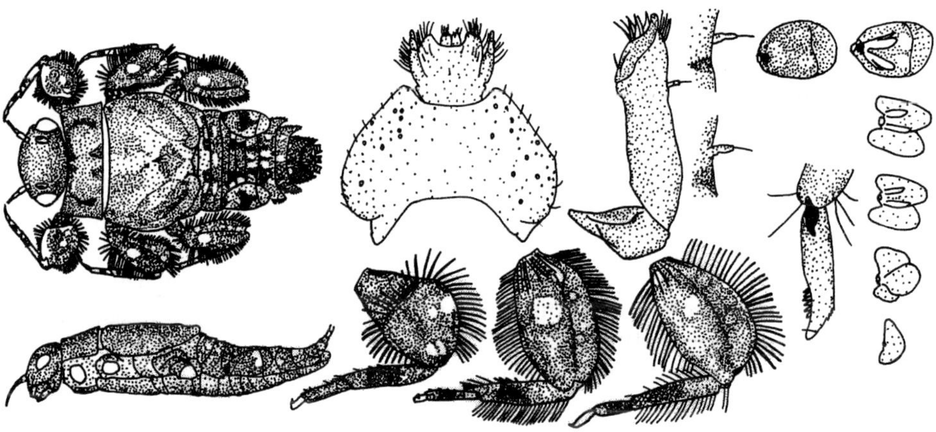

Fig. 2.124 *Tricorythopsis yacutinga* larva: habitus in dorsal view and a lateral view with the legs removed (left, above and below, respectively); labium (upper left center); maxilla with enlargements of the palps from two specimens (upper right center); the fore, middle, and hind legs and an enlarged tarsal claw (lower center, left to right); the gill on the second abdominal segment in dorsal view (upper right center); the gills on the second through sixth abdominal segments in ventral view (right, top to bottom).

4. There is a prominent dark median mark near the anterior margin of the mesonotum and a dark midline stripe on the pronotum bordered by paired light markings (**Fig. 2.126**). Length of male larva: 1.5 to 1.7 mm; female larva: about 2 mm. Color: whitish or yellowish with black markings.
....*Tricorythopsis chiriguano* Molineri, 2002 (Bolivia).

- There is no prominent dark median mark near the anterior margin of the mesonotum. If the median stripe on the pronotum is dark, then it is not bordered by paired light markings. The abdomen is generally pale with a dark pattern (**Figs. 2.128** and **2.129**). 5

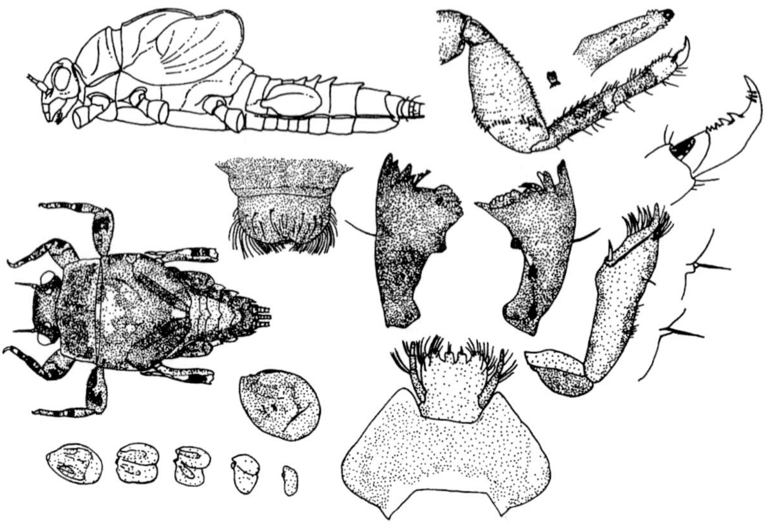

Fig. 2.125 *Tricorythopsis gibbus* larva (upper row, left to right): outline of a final instar larva without appendages in lateral view, the fore-leg with a seta from the femur enlarged, and enlargements of two aspects of the tarsal claw; (middle row, left to right): habitus without cerci, labrum, mandibles, maxilla, and enlarged maxillary palps from two specimens; and (lower row, left to right): the gills in ventral view from anterior to posterior, the first abdominal gill in dorsal view enlarged (above row of gills), and the labium.

5. There is a narrow dark stripe along the midline of the pronotum. There are pale ellipses surrounded by dark pigment on the operculate gills. A pale area runs down the centers of the first three abdominal tergites and onto the anterior edge of the fourth. The sixth and tenth tergites are almost entirely pale, as are the central areas of the seventh and ninth tergites. The tarsal claws bear six to eight marginal and a row of five to seven submarginal denticles near the apex (**Figs. 2.110** and **2.128**). Length: 3 to 4 mm. Color: pale with brown markings. Dark markings on the head include paired brown maculae on the occiput between the compound eyes and a generally dark frons except for light areas around the middle ocellus and antennal bases. The pattern on the abdomen is shown in **Figure 2.128**.
....*Tricorythopsis artigas* Traver, 1958 (Argentina, Uruguay, Rio Grande do Sul). Syn: *Leptohyphes tinctus* Allen, 1973; *Allenhyphes tinctus* (Allen, 1973); *Tricorythopsis fictilis* Molineri, 1999.

- There is a broad pale stripe along the midline of the pronotum. The abdominal tergites are brown with both black and pale markings, which do not include a distinct pale area in the center of the first three tergites or a sixth tergite paler than the adjacent tergites (**Fig. 2.129**). The head is brown with a pale transverse band between the compound eyes on the vertex. The tarsal claws bear five to six marginal denticles and a row of five or six denticles near the apex. The length is 2 to 4 mm. Color: brown with dark brown and black markings.
....*Tricorythopsis minimus* (Allen, 1973) (Rio Grande do Sul). Syn: *Leptohyphes minimus* Allen, 1973, and *Leptohyphes viriosus* (Allen, 1973). Although Molineri (1999a) placed *Leptohyphes viriosus* in synonymy with this species, there are some discrepancies in the original description, including the total length and head markings.

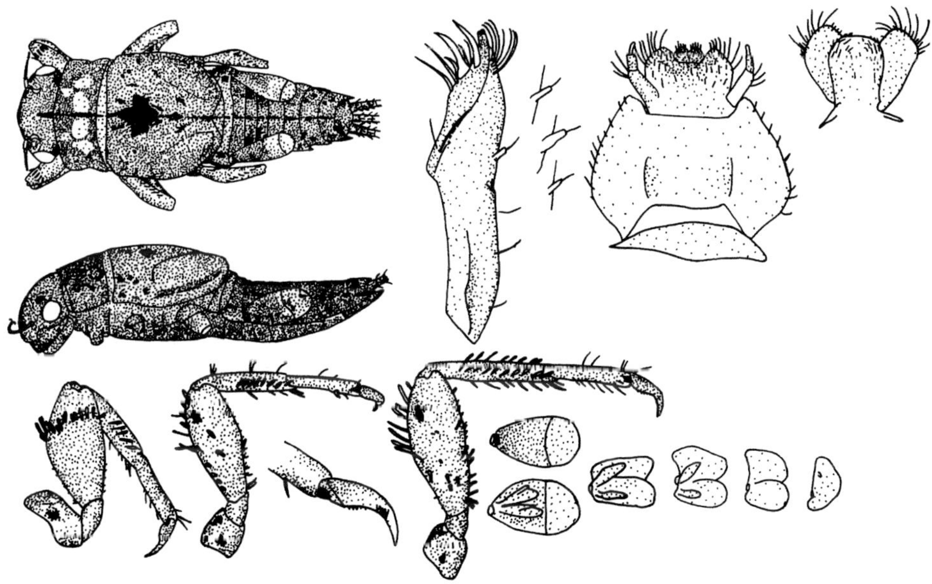

Fig. 2.126 *Tricorythopsis chiriguano* larva: habitus in dorsal view and a lateral view with the legs removed (upper and middle left, respectively); maxilla with enlargements of the palps from three specimens (upper left center); labium (upper right center); hypopharynx (upper right); and (lower row, left to right) fore and middle legs, enlarged tarsal claw, hind leg; the gill on the second abdominal segment in dorsal (above) and ventral (below) view, and the gills on the second through sixth abdominal segments in ventral view.

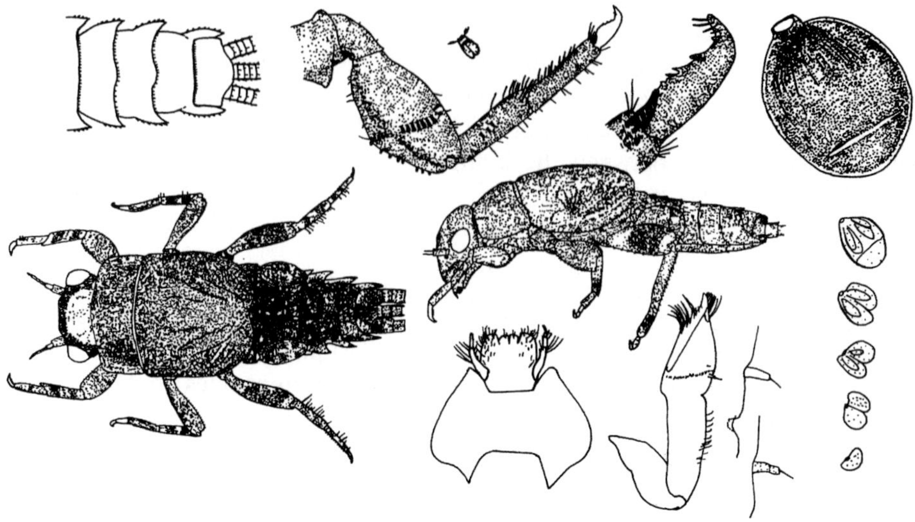

Fig. 2.127 *Tricorythopsis undulatus* larva (upper row, left to right): dorsal outline of the seventh through tenth abdominal segments, the fore-leg with a seta from the femur enlarged, an enlarged tarsal claw, and the first gill in dorsal view; habitus in dorsal (lower left) and lateral view (center); the labium, maxilla, and enlarged maxillary palps from two specimens (lower center, left to right); the gills in ventral view (lower right, top to bottom = anterior to posterior).

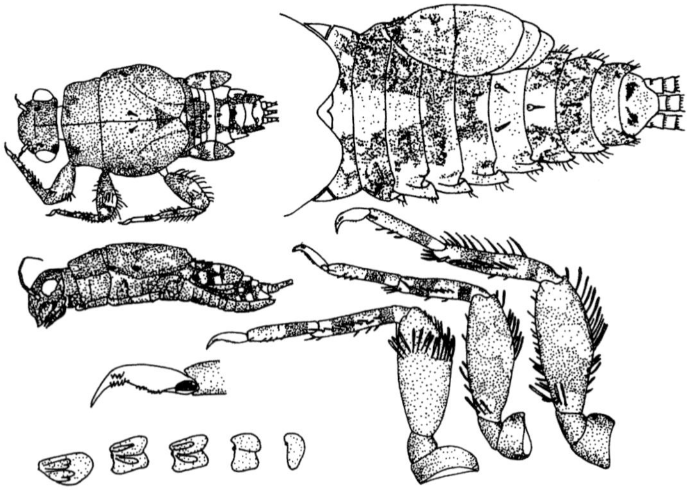

Fig. 2.128 *Tricorythopsis artigas* larva: habitus in dorsal and lateral view (upper left); dorsal surface of the abdomen with three of the gills (upper right); the fore, middle, and hind-legs (lower right, left to right) with an enlarged tarsal claw (lower middle left); the five gills in ventral view (lower left, left to right = anterior to posterior). The illustrations of Molineri (2002a) differ from those shown in **Figure 2.110**, based on illustrations by Traver (1958) and Allen (1973). Traver (1958) takes priority for identification of the species, even though his figures are of poorer quality than later illustrations.

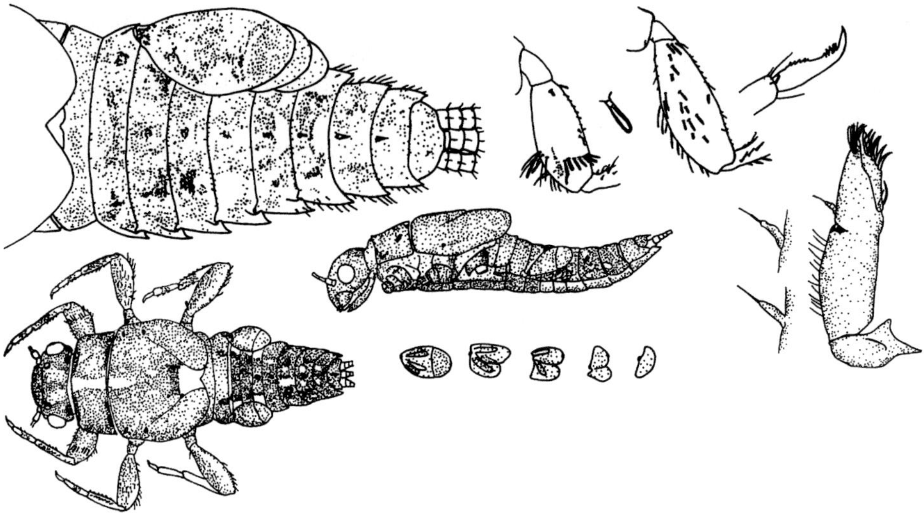

Fig. 2.129 *Tricorythopsis minimus* larva: the abdomen in dorsal view with three gills still attached (upper left); habitus without cerci in dorsal (lower left) and lateral view (center); femur of the fore-leg, a modified spine from the femur, the hind femur, and a tarsal claw (upper right, left to right); the maxilla with the enlarged maxillary palps of two specimens to its left (lower right); the gills in ventral view (lower center, left to right is from anterior to posterior).

Key to the species of *Tricoryhyphes* larvae reported from South America

Information for the key was provided by Allen (1967) and Domínguez (1982). The only adults in this genus from South America that have been described are those of *Tricoryhyphes popayanicus* (**Fig. 2.98**).

1. The claw on the fore-tarsus has three short basal denticles and two longer distal ones. There is a small black mark near the articulation at the base of each gill (**Fig. 2.130**). Body length: 3 to 4 mm. Length of the cerci: 1.5 to 2.5 mm. Coloration: generally pale yellow with blackish markings.
....*Tricoryhyphes popayanicus* (Domínguez, 1982) (Argentina). Syn: *Tricorythodes popayanicus* Domínguez, 1982.
- The tarsal claws each have one denticle (**Fig. 2.131**). 2
2. The head is light brown. The thorax is light brown with black markings. The abdomen is entirely light brown with the exception of an irregular black marking on the seventh abdominal tergite and one central black marking on the eighth and tenth tergites (**Fig. 2.111**). Body length: 4 to 5 mm; length of caudal filaments: 1 to 2 mm. Color: light brown with black markings.
.....................................*Tricoryhyphes ocellus* (Allen and Roback, 1969) (Peru). Syn: *Tricorythodes ocellus* Allen and Roback, 1969.

- The head is yellow with a black marking on the occiput and a black triangle on the frons between the antennae. The thorax and abdomen are black with yellow markings (**Fig. 2.131**). Length: 7 to 8 mm; caudal filament length: 3.5 to 4.5 mm.
....*Tricoryhyphes barbus* (Allen, 1967) (Minas Gerais). Syn: *Tricorythodes barbus* Allen, 1967.

Fig. 2.130 *Tricoryhyphes popayanicus* larva (left to right): the head and anterior part of the pronotum, fore-leg, enlargement of its tarsal claw, and a gill.

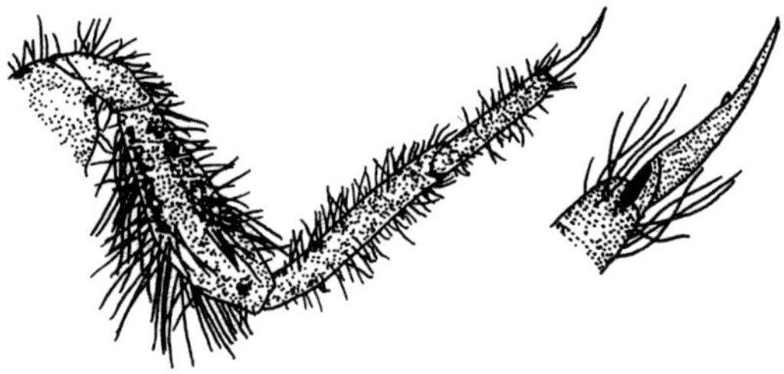

Fig. 2.131 The fore-leg with an enlarged tarsal claw of *Tricoryhyphes barbus.*

Key to the species of known subadult and adult *Tricorythodes* in South America.

Information for the key was provided by Traver (1943, 1958, 1959a), Domínguez (1982), and Molineri (2002b). The larva of *Tricorythodes hiemalis* Molineri, 2002, is the only one that has been described (**Fig. 2.112**). Larvae of other South American species formerly assigned to this genus were transferred to *Haplohyphes* (Domínguez *et al.*, 2001), although Molineri (2002b) still regards them as species of *Tricorythodes*.

1. The head and thorax are mainly dark brown or dark reddish brown. The abdomen is yellow or brown with black markings. 2

- The ground colors of the head and thorax are whitish, yellowish, pale flesh color, or light brown; sometimes black or gray markings are present. 3

2. The head of the female is dark grayish brown heavily marbled with black dorsally, except on the front. The thorax is brownish with black markings, including a median stripe on the pale reddish brown mesonotum. The abdomen is yellowish with a reddish tinge and dark markings on the ninth and tenth tergites. The entire basal portion of the wing has a purplish gray tinge (**Fig. 2.132**). Length: 2.7 to 2.8 mm; fore-wing length: 3.7 to 3.8 mm. The male has not been described.

....*Tricorythodes sanitaria* Traver, 1959 (Uruguay).

- The head is very dark reddish brown without black markings. The pronotum is dull blackish brown. The mesonotum is shiny dark reddish brown. The abdomen is pale yellowish or reddish brown with black shading on the basal and apical tergites. Length: 3 to 4 mm. Fore-wing length of male: 3.5 to 4.5 mm; of female: 5.5 mm.

....*Tricorythodes lichyi* Traver, 1943 (Venezuela).

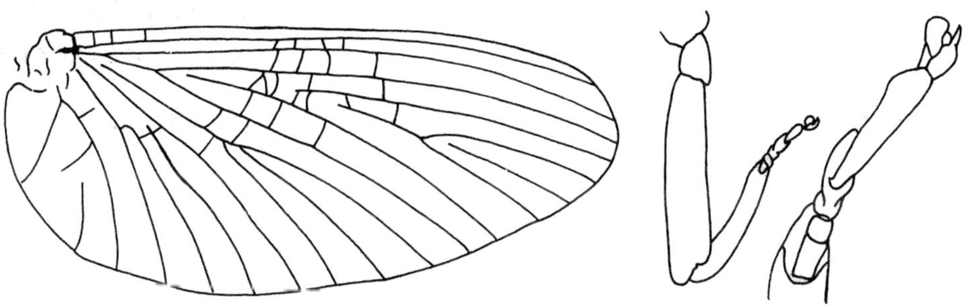

Fig. 2.132 Fore-wing (left) and the hind leg, lacking its color pattern, with an enlargement of the tarsal segments and claws (right) of *Tricorythodes sanitaria.*

3. The length reported by Traver (1958) is 2 to 3 mm, and the fore-wing length is 2 to 3 mm. Banks (1913) reported the "expanse" of Brazilian specimens to be 8 mm. The ground color is pale yellowish. The head has pale crescent-shaped marks on the vertex near each eye. Black markings on the abdomen include lateral marks that extend entirely across the posterior margins of some tergites. The femora and tibiae of the fore and hind legs of the male are all subequal in length. The legs are particularly long and thin (**Fig. 2.99**).

....*Tricorythodes australis* (Banks, 1913) (Guyana, Rondônia). Syn: *Tricorythus australis* Banks, 1913; *Leptohyphodes australis* (Banks, 1913).

- The length is greater than 4 mm, and the fore-wing is at least 4.5 mm long.

....4

4. The lower half of the penis of the subadult is bulbous with its widest point at midlength. The excavation at the base of the forceps is shallow (**Fig. 2.133**). The head of the male subadult is pale flesh colored with a black posterior margin and pale antennae. The pronotum is yellowish with narrow black lines along the anterior and lateral margins and a narrow black middorsal line. A black line extends from the fore-coxa upward to brown lateral, triangular patches. The abdomen is yellowish white with pale gray transverse bands on the first and last tergites, narrow dark dorsolateral lines, and other black markings, including a large patch on the third tergite that extends onto the fourth. Length: 4.5 mm; fore-wing length: 4.5 to 5 mm. It is assumed that the adults have similar patterns, but this cannot be reported with certainty until specimens are obtained and described.
....*Tricorythodes arequita* Traver, 1959 (Uruguay).

- The penis is widest at its base and is narrow at midlength. The base of the forceps is shallow and bordered by small tooth-like processes (**Fig. 2.134**). The head of the adult is whitish yellow with a pair of circular blackish markings between the median and lateral ocelli and V-shaped marks in the occipital region posterior to the compound eyes and between the posterior margin of the head and the lateral ocelli. The pronotum is whitish with a grayish black wash and two brownish gray lateral stripes. The coxae and trochanters are light brown. There are dark stripes along the femora: brown on the fore-leg and dark yellow on the middle and hind legs. The dorsal surface of the hind femur has a gray V-shaped mark near the apex. The ground color of the first seven segments of the male abdomen is whitish yellow, while the eighth through tenth are darker yellow. The ground color of the female abdomen is light brownish yellow. The tergites of both sexes are all darkened by a grayish black wash, which is not very intense on the third through sixth segments. The posterior part of the tenth tergite of the male is light brown. Length: 4.1 to 4.4 mm; fore-wing length of male: 5.4 to 5.6 mm; of female: 6.2 mm.
....*Tricorythodes hiemalis* Molineri, 2002 (Argentina).

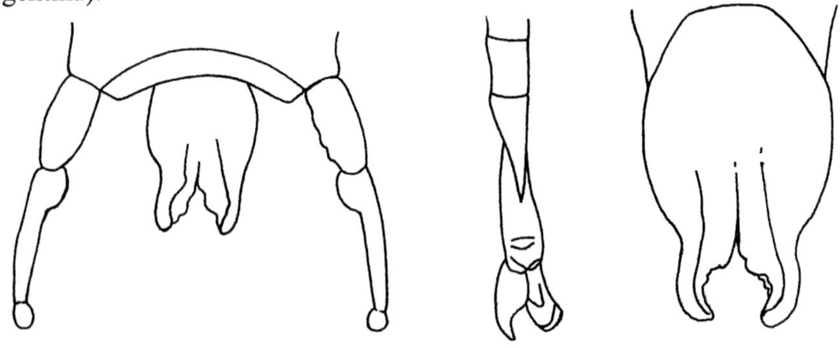

Fig. 2.133 *Tricorythodes arequita* male (left to right): the genitalia, the apical tarsal segments and claw of the fore-leg, and the penis lobes.

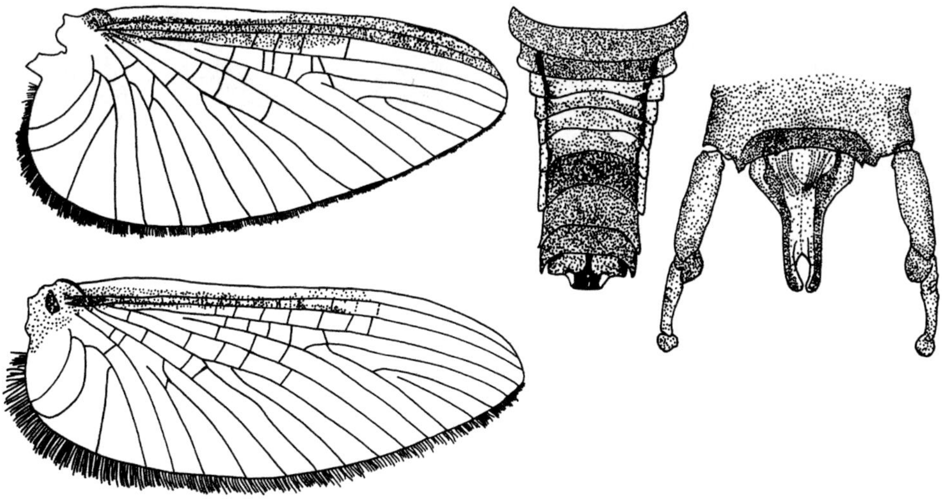

Fig. 2.134 *Tricorythodes hiemalis:* the wing of a male (upper left) and a female (lower left), the abdomen of a female (center), and the male genitalia in ventral view.

Key to the larvae in the genus *Epiphrades* reported from South America

Information for the key was provided by Allen (1967) and Wiersema and McCafferty (2000). No adults of South American species in this genus have been described.

1. The tarsal claws have 16 to 18 denticles (**Fig. 2.109**). The general coloration is black with lighter markings on the abdomen and brown caudal filaments. The head is almost entirely black. Length: 5.5 to 6.5 mm; caudal filament length: 5 to 6 mm.
....*Epiphrades cristatus* (Allen, 1967) (Serra do Mar, presumably Rio de Janeiro but not reported precisely). Syn: *Tricorythodes cristatus* Allen, 1967.
- The tarsal claws have 8 to 11 denticles. The head is brown with a black marking on the vertex. The thorax and abdomen are brown with black or dark brown markings (**Fig. 2.135**). Length: 3.5 to 4.5 mm; caudal filament length: 1.5 to 2.5 mm.
....*Epiphrades bullus* (Allen, 1967) (Santa Catarina). Syn: *Tricorythodes bullus* Allen, 1967.

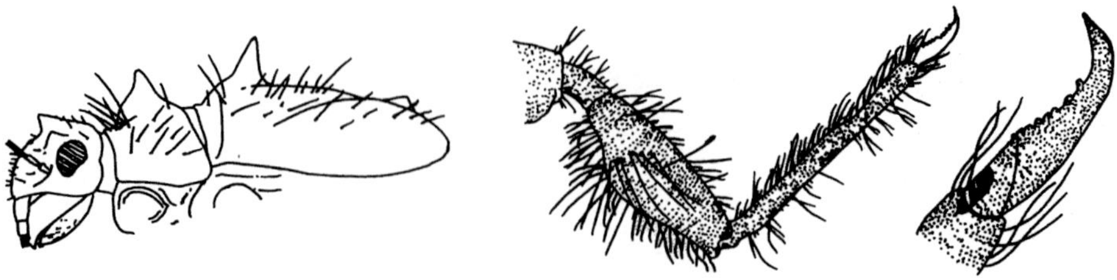

Fig. 2.135 A profile of the head and thorax and the fore-leg with an enlarged tarsal claw of *Epiphrades bullus.*

Key to the species of adult *Haplohyphes* in South America

Information for the key was taken from Allen (1966), Domínguez (1984), and Molineri (1999b). The larva of *Haplohyphes baritu* is the only one of the genus that has been described (**Fig. 2.108**).

1. The hind wings have fewer than three longitudinal veins (**Fig. 2.13, 2.102**). 2
- The hind wings have three longitudinal veins (**Fig. 2.136**). 3

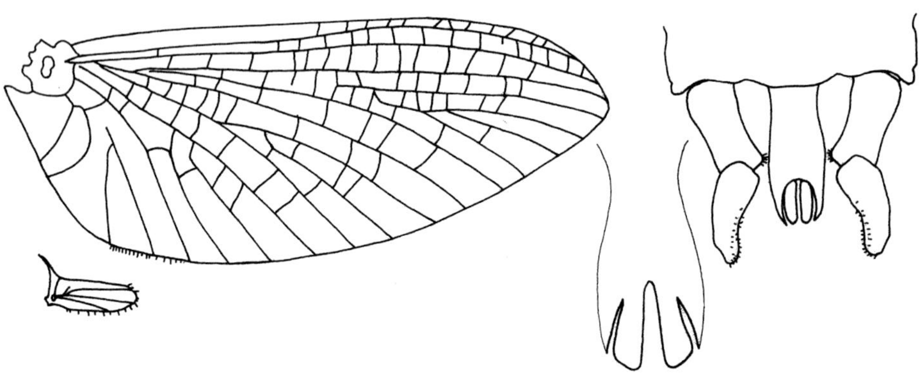

Fig. 2.136 Fore and hind wing (left), male genitalia (right), and the penis (lower right center) of *Haplohyphes furtiva.*

2. Veins A and CuP join at the posterior edge of the fore-wing (**Fig. 2.13**). Length of male: 4 to 5 mm; female: 6 to 6.5 mm. Fore-wing length of male: 4 to 5 mm; female 6 mm. Color: the male is brown with dark brown and black markings; the female is black with paler brown markings. ....*Haplohyphes huallaga* Allen, 1966 (Peru).

- Veins A and CuP are well separated at the edge of the wing (**Fig. 2.102**).
....*Haplohyphes mithras* Allen, 1966
(Costa Rica, not yet found in South America).
3. The claws on the fore-tarsus of the male are different; one is acutely pointed, and the other is expanded and rounded. The penis is widest at the base and tapers slightly toward the apex. The apical segment of the paraproct has a deep excision in the middle of the inner margin (**Fig. 2.137**). Length of male: 4 to 5 mm; female: 6.5 to 6.7 mm. Length of fore-wing of male: 7.2 mm; of female: 8.5 to 8.7 mm. Hind wing length of male: 1.2 mm; of female: 1.1 to 1.2 mm. Coloration: yellowish brown with dark markings and a grayish abdomen.
....*Haplohyphes baritu* Domínguez, 1984
(Argentina).
- The claws on the fore-tarsus are alike, both expanded and rounded at the apex (**Fig. 2.136**). 4

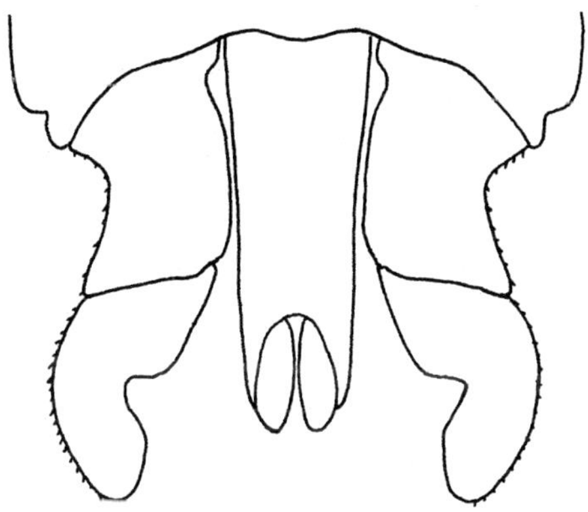

Fig. 2.137 The male genitalia of *Haplohyphes baritu*.

4. The penis is widest at the base and is divided in the apical 2/5 of its length; each lobe is roughly rectangular at its apex (**Fig. 2.138**). Length of male: 4.5 to 4.9 mm. Length of fore-wing of male: 5.0 to 5.2 mm. Hind wing length of male: 0.8 to 0.9 mm. Coloration: whitish; the abdomen is darkened with gray. The female has not been described.
....*Haplohyphes dominguezi*, Molineri, 1999
(Ecuador).
- The penis is widest near the middle and is divided only in the apical 1/7 of its length; each lobe is roughly rectangular at its apex (**Fig. 2.136**). Length of male: 5 to 5.5 mm; female: 5 to 5.2 mm. Length of fore-wing of male: 6.2 to 6.5 mm; of female: 7 to 7.1 mm. Hind wing length of male: 1.0 to 1.2 mm; of female: 0.9 to 1.0 mm. Coloration: brown with black markings and a grayish abdomen.
....*Haplohyphes furtiva* Domínguez, 1984
(Argentina).

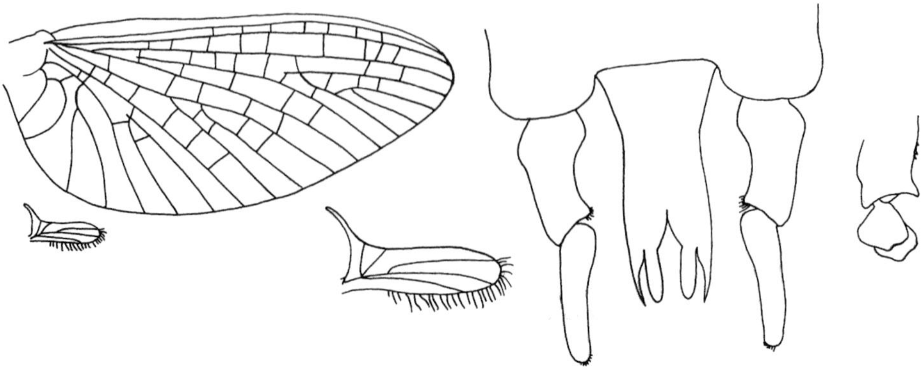

Fig. 2.138 *Haplohyphes dominguezi:* fore and hind wing drawn to scale (left), hind wing enlarged (lower center), male genitalia (right center), and the tarsal claw from the fore-leg (right).

Key to the adults in the genus *Traverhyphes* reported from South America

Information for the key was provided by Molineri (2001b). The larva of *Traverhyphes indicator* is the only one in this genus that has been described (**Fig. 2.106**).

1. In the male subadult, the dorsal coloration of the abdomen is uniformly gray except for two darker longitudinal stripes near the midline of the first through sixth tergites. The basal half of the penis is narrower than the apical half. The spines on the penis are inserted on the dorsolateral surface (**Fig. 2.139**). Length of male subadult: 3.4 to 3.5 mm. Fore-wing length of male subadult: 3.3 to 3.6 mm. Color: mainly yellowish orange, shaded with gray on the pronotum and abdomen. Only the male subadult has been described.
....*Traverhyphes pirai* Molineri, 2001 (Rio de Janeiro).

- The abdomen of the male subadult and adult has black shading only along the lateral margins, while the middle parts of the tergites are pale, forming a broad, pale, median longitudinal stripe. The penis lobes of the male are parallel and acute at the tips. The spines on the penis are inserted on the dorsal surface (**Fig. 2.103**). Length of male: 3.4 to 3.8 mm; female: 4.0 to 5.0 mm. Fore-wing length of male: 3.3 to 3.9 mm; of female: 4.3 to 3.9 mm. Hind wing length of male: 0.6 to 0.8 mm. Caudal filament length: 5 mm. Color: mainly yellowish with black shading in places.
...................................*Traverhyphes indicator* (Needham & Murphy, 1924) (Argentina, Uruguay).

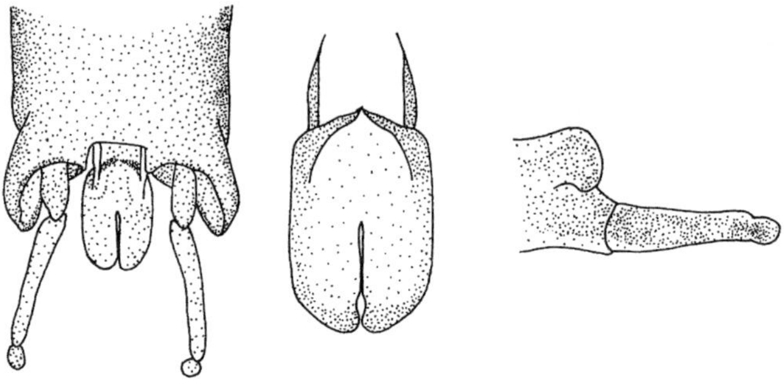

Fig. 2.139 *Traverhyphes pirai* (left to right): male genitalia in ventral view, enlargement of the penis, and the genitalia in lateral view.

Key to the adults in the genus *Yaurina* reported from South America

Information for the key was provided by Molineri (2001a).

1. In male adults, the margins and an irregular sublateral band on the pronotum are black. The head is whitish with a heavy suffusion of gray on the dorsal and lateral surfaces. The middle section of the penis is narrower than the basal and apical parts, and the ventral spine-like processes are about 0.7 times as long as the penis. The apices of the processes are acutely pointed and curve laterad (**Fig. 2.140**). Length of adult male: 3.0 to 3.2 mm; of female subadult: 3.7 to 4.5 mm. Fore-wing length of adult male: 3.1 to 3.3 mm; of female subadult: 4.3 mm. Hind wing length of male: 0.40 to 0.45 mm. Color: mainly yellowish orange with a whitish translucent abdomen in the male and an abdomen yellowish from the egg content in females. The adult female has not been described.
....*Yaurina yuta* Molineri, 2001 (Argentina).

- There are no well-defined black markings on the prothorax of the male, although there may be a suffusion on gray on a lighter background. The head is suffused with brown or has gray shading only around the ocelli and antenna bases. The ventral processes are longer than the penis, or, if they are shorter than the penis, then they are blunt at the apices (**Fig. 2.97, 2.141**). 2

2. The pronotum is brown except for three pale round marks on each side of the midline. The spine-like appendages on the ventral side of the penis are blunt and only slightly more than half as long as the penis (**Fig. 2.97**). Length of adult male: 3.2 to 3.6 mm; of female subadult: 3.7 to 4.0 mm. Fore-wing length of adult male: 3.6 to 3.8 mm; of female subadult: 4.3 to 4.5 mm. Hind wing length of male: 0.5 mm. Color: mainly whitish brown.
....*Yaurina mota* Molineri, 2001 (Argentina).

- The pronotum of the adult male is whitish and translucent with slight gray shading that forms oblique bands near the lateral margins. The penis is narrowest at midlength, and its ventral processes are coiled and longer than the penis (**Fig. 2.141**). Length of adult male: 3.0 mm. Fore-wing length of male adult: 2.9 mm. Hind wing length of male: 0.4 mm. Color: yellowish with a whitish abdomen. The female has not been described. .... *Yaurina yapa* Molineri, 2001 (Ecuador).

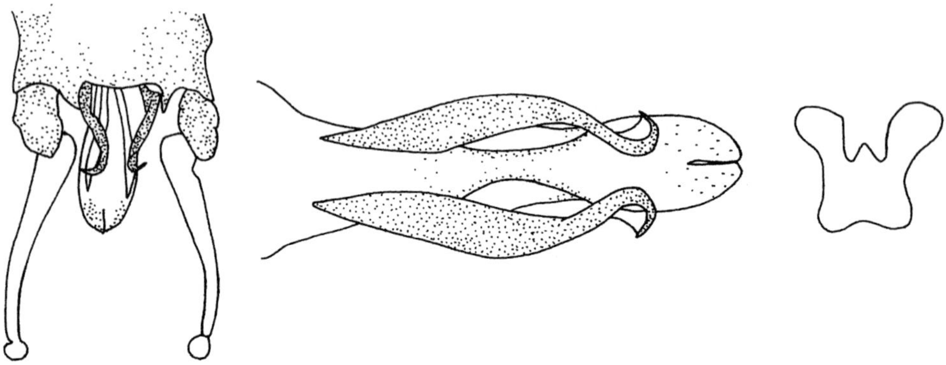

Fig. 2.140 *Yaurina yuta* male (left to right): genitalia and penis in ventral view and a transverse section of the penis.

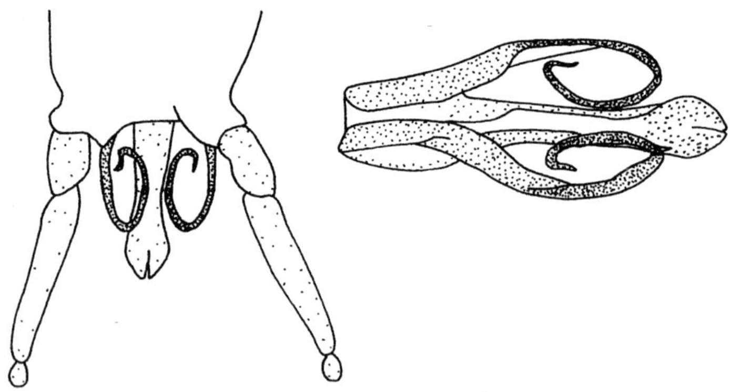

Fig. 2.141 The male genitalia (left) and penis of *Yaurina yapa* in ventral view.

Key to the larvae in the genus *Yaurina* reported from South America

Information for the key was provided by Molineri (2001a). He stated that a positive identification of the larvae is not yet possible, making this key tentative, and no information is available about the larva of *Yaurina yapa*.

1. The opercular gill on the second abdominal segment is yellowish white surrounded by gray except on the distal part of the lateral margin. The other

gills are white with a gray wash at the base (**Fig. 2.113**). Length of the final instar male larva: 3.4 mm; female larva: 4.3 to 4.5 mm. Cerci of male: 1.8 mm long. Color: mainly whitish yellow.
....*Yaurina yuta* Molineri, 2001 (Argentina).
- The opercular gill on the second abdominal segment is translucent yellowish brown in the center surrounded by hyaline margins. The other gills are white (**Fig. 2.142**). Length of the final instar male larva: 3.0 to 3.2 mm. Cerci: 2.4 mm long. Median caudal filament: 2.7 mm. Color: yellowish white or light yellowish brown with grayish markings.
....*Yaurina mota* Molineri, 2001 (Argentina).

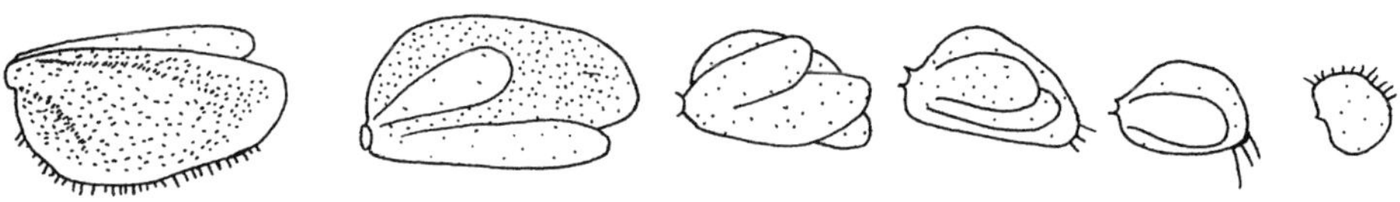

Fig. 2.142 *Yaurina mota* larva (left to right): the gill from the second abdominal segment in dorsal view, and the gills from the second, third, fourth, fifth, and sixth segments in ventral view.

Key to the adults in the genus *Leptohyphes* reported from South America

Information for the key was provided by Needham and Murphy (1924) and Navás (1920b, 1931b), Demoulin (1952b). The adults of many species described as larvae by Allen (1967, 1973) and Mayo (1968a) have never been accounted for, but many of these species have since been transferred to other genera. The larvae of the species in the key to adults, except for those of *Leptohyphes petersi, L. setosus,* and *L. maculatus,* have not been described. Confusion will persist until all life stages are accounted for.

1. The wing is about 8 mm long (**Fig. 2.143**). The color is mainly black.
....*Leptohyphes eximius* Eaton, 1882 (Argentina).
- The fore-wing is 3 to 6 mm long. 2
2. The dorsal coloration is mainly dark, sometimes with pale markings and appendages. The anal veins in the fore-wing curve evenly to the wing margin (**Fig. 2.144**). 3

- The color is mainly yellowish or brown, sometimes with black markings, or even with some of the abdominal tergites black (**Fig. 2.145**). 4

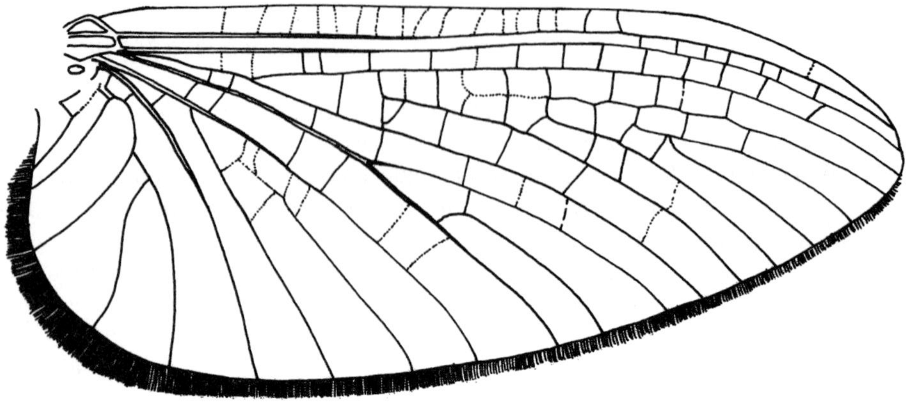

Fig. 2.143 The wing of *Leptohyphes eximius.*

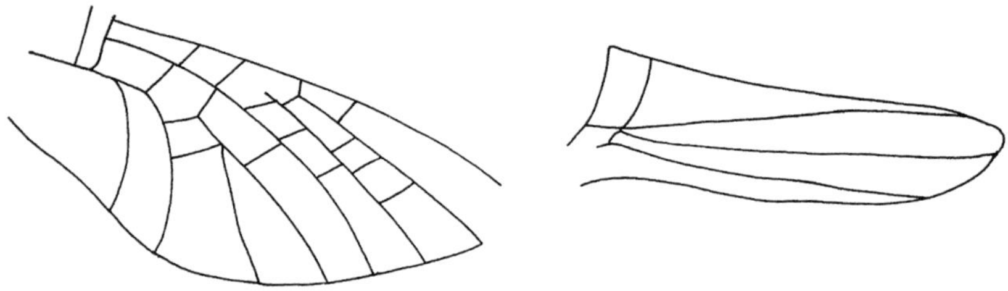

Fig. 2.144 The veins of the inner, posterior part of the fore-wing (left) and the hind wing of a male *Leptohyphes niger.*

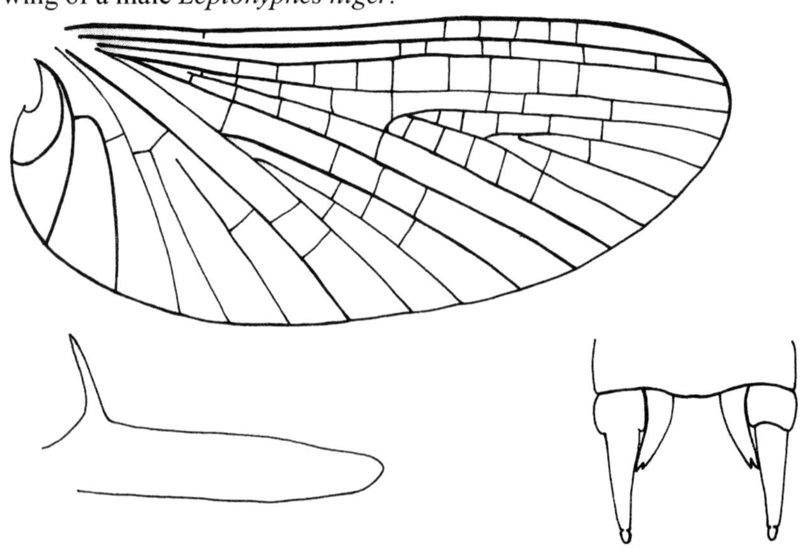

Fig. 2.145 *Leptohyphes mollipes:* fore and hind wings (left) and genitalia of a male (lower right).

3. The hind wing has three long veins connected only near the base (**Fig. 2.144**). The head, eyes, and thorax are shiny black, and the abdomen is mainly dark. The antennal flagellum and cerci are whitish. The legs are pale except for dark stripes on the femora. Length: 3.6 mm; fore-wing length: 5.5 mm. Only the male has been described.
....*Leptohyphes niger* (Navás, 1920) (Argentina). Syn: *Bruchella nigra* Navás, 1920
- The hind wing has only one long vein, which fades out in the proximal half (**Fig. 2.146**). The head, eyes, and thorax are dark dorsally but not very shiny. The dorsal surface of the abdomen is black with pale posterior margins of the tergites, and its ventral surface is light yellow. The legs are light yellow with some darker infuscation on the femora. Length: 4.3 mm; fore-wing length: 5 mm.
....*Leptohyphes bruchi* Navás, 1931 (Argentina).

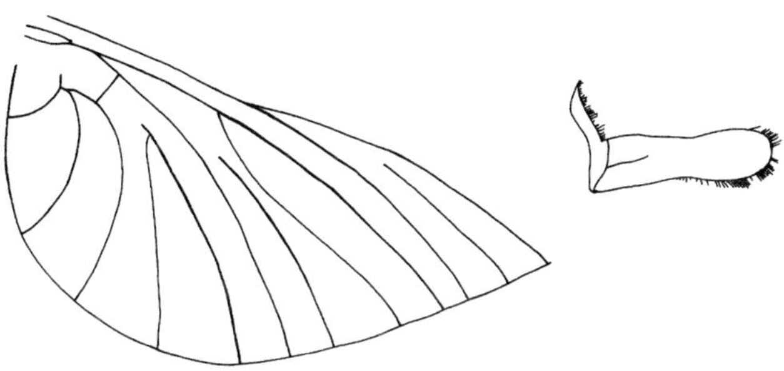

Fig. 2.146 The veins of the inner, posterior part of the fore-wing (left) and the hind wing of a male *Leptohyphes bruchi.*

4. There is a small black mark in the middle of the second through tenth abdominal tergites of the subadult; the ground color of the tergites is light brown suffused with black. Length: 4 mm. Fore-wing length: 4 mm.
....*Leptohyphes maculatus* Allen, 1967 (Peru).
- If there are black markings on the abdominal tergites, they are lateral to a light longitudinal stripe in the center (**Fig. 2.147**). 5
5. The first four abdominal tergites of the male subadult are black with a pale midline stripe. The fifth through ninth tergites are light in color with four longitudinal black markings, one pair beside the pale midline and the other pair sublateral (**Fig. 2.147**). The hind wing has a long costal projection. The veins in both wings are pale, and the cross veins are indistinct. The head is light brown with a wide transverse band between the ocelli and compound eyes. The thorax is light in color with black markings. Length: 4 to 5 mm. Fore-wing length: 4.5 to 5.5 mm. The adults have not been described, and the subadults

were associated with the larvae according to occurrence in the same streams and similarity of the color patterns. ....*Leptohyphes setosus* Allen, 1967 (Peru).
- The first four abdominal segments are not predominantly black. 6

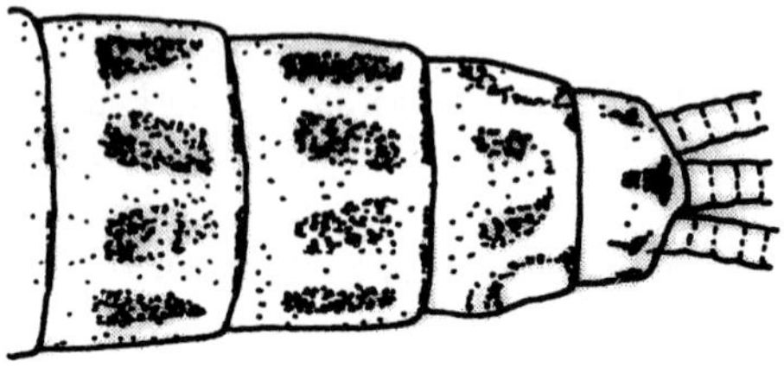

Fig. 2.147 The color pattern on the seventh through tenth abdominal tergites of a subadult male *Leptohyphes setosus.*

6. The hind wing is veinless, or the long veins are evident only as traces in the wings of the subadult; its costal process is fairly long but only slightly recurved (**Fig. 2.145**). 7
- The hind wings have two distinct long veins; its costal process is long and recurved (**Fig. 2.148**). 8

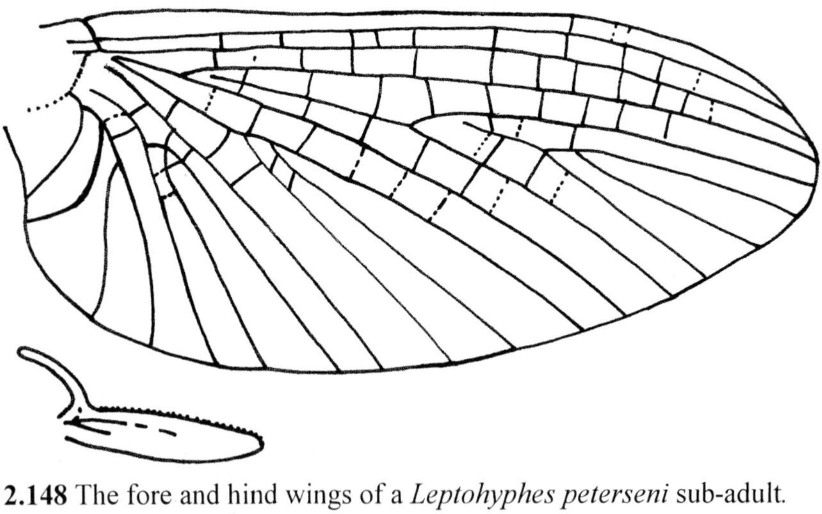

Fig. 2.148 The fore and hind wings of a *Leptohyphes peterseni* sub-adult.

7. There is a trace of long veins evident in the hind wing of the subadult (**Fig. 2.149**). The mesonotum and metanotum are bright reddish brown and lack black markings, although the pronotum has extensive black markings. Length: 4 mm.

Caudal filament length: 10 mm. Color: the head is gray above and white below; the second through sixth segments of the abdomen are yellowish white, and the others are light reddish brown with a yellow tinge. The sternites are the same colors. Caution is required in distinguishing the two species in this couplet because this species is described only from subadults, while the other is described only from adults.
....*Leptohyphes nigripunctum* Traver, 1943 (Venezuela).
- There is no trace of veins in the hind wing of the adult (**Fig. 2.145**). There are narrow blackish lines on some of the more prominent carinae and two pairs of blackish marks beside the dorsal crest of the mesonotum. Length: 3 to 4.5 mm. Fore-wing length: 3.5 to 4.7 mm. Caudal filament length: 4 mm. Color: pale brown with blackish head and whitish abdomen and caudal filaments.
....*Leptohyphes mollipes* Needham & Murphy, 1924 (Minas Gerais).

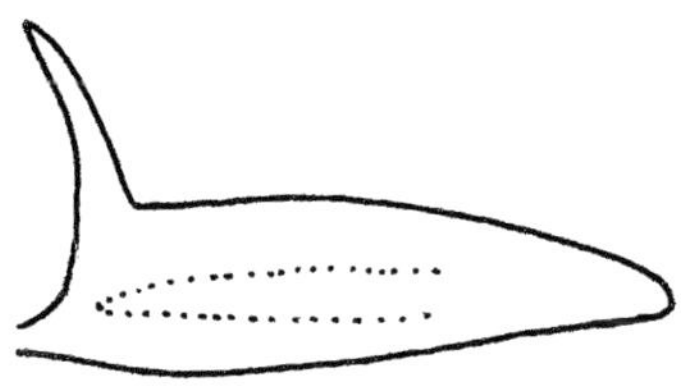

Fig. 2.149 The hind wing of a subadult *Leptohyphes nigripunctum.*

8. The costal process of the hind wing is very strongly recurved and blunt at the apex (**Fig. 2.148**). The penis lobes of the male are divaricated and blunt at the tips. Length of male: 2.5 to 3 mm; of female: 3 to 5 mm. Fore-wing length of male: 3 to 4 mm; of female: 3.5 to 6 mm. Color: yellowish brown with black markings, including black marks on the abdominal tergites on either side of a light median stripe.
....*Leptohyphes peterseni* Ulmer, 1920 (Argentina, Bolivia, Santa Catarina).
- The costal process of the hind wing is moderately recurved and drawn out into an acute tip. The penis lobes of the male are divaricated and bear acute processes at the apex (**Fig. 2.104**). Length of the male and female: 3.5 to 4.5 mm. Fore-wing length of the male: 3.5 to 4.5 mm; of the female: 4.5 to 5.5 mm. Color: pale to light brown on the head with a black mark between the ocelli and black rings around the antennal bases; the pronotum is light brown with a black margin and sublateral black streaks. The mesonotum and pleurae are light brown with black anterior to the base of the fore-wing. The thoracic sternum is black. The abdomen is pale with scattered black pigment and black parallel longitudinal dashes on either side of the light median stripe on the second through sixth tergites.
....*Leptohyphes petersi* Allen, 1967 (Peru).

Key to the species of larval *Leptohyphes* in South America.

Information for the key was provided by Allen (1967, 1973), Allen and Roback (1960), Mayo (1968a), and Wang *et al.* (1998), all of whom described few adults. As a result, the adults of most species have never been described.

1. The hind femur bears a large projection from the distal end (**Fig. 2.150**). 2
- There is no large projection from the hind femur (**Fig. 2.151**). 3

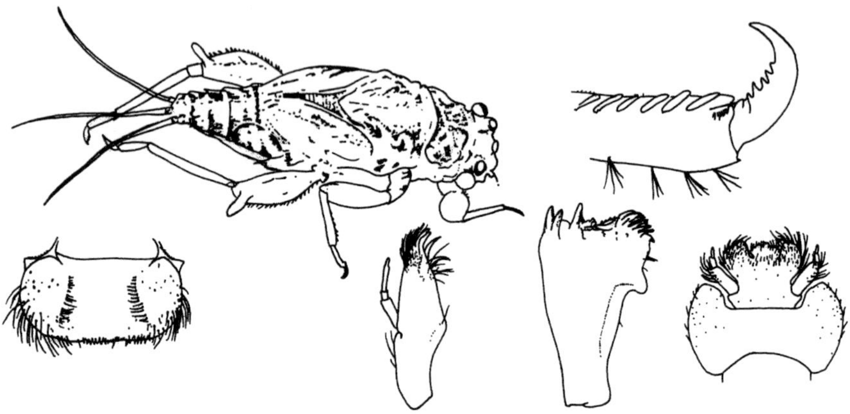

Fig. 2.150 *Leptohyphes ecuador* larva: habitus (upper left), the apex of a tarsus and claw (upper right) and (below, left to right): the labrum, maxilla, mandible, and labium.

Fig. 2.151 *Leptohyphes carinus* larva (left to right): the femur of the fore-leg, a modified bristle from the femur, the femur from the hind leg, and a tarsal claw.

2. The fifth through ninth abdominal tergites bear round median tubercles. The head and thorax lack tubercles. There are four or five small marginal denticles on the tarsal claws (**Fig. 2.152**). Length: 4 to 5 mm. Caudal filament length: 3.5 to 4.5 mm. Color: the head is yellow with a black line on the posterior margin form the eyes to the epicranial suture and a black design on the frons. The

thorax and legs are yellow with variable black markings. The five anterior abdominal tergites are yellow with black transverse bands.
....*Leptohyphes tuberculatus* Allen, 1967 (Peru).
- There are no tubercles on the abdominal tergites. The head and thorax bear small tubercles (**Fig. 2.150**). Length of body: 4 mm; length of cerci: about 2 mm. Color: amber with dark brown markings and a blackish head.
....*Leptohyphes ecuador* Mayo, 1968 (Ecuador).

Fig. 2.152 Profile of the abdomen and the fore-leg with enlargements of a seta and tarsal claw of *Leptohyphes tuberculatus*.

3. The maxillary palp consists of two segments (**Fig. 2.153**). The tarsal claws bear only five marginal denticles. The head is pale yellow, except for the ocelli. The operculate gills are smoky at the base and pale distally. Length: 5 mm; caudal filaments: 3.5 mm. Color: pale yellow with darker markings. The head is uniform yellow without any distinct markings.
....*Leptohyphes albus* Mayo, 1968 (Ecuador).
- The maxillary palp consists of three segments (**Fig. 2.154**). 4

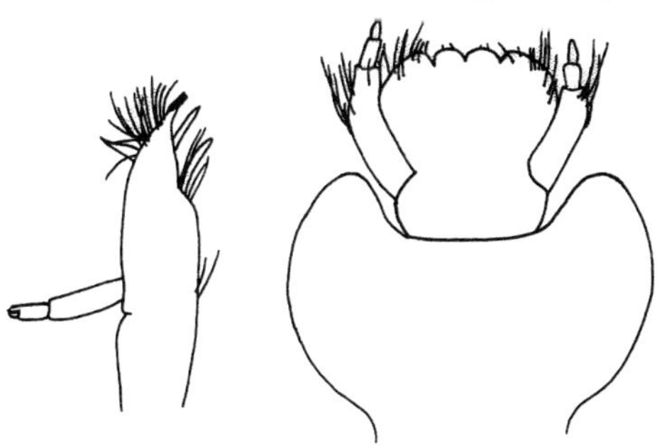

Fig. 2.153 *Leptohyphes albus* larva: the apex of the maxilla with the palp (left) and the labium (right).

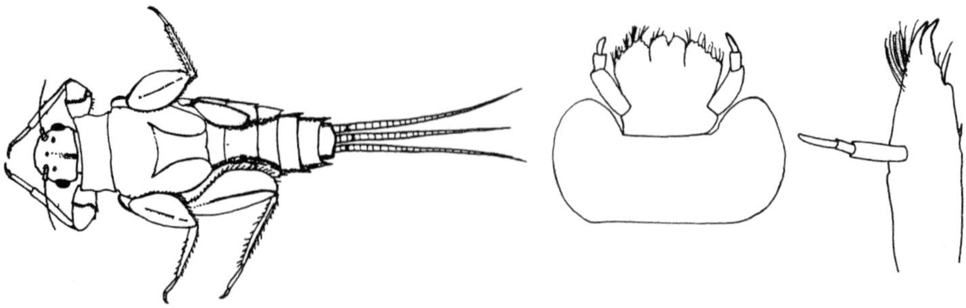

Fig. 2.154 *Leptohyphes tacajalo* larva (left to right): habitus with the right hind leg removed to show the flattened depression in the abdominal segments, the labium, and maxilla.

4. The fourth through ninth abdominal tergites have round median tubercles. The dorsal and ventral margins of the middle and hind femora bear numerous moderately long spines in elevated sockets, but there is no tubercle at the apex of the hind femur. Near the apex of the tarsal claws, there are four blunt marginal denticles (**Fig. 2.151**). Length: 3.5 to 4.5 mm. Color: brown with black markings. The head has a narrow black stripe from the posterior margin of the compound eyes to the posterior margin of the head. The thorax has irregular black markings, while the legs are uniform brown. The first through third abdominal tergites each bear transverse black bands, while tergites four through ten are uniformly brown.
....*Leptohyphes carinus* Allen, 1967 (Peru).

- The abdominal tergites do not bear prominent tubercles in the middle (**Fig. 2.154**). 5

5. The tarsal claws each bear three to five marginal and five or six submarginal denticles near the apex. The first four abdominal tergites are black with a pale midline stripe. The fifth through ninth are light in color with four black longitudinal markings (**Fig. 2.155**). Each operculate gill is dark at the base and pale at the apex. Length: 5 to 6 mm. Caudal filament length: 3 to 4 mm. Coloration: generally yellow with black markings. The caudal filaments are pale.
....*Leptohyphes setosus* Allen, 1967 (Peru).

- The tarsal claws bear only a single group of denticles, which usually number six or fewer, although in two cases, a single subapical denticle may also be present. The operculate gill lacks a lateral flap (**Fig. 2.157**). 6

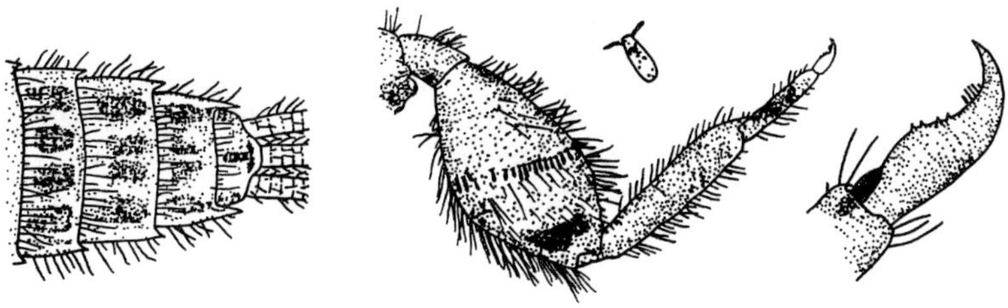

Fig. 2.155 *Leptohyphes setosus* larva (left to right): the color pattern on the seventh through tenth abdominal tergites, the right fore-leg with one of its modified spines enlarged, and an enlargement of the tarsal claw.

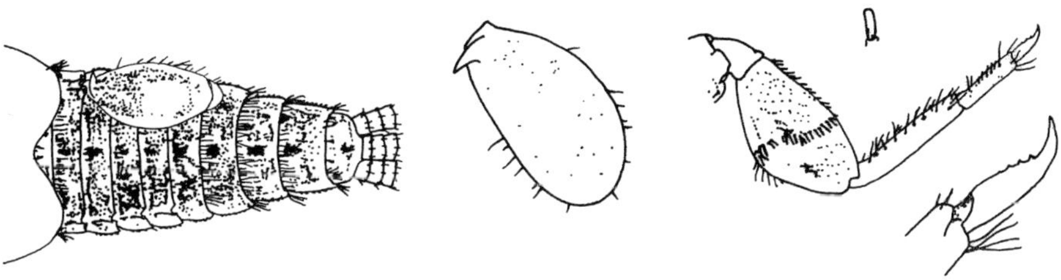

Fig. 2.156 *Leptohyphes maculatus* (left to right): the abdomen in dorsal view with all but one of the operculate gills and the apical segments of the cerci removed, a gill, the right fore-leg with one of the spines enlarged above it, and an enlargement of the tarsal claw.

6. The general color is light or reddish brown without extensive darker markings (**Fig. 2.154**). 7
 - The body has extensive dark markings (**Fig. 2.157**). 8

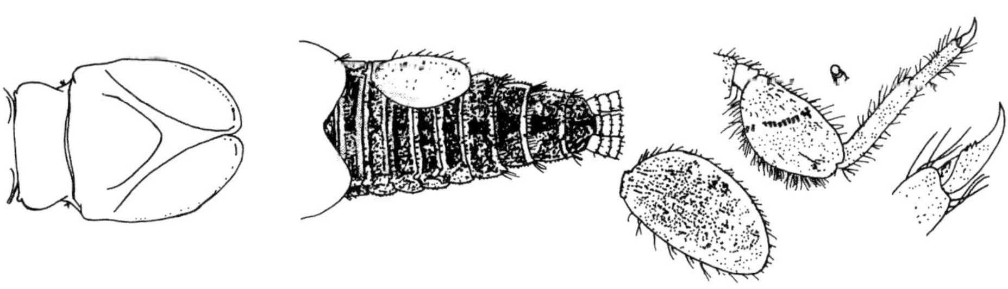

Fig. 2.157 *Leptohyphes myllonotus* larva (left to right): the thorax in dorsal view without appendages, the abdomen slightly enlarged in dorsal view with all but one of the operculate gills and the apical segments of the cerci removed, a gill, the right fore-leg with one of the spines enlarged above it, and an enlargement of the tarsal claw.

7. The margins of the hind tibia are not lined with dense rows of hair-like setae. The apex of the hind femur is not armed with spurs. The abdomen is narrowed and flattened to receive the large curved hind femora, and the flattened areas are hyaline. The lateral margins of the tergites are extended as spines. The caudal filaments have narrow dark brown bands near the base. The head and thorax bear tiny spinules. There are lateral protuberances near the anterior border of the pronotum (**Fig. 2.154**). Length: 4.5 to 5 mm; length of caudal filaments: 3 mm.
....*Leptohyphes tacajalo* Mayo, 1968 (Ecuador).
- The margins of the hind tibia are lined with dense rows of hair-like setae. The apex of the hind femur is armed with short, strong spurs (**Fig. 2.158**). Length: 4.5 to 6.0 mm; length of caudal filaments: 3 to 4 mm. Color: reddish to light brown with darker markings.
....*Leptohyphes liniti* Wang. Sites, and McCafferty, 1998 (Ecuador).

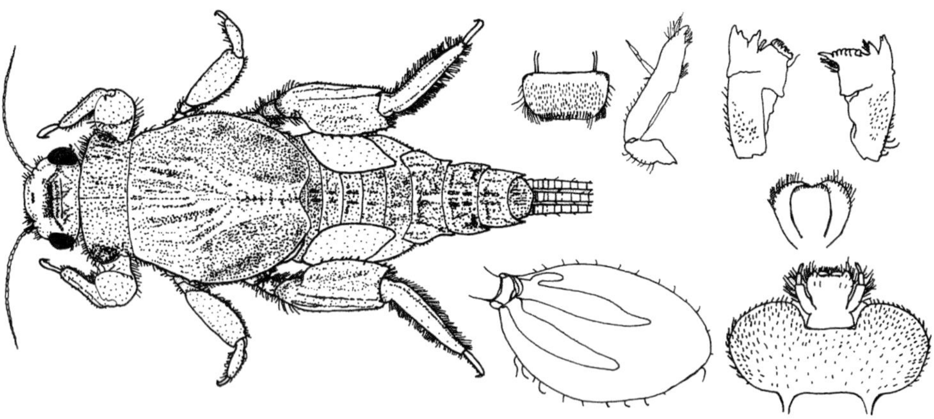

Fig. 2.158 *Leptohyphes liniti* larva: habitus with only the bases of the cerci and median caudal filament (left), labrum, maxilla, mandibles (upper right, left to right), hypopharynx (middle right), labium (lower right), and the second gill (lower center).

8. The legs are uniform dark or chestnut brown, sometimes with scattered black pigment on the femora (**Fig. 2.157**). 9
- The legs have distinct light or dark markings (**Fig. 2.159**). 12

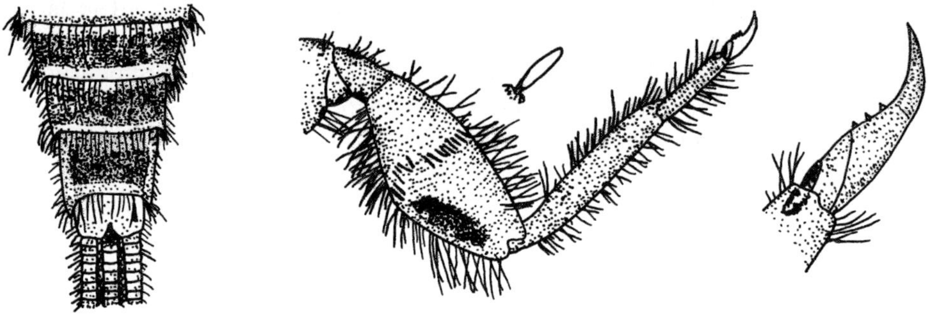

Fig. 2.159 The posterior abdominal tergites and a fore-leg with an enlarged seta and tarsal claw of *Leptohyphes comatus.*

9. There are parallel black dashes on either side of a lighter median line on the second through sixth abdominal tergites. The tarsal claws bear three to five marginal denticles (**Fig. 2.160**). Length: 4 to 5 mm. Caudal filament length: 3.5 to 4.5 mm. Color: brown to chocolate brown with dark brown and black markings. The head is brown with black on the vertex and on the frons between the compound eyes. The legs are uniform brown except for some black pigment scattered on the femora. The abdominal tergites and sternites are dark brown to black.
....*Leptohyphes petersi* Allen, 1967 (Peru).
- There are no distinct parallel black dashes on either side of a lighter median line on the second through sixth abdominal tergites (**Fig. 2.157**). 10

Fig. 2.160 *Leptohyphes petersi* larva (left to right): the fore-leg with a seta from the femur enlarged, an enlargement of the tarsal claw (right), the pattern on the fourth and fifth abdominal tergites, and a gill.

10. The compound eyes of the last instar male larva are divided. The tarsal claws each bear two to four marginal denticles. The marginal spines on the hind femora are inserted in elevated sockets (**Fig. 2.161**). Length: 4 to 5 mm. Caudal filament length: 4 to 5 mm. Color: brown with dark brown markings. There are dark brown transverse bands on the posterior margins of the sixth to eighth abdominal tergites.
....*Leptohyphes populus* Allen, 1973 (Amazonas).
- The compound eyes of the last instar male larva are not divided. The sockets of the marginal spines on the hind femora are not elevated, although those on the fore-femur may have elevated bases (**Fig. 2.157**). 11

Fig. 2.161 *Leptohyphes populus* larva (left to right): the head of a male larva, femur of the fore-leg, a modified bristle from the femur, the femur from the hind leg, a tarsal claw (above), and a gill (below).

11. The abdomen is dark brown with a whitish median line and darker posterior margins of the tergites. The head is brown with a black transversal line between the eyes and irregular dark markings on the vertex. The thorax is brown with a whitened median line and a few darker markings. The tarsal claws bear three or four marginal denticles (**Fig. 2.162**). Length: 5.6 to 6.4 mm; length of the caudal filaments: 6 to 7 mm. Color: dark brown with a black band between the eyes and irregular dark markings on the vertex. The whitish median line runs from the thorax through the abdominal tergites.
....*Leptohyphes pereirae* Silva, 1993 (Rio de Janeiro). The reported differences between this and the next species are minor, and examination of adults should be undertaken to determine whether these species are really distinct.
- The head is brown with a black band across the frons between the compound eyes. The abdomen is dark brown with darker median markings on the seventh through tenth tergites. The opercular gills are dark brown with pale margins and lack apical spines. There are three or four blunt marginal denticles on the tarsal claws (**Fig. 2.157**). Length: 3.5 to 4.5 mm. Length of caudal filaments: 4 to 5 mm.
....*Leptohyphes myllonotus* Allen & Roback, 1969 (Peru).

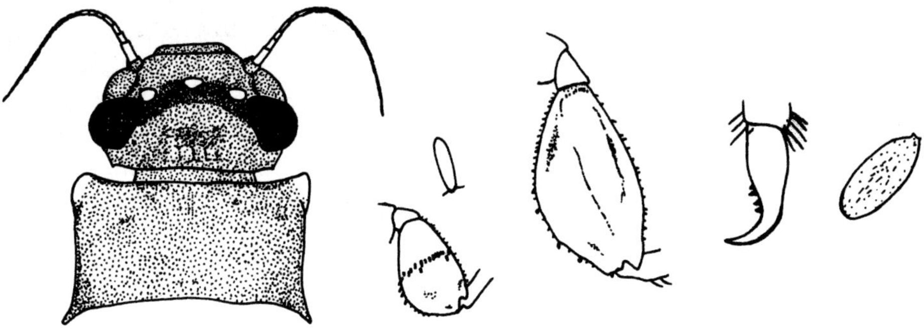

Fig. 2.162 *Leptohyphes pereirae* larva (left to right): the head and pronotum, fore-leg with an enlarged femoral spine, hind leg, tarsal claw, and a gill.

12. The opercular gills lack a spine near the base, or they are uniformly pale and not darker at the base than at the apex (**Fig. 2.156**). 13
- Each opercular gill is brown at the base and pale at the apex and bears a spine near its point of attachment to the tergite (**Fig. 1.163**). Each tarsal claw bears three or four small marginal denticles. The length of the last larval instar is no more than about 5 mm. 19

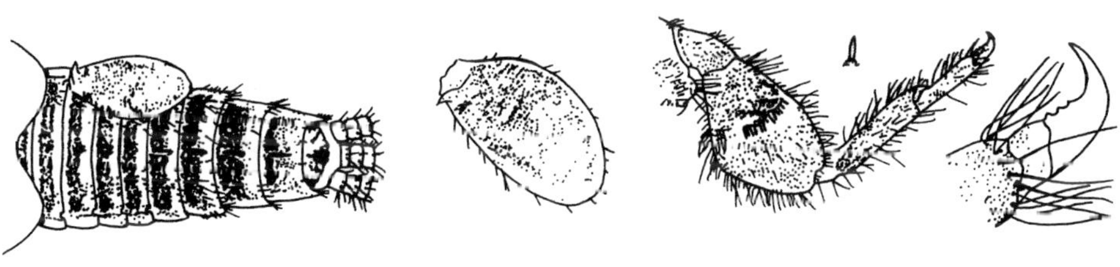

Fig. 2.163 *Leptohyphes hirsutus* (left to right): the abdomen in dorsal view with all but one of the operculate gills and the apical segments of the cerci removed, a gill, the right fore-leg with one of the spines enlarged above it, and an enlargement of the tarsal claw.

13. There is a spine near the base of the opercular gill, but the gill is uniformly pale and lacks a brown base or apex. The second through tenth abdominal tergites are brown with scattered black pigment and distinct small black markings in the center. In addition to three to five marginal denticles, most tarsal claws bear a single subapical denticle (**Fig. 2.156**). Length: 4 to 5 mm. Caudal filament length: 4 to 5 mm. Color: mainly brown with dark brown and black markings.
....*Leptohyphes maculatus* Allen, 1967 (Peru).

- The abdominal tergites are brown without central black markings or have extensive dark markings lateral to the center, which is sometimes pale (**Fig. 2.159**). 14

14. The color is yellow with black markings. The head is yellow with a black band across the frons between the compound eyes. The first nine abdominal tergites are black with yellow anterior and posterior margins; the tenth tergite is yellow with paired sublateral markings. There are three or four teeth on each tarsal claw (**Fig. 2.159**). Length: 6 to 7 mm.
....*Leptohyphes comatus* Allen, 1967 (Peru).

- The general color is brown with dark brown or black markings or black with lighter markings. 15

15. The general color is black with lighter markings. The legs are pale with extensive black markings. There are paired tubercles on the occiput, pronotum, and mesonotum. The fore-femur has a sub-apical enlargement bearing spines (**Fig. 2.164**). The gills are entirely black, and the abdomen is black with a pale mid-dorsal stripe on the third through ninth tergites. Length: 3 mm. Caudal filament length: 3 mm.
....*Leptohyphes cornutus* Allen, 1967 (Santa Catarina).

- The general color is brown with dark brown or black markings (**Fig. 2.165**).
....16

Fig. 2.164 The posterior part of the head showing occipital tubercles and a fore-leg with an enlarged seta and tarsal claw of *Leptohyphes cornutus*.

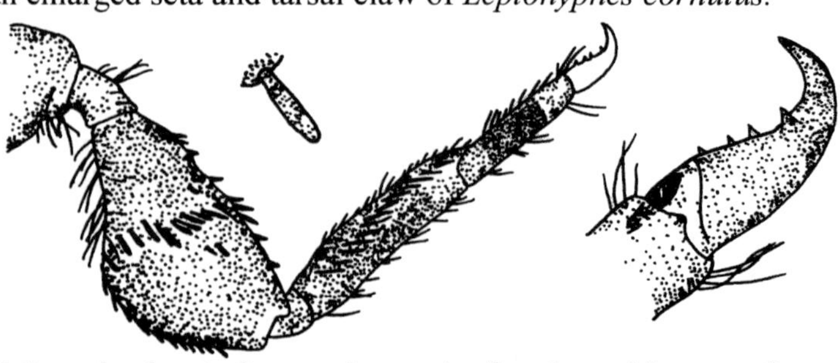

Fig. 2.165 *Leptohyphes jodiannae* larva: the fore-leg with a seta from the femur enlarged (left), and an enlargement of the tarsal claw (right).

16. Each tarsal claw bears only marginal denticles and lack a single one near the apex (**Fig. 2.165**). The abdomen usually has distinct light or dark markings. 17

- Each tarsal claw bears two or three marginal denticles and a single submarginal denticle near the apex (**Fig. 2.166**). The abdomen is light brown without prominent markings and bears long spines along the posterior margins of the second through ninth tergites; the operculate gills are light brown with dark apices. Length: 2.5 to 3.5 mm. Caudal filament length: 2 to 3 mm. Color: light brown with darker brown markings. ....*Leptohyphes invictus* Allen, 1973 (Peru).

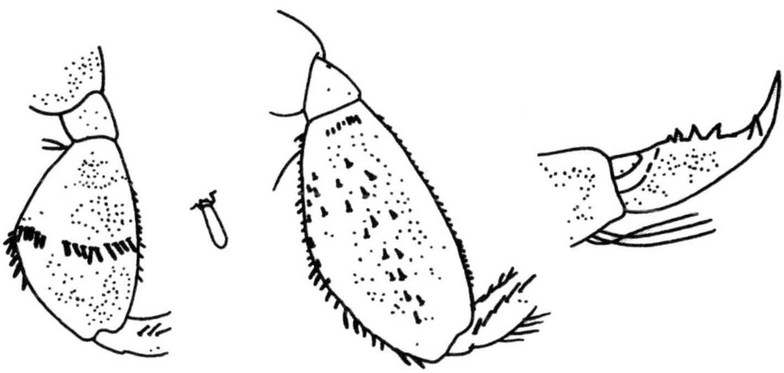

Fig. 2.166 *Leptohyphes invictus* larva (left to right): the femur of the fore-leg, a modified bristle from the femur, the femur from the hind leg, and a tarsal claw.

17. The head is brown with narrow black markings, including lines running from the compound eyes to the posterior margin of the head. The thorax is brown with variable black markings. The legs are brown with black longitudinal lines on the femora. The nine anterior abdominal tergites are brown with transverse black bands, and the tenth tergite is brown with a median and paired lateral markings. The tarsal claws bear three or four marginal denticles (**Fig. 2.167**). Length: 5.5 to 6.5 mm. Length of caudal filaments: 4 to 5 mm. ....*Leptohyphes illiesi* Allen, 1967 (Peru).

- The head and thorax are brown with irregular dark brown or black markings, including a black band between the compound eyes (**Fig. 2.116**). 18

18. The legs are brown; the femora suffused with black and with dark brown bands on the tarsi. The second through ninth tergites are often black with a pale median stripe running down the sixth through ninth. The tenth tergite often has a median and paired sublateral markings. There are three to five marginal denticles on each tarsal claw (**Fig. 2.116**). Length: 5 to 7 mm. Caudal filament length: 5.5 to 6.5 mm. ....*Leptohyphes plaumanni* Allen, 1964 (Santa Catarina).

- The legs are dark brown with pale markings; the tibiae have markings at each end, and the tarsi are pale at the base and apex. The tergites are brown or dark brown, sometimes with parallel black submedian longitudinal stripes. The tarsal claws bear four or five marginal denticles (**Fig. 2.165**). Length: 4 to 5 mm. Caudal filament length: 4 to 5 mm.
....*Leptohyphes jodiannae* Allen, 1967 (Peru).

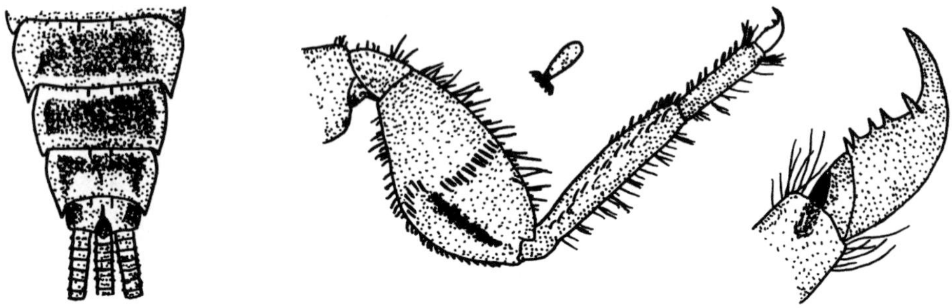

Fig. 2.167 The posterior abdominal tergites and a fore-leg with an enlarged seta and tarsal claw of *Leptohyphes illiesi.*

19. The tarsal claws are sharply hooked almost 90° from the axis of the tarsus (**Fig. 2.163**). The head is brown with dark brown markings near the middle of the occiput. Length: 4 to 5 mm. Caudal filament length: 4 to 5 mm.
....*Leptohyphes hirsutus* Allen & Roback, 1969 (Peru).
- The tarsal claws are slightly curved less than 30° from the axis of the tarsus (**Fig. 2.168**). The head is light brown with a dark band on the frons between the compound eyes. Length: 4 to 5 mm. Caudal filament length: 3.5 to 4.5 mm.
....*Leptohyphes echinatus* Allen & Roback, 1969 (Peru). That the two species in this couplet are distinct seems very doubtful.

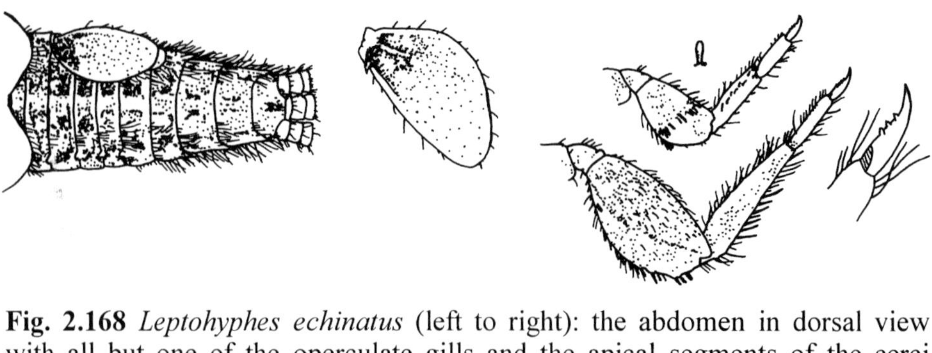

Fig. 2.168 *Leptohyphes echinatus* (left to right): the abdomen in dorsal view with all but one of the operculate gills and the apical segments of the cerci removed, a gill, the right fore-leg with one of the spines enlarged above it and the right hind leg below, and an enlargement of a tarsal claw.

Key to the species of larvae in the genus *Allenhyphes* from South America

Information for the key was provided by Needham and Murphy (1924), Allen (1967, 1973), Mayo (1968a), and Wiersema and McCafferty (2000). The adult of only one species, *Allenhyphes flinti,* has been described (**Fig. 2.105**).

1. The maxillary palp consists of two segments. 2
- The maxillary palp consists of three segments (**Fig. 2.115**). 3
2. The head is uniform brown without any distinct markings. The tarsal claws each bear four to six marginal denticles and a row of six to eight submarginal denticles near the apex (**Fig. 2.169**). The operculate gills are black at the base and brown at the apex. Length: 3.5 to 4.5 mm. Caudal filament length: 2 to 3 mm. Color: brown with few distinct black markings.
....*Allenhyphes rallus* (Allen, 1967) (Peru). Syn: *Leptohyphes rallus* Allen, 1967.
- The head has contrasting light brown and black markings. The tarsal claws bear six to seven marginal and a row of four or five submarginal denticles near the apex (**Fig. 2.170**). The thorax is light brown dorsally with variable black markings. The first through ninth abdominal tergites have dark brown markings separated by a brown median longitudinal stripe. The first six tergites have black lateral markings dorsal to the gill insertions. Length: 4 to 5 mm; length of the unbroken cerci: 3 to 4 mm.
....*Allenhyphes edmundsi* (Allen, 1973) (Paraná, Santa Catarina, Rio Grande do Sul). Syn: *Leptohyphes edmundsi* Allen, 1973.

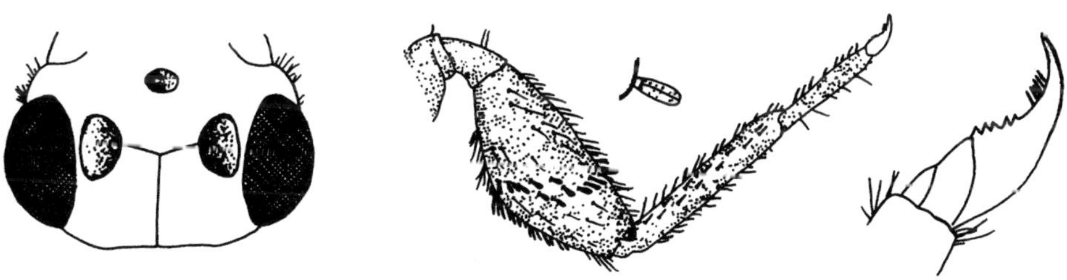

Fig. 2.169 *Allenhyphes rallus* larva (left to right): the posterior portion of the head, the fore-leg with an enlarged bristle from the femur, and an enlarged tarsal claw.

3. The tarsal claws bear only 2 or 3 small marginal denticles (**Fig. 2.171**). The caudal filaments are yellow without distinct darker bands, and the legs are uniform yellow with long spines on the femora. The operculate gill lacks a lateral flap. The gills are black with pale apices. Length: 4 to 5 mm; length of

the unbroken cerci: 3 to 4 mm. Color: mainly yellow with black markings; the thorax is light brown with scattered black marks dorsally.
....*Allenhyphes asperulus* (Allen, 1967) (Peru). Syn: *Leptohyphes asperulus* Allen, 1967.
- The tarsal claws each bear three to eight marginal denticles and four to seven submarginal denticles near the apex (**Fig. 2.172**). 4

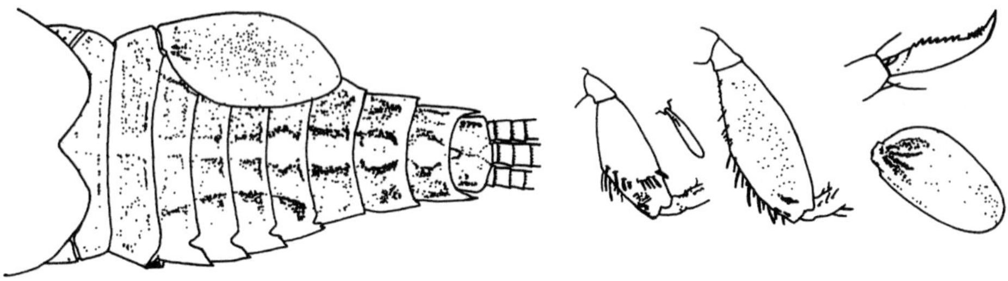

Fig. 2.170 *Allenhyphes edmundsi* larva (left to right): the color pattern on the dorsal surface of the abdomen showing one of the gills, the fore-femur and one of its spines enlarged, the hind femur, a tarsal claw (above), and a gill.

Fig. 2.171 The fore-leg with enlargements of a seta and tarsal claw of *Allenhyphes asperulus*

4. The legs are brown with black and dark brown markings. The cerci are uniformly pale, and their length is about 50% of the total length of the body without cerci or antennae. The tarsal claws each bear six to eight marginal denticles and four to seven submarginal denticles near the apex. Each operculate gill has a lateral flap and is dark at the base and pale at the apex. The abdominal tergites have black markings that form two longitudinal bands (**Fig. 2.172**). The abdominal sternites are pale with black markings along the midline and laterally. Length: 4 mm; length of caudal filament: 2 mm.
....*Allenhyphes spinosus* (Allen & Roback, 1969) (Peru). Syn: *Leptohyphes spinosus* Allen & Roback, 1969.
- The legs are uniformly yellow. The cerci are brown, and their length is at least 80% of the total length of the body without cerci or antennae. The tarsal

claws each bear three to five marginal denticles and a row of four or five submarginal denticles near the apex (**Fig. 2.115**). The abdominal tergites are pale or brown with a suffusion of black. The abdominal sternites are brown. Length: 3 to 4 mm; length of caudal filament: 2.5 to 3.5 mm.
....*Allenhyphes flinti* (Allen, 1967) (West Indies, Venezuela). Syn: *Leptohyphes flinti* Allen, 1967.

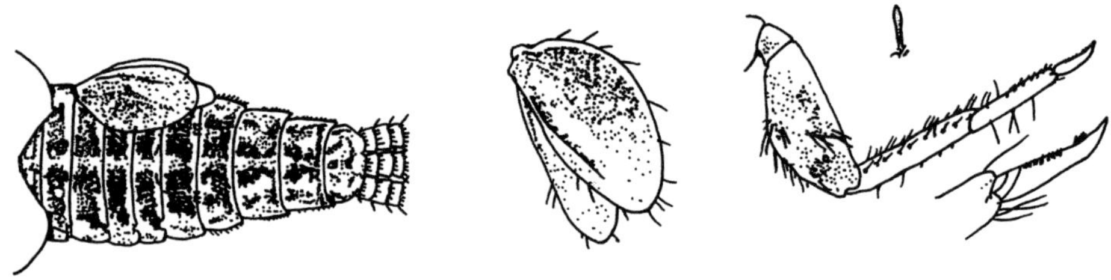

Fig. 2.172 *Allenhyphes spinosus* (left to right): the abdomen in dorsal view with all but one of the operculate gills and the apical segments of the cerci removed, a gill, the right fore-leg with one of the spines enlarged above it, and an enlargement of the tarsal claw.

Oligoneuriidae

Key to the South American genera of adults

Information for the key was taken from Traver (1944), Demoulin (1952c, 1955a), Pescador and Peters (1980), and Pescador and Edmunds (1994).

1. Two posterior filaments are present (**Fig. 2.173**). 2
- Three posterior filaments are present. 3

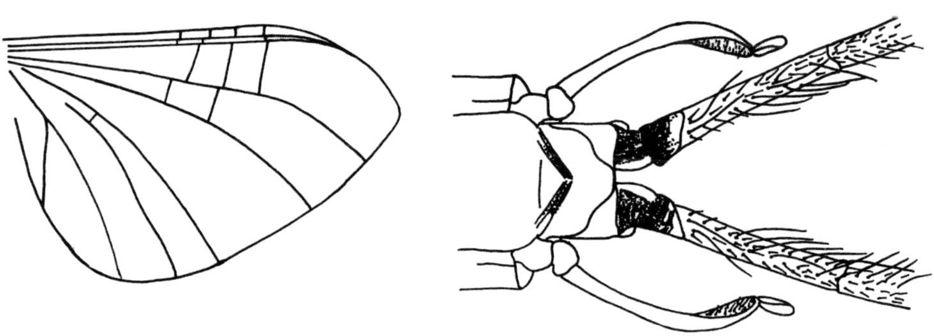

Fig. 2.173 *Spaniophlebia assimilis:* the fore-wing (left) and apex of the abdomen of a male showing the bases of the cerci (right).

2. There are three prominent, long longitudinal veins between veins R and Cu that arise from the base of the fore-wing. The branches of vein M are very long. There are at least several cross veins in the costal region, three or four in the radial region, and two in the next region (**Fig. 2.173**).
....*Spaniophlebia* Eaton, 1881..p. 149
- There are only two prominent, long longitudinal veins between veins R and Cu that arise from the base of the fore-wing. The branches of vein M are not very long. There are several cross veins in each region (Fig. **2.174**).
....*Lachlania* Hagen, 1868..p. 150

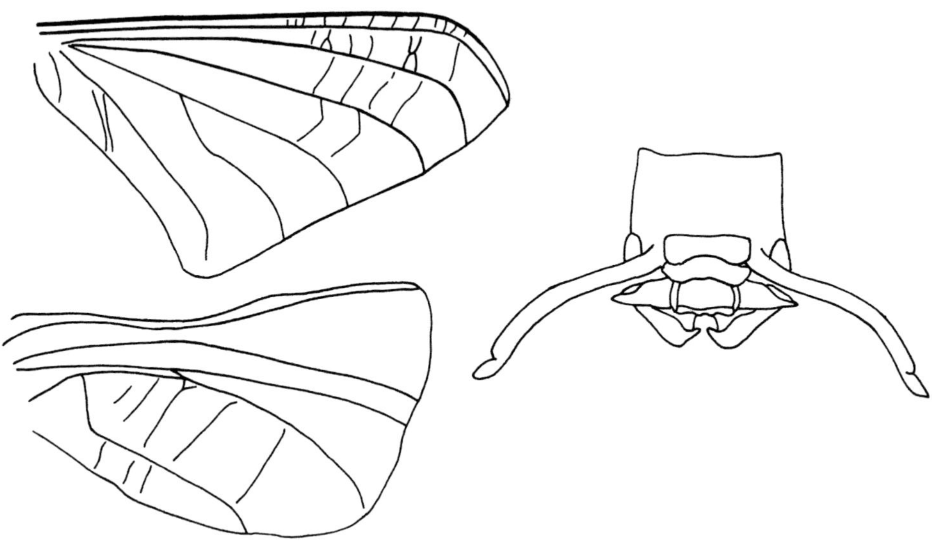

Fig. 2.174 *Lachlania dominguezi*: fore and hind wings of a male (left) and the male genitalia in ventral view (right).

3. Vein Cu of the fore-wing is not bifurcate. Vein R_3b in the fore-wing arises from the wing base (**Fig. 2.175**).
....*Oligoneurioides* Demoulin, 1955
The only species in this genus is *Oligoneurioides amazonicus* Demoulin, 1955, known from Amazonas.
- Vein Cu of the fore-wing is bifurcate (**Fig. 2.176**). 4
4. Cross veins are absent from both fore and hind wings (**Fig. 2.176**).
....*Homoeoneuria* Eaton, 1881
The only known South American representative of this genus is *Homoeoneuria (Notochora) fittkaui* Pescador and Peters, 1980, a brownish and yellow species from Amazonas, 7 mm long.
- Cross veins are present on both fore and hind wings. Vein R_{3b} in the fore-wing arises distal to the wing base. Vein IMP is present (**Fig. 2.12**). 5

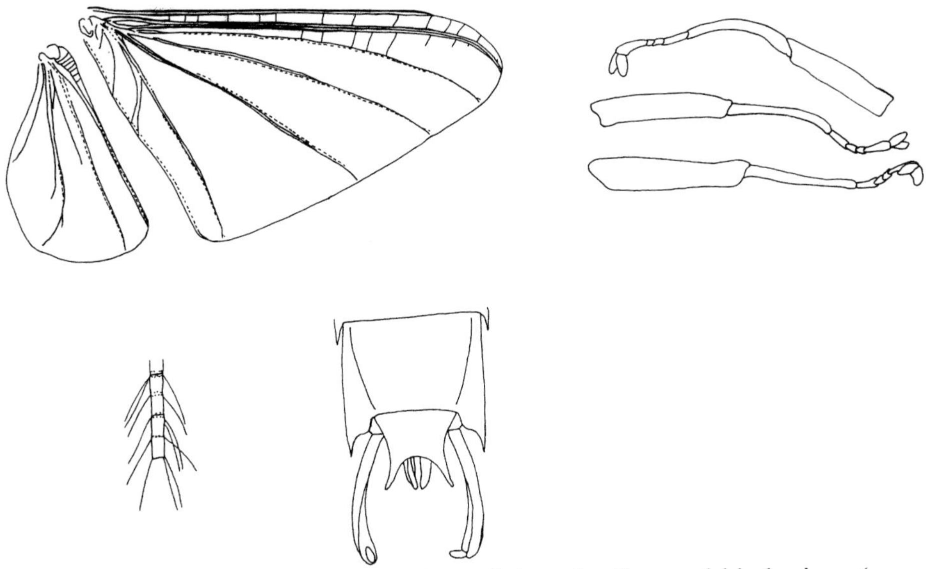

Fig. 2.175 *Oligoneurioides amazonicus* adult male: fore and hind wings (upper left); middle portion of a lateral cercus (lower left); fore, middle, and hind legs (right, top to bottom), and genitalia in ventral view (lower center).

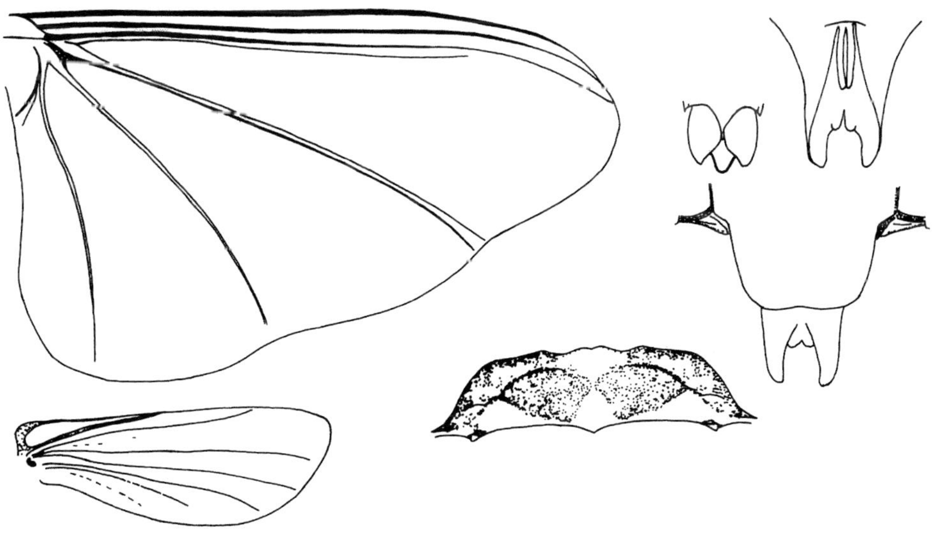

Fig. 2.176 The fore and hind wing of *Homoeoneuria cahabensis* (left), a species not known from South American, and *Homoeoneuria fittkaui:* pronotum (lower right center), male genitalia (lower right), penis (upper right), and the apex of the abdomen of a female (upper right center).

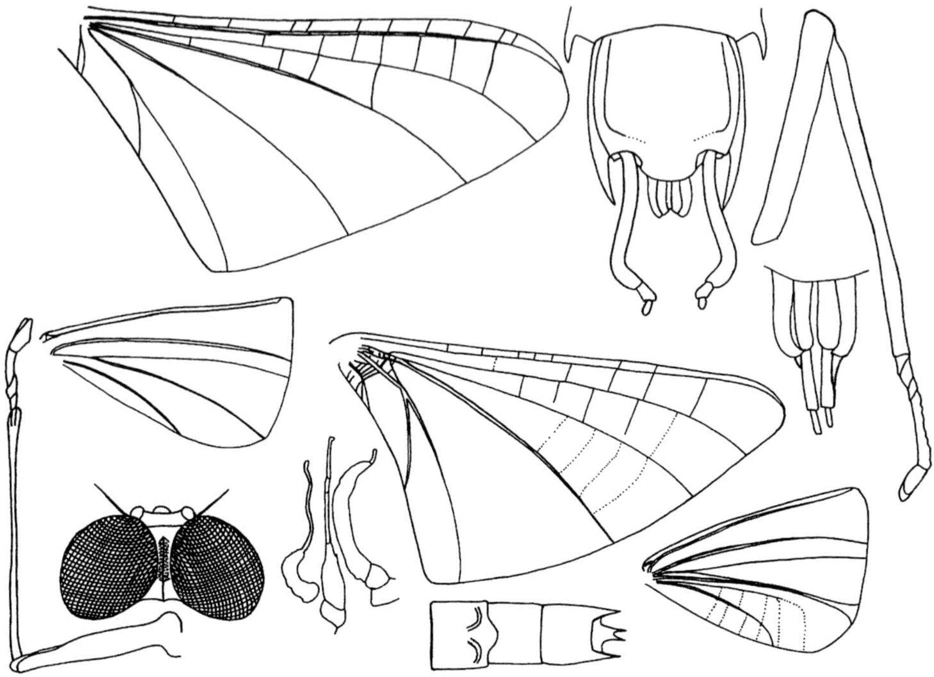

Fig. 2.177 *Oligoneuria anomala:* fore and hind wing of a male (upper left) and a female (lower right); head above the fore-leg of a male (lower left); hind leg of a male (upper right); male genitalia in dorsal view (upper right center) and a ventral view of the penis lobes and paraprocts (middle right); apex of the abdomen of a female in ventral view (lower center), and the vestigial fore, middle, and hind legs of a female (lower left center, left to right).

5. There is a median ridge on the membranous epicranium. In the fore-wing, vein IMP_2 arises from a fork in MP_1 near the fork giving rise to MP_2. The fore-legs are only about half as long as the middle and hind legs, and the tibiae of the middle and hind legs are about three times as long as the femora and are membranous. Genital forceps are absent, and the penis is bulbous and short(**Fig. 2.12**). The adult male *Fittkauneuria carina* is 13 mm long with a 12 mm fore-wing. Its color is mainly yellow with a brown wash on the head and dark brown on the dorsal part of the thorax with yellow furrows and a cross-band forming a yellow H-shaped mark on the mesonotum. The tergites are brown with yellow markings. The adults of other South American species have not been described. ....*Fittkauneuria* Pescador and Edmunds, 1994..p. 155

- There is no prominent median ridge on the epicranium. Vein R_{3b} in the fore-wing arises distal to the wing base. Vein IMP is present but weak. There are semi-circular sclerotizations on each side of the midline in the posterior half of the seventh sternite of the female of the only South American species. The apical process in the middle of the tenth segment of the female is deeply emarginated in this species (**Fig. 2.177**). A more detailed description of the male is urgently needed to establish whether this genus really occurs in South America.
....*Oligoneuria* Pictet, 1845

This genus is represented in South America only by the species *Oligoneuria anomala* Pictet, 1845, with a body length of 10.5 mm and a fore-wing length of 9 mm, known from Surinam, Paraguay, and Brazil. It is likely that the species will eventually be placed in a different genus.

Key to the South American genera of larvae

Information for the key was taken from Demoulin (1955a). The larva of *Oligoneuria anomala* has never been described well enough to place *Oligoneuria* in the key. *Oligoneuria* larvae are assumed to have three caudal filaments, but other features of *O. anomala* are unknown. In other species of the genus, gills are present on the first seven abdominal segments and at the bases of the maxillae. The problem with using larval features of species from other regions, however, is the fact that the South American species may actually belong to another genus. Identification of the last instar larva may be possible by dissecting the wings from the wing buds and examining the veins using the key to the adults.

1. There are only two caudal filaments present. The gills consist of a large ventral plate with a dorsal tuft of filaments (**Fig. 2.178**).
....*Lachlania* Hagen, 1868..p. 150
- Three caudal filaments are present (**Fig. 2.179**). 2
2. The anterior part of the epicranium tapers and is covered thickly with marginal hair-like and peg-like setae. The antennae are attached near the anterior margins of the eyes and extend double the length of the head. The middle and hind tarsi bear two rows of spines. The tarsal claws bear denticles. The second through fifth gill lamellae are broad and bear fringes of hair-like and blade-like setae. The eighth and ninth abdominal segments bear prominent posterolateral spines (**Fig. 2.179**). The median terminal filament is only half as long as the cerci.
....*Fittkauneuria* Pescador and Edmunds, 1994..p. 155
- The anterior part of the epicranium is generally broadly rounded and not tapering with a thick covering of setae. Gills on the second through seventh abdominal segments are lanceolate without a tuft of filaments, or they lack fringes of hair-like and blade-like setae (**Fig. 2.26**). 3

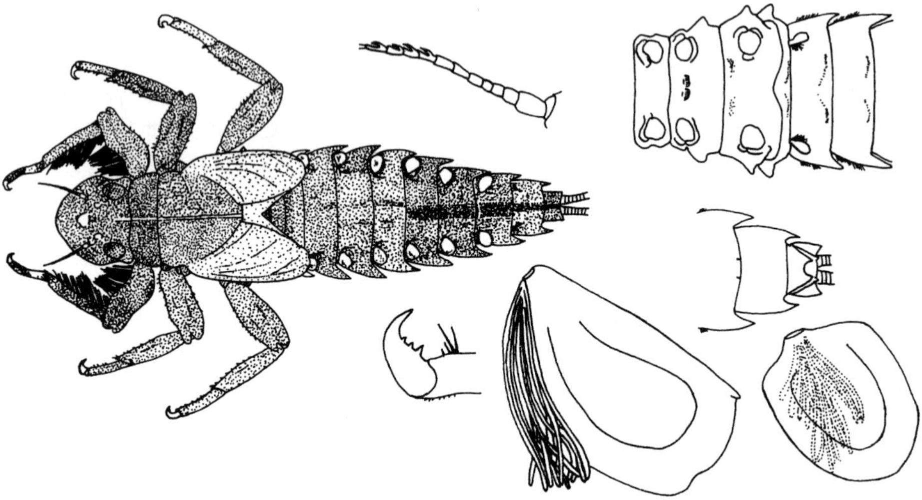

Fig. 2.178 *Lachlania boanovae* larva: habitus (left), right antenna (upper center), thorax and first two abdominal segments in ventral view with the legs removed (upper right), tarsal claw enlarged (lower center), apex of the abdomen with the basal segments of the cerci in ventral view (middle right), and two gills showing the locations of the filaments in dorsal and ventral view (lower right).

3. The laminar portions of the gills on the second through seventh abdominal segments are lanceolate without a dorsal tuft of filaments. The ventral gill on the first abdominal segment is much larger than the others. The hind coxa is larger than the femur.

....*Homoeoneuria* Eaton, 1881

The only known South American representative of this genus is *Homoeoneuria (Notochora) fittkaui* Pescador and Peters, 1980, a brownish and yellow species from Amazonas with a body length of 7 mm.

- The gills on the anterior segments of the abdomen bear a tuft of filaments (**Fig. 2.26**).

....*Spaniophlebia* Eaton, 1881..p. 149

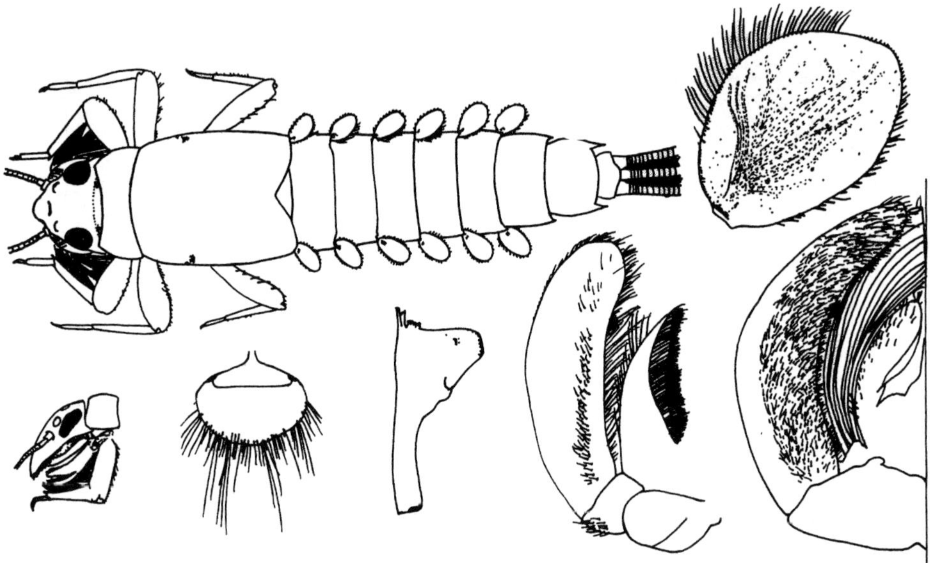

Fig. 2.179 *Fittkauneuria carina* larva: habitus with apical sections of the cerci, terminal filament, and antennae removed (upper left); a gill lamella from the fourth segment of the abdomen with the tuft of filaments on the opposite side showing through in outline (upper right); and (lower row, left to right): the head and pronotum in lateral view, labrum, left mandible, right maxilla, and one side of the labium in dorsal view.

Key to the species of adult *Spaniophlebia* in South America

Information for the key was taken from Banks (1913) and Cockerell (1926). Descriptions of the larvae are not available.

1. The wings have dark markings (**Fig. 2.180**).
....*Spaniophlebia trailiae* Eaton, 1881 (Argentina, São Paulo). Syn: *Spaniophlebia traili,* misspelled by Banks (1913).
- The wings lack dark markings (**Fig. 2.173**). 2
2. The thorax is a pale rust color, and the abdomen is medium brown. Length of body: c. 13 mm; of fore-wing: c. 19 mm.
....*Spaniophlebia escomeli* Cockerell, 1926 (Peru).
- The head and thorax are dark brown, and the abdomen and legs are lighter brown (**Fig. 2.173**). The "expanse" reported by Banks (1913) was 16 mm, but he did not report how he measured it.
....*Spaniophlebia assimilis* Banks, 1913 (Rondônia).

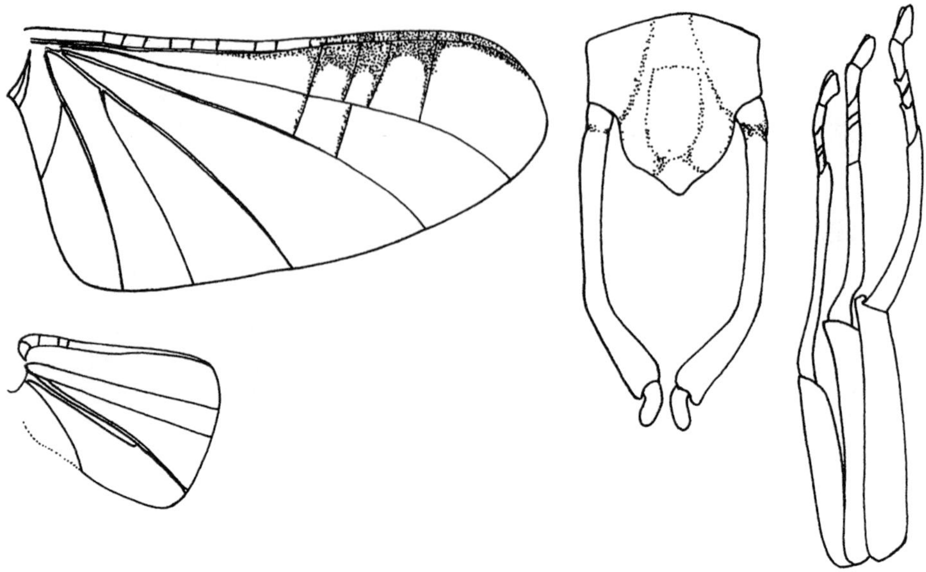

Fig. 2.180 *Spaniophlebia trailiae:* the fore-wing and hind wing (left), apex of the abdomen of a male (center), and the fore, middle, and hind legs (right, left to right).

Key to the species of adult *Lachlania* in South America

Information for the key was taken from Navás (1912, 1926b), Ulmer (1943), Pereira (1987, 1989), and da Silva and Pereira (1993). Better descriptions of the species are needed. Pereira (1987) distinguished species mainly by the number of cross veins in the wings without reporting possible sexual differences.

1.There are three strongly arced veins running close together from A to the posterior margin of the fore-wing (**Fig. 2.181**).
....*Lachlania cacautana* (Needham, 1932) (Colombia). Syn: *Alloydia cacautana* Needham, 1932.
- There are no more than two veins joining A to the posterior margin of the fore-wing, and if as many as two are present, they are straight or sigmoid (**Fig. 2.182; 2.183**). 2
2. Some of the cross veins between R_2 and R_3 in the fore-wing of the male are branched or joined to each other; they number about seven. There are four cross veins between IR_3 and R_{4+5}, and two between MA_1 and MA_2 (**Fig. 2.174**). Length: 14 mm. Length of fore-wing: 15 mm; hind wing: 7 mm. Color: light reddish with black joining the compound eyes and the ocelli. The mesonotum has three dark longitudinal stripes, which join at their posterior ends. A poorly

described species distinguished only through illustrations of the wing veins of the male.
…………………………...............................*Lachlania dominguezi* Pereira, 1989 (Argentina).

- The cross veins between R_2 and R_3 in the fore-wing of the male are not branched or joined to each other (**Fig. 2.182; 2.183**). …………………………......3

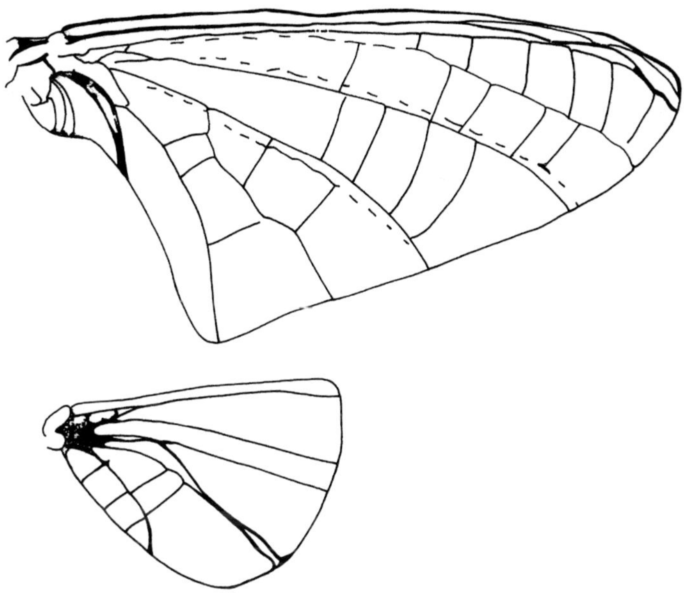

Fig. 2.181: Fore and hind wings of *Lachlania cacautana.*

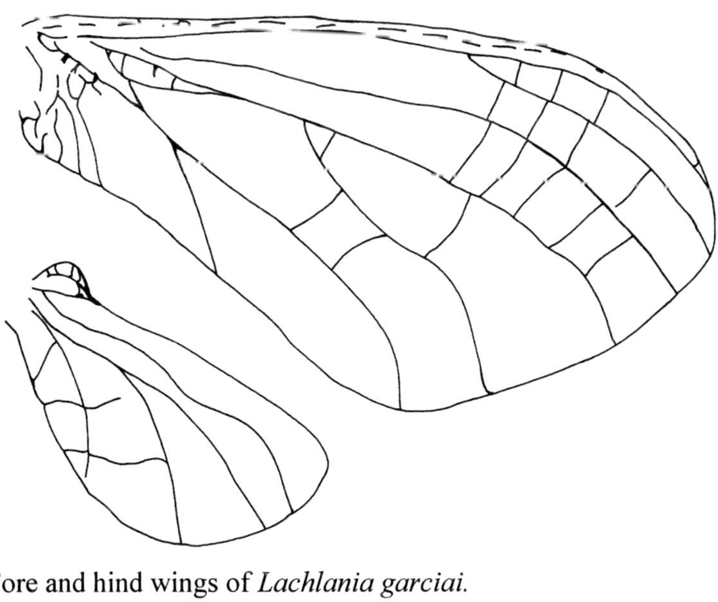

Fig. 2.182 Fore and hind wings of *Lachlania garciai.*

3. The wings are grayish with yellow veins. There are only about two cross veins between R_{1+2} and R_3 in the fore-wing, and vein IR_3 is absent (**Fig. 2.183**). Length of female: 12 mm. Fore-wing length of female: 13 mm. Hind wing length of female: 7 mm. Length of cerci: 11 mm. Color: generally brown to dark brown.
....*Lachlania boanovae* da Silva and Pereira, 1993 (Rio de Janeiro).
- The wings have grayish, blackish, or brownish veins (**Fig. 2.184**). 4

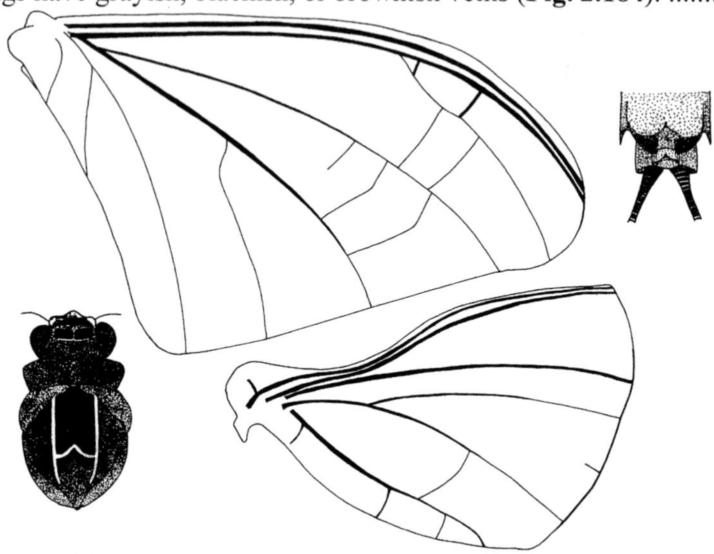

Fig. 2.183 *Lachlania boanovae:* pattern on the head and thorax (lower left), fore and hind wing (upper left to lower right), and the apex of the abdomen (upper right).

4. The caudal filaments are dark brown or blackish. 5
- The caudal filaments are white or yellowish brown. The wings are hyaline or smoky gray with brown veins. There are more than two cross veins between R_2 and R_3, and vein IR_3 is present (**Fig. 2.185**). 6
5. There are three cross veins between R_2 and R_3, four between IR_3 and R_{4+5}, and five between MA_1 and MA_2. The caudal filaments are dark brown. The dorsal portion of the sternum is blackish. The abdomen and legs are blackish brown. The wings are smoky gray, darker at the base. The wing veins are blackish brown (**Fig. 2.182**). Length: 13 mm, fore-wing length: 19 mm.
....*Lachlania garciai* (Navás, 1912) (Colombia, Ecuador, Bolivia). Syn: *Neophlebia garciai* (Navás, 1912); *Noya garciai* (Navás, 1912).
- There are four cross veins between R_2 and R_3, one between IR_3 and R_{4+5}, and one between MA_1 and MA_2. The caudal filaments are blackish. The dorsal portion of the sternum is blackish. The legs are reddish with whitish markings.

The first six segments of the abdomen are light reddish; the seventh to tenth become progressively darker. The wings are grayish hyaline or grayish (**Fig. 2.184**). Only females have been described. Length: 13 mm, fore-wing length: 10 mm.
....*Lachlania santosi* Pereira, 1987 (Rio de Janeiro).

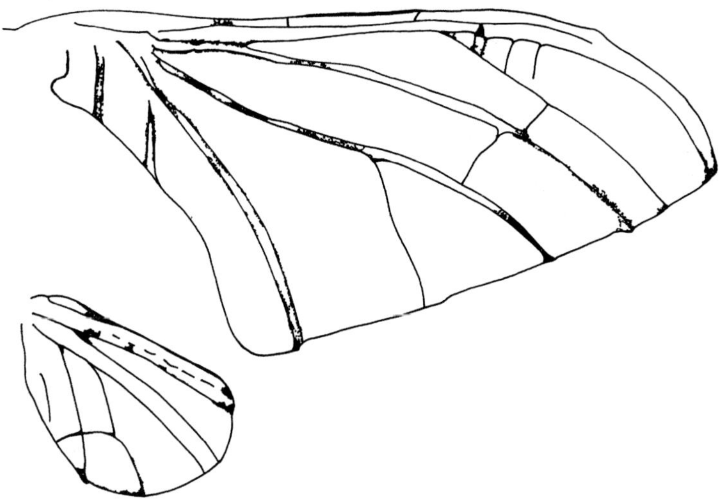

Fig. 2.184: Fore and hind wing of *Lachlania santosi.*

6. The caudal setae are yellowish brown. The sternum and dorsal portion of the abdomen are rust or cinnamon colored. The wings are hyaline with greenish or coppery iridescence. The wing veins are yellowish brown. There are at least eight cross veins between R_2 and R_3 and about five and a half between IR_3 and R_{4+5} (**Fig. 2.185**). Length of female: 9 mm, length of the fore-wing of the female: 18 mm. The male has not been described, but its description will be necessary before the validity of this species can be assured.
....*Lachlania radai* (Navás, 1926) (Peru). Syn: *Noya radai* Navás, 1926.
- The caudal setae are whitish. The sternum is light brown, and the mesonotum and dorsal part of the abdomen are darker, sometimes blackish. The legs are dull brownish yellow with the tibiae near the tips and the tarsi darkened. The wings are clouded smoky gray. The wing veins are brown. There are about four cross veins between R_2 and R_3 and about three between IR_3 and R_{4+5} (**Fig. 2.186**). Length of male: 12 to 17 mm. Only the male has been described; it can therefore not be logically ruled out that the two nominal species in this couplet are, in fact, male and female of the same species.
....*Lachlania pallipes* (Eaton, 1883) (Ecuador, Argentina). Syn: *Spaniophlebia pallipes* Eaton, 1883; *Noya pallipes* (Eaton, 1883).

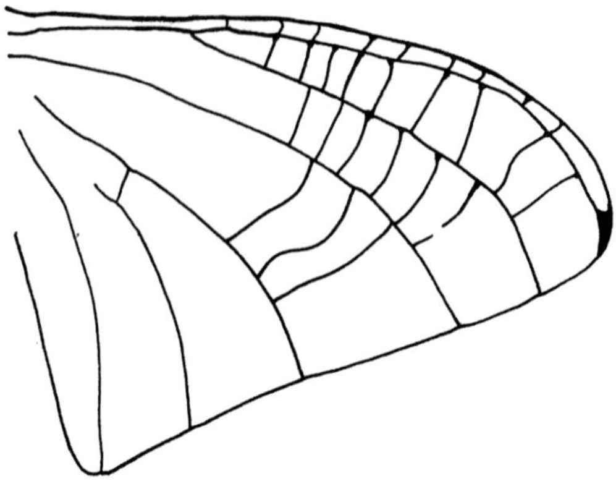

Fig. 2.185: The fore-wing of a female *Lachlania radai.*

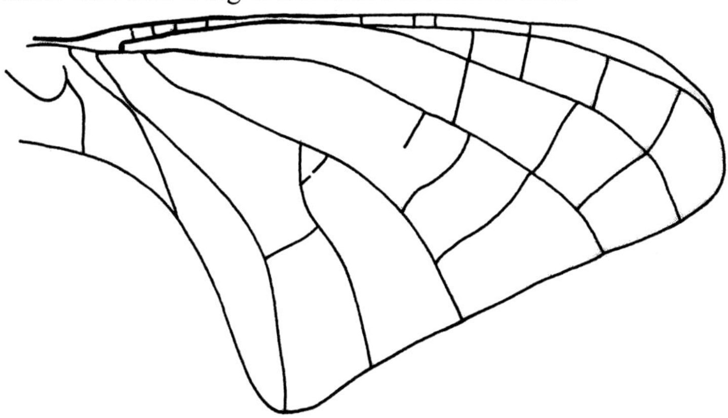

Fig. 2.186: Fore-wing of a male *Lachlania pallipes*.

Key to the species of larval *Lachlania* in South America

Information for the key was taken from Eaton (1883) and da Silva and Pereira (1993). Descriptions of the larvae of only two South American species are available. From the original descriptions of the larvae of these two species, it is not possible to distinguish them with certainty.

1\. The denticles are confined to the basal half of the tarsal claws (**Fig, 2.187**).
....*Lachlania pallipes* (Eaton, 1883) (Ecuador, Argentina). Syn: *Spaniophlebia pallipes* Eaton, 1883; *Noya pallipes* (Eaton, 1883).
\- At least the apical denticle is located in the apical half of the tarsal claw (**Fig. 2.178**).
....*Lachlania boanovae* da Silva and Pereira, 1993 (Rio de Janeiro).

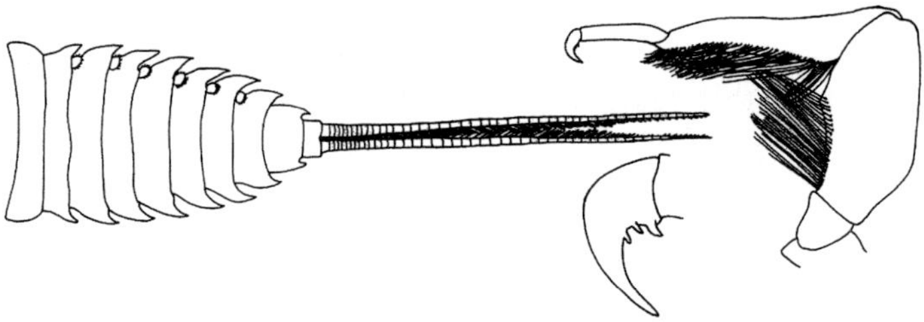

Fig. 2.187: *Lachlania pallipes:* the abdomen and cerci in dorsal view (left) and a fore-leg with an enlargement of the tarsal claw (right).

Key to the species of *Fittkauneuria* larvae in South America

Information for the key was taken from Pescador and Edmunds (1994). The male adult *Fittkauneuria carina* Pescador and Edmunds, 1994, has been described (**Fig. 2.12**), while *Fittkauneuria adusta* Pescador and Edmunds, 1994, is known only as a larva.

1. Only vestigial gill tufts are present on the ovate gill lamellae (**Fig. 2.188**). On the ninth abdominal segment, the posterolateral spines are about 2/5 as long as the maximal segment width. The head is extensively marked with a brown coloration, with covers the vertex and extends between the bases of the antennae. Length of fully grown larva: 14 to 18 mm plus 10 to 14 mm caudal filaments. Coloration: mainly brown with dark brown and yellowish markings and black eyes.
................................*Fittkauneuria adusta* Pescador and Edmunds, 1994 (Venezuela, Amazonas).
 - There is a fully developed tuft of filaments on the lamellae attached to the second through sixth abdominal segments. The lamellae on the second through seventh segments are oval. On the ninth abdominal segment, the posterolateral spines are less than 1/3 as long as the maximal segment width (**Fig. 2.179**). The head is marked with brown only between the ocelli and the compound eyes. Length of fully grown larva: 13 to 18 mm plus 10 to 15 mm caudal filaments. Coloration: yellow and brown with black eyes.
................................*Fittkauneuria carina* Pescador and Edmunds, 1994 (Venezuela).

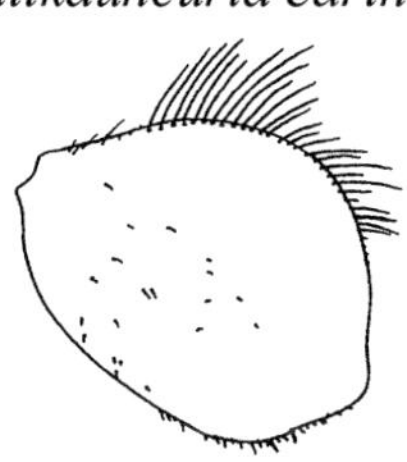

Fig. 2.188 A gill lamella from the fourth abdominal segment of *Fittkauneuria adusta.*

Siphlonuridae

The only genus in this family known from South America is *Metamonius* Eaton, 1885.

Key to the species of adult *Metamonius* in South America

Information for the key was obtained from Navás (1936a) and Demoulin (1955e). The adult of *Metamonius fuegiensis* has not been described.

1. The abdomen is yellow with dark markings at the apices of the tergites that are expanded to form dorsal and lateral stripes. Many of the costal cross veins in the stigmatic area of the fore-wing are bifurcated (**Fig. 2.189**). Length: 11.5 to 12.4 mm. Fore-wing length: 11.5 to 13 mm. Hind wing length: 5.2 to 5.6 mm. Color: head and thorax mainly brownish with darker markings.
....*Metamonius hollermayeri* Navás, 1936 (Chile).

- The abdomen is deep wine-red with lighter markings on the anterolateral parts of the second through seventh tergites and at the lateral margin of the sternites. In the fore-wing, there are few, if any, bifurcated cross veins in the costal space of the stigmatic area (**Fig. 2.14**). Length: 14 mm. Fore-wing length: 18 mm. Hind wing length: 7 mm. Color: head and thorax light brown with ivory markings.
....*Metamonius anceps* (Eaton, 1885) (Chile, Argentina). Syn: *Siphlurus anceps* Eaton, 1885.

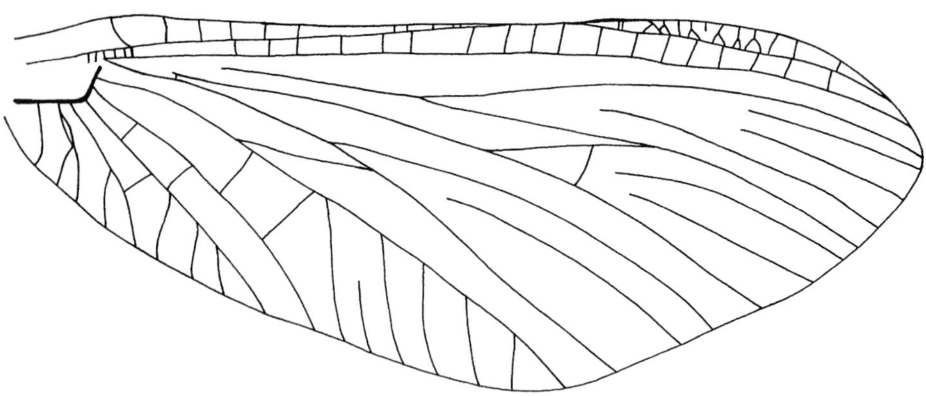

Fig. 2.189 The fore-wing of a female *Metamonius hollermayeri.* According to Navás (1936a), the illustration on which this figure is based is "schematic," suggesting that various cross veins may have been omitted.

Key to the species of *Metamonius* larvae in South America

Information for the key was obtained from Lestage (1935), Ulmer (1904, 1938), and Demoulin (1955). The larva of *Metamonius hollermayeri* has not been described.

1\. The second segment of the labial palp shows no sign of a process on the inner side at its apex. The outer apical projection of the right mandible is about three times as long as its width at the apex. The margins of the gills are shallowly but clearly concave on one side (**Fig. 2.190**). Length: c. 13.5 mm. Length of cerci: 5 mm.
....*Metamonius fuegiensis* (Lestage, 1935) (Chile, Argentina, including Tierra del Fuego). Syn: *Ameletoides fuegiensis* (Lestage, 1935), *Ameletus fuegiensis* Lestage, 1935.

\- The second segment of the labial palp has a short, broad process on the inner side at its apex. The outer apical projection of the right mandible is only slightly longer than its width at the apex. The margins of the gills are straight or concave of both sides, except at the base (**Fig. 2.33**). Length: 9 to 12 mm. Length of cerci: 3 to 6 mm.
....*Metamonius anceps* (Eaton, 1885) (Chile, Argentina). Syn: *Siphlurus anceps* Eaton, 1885.

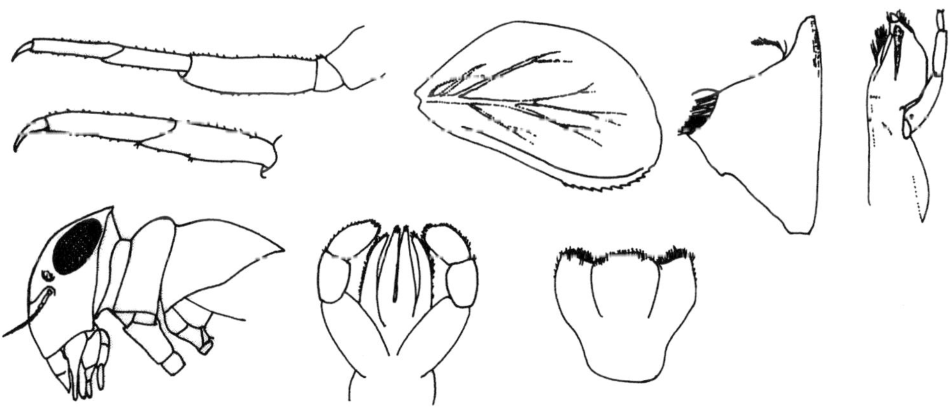

Fig. 2.190 *Metamonius fuegiensis* larva: the head pronotum, and mesonotum in lateral view (lower left); a leg with an enlargement of the tibia and tarsus (upper left); a gill (upper center); right mandible (upper right center); apical section of maxilla (upper right); labium (lower left center); hypopharynx (lower right).

Ameletopsidae

Key to the genera of adults in South America

Information for the key was taken from Demoulin (1952d, 1955c). Descriptions of the South American larvae of the *Chaquihua* species are not available, precluding the preparation of a key to the genera of larvae.

1. The cross veins in the stigmatic region of the costal space of the fore-wing do not form an anastomosing network. There are 15 or more cross veins in the subcostal space of the hind wing (**Fig. 2.191**).
....*Chiloporter* Lestage, 1931
- The cross veins in the stigmatic region of the costal space of the fore-wing form an anastomosing network. There are 10 or fewer cross veins in the subcostal space of the hind wing (**Fig. 2.15**).
....*Chaquihua* Demoulin, 1955..p. 159

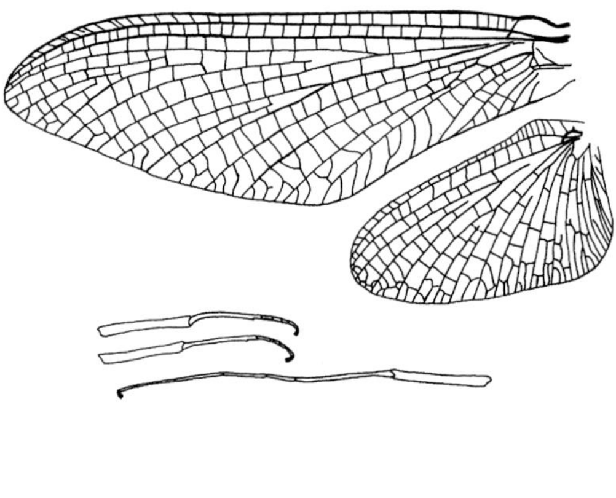

Fig. 2.191 *Chiloporter penai:* fore and hind wings (above); male genitalia (lower right); and hind, middle, and fore-leg (top to bottom, lower left).

Key to the species of *Chiloporter* larvae in South America

Information for the key was taken from Lestage (1931) and Demoulin (1952d, 1955c). The adult *Chiloporter eatoni* has never been described, so the species can only be distinguished from the larvae.

1. The antenna consists of 20 segments. The maxillary palp consists of 11 segments, and the labial palp consists of 14. No protective lamella covers each abdominal gill of the first pair (**Fig. 2.31**).
....*Chiloporter eatoni* Lestage, 1931
(Chile, Argentina). Syn: *Ameletopsis eatoni* (Lestage, 1931).

- The antenna consists of 17 segments. The maxillary palp consists of 13 segments, and the labial palp consists of 11. There is a protective lamella covering the gill on all pairs of abdominal gills (**Fig. 2.192**). Length: 15 mm. Length of the cerci: 4 mm. Color: brown, lighter on the underside and lateral parts of the abdomen. The lamellae covering the gills are orange, and the cerci are brown, becoming lighter toward the apex.
....*Chiloporter penai* Demoulin, 1955 (Chile).

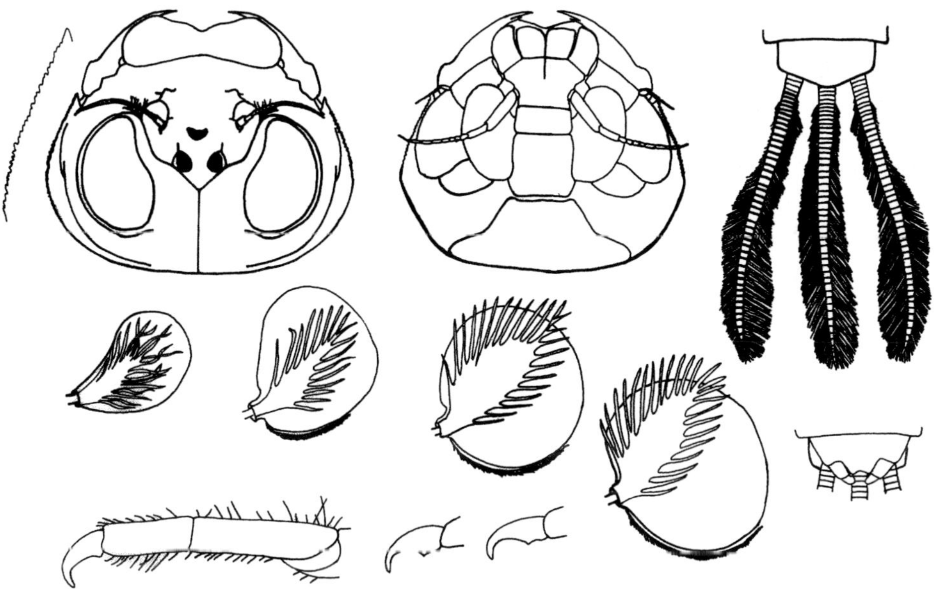

Fig. 2.192 *Chiloporter penai* larva (upper row, left to right): the head in dorsal and ventral view with an enlargement of the cheek margin to the left of the dorsal view and the apex of the abdomen with the cerci; (middle row, left to right): the gills of the first, second, third, and seventh abdominal segments and the anlage of the male genital structures at the apex of the abdomen of the last instar larva; (bottom row, left to right): the tarsal segments and claw of the fore leg and the claws on the middle and hind legs.

Key to adults of *Chaquihua* in South America

Information for the key was provided by Navás (1930b) and Demoulin (1955). The larvae have not been described.

1. The head is creamy white with a brownish shade on the vertex and occiput and grayish ocellae. The antennae are brown. The middle portions of the pronotum and the anterior part of the mesonotum and metonotum are darkened. The legs are entirely whitish in color. There are no branches in the

posteriormost anal vein (**Fig. 2.15**). Length: 13 mm; fore-wing length: 16 mm; hind wing length: 6 mm.
....*Chaquihua penai* Demoulin, 1955 (Chile). This species is very similar to the other South American species in this genus, and the differences between them, if they are both actually valid, must be better defined.
- The head is yellow with a dark band between the ocelli and dark compound eyes. The scape and pedicel of the antenna are yellow, and the flagellum is dark. The middle portions of the pronotum and metanotum and two median areas at the anterior and posterior parts of the mesonotum are darkened. The legs are light yellow with the apices of the tibiae and narrow rings around the tarsal joints darkened. There are two branches in the posteriormost anal vein in the fore-wing (**Fig. 2.193**). Length: 14.3 mm; fore-wing length: 18 mm; hind wing length: 7.5 mm.
....*Chaquihua bullocki* (Navás, 1930) (Chile). Syn: *Euthyplocia bullocki* Navás, 1930.

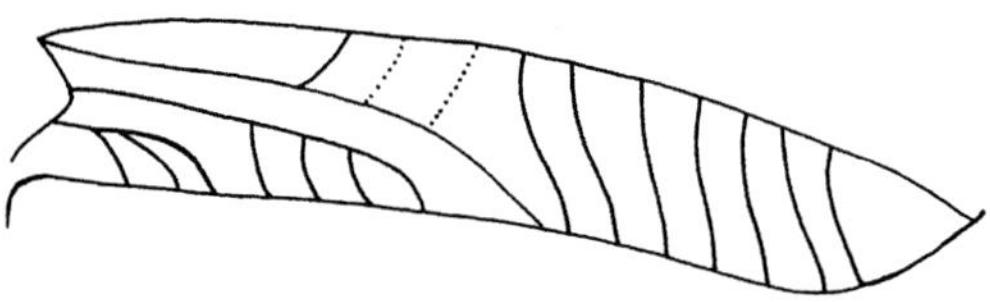

Fig. 2.193 The veins of the inner posterior section of the fore-wing of *Chaquihua bullocki.*

Baetidae

Key to the genera of adults in South America

Information taken from Ulmer (1920a, b, 1930), Traver (1947b), Day (1955, 1963), Mayo (1973a, b), Flowers (1985), Lugo-Ortiz and McCafferty (1994, 1995, 1996a, b, 1997, 1999a), Lugo-Ortiz *et al.* (1994), and Waltz and McCafferty (1985, 1987 a, b, c, 1999). Adults of the genera *Adebrotus, Tomedontus, Waltzoyphius, Bernerius, Harpagobaetis, Nanomis, Prebaetodes, Cryptonympha, Rivudiva, Spiritiops, Varipes,* and *Zelusia* remain undescribed. Characteristics of adult *Apobaetis* are known only from North American species. Due to the lack of descriptions of all stages of each species, the taxonomy of South American baetids is chaotic. First, a recent trend to describe each new taxon as a new genus based only on the examination of larvae makes it impossible to determine whether the morphology of the adults is similar to other known genera. There is no way even to rule out that species described only from larvae were not previously described as adults belonging to other genera. It would be more responsible to obtain both adults and larvae, and also subadults if possible, before making major modifications in the taxonomy of the family.

The reckless rush to add new species and genera without undertaking necessary taxonomic revisions of the family leaves everyone interested in identifying South American mayflies with a choice between a controversial new systematic arrangement of the species in newly named genera and the traditional arrangement without any yardstick on which to make a rational choice. Another major problem related to basing taxonomy on the description of larvae is the lack of information on the male genitalia, which provide the definitive characteristics on which to delimit species. It is concluded that a major revision of this family in South America is urgently needed, and this revision should be based primarily upon descriptions of adults with evidence to justify the findings obtained through comparisons of the larvae. Until such a revision is finished, the following keys can be used to identify those species described to date. However, a considerable number of changes in generic classification will be inevitable, probably resulting in the reduction of many, if not most of the genera in this key to the status of synonyms.

1. Hind wings absent (**Fig. 2.194**). 2
- Hind wings present (**Fig. 2.195**). 7

2. One short, free intercalary vein is present in each space between adjacent longitudinal veins along the outer margin of the wing. ....*Cloeon* Leach, 1815

No species of *Cloeon* are presently known to occur in South America. However, many South American species were assigned to this genus in the past, so the generic name may be encountered frequently in older literature.

- The marginal intercalary veins are arranged in pairs (**Fig. 2.194**). 3

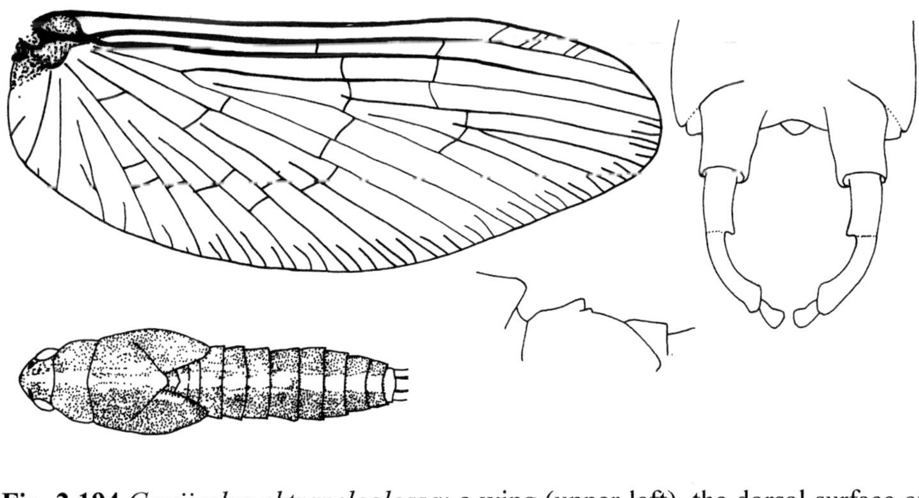

Fig. 2.194 *Guajirolus ektrapeloglossa:* a wing (upper left), the dorsal surface of the metanotum in lateral view showing the typical small tubercle (lower right center), the male genitalia (right), and the color pattern of the final larval instar and the location of its single pair of wing buds (lower left).

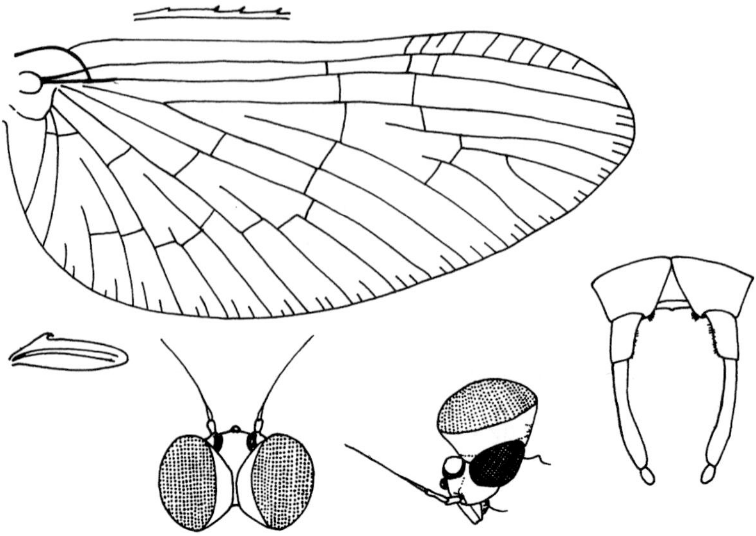

Fig. 2.195 *Cloeodes hydation* adult male: fore and hind wing drawn to scale with an enlargement of the spines on the anterior margin of the fore-wing (above and left), head in dorsal (lower left) and lateral view (lower right center), and the genitalia in ventral view (right).

3. There is an acutely pointed conical process on the distal end of the subgenital plate between the genital forceps. The total length of the adult is less than about 4 mm.
....*Apobaetis* Day, 1955

The only known South America representative of this genus is *Apobaetis signifer* Lugo-Ortiz and McCafferty (1997), from Paraguay and Pará. Also see the key to the South American species of *Paracloeodes* Day, 1955, below. The characteristics used in the key are for the known North American species because only the larva of the South American species has been described. Confusion is sure to result from the fact that some species of *Paracloeodes* lack hind wings and are indistinguishable from *Apobaetis* as adults. Of the four *Paracloeodes* species known from South America, only one probably lacks hind wings, based on the lack of larval hind wing buds. However, only the larva of this species has been described. Since the main distinction between the larvae of these two genera is a minor difference in the shape of the labial palps, the judgement of Day (1955) in declaring them to be distinct genera might be questioned. Not a single adult male of any South American species has yet been described.

- If an acutely pointed conical process is present on the distal end of the subgenital plate between the genital forceps, then the total length of the adult is much longer than 4 mm. 4

4. The intercalary veins are shorter than the space between the long veins in which each pair is located. The cross veins in the region of the stigma are sometimes oblique or partially anastomosed **(Fig. 2.196)**. 5
- Most intercalary veins are longer than the space between the long veins in which each pair is located. The cross veins in the region of the stigma are not anastomosed **(Fig. 2.194)**.
....6

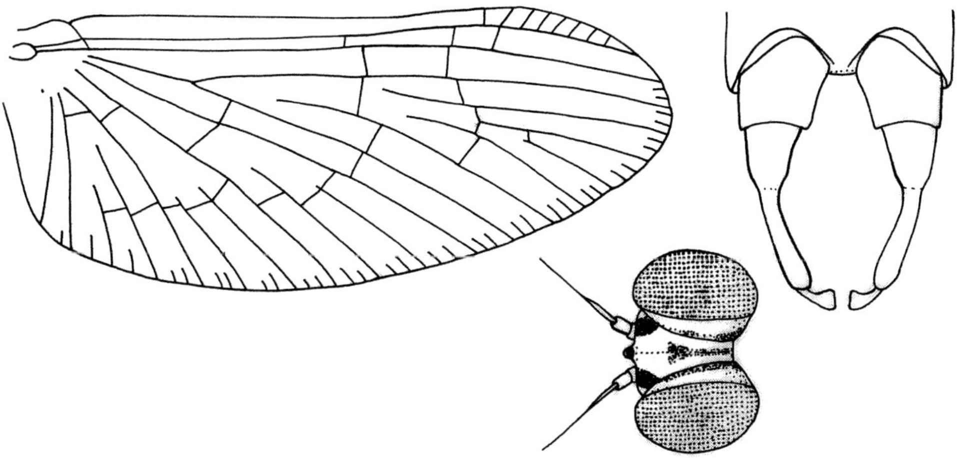

Fig. 2.196 *Americabaetis robacki:* the wing (left), head of a male in dorsal view (lower right center), and the genitalia of a male in ventral view (right).

5. The second segment of the labial palp of the known species bears a moderately developed or a thumb-like process. Each forceps arm consists of four segments, although in some cases, the divisions between the segments may be faint **(Fig. 2.196)**.
....*Americabaetis* Kluge, 1992..p. 188
Species of this genus had been placed in *Acerpenna* by Waltz and McCafferty (1987c), but all known Neotropical species with the characteristics described here have since been assigned to *Americabaetis* by Lugo-Ortiz and McCafferty (1996a). However, most South American species in these controversial genera are known only from larvae. Adults of presumed congeneric species elsewhere have hind wings. The validity of this genus, like several other newly described genera from the northern Neotropical Region, seems extremely doubtful, and the taxonomy of the South American baetids is in a state of rapid flux.
- The second segment of the labial palp is presumed not to bear a moderately developed or thumb-like process, although this has not been established for most South American species, which are known only from larvae. Each arm of the forceps usually consists of three visible segments, and the basal half of the second segment is swollen **(Fig. 2.197)**.
....*Baetodes* Needham and Murphy, 1924..p. 195

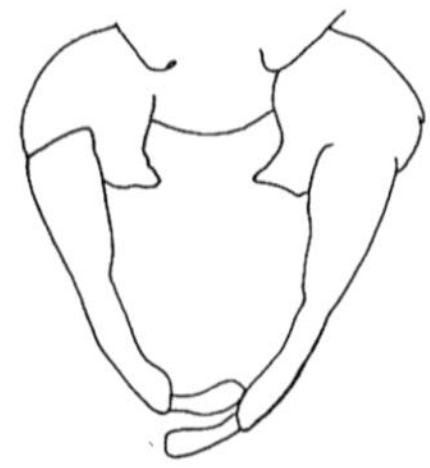

Fig. 2.197 The male genitalia of *Baetodes arawak*.

6. There is a small projection between the genital forceps. The second segment of the genital forceps has a large internal projection (**Fig. 2.194**).
....*Guajirolus* Flowers, 1985

The only species yet described in this genus is *Guajirolus ektrapeloglossa* Flowers, 1985, found in Panama and Colombia. Its body length is 4.9 to 5.8 mm, and its fore-wing length is 4.5 to 5.7 mm. Color: generally orange brown with darker markings and lighter appendages.

- There are no projections between the genital forceps or on the internal side of its second segment.
....*Pseudocloeon* Klapálek, 1905

Keffermüller (1979) proposed that the genus *Pseudocloeon* be eliminated and the species be included in *Baetis*. In the past, many South American species have been assigned to this genus, but McCafferty (2000) transferred the last of them to other genera, although two *nomena dubia* are still maintained in the list of Domínguez *et al*. (2001).

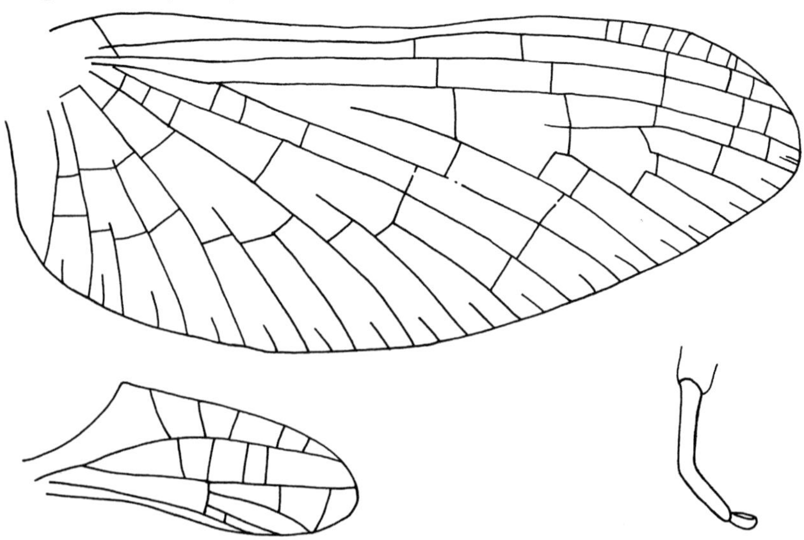

Fig. 2.198 *Callibaetis pollens:* fore and hind wing (left) and the forceps segments of a male (lower right).

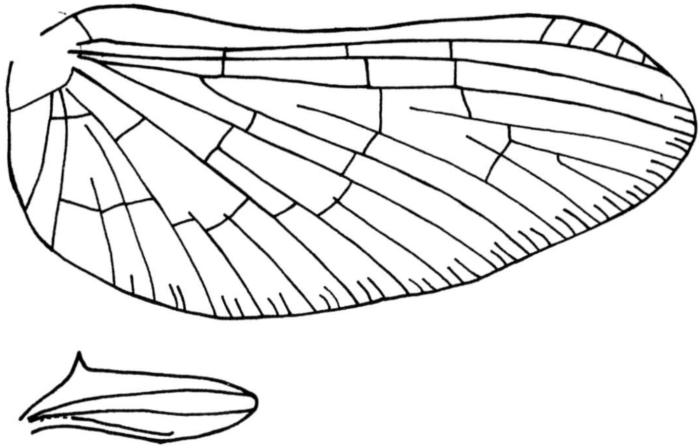

Fig. 2.199 The fore and hind wings of *Fallceon murphyae.*

7. The body is covered by small reddish brown spots. Fore and hind wings have many cross-veins in the costal space (**Fig. 2.198**).
....*Callibaetis* Eaton, 1881..p. 202
- The body lacks small reddish brown spots. The hind wing usually does not have many cross veins in the costal space (**Fig. 2.199**). 8
8. The hind wing is well developed and, although considerably smaller than the fore-wing, is not minute. In the known South American species, it has three longitudinal veins, the first two of which arise from a common fork or at least very close together (**Fig. 2.199**). 9
- The hind wings are minute in comparison with the fore-wing, or, if not, then they each have only two longitudinal veins (**Fig. 2.195**). 10
9. The hind wing is ovoid or lancet-shaped with an acutely pointed costal prominence, which is sometimes hooked. The middle veins of the two known South American species are not forked (**Fig. 2.199**), but they fork in some species found to the north. In South America, the insects belonging to the known species are not longer than about 7 mm.
....*Fallceon* Waltz and McCafferty, 1987..p. 210
- The hind wing is roughly trapezoidal with a broadly rounded apex and a low costal prominence; it has three longitudinal veins, the middle of which is forked (**Fig. 2.200**). The hind wings are obviously smaller than the fore-wing but not minute. The insects are about 10 to 12 mm long.
.....................................*Moribaetis* Waltz and McCafferty, 1985..p. 212
10. The hind wings have two longitudinal veins and a costal projection but are not long and narrow in shape and minute in comparison with the fore-wing (**Fig. 2.200**). 11
- The hind wings are greatly reduced in size in comparison with the fore-wing, sometimes long and narrow (**Fig. 2.195**), and there may sometimes be more than two longitudinal veins present. 13

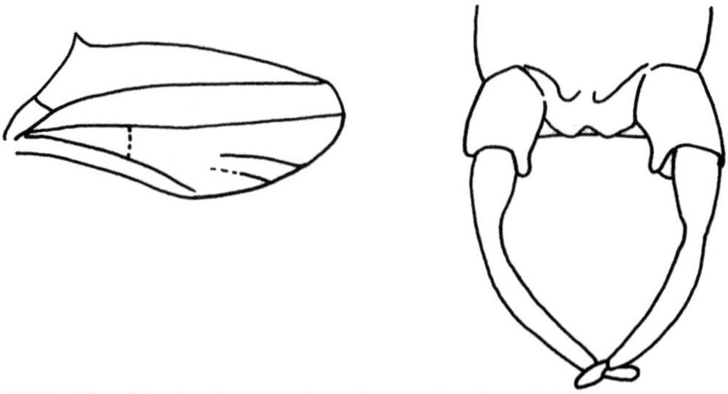

Fig. 2.200 The hind wing and male genitalia of *Moribaetis aneto*.

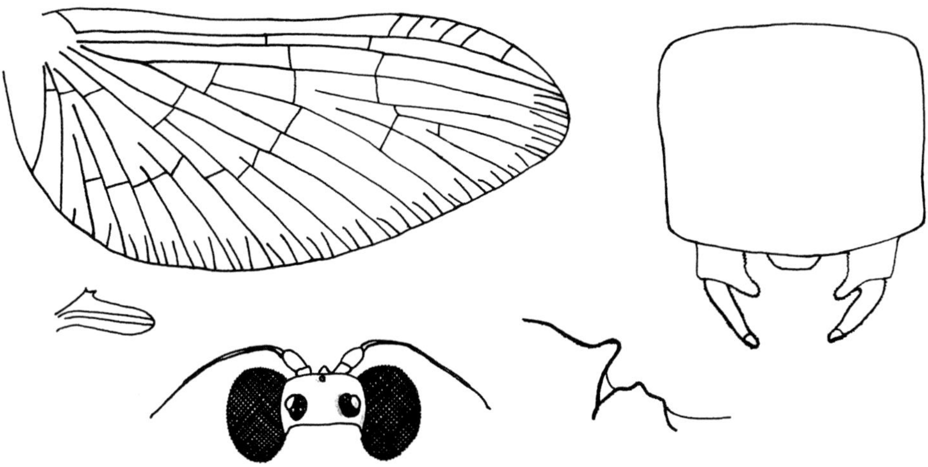

Fig. 2.201 *Aturbina georgei*: fore and hind wings of adult male showing double-peaked costal projection of the hind wing (upper left), head of a male in dorsal view (lower left center), metanotum with small median hump (lower right center), male genitalia (right).

11. The costal projection on the hind wing has two apical peaks. There is a small median hump on the metanotum. The eyes of the male are not turbinate (**Fig. 2.201**).
..................................*Aturbina* Lugo-Ortiz and McCafferty, 1996..p. 213
- The costal projection on the hind wing has only a single apical peak. There is no small median hump on the metanotum. The eyes of the male are turbinate (**Fig. 2.202**). 12

12. There are brown clouds on the fore-wing of the male. The only known South American species has posterior cross veins connecting the posterior long vein with the posterior costal margin (**Fig. 2.202**). The body length is 10 to 12 mm.
....*Mayobaetis* Waltz and McCafferty, 1985

Only one of the species presently assigned to this genus occurs in South America: *Mayobaetis ellenae* (Mayo, 1973) from Costa Rica, Ecuador, and Peru. This species has been referred to as *Baetis ellenae* Mayo (1973) and *Moribaetis (Mayobaetis) ellenae* (Mayo, 1973).

- The fore-wing lacks cross veins along the costal vein except near the stigma, and hind wings lack cross veins reaching the costal vein (**Fig. 2.203**).
....*Camelobaetidius* Demoulin, 1967..p. 214

The name *Dactylobaetis* Traver and Edmunds, 1968, is now generally regarded to be a junior synonym of *Camelobaetidius,* which had been described exclusively from larvae. Recently, several species known only from adults and formerly considered to be species of *Baetis* have been transferred to this genus (Domínguez *et al*., 2001).

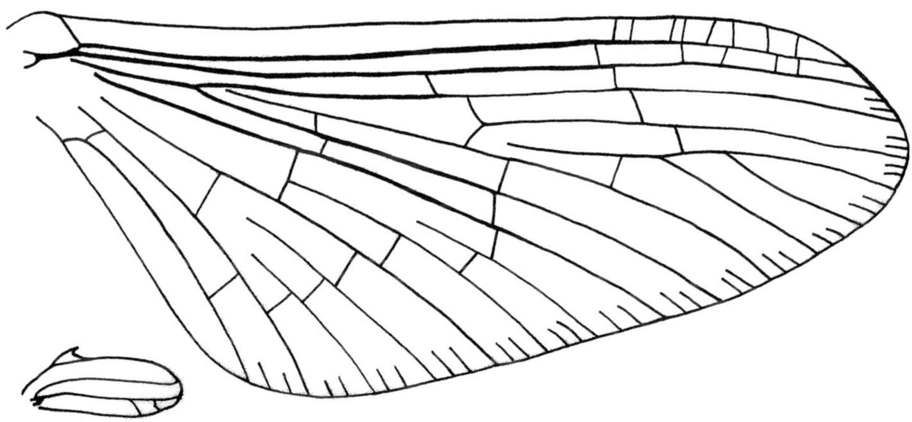

Fig. 2.202 The fore and hind wing of *Mayobaetis ellenae* with the hind wing enlarged relative to the fore-wing.

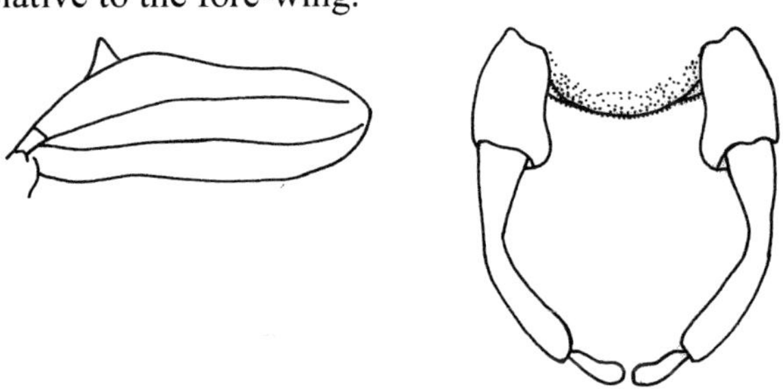

Fig. 2.203 Hind wing (left) and male genitalia (right) of *Camelobaetidius penai.*

13. There are two longitudinal veins that arise from a single fork near the base of the hind wing. Its costal prominence is short, broadly obtuse, truncate at the apex, and located in the basal ¼ of the wing (**Fig. 2.204**). The adult male has not been described.

....*Paracloeodes* Day, 1955, pars..p. 222

The adult of only one species in South America has been described, the female *Paracloeodes eurybranchus* Lugo-Ortiz and McCafferty, 1996, from Argentina and Rio Grande do Sul. It is 4.3 to 4.4 mm long, has a fore-wing 4.1 to 4.2 mm and a hind wing 0.8 to 0.9 mm long. The other species, including one apparently lacking a hind wing, can only be distinguished as larvae.

- The hind wings do not have only two longitudinal veins arising from a fork in a single basal vein, or, if they do, the costal prominence is not short, broadly obtuse, or truncate at the apex, or it is not located in the basal ¼ of the wing (**Fig. 2.195**). 14

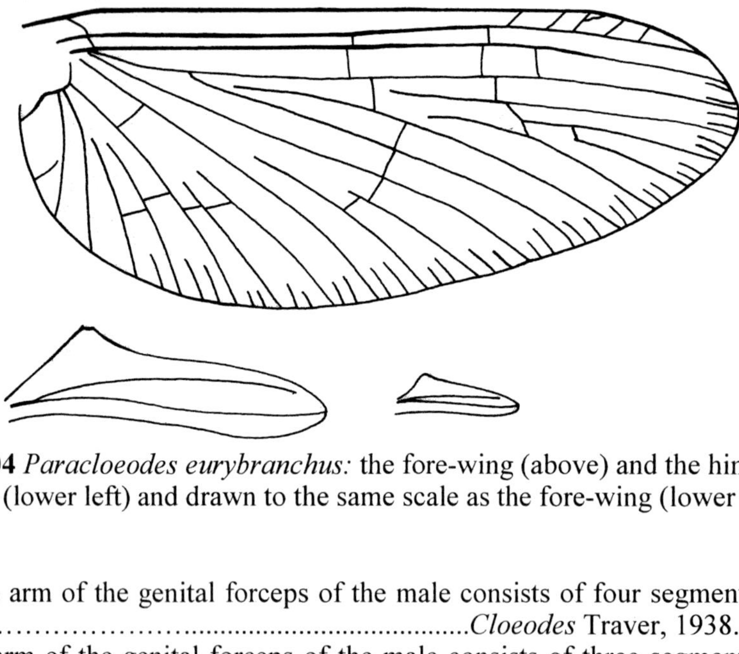

Fig. 2.204 *Paracloeodes eurybranchus:* the fore-wing (above) and the hind wing enlarged (lower left) and drawn to the same scale as the fore-wing (lower right).

14. Each arm of the genital forceps of the male consists of four segments (**Fig. 2.195**). ....*Cloeodes* Traver, 1938..p. 224

- Each arm of the genital forceps of the male consists of three segments (**Fig. 2.205**). 15

15. The eyes of the adult male are not ellipsoid in shape and lack broad bases, or the costal process of the hind wing, if present, is not small or not located in the basal third of the wing, or the hind wing does not have three longitudinal veins with the middle one forked.

....*Baetis* Leach, 1815

All 32 of the South American species formerly assigned to the genus *Baetis* have been transferred to other genera (McCafferty, 2000). Because the taxonomy of the family Baetidae has been in a state of flux during recent years,

it is important to keep in mind that the "disappearance" of *Baetis* from the South American fauna may not be permanent.
- The eyes of the adult male are ellipsoid in shape and have broad bases. The small costal process on the hind wing is located in the basal third of the wing. There are three longitudinal veins in the hind wing, and the middle one is forked. The apical segment of the genital forceps has an elongate ellipsoidal shape (**Fig. 2.205**). 16

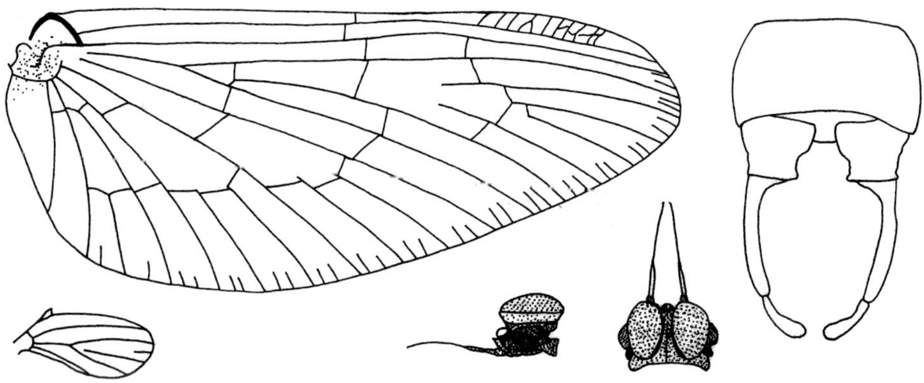

Fig. 2.205 *Andesiops peruvianus:* fore and hind wing (left), the head of the male in lateral and dorsal view (lower right center), and the male genitalia.

16. The longest diameter of the ellipsoidal turbinate eyes of the male adult is at least twice that of its height, including the stalk (**Fig. 2.205**). Length of only known species: 5.4 to 5.8 mm; fore-wing length: 5.8 to 6.3 mm; hind-wing length: 1.4 to 1.6 mm; caudal filament length: 15 to 17 mm. Color: head yellowish or brown, thorax mainly brown to dark brown with yellow brown markings; abdomen brown with the second through sixth segments translucent. ....*Andesiops* Lugo-Ortiz and McCafferty, 1999
The only known species in this genus is *Andesiops peruvianus* (Ulmer, 1920) from Venezuela, Colombia, Ecuador, Peru, Bolivia, Chile, and Argentina. Syn: *Baetis peruvianus* Ulmer, 1920; *Baetis weiseri* Navás, 1922.
- The longest diameter of the turbinate eyes of the male adult is about equal to or only slightly greater than its height, including the stalk (**Fig. 2.206**). ....*Deceptiviosa* Lugo-Ortiz and McCafferty, 1999..p. 230

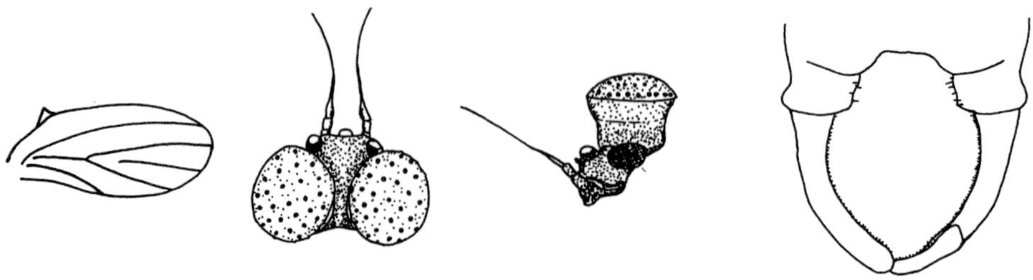

Fig. 2.206 *Deceptiviosa torrens* male (left to right): hind wing, head in dorsal and lateral view, and male genitalia.

Key to the genera of larvae of Baetidae in South America

Information for the key was provided by Day (1955), Traver and Edmunds (1968), and Flowers (1985), Lugo-Ortiz and McCafferty (1996a, b, c, d, 1997, 1998), Lugo-Ortiz *et al.* (1994), and Waltz and McCafferty (1985, 1987a, b, c, 1995). Unfortunately, this key is not easy to use, not only because of the minutiae used to distinguish the genera but also because many of the original descriptions are sketchy. *Cloeon* is not included (see key for adults).

1. The claws are spatulate (**Fig. 2.207**).
....*Camelobaetidius* Demoulin, 1967..p. 214
The name *Dactylobaetis* Traver and Edmunds, 1968, is now generally regarded to be a junior synonym of *Camelobaetidius,* which had been described exclusively from larvae. Recently, several species known only from adults and formerly considered to be species of *Baetis* have been transferred to this genus (Domínguez *et al.*, 2001).
- The claws are pointed (**Fig. 2.208**). 2
2. Gills are present only on the first five segments. The lateral caudal filaments might bear a few inconspicuous setae, which may be absent in some species (**Fig. 2.208**).
....*Baetodes* Needham and Murphy, 1924..p. 195
- Gills are present on the first seven or the second to seventh segments (**Fig. 2.209**). 3
3. Double gill lamellae, sometimes consisting only of a recurved flap, are present on at least the first six segments. The larvae of the South American species have not yet been described.
....*Callibaetis* Eaton, 1881..p. 202
- Single gill lamellae are present at least on the second through seventh abdominal segments (**Fig. 2.209**). 4

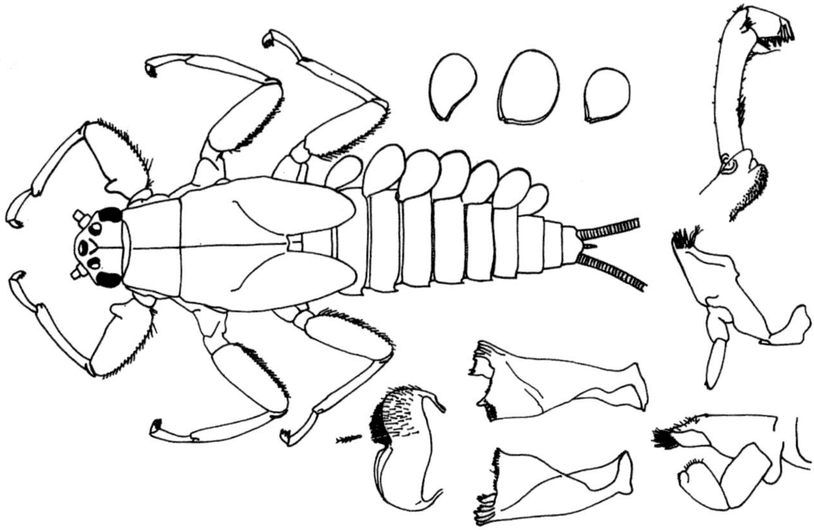

Fig. 2.207 *Camelobaetidius leentvaari* larva: habitus without the apical portions of the antennae and cerci and the gills on the left side (left); the labrum (lower center); mandibles (lower right center); half of the labium (lower right); maxilla (middle right); the apex of a tarsus and claw (upper right); and the gills on the first, fourth, and seventh abdominal segments (upper right center, left to right).

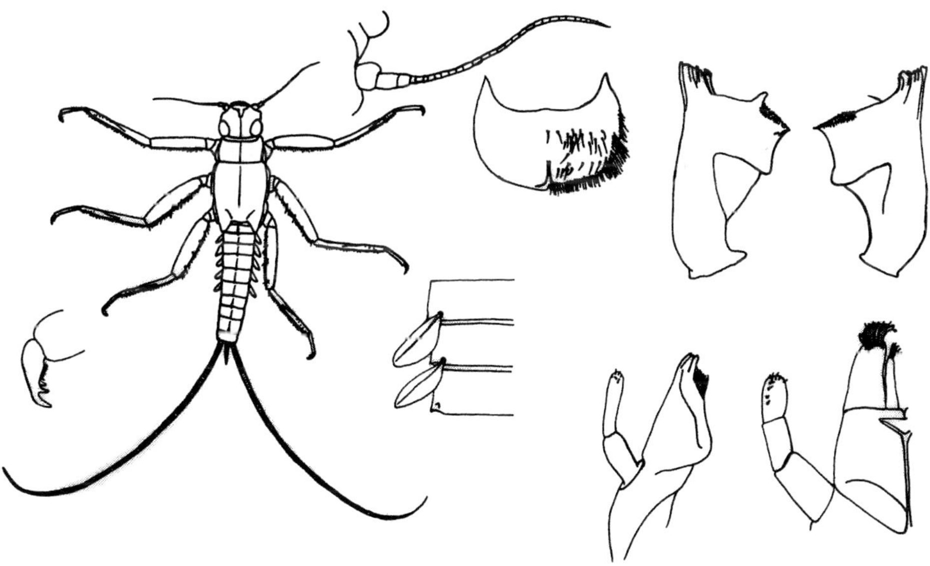

Fig. 2.208 *Baetodes itatiayanus* larva: habitus with enlargements of the hind claw and antenna (left), labrum (upper center), mandibles (upper right), gills on two abdominal segments (lower center), left maxilla (lower right center), and half of the labium (lower right).

4. The claw has a large subapical tooth followed by a row of fine teeth. The second segment of the labial palp is greatly enlarged on the side toward the mid-line (**Figs. 2.210**). The first abdominal segment usually lacks gills.
....*Guajirolus* Flowers, 1985

The only species yet described in this genus is *Guajirolus ektrapeloglossa* Flowers, 1985, found in Panama and Colombia. The late instar larvae are 3.6 to 5.3 mm.

- A relatively large subapical tooth is not present on each claw (**Fig. 2.211**), or, if one is present, it is followed by a row of teeth decreasing gradually in size toward the base of the claw rather than by a row of fine teeth, or it is followed by fairly uniform teeth or setae that are not fine, usually arranged in one or more rows.
....5

5. Apically rounded gills are present only on the second through seventh abdominal segments (**Fig. 2.211**). Hind wing pads are absent. 6

- Single gill lamellae are present on all abdominal segments through the seventh, and one or more of them may not be rounded apically (**Fig. 2.209**). 7

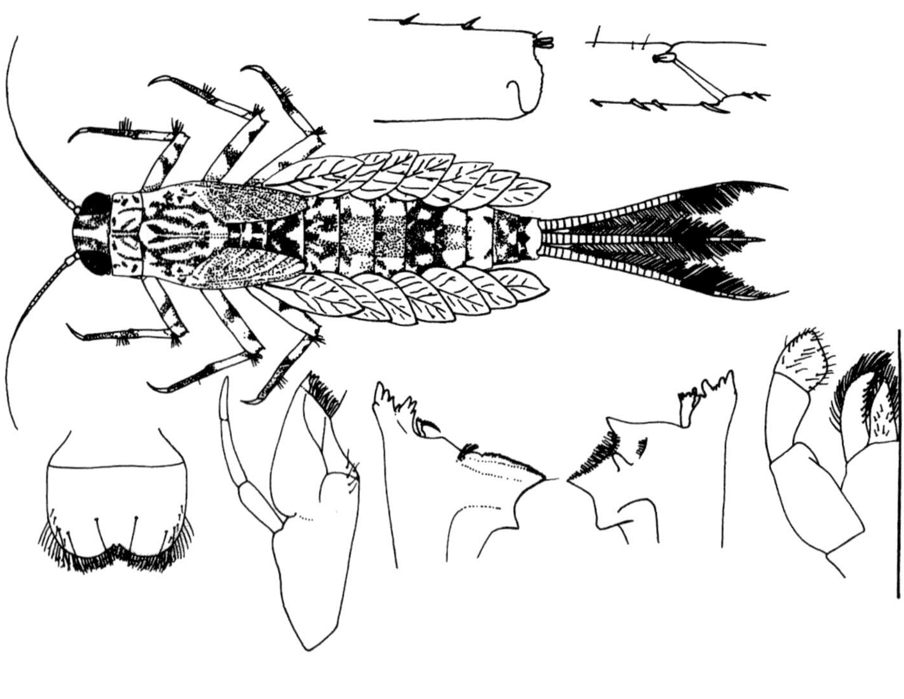

Fig. 2.209 *Cloeodes hydation* larva: habitus (above left) below enlargements of the apices of the femur and tibia on the fore-leg (upper right, left and right, respectively), and mouthparts (left to right): the labrum, maxilla, apical parts of the mandibles, and one side of the labium in ventral view.

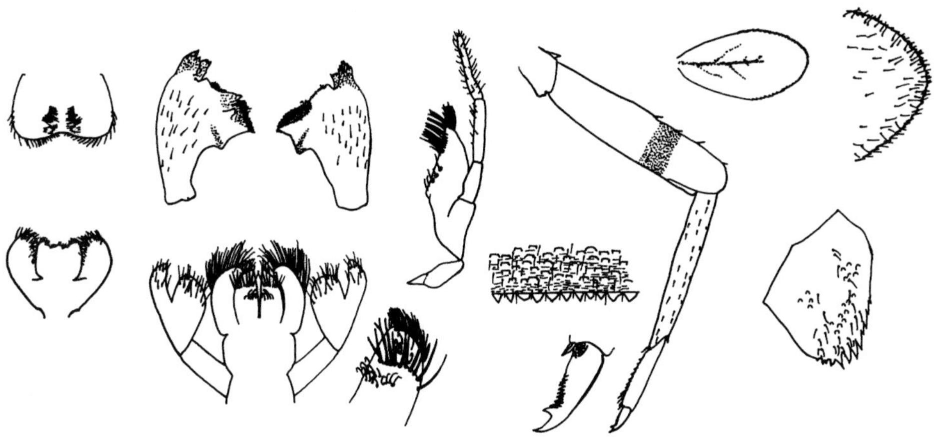

Fig. 2.210 *Guajirolus ektrapeloglossa* larva: labrum in ventral view (upper left), mandibles (upper left center), maxilla (upper center), hypopharynx (lower left), labium in dorsal view with an enlargement of the apex of a glossa (lower left center), the middle leg with an enlargement of a tarsal claw (right center), posterior margin of the seventh abdominal tergite (between maxilla and leg), a gill with an enlargement of its apex (upper right), and a paraproct (lower right).

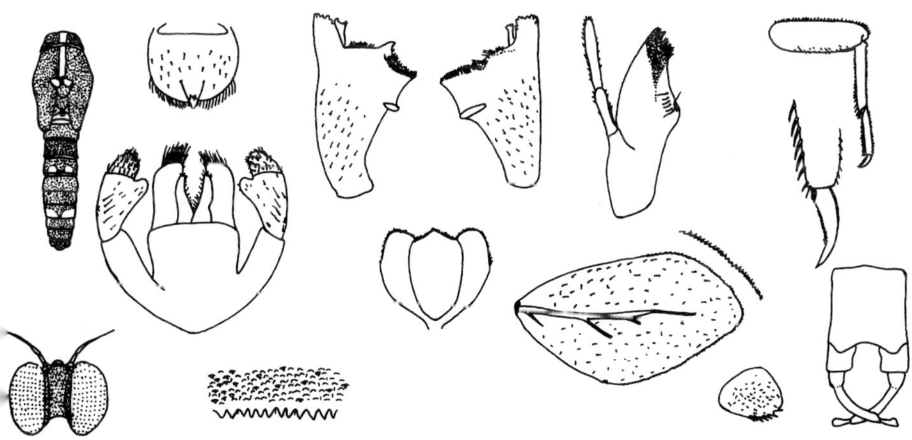

Fig. 2.211 *Americabaetis alphus* (larva): the thorax and abdomen of a male larva in dorsal view (upper left); the head of a male adult showing the turbinate eyes (lower left); the labrum, mandibles, maxilla, and left fore-leg with an enlargement of the tarsal claw (upper row, left to right from thorax); the labium in ventral view (below labrum), the hypopharynx, gill pad from the fourth abdominal segment with an enlargement of its margin, paraproct, apex of the abdomen of a male adult (left to right from labium), and details of the posterior margin of the fourth tergite (lower left center).

6. The pre-apical segment of the labial palp is at least moderately produced along the inside margin at the apex. The prostheca on the right mandible is robust and has a molar-like apical surface (**Fig. 2.211**). The frontal surface of the labrum is not at all raised in the middle.

....*Americabaetis* Kluge, 1992..p. 188

- The labial palp is relatively thick, and its pre-apical segment is not produced along the inside margin at the apex. The prostheca on the right mandible is reduced to a thin, bifid appendage without any trace of a molar surface at the apex (**Fig. 2.212**). The frontal surface of the labrum is raised in the middle, as best seen in profile.

....*Zelusia* Lugo-Ortiz and McCafferty, 1998

The only known species in this genus is *Zelusia principalis* Lugo-Ortiz and McCafferty, 1999, from Amazonas. Its last larval instar is 3.8 to 4.2 mm long, and the length of its caudal filament is 1.5 to 1.7 mm. It is yellowish brown and lacks a distinct color pattern. The distinctions between this genus and *Americabaetis* are very minor.

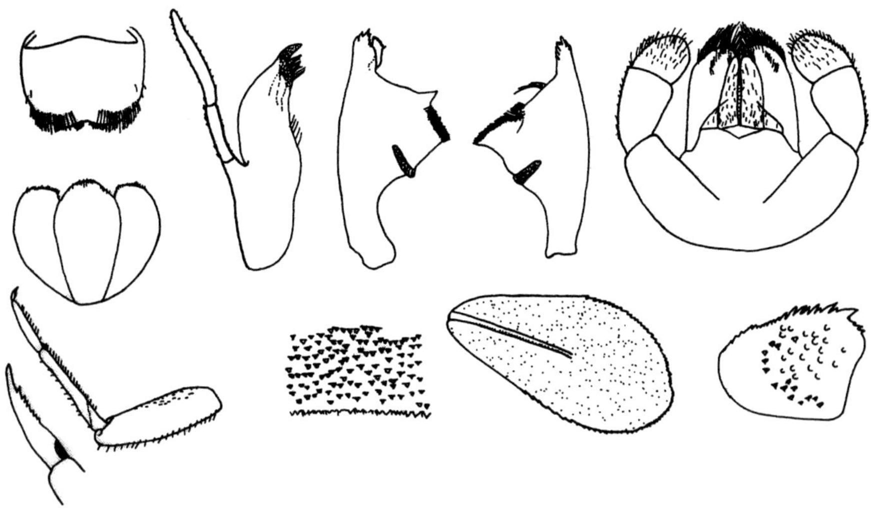

Fig. 2.212 *Zelusia principalis* larva (upper row, left to right): labrum (above), hypopharynx (below), maxilla, mandibles, labium, and (lower row, left to right): fore-leg above an enlargement of the tarsal claw, an enlargement of the denticles and sculpture on the fourth abdominal tergite, the fourth gill, and a paraproct.

7. The labial palp consists of two apparent segments because the second and third are fused. There is a distinct subapical process on the inner margin of the palp, which represents an apical process of the fused second segment. The claw is narrow, and if teeth are present, they are very small (**Fig. 2.213**). 8

- The labial palps consist of three distinct segments, or the apical segment does not bear an inner, thumb-like process (**Fig. 2.209**). 9

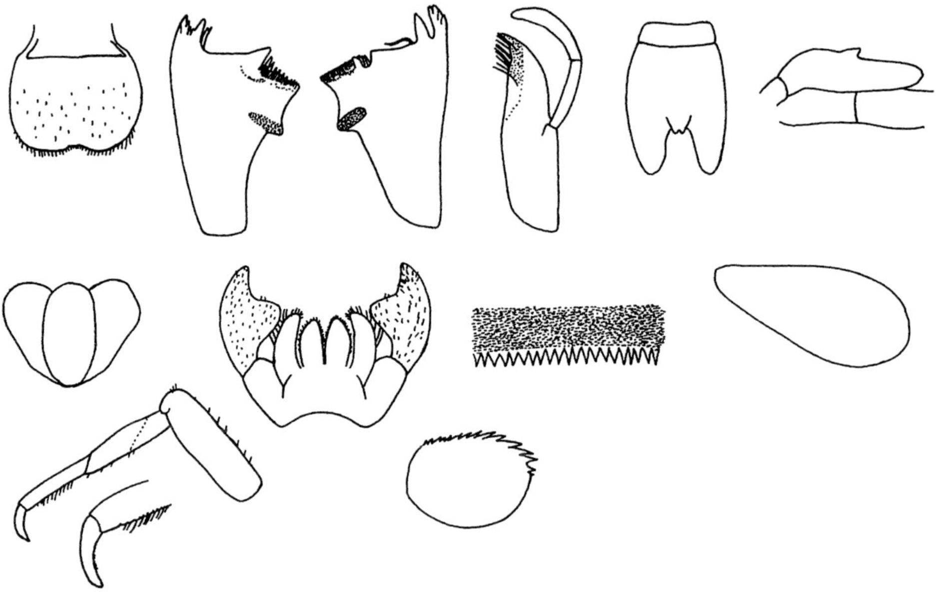

Fig. 2.213 *Paracloeodes binodulus* larva (upper row, left to right): the labrum, mandibles, maxilla, profile of the thorax in dorsal and lateral view, and (middle row, left to right) the hypopharynx, labium, an enlargement of the posterior margin and adjacent surface of the fourth abdominal tergite, the fourth gill, and (lower row, left and right) the fore-leg with an enlargement of the apex of the tarsus and tarsal claw, and a paraproct.

8. The tarsal claw is 75% to 80% as long as the apical tarsal segment. The anterior margin of the labrum is not noticeably emarginated in the middle. Hind wing buds are absent. The surfaces of the tergites and paraprocts are covered by minute spines (**Fig. 2.214**). The maximum length of the final instar larva is less than 4 mm.

....*Apobaetis* Day, 1955

The only known South America representative of this genus is *Apobaetis signifer* Lugo-Ortiz and McCafferty, 1997, from Paraguay and Pará.

- The tarsal claw is less than half as long as the apical tarsal segment. The anterior margin of the labrum is obviously emarginated (**Fig. 2.213**).

....*Paracloeodes* Day, 1955, pars..p. 222

Larva of the genus *Pseudocloeon* might be confused with those of *Paracloeodes*, but the South American species formerly assigned to *Pseudocloeon* have been assigned to other genera, and the genus is thought to be absent from South America.

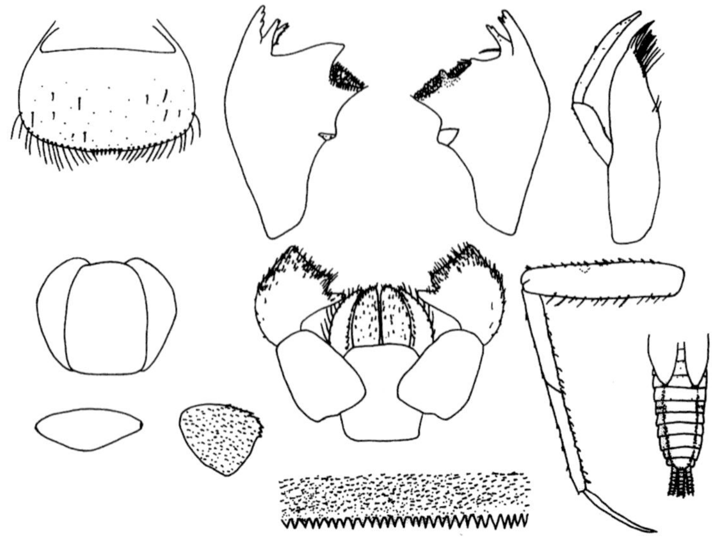

Fig. 2.214 *Apobaetis signifer* larva: the labrum (upper left), hypopharynx (middle left), mandibles (upper center), right maxilla (upper right), labium in ventral view (center), fore-leg (lower right center), dorsal pattern on the abdomen (lower right), posterior margin of the fourth tergite (lower center), gill from the fourth abdominal segment (lower left), and paraproct (lower left center).

9. The lateral margins of the labrum are notably expanded and its base is narrow, making it broadly rounded laterally. The labrum has a median emargination (**Fig. 2.215**). 10
- The labrum is only moderately rounded on the lateral margin or its anterior margin is not emarginated (**Fig. 2.209**). 11
10. The hypopharynx bears wing-like lateral processes. The labrum lacks long setae. It is emarginated in the middle of its anterior margin, and there is a rounded prominence in the middle of the emargination. The molar region of each mandible protrudes. The maxillary palps and apical section of the labial palps are slender. The tarsal claw bears two parallel rows of denticles that increase in size from the base to the apex of the claw (**Fig. 2.215**). Both fore and hind wing buds are present.
....*Harpagobaetis* Mol, 1986

The only species in this genus from South America is *Harpagobaetis gulosus* Mol, 1986, from Surinam. The fully-grown larvae are 7.2 mm long. The length of the cerci is about 3.1 mm, and the medial caudal filament is about 2.8 mm long. The larvae are pale yellow with light brown markings.

- The hypopharynx does not bear wing-like extensions; it has narrow, elongate superlinguae. The anterior median emargination is relatively shallow and lacks a rounded prominence extending from the middle. The elongate tarsal claws bear a single row of denticles and a subapical seta. Hind wing pads are present but appear vestigial. The posterior margins of the abdominal tergites bear triangular spines (**Fig. 2.216**). The middle caudal filament is 1/5 to 1/4 as long as the cerci.

.....................................*Prebaetodes* Lugo-Ortiz and McCafferty, 1996

The only species in this genus is *Prebaetodes sitesi* Lugo-Ortiz and McCafferty, 1996, known from high mountainous regions of Ecuador and Colombia. Its final larval instar is 5.5 to 6.3 mm long, and its caudal filament length 4.2 to 5.0 mm.

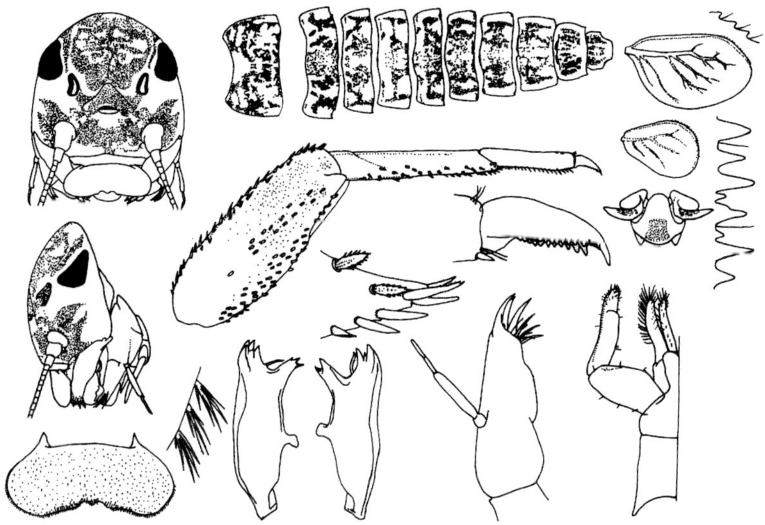

Fig. 2.215 *Harpagobaetis gulosus* larva: the head in frontal (upper left) and lateral view (middle left), the pattern on the pronotum and the abdominal segments (upper center), the gills from the second (above) and seventh (below) abdominal segments with an enlargement of the margin of the second gill (upper right), the posterior margin of a tergite enlarged (middle right), a fore-leg above enlargements of the apex of its tibia and its tarsal claw (center), the hypopharynx (below gills), labrum with an enlargement of the setae along its lateral margin (lower left), mandibles (lower left center), maxilla (lower right center), and half of the labium in ventral view (lower right).

11. A prostheca is not developed on the right mandible, and only a seta is present in its place. There are two curving rows of denticles on each tarsal claw, which nearly meet at the apex to form a U with the open end toward the base. The posterior margins of the tergites are zigzag with triangular processes (**Fig. 2.217**). The caudal filament is approximately as long as the cerci.

....*Varipes* Lugo-Ortiz and McCafferty, 1998

The only known species in this genus is *Varipes lasiobrachius* Lugo-Ortiz and McCafferty, 1998, from Ecuador and Colombia. It is 4.3 to 5.8 mm long and has a caudal filament 1.9 to 2.4 mm long. It is generally dark yellowish brown to medium brown with some light and dark markings.

- The prostheca is developed on the right mandible and usually bears a seta (**Fig. 2.218**). 12

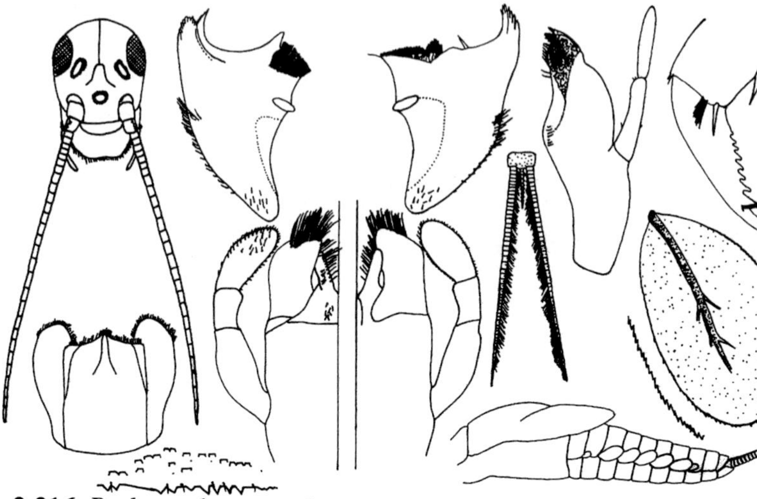

Fig. 2.216 *Prebaetodes sitesi* larva (upper row, left to right): head in frontal view, mandibles, apical segment of the abdomen showing the cerci and median caudal filament, maxilla, tarsal claw, and (lower row, left to right): hypopharynx, posterior margin of a tergite (below), half of the labium in ventral and dorsal view, outline of the thorax and abdomen in lateral view below the fourth gill with an enlargement of the denticles on the gill margin.

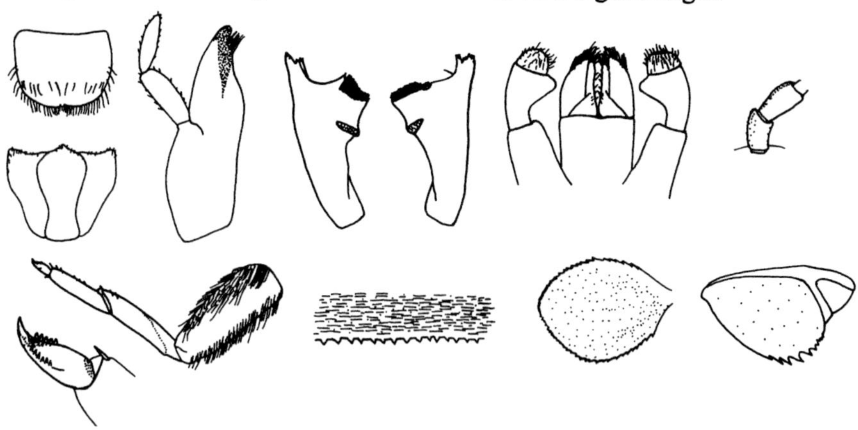

Fig. 2.217 *Varipes lasiobrachius* larva (upper row, left to right): labrum (above), hypopharynx (below), maxilla, mandibles, labium, scape and pedicel at the base of the antenna, and (lower row, left to right): fore-leg above an enlargement of the tarsal claw, an enlargement of the denticles and sculpture on the fourth abdominal tergite, the fourth gill, and a paraproct.

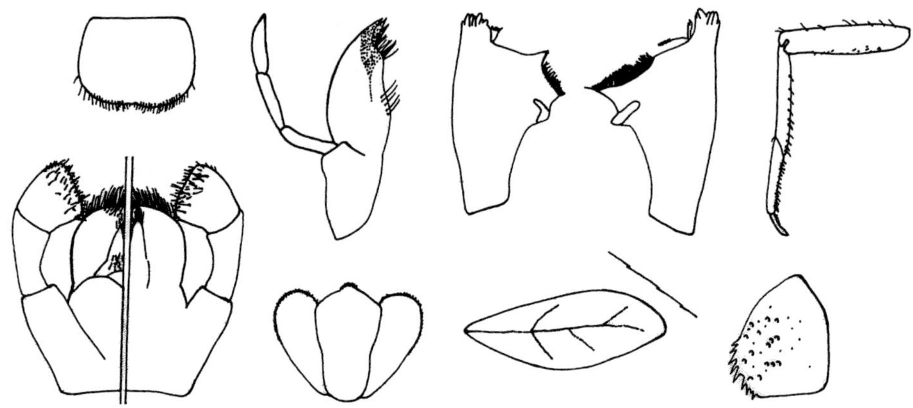

Fig. 2.218 Larva of *Aturbina georgei* (upper row, left to right): labrum in dorsal view, left maxilla, left mandible, right mandible, right fore-leg; (lower row, left to right): labium in ventral (left) and dorsal (right) view, hypopharynx, gill from the fifth abdominal segment with details of the margin above, and the paraproct.

12. The prostheca of the right mandible bears a sclerotized, slender process with an elongate, distally branched seta arising from the base of the process (**Fig. 2.218**). There are small denticles at the bases of the claws.
..................................*Aturbina* Lugo-Ortiz and McCafferty, 1996..p. 213
- There is no slender sclerotized process bearing a branched seta on the prostheca of the right mandible, or the seta does not arise from the base of the process (**Fig. 2.209**). 13

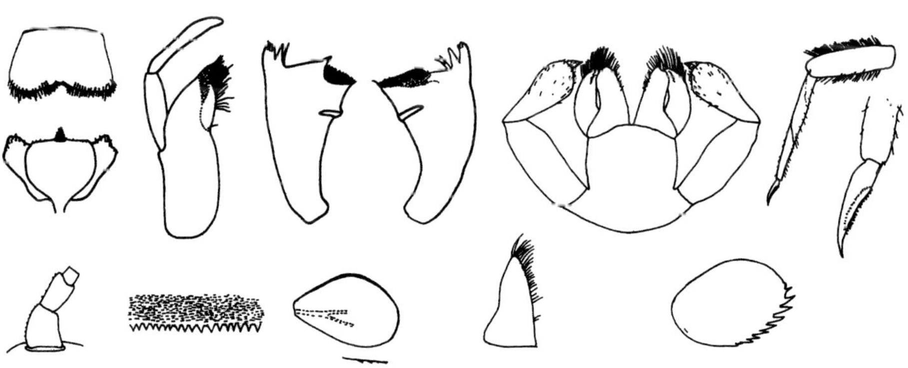

Fig. 2.219 *Rivudiva minantenna* larva (upper row, left to right): labrum (above) and hypopharynx (below), maxilla, mandibles, labium, fore-leg with an enlargement of the claw, and (lower row, left to right): the scape and pedicel of the antenna, an enlargement of the denticles and sculpture on the fourth abdominal tergite, the fourth gill with an enlargement of its margin, an enlarged glossa from the labium, and a paraproct.

13. The prostheca of the right mandible is bifid (**Fig. 2.219**). 14
- The prostheca of the right mandible is not bifid (**Fig. 2.209**). 18
14. The median caudal filament is about the same length as the cerci. The hypopharynx is wider than long (**Fig. 2.219**). All species thus far described come from lowland water courses in eastern South America.
......................................*Rivudiva* Lugo-Ortiz and McCafferty, 1998..p. 229
- The median caudal filament is shorter than the cerci, not exceeding 85% of the length of a cercus (**Fig. 2.220**). The species in these genera thus far described inhabit cool streams high in the Andes or in the cool southern part of the continent. 15

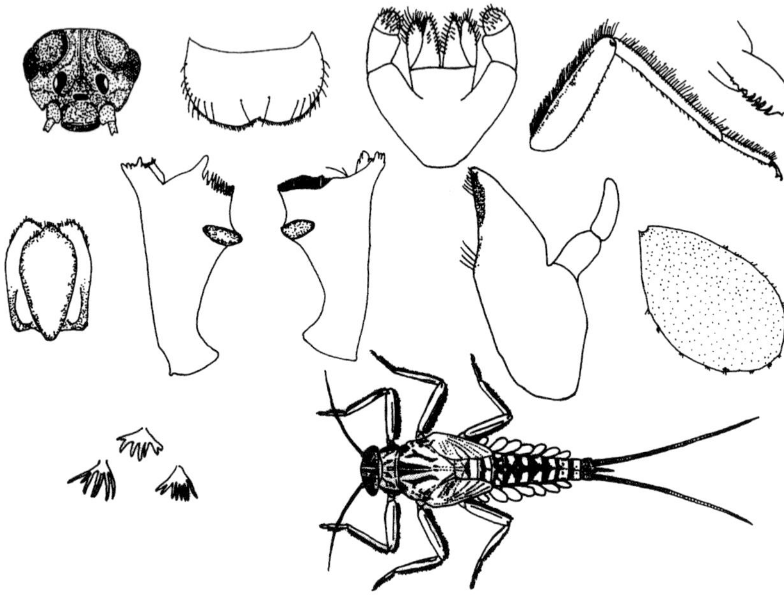

Fig. 2.220 *Deceptiviosa torrens* larva (upper row, left to right): head in frontal view, labrum, labium in ventral view, fore-leg, and enlargement of the tarsal claw; (middle row, left to right) hypopharynx, mandibles, maxilla, and the gill from the fourth abdominal segment; (lower row) structures on the fourth tergite measuring just under 10 µm across (left) and the habitus (right).

15. The dorsal margins of the femora, tibiae, and tarsi are all lined with long, fine setae. The length of the caudal filament is from 15% to 60% of the length of the cerci. The subapical denticle on the tarsal claws is minute. There are creases on the surfaces of the tergites, which have triangular processes along the posterior margins (**Fig. 2.220**).
..................................*Deceptiviosa* Lugo-Ortiz and McCafferty, 1999..p. 230
- The dorsal margin of the tibiae and tarsi are not lined by long, fine setae (**Fig. 2.221**). The length of the caudal filament is at least 65% of the length of the cerci. 16

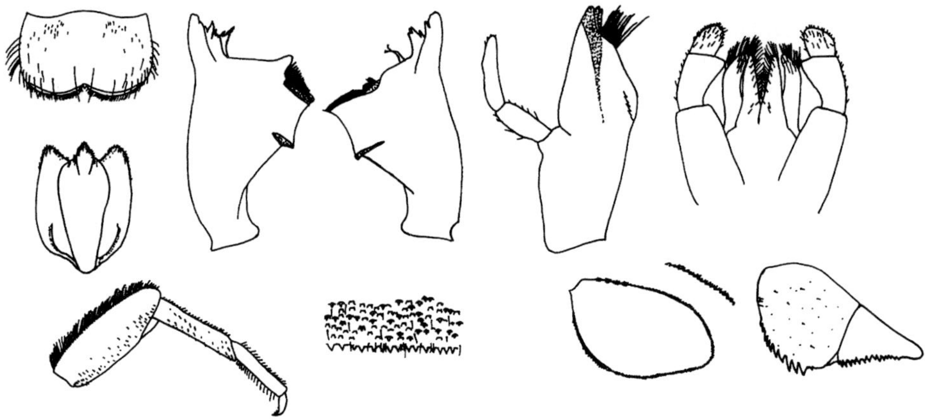

Fig. 2.221 *Andesiops peruvianus* larva (upper row, left to right): the labrum (above) and hypopharynx (below), mandibles, maxilla, labium in ventral view; (lower row, left to right): fore-leg, posterior margin of the fourth tergite, gill from the fourth abdominal segment with an enlargement of its margin, and a paraproct.

16. The apical segment of the maxillary palp is broadly rounded. The second segment of the labial palp is produced on the inner side at the apex. There are long, fine setae on the dorsal margins of the femora. The posterior margins of the tergites are zigzag with triangular processes. The caudal filament is 65% to 70% the length of the cerci. There is no minute paddle-like seta near the apex of the hind tibia. The hypopharynx is distinctly narrower than long and has a distinct median process covered by minute setae at the apex (**Fig. 2.221**). The only species described reaches a length of about 7 to 9.5 mm and bears caudal filaments from 3 to 4 mm. The larva is cream or brown without a distinct pattern.
....*Andesiops* Lugo-Ortiz and McCafferty, 1999

The only known species in this genus is *Andesiops peruvianus* (Ulmer, 1920) from Venezuela, Colombia, Ecuador, Peru, Bolivia, Chile, and Argentina. Syn: *Baetis peruvianus* Ulmer, 1920; *Baetis weiseri* Navás, 1922.

- The apical segment of the maxillary palp is acutely pointed. The second segment of the labial palp is not or only slightly produced on the inner side at the apex The hypopharynx is nearly as wide as long or wider, and if a median projection is present, it is small and not covered by minute setae (**Fig. 2.222**). The caudal filament is 80% to 85% the length of the cerci. 17

17. There are no long, fine setae on the dorsal margins of the femora, but a few short thick ones may be present. The posterior margins of the tergites are straight and smooth without triangular processes. Scales are abundant on the body surfaces. There is a minute paddle-like seta near the apex of the hind tibia

(**Fig. 2.222**). The only species described reaches a length of about 7.5 to 9 mm and bears caudal filaments from 4.5 to 5 mm. The head is cream or brown, and the rest of the body is yellowish brown or brown with variable markings.
....*Nanomis* Lugo-Ortiz and McCafferty, 1999

The only known species in this genus is *Nanomis galera* Lugo-Ortiz and McCafferty, 1999, from Ecuador, Peru, Colombia, and Argentina.

- There is a dense row of fine setae lining the dorsal margins of the femora. The posterior margins of the tergites are zigzag with triangular processes. The caudal filament is almost 80% the length of the cerci. The right mandible has a long, strong, pilose seta that arises at about midlength of the prostheca and a dense group of simple setae between the molar area and the prostheca. A long seta is located near the base of the two relatively large subapical denticles on each tarsal claw (**Fig. 2.223**). There is no minute paddle-like seta near the apex of the hind tibia. The only species known is 4.0 to 5.1 mm long and has caudal filaments 2.3 to 2.9 mm long.
....*Spiritiops* Lugo-Ortiz and McCafferty, 1998

The only known species in this genus is *Spiritiops silvudus* Lugo-Ortiz and McCafferty, 1998, from French Guiana and Amazonas. It is yellowish brown to medium brown with faint markings.

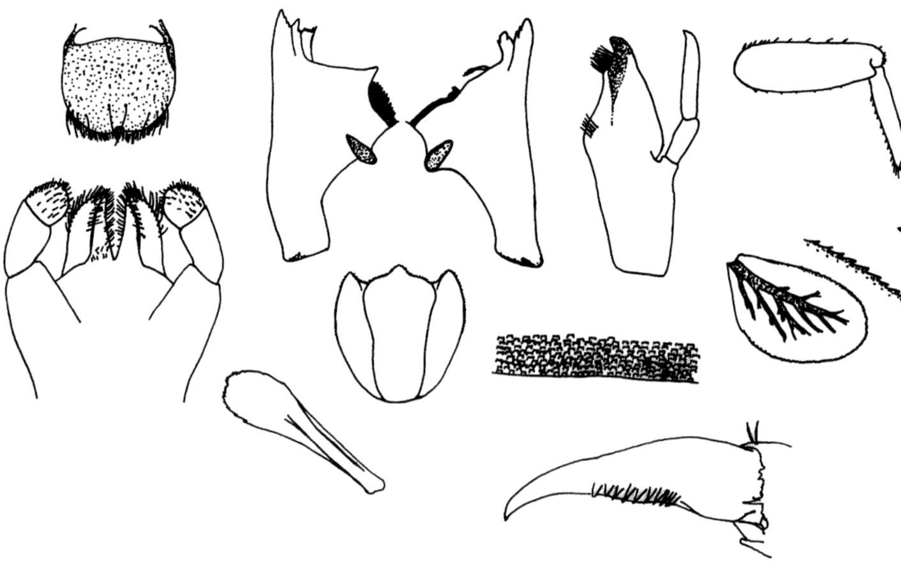

Fig. 2.222 *Nanomis galera* larva (upper row, left to right): labrum, mandibles, maxilla, fore-leg; (middle row, left to right) labium in ventral view, hypopharynx, posterior margin of the fourth tergite, the gill from the fourth abdominal segment with an enlargement of its margin; (below): a paddle-like seta from the subapical region of the hind tibia measuring between 20 and 30 μm (left) and a tarsal claw (right).

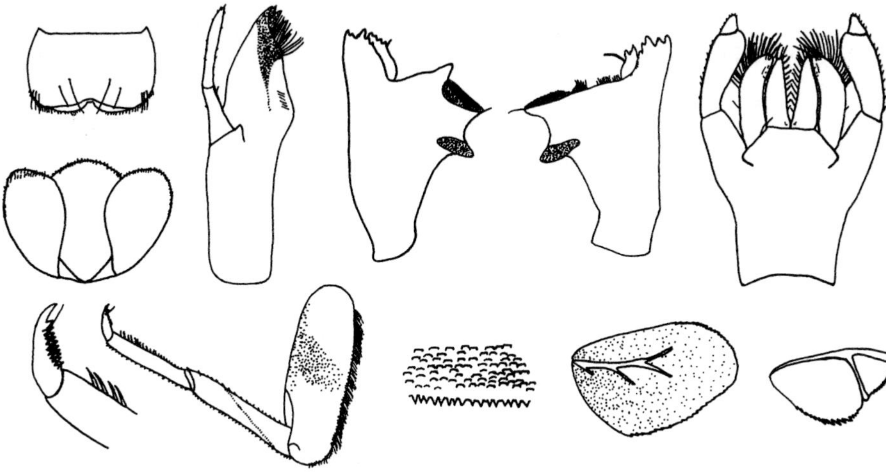

Fig. 2.223 *Spiritiops silvudus* larva (upper row, left to right): labrum (above) and hypopharynx (below), maxilla, mandibles, labium, and (lower row, left to right): fore-leg with an enlargement of the tarsal claw, an enlargement of the denticles and sculpture on the fourth abdominal tergite, the fourth gill, and a paraproct.

18. There is a spur in the middle of the mesosternum. The maxillae are elongate and robust; they have only sharp, slender denticles on the galealaciniae. The second segment of the maxillary palp is slender and nearly 1.5 times as long as the first segment. The palp itself is set on a prominent, extended base (**Fig. 2.224**).
......................................*Waltzoyphius* Lugo-Ortiz and McCafferty, 1995
The only species yet described in this genus is *Waltzoyphius fasciatus* Lugo-Ortiz and McCafferty, 1995, from Paraguay, Pará, and Amazonas. The body length of its type specimens ranges from 4.1 to 5.0 mm. Its caudal filaments are 2.1 to 2.3 mm. It is yellowish brown without a distinguishable pattern except for brown markings on the second through ninth tergites.
- There is no spur in the middle of the mesosternum, and the second segment of the maxillary palp is not nearly 1.5 times as long as the first or nearly twice as long (**Figs. 2.225, 2.226**). 19
19. The labrum has the approximate shape of a trapezoid and lacks a medial emargination on the anterior edge. The fused mandibular incisors are armed with minute lateral denticles (**Fig. 2.225**).
......................................*Tomedontus* Lugo-Ortiz and McCafferty, 1995
The only species yet described in this genus is *Tomedontus primus* Lugo-Ortiz and McCafferty, 1995, from Pará and Amazonas. The body length of its type specimens ranges from 4.8 to 5.1 mm. The caudal filaments are 5.8 to 6.0 mm. It is yellowish brown without any distinctive pattern.

- The labrum is not trapezoidal in shape, or an emargination is present in the center (**Fig. 2.209**). 20

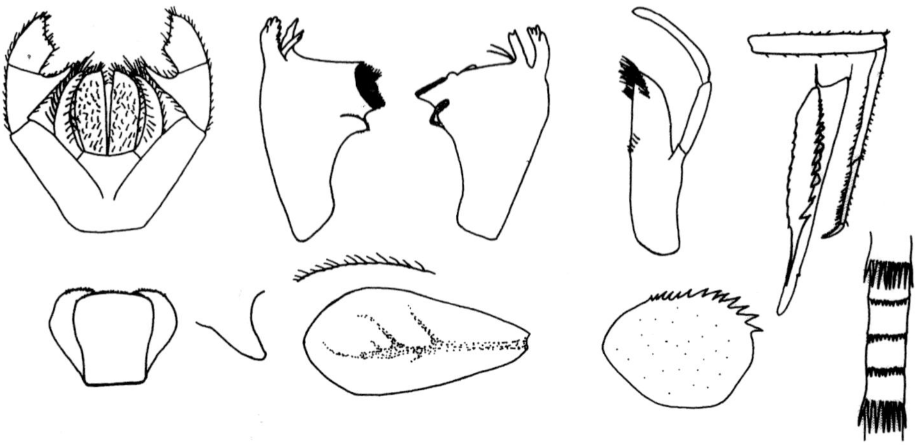

Fig. 2.224 *Waltzoyphius fasciatus* larva (upper row, left to right): labium in ventral view, mandibles, right maxilla, fore-leg with an enlargement of the tarsal claw in ventral view; (lower row, left to right): hypopharynx, outline of spur between the middle legs on the mesosternum, a gill from the sixth pair with an enlargement of the margin, a paraproct, and details of the cercal segments.

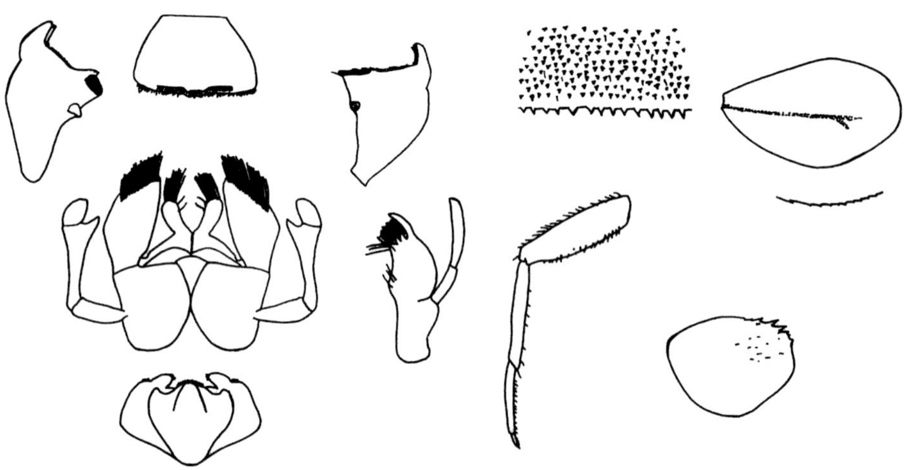

Fig. 2.225 *Tomedontus primus* larva: labrum (upper left center); left and right mandibles to the left and right of the labrum, respectively; labium in ventral view (middle left); hypopharynx (lower left); right maxilla (right of labium); surface of the fourth tergite (upper right center); foreleg (lower right center); the gill plate on the fifth abdominal segment with an enlargement of its edge (upper right); a paraproct (lower right).

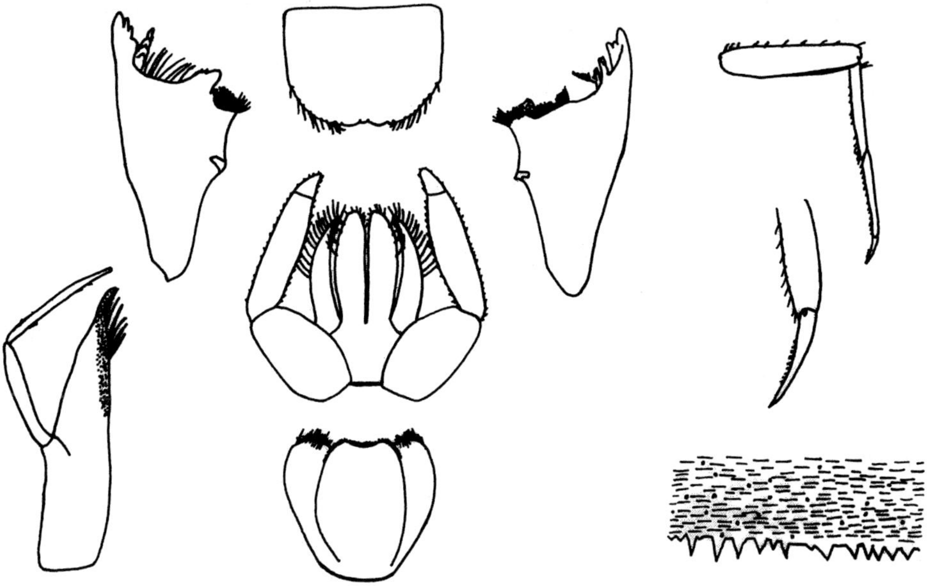

Fig. 2.226 *Adebrotus amazonicus* larva: labrum (upper left center), labium (left center), hypopharynx (lower left center), left and right mandibles (left and right of labrum, respectively), right maxilla (lower left), right foreleg with an enlargement of the tarsal claw (upper right), and the surface of a tergite (lower right).

20. There are small lobes and no fine setae in the middle of the anterior margin of the labrum. The inner row of incisors on the left mandible is set perpendicular to the outer row (**Fig. 2.226**).
....*Adebrotus* Lugo-Ortiz and McCafferty, 1995
The only known species in this genus is *Adebrotus amazonicus* Lugo-Ortiz and McCafferty, 1995, from Amazonas.
- There are no small lobes, and there are usually fine, simple setae on the anterior margin of the labrum. The inner and outer rows of incisors are not set perpendicular to one another (**Fig. 2.209**). 21
21. The inner margin of each claw bears a row of denticles followed by a pair of large teeth and long setae. The labrum is narrowed basally and bears long, plumose marginal setae. The gill on the first abdominal segment is smaller than the others. The terminal filament is well developed (**Fig. 2.23**). Last instar larvae are usually 9 to 12 mm long. 22
- The claws lack long sub-apical setae at the end of a row of denticles (**Fig. 2.227**). 23

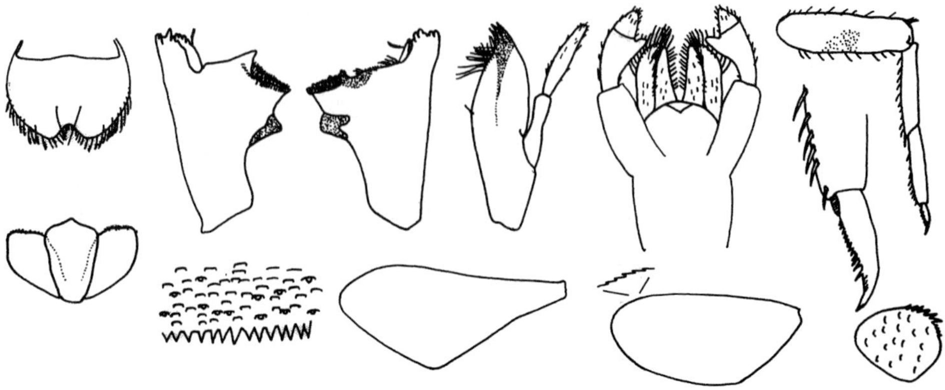

Fig. 2.227 *Cryptonympha copiosa* larva (upper row, left to right): labrum, mandibles, maxilla, labium, fore-leg with enlarged tarsal claw below; (lower row, left to right): hypopharynx, posterior part of an abdominal tergite, first gill, fourth gill with an enlargement of its margin, and the paraproct.

22. The head and antennae are about equally long. The median caudal filament is half as long as the cerci or shorter. There is a row of fine setae on the outer margin of the scape and pedicel, the two basal segments of the antenna. The pedicel is broadened, and there is a row of blunt, scale-like tuberculations on its dorsal side. There is no raised area on the mediobasal side of the labrum. The maxillary palp has a weakly developed third segment **(Fig. 2.23)**.
....*Mayobaetis* Waltz and McCafferty, 1985

Only one of the species presently assigned to this genus occurs in South America: *Mayobaetis ellenae* (Mayo, 1973) from Costa Rica, Ecuador, and Peru. This species has been referred to as *Baetis ellenae* Mayo (1973) and *Moribaetis (Mayobaetis) ellenae* (Mayo, 1973).

- The antenna is much longer than the head. The median caudal filament is much longer than half the length of the cerci. The scape and pedicel bear neither a row of fine setae nor scale-like tuberculations. There is a raised area on the mediobasal side of the labrum. The maxillary palp clearly consists of two segments. The larvae of the known South American species have not been described.
....*Moribaetis* Waltz and McCafferty, 1985..p. 212

23. Hind wing pads are present **(Fig. 2.209)**. 24

- Hind wing pads are absent **(Fig. 2.228)**. 26

24. A "villopore," which is a small patch of small setae, is present on the ventral side of each femur near its base. If the median caudal filament is present at all, it is considerably shorter than the cerci.

....*Baetis* Leach, 1815

Most South American species formerly assigned to *Baetis* have been transferred to other genera. Because the larvae of the species still listed under *Baetis* are unknown, it is uncertain whether they would fit in here or show characteristics of other genera, as presented in the key. Lugo-Ortiz and McCafferty (1998) stated bluntly that *Baetis* does not occur in South America. This seems to imply that they already know all species in the South American mayfly fauna, which would certainly be a premature claim.

- There is no patch of small setae near the base of each femur. 25

25. The tarsal claws of the known South American species are free of denticles, or, if present, the denticles are few and inconspicuous. The median caudal filament is more than half as long as the cerci **(Fig. 2.209)**.

....*Cloeodes* Traver, 1938 pars..p. 224

- The tarsal claws of species not found in South America usually bear nine or more denticles.

....*Fallceon* Waltz and McCafferty, 1987..p. 210

The larvae of the two known South American species have never been described. From the information available, the distinction of this nominal genus from *Cloeodes* seems dubious, at best.

26. There are long, strong feathered setae on the prostheca of the mandible and along the apical margin of the labrum. The maxillary palp consists of two segments. The basal section of the first gill is narrow **(Fig. 2.227)**.

....*Cryptonympha* Lugo-Ortiz and McCafferty, 1998

The only species in this genus is *Cryptonympha copiosa* Lugo-Ortiz and McCafferty, 1998, from Pará, Amazonas, Paraná, Santa Catarina, and the Territory of Acre. It reaches a length of 4.3 to 5.5 mm and has a caudal filament from 1.8 to 2.3 mm long. The color is mainly yellowish brown with a pattern of vermiform markings.

- The setae on the prostheca of the mandible and along the margin of the labrum are not obviously feathered **(Fig. 2.228)**. 27

27. The long setae present on most abdominal sternites (ventral plates) are arranged in distinct tufts. The abdomen of the only South American species of the genus without hind wing buds, *Cloeodes redactus*, is dark with a pattern of circular light markings **(Fig. 2.229)**.

....*Cloeodes* Traver, 1938 pars..p. 224

- The long setae present on most abdominal sternites (ventral plates) are not arranged in distinct tufts. The thumb of the left mandible is elevated above the plane of the incisor bases **(Fig. 2.228)**.

....*Bernerius* Waltz and McCafferty, 1987

The only known species in this genus is *Bernerius incus* Waltz and McCafferty, 1987, from Peru, which reaches a length of 10 to 11 mm. Considering the relatively minor differences between the larvae of this species and those of the known species of *Cloeodes* and the fact that the adult of *B. incus* is unknown, placing this species in a new genus seems premature.

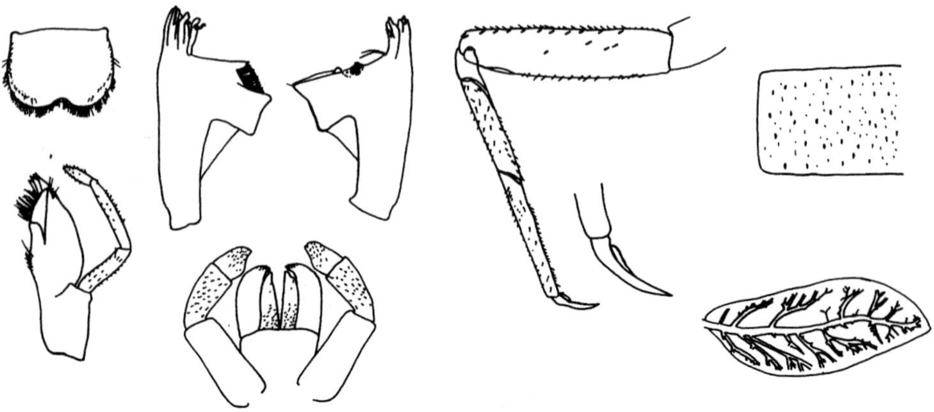

Fig. 2.228 *Bernerius incus* larva (upper row, left to right): the labrum in ventral view, mandibles, fore-leg with an enlargement of the tarsal claw, and a sternite with scattered scales and small setae but without a tuft of long setae; (lower row, left to right) a maxilla, labium in ventral view, and fourth abdominal gill.

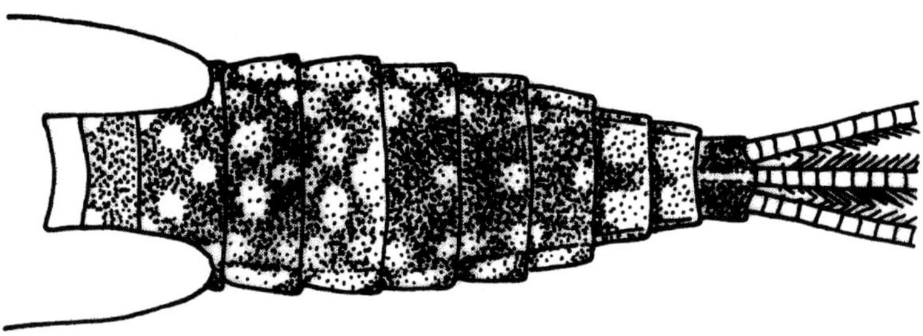

Fig. 2.229 The color pattern on the abdominal tergites of a *Cloeodes redactus* larva. The spines on the segments are not shown.

Key to the known species of adult and sub-adult *Americabaetis* in South America

Information for the key was taken from Weyenbergh (1883), Esben-Petersen (1909), Navás (1922b, 1923a, 1926b, 1932), Hubbard (1974), Lugo-Ortiz and McCafferty (1994), Waltz and McCafferty (1999), and McCafferty (2000). Adult *Americabaetis labiosus, A. longetron, A. maxifolium,* and *A. titthion* remain undescribed.

1. The length is 6 to 6.5 mm, and the fore-wing is 5 to 6.5 mm long. The anterior part of the first abdominal segment of the adult male is brown with a

dark W-shaped marking. The rest of the abdomen is white, with a reddish tint dorsally (**Fig. 2.230**). The abdomen of the male sub-adult is brownish yellow dorsally and grayish yellow ventrally. The abdomen of the female is brownish yellow dorsally and ventrally, and that of the female sub-adult is reddish brown on the dorsum and paler ventrally.
....*Americabaetis peterseni* (Hubbard, 1974) (Argentina). Syn: *Cloeon peterseni* Hubbard, 1974; *Pseudocloeon peterseni* (Hubbard, 1974); *Cloeon brunneum* Esben-Petersen (non Rambur), 1909.
- The length is not greater than 5.5 mm. If it approaches 6 mm in length, then the abdomen is grayish brown dorsally. 2

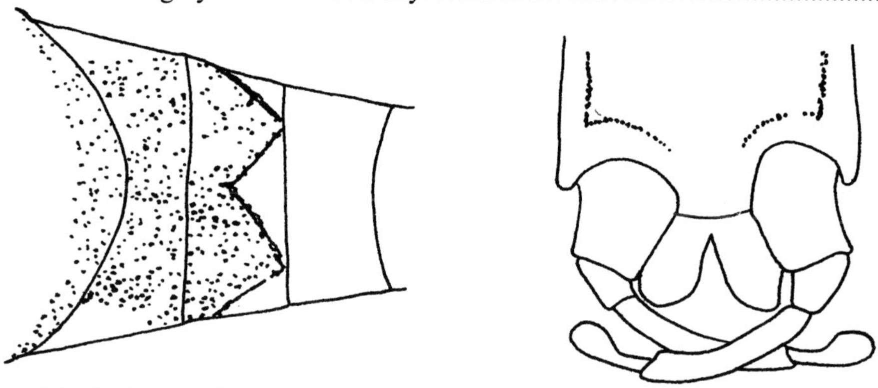

Fig. 2.230 *Americabaetis peterseni* adult male: the metanotum and first two abdominal tergites showing the pattern (left) and the genitalia (right).

2. The intermediate segments of the abdomen are yellow, while the first and last two are dark. The head is dark, and the thorax is blackish dorsally and paler ventrally. The wing veins are pale and faint. There are six or seven oblique cross veins in the stigmatic area. Length: 2.5 mm. Wing length: 3.6 mm.
....*Americabaetis bruchi* (Navás, 1926) (Argentina). Syn: *Pseudocloeon bruchi* Navás, 1926. Because of its poor description, this name is considered a *nomen dubium* (Domínguez *et al*., 2001).
- The abdomen is uniform grayish or browning dorsally, sometimes with darkened posterior margins of the tergites, or its second through sixth abdominal tergites have unpigmented posterior margins with the unpigmented area increasing toward the midline. 3
3. There is an unpigmented area along the posterior margins of the second through sixth abdominal tergites that extends increasingly farther anteriad toward the midline. The anterior and posterior parts of the turbinate eyes of the male are separated an equal distance from each other.
.... *Americabaetis alphus* Lugo-Ortiz and McCafferty, 1996 (Bolivia, Paraguay, Argentina, Brazil).
- The abdomen is almost entirely uniform grayish or brownish dorsally, usually rather dark and sometimes with darkened posterior margins. 4

4. The head is a shiny dark brown; the thorax is grayish. The abdomen is grayish brown. The wing apparently has a long, faint intercalary vein between Sc and R (**Fig. 2.231**). Length: 5 to 5.5 mm. .... *Americabaetis oldendorffii* (Weyenbergh, 1883) (Argentina, Brazil). This species, originally described as *Oxycypha oldendorffii* Weyenbergh, 1883; *Pseudocloeon oldendorffii* (Weyenbergh, 1883); *Cloeon oldendorffii* (Weyenbergh, 1883); *Baetis oldendorffii* (Weyenbergh, 1883).
- The thorax is not grayish, and the abdomen is not grayish brown. 5

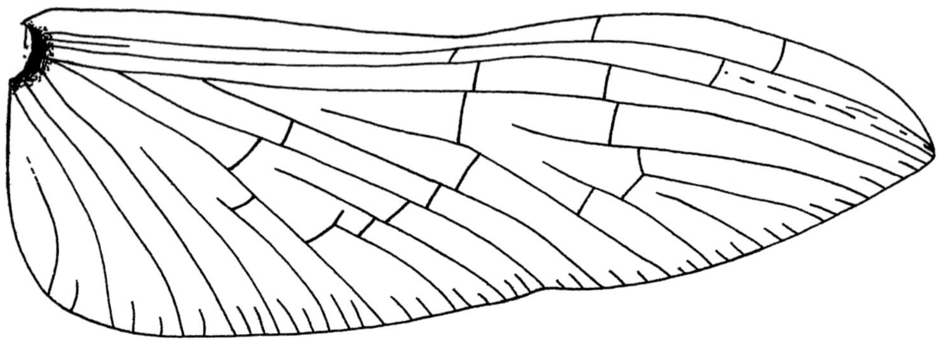

Fig. 2.231 The wing of *Americabaetis oldendorffii.*

5. The dorsal coloration of the thorax is dark brownish with yellow borders of the individual thoracic segments. The head is dark reddish brown with a blackish vertex and dark basal antennal segments. The apical segments of the antenna are whitish. The abdomen is reddish brown dorsally with blackish brown borders of the segments. The underside is yellowish brown. The legs are pale yellow with whitish tibiae. The wings are hyaline with whitish veins. Length: 4.2 mm. Fore-wing length: 4.1 mm. .... *Americabaetis albinervis* (Navás, 1923) (Chile). Syn: *Pseudocloeon albinerve* Navás, 1923.
- The thorax and abdomen, and sometimes the head, as well, are almost entirely brownish. The bases of the male forceps in all species with known males are short and cylindrical or tapering, but the males of all species have not yet been described (**Fig. 2.232**). 6
6. The thorax is reddish brown with a dark midline stripe. The head is reddish brown with darker eyes. The abdomen is yellowish brown with darker bands across the posterior margins of the segments dorsally and yellow ventrally. Most wing veins are white, and three to five of them in the stigmatic area are somewhat thickened. Length: 4 mm. Wing length: 4.5 mm. .... *Americabaetis bridarolli* (Navás, 1932) (Argentina). Syn: *Pseudocloeon bridarolli* Navás, 1932.

- The thorax is either gray or brown without a longitudinal stripe, and the head and abdomen are almost entirely brownish dorsally, although the posterior margins of some abdominal tergites may be darkened. 7

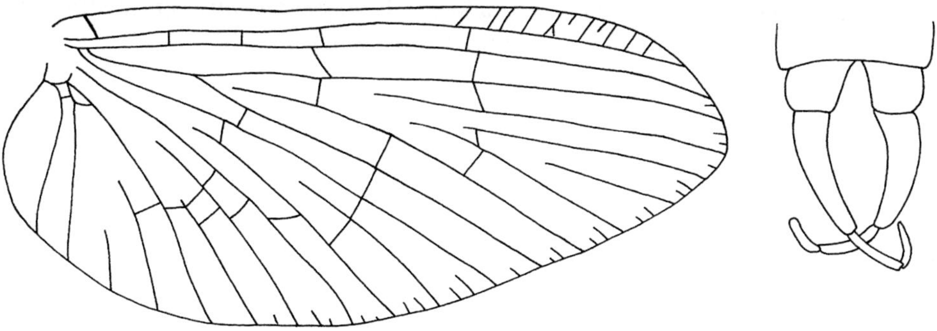

Fig. 2.232 *Americabaetis jorgenseni:* the wing with the veins much darker than they appear in a life (left) and the genitalia of a male (right).

7. The length of both males and females is 2.5 to 3 mm, and the fore-wing is 3.3 to 3.5 mm long. The male has yellowish ellipsoidal turbinate eyes, the posterior parts of which are closer together that the anterior parts (**Fig. 2.196**). The thorax and abdomen are uniform light to medium brown dorsally. The abdomen is light yellow ventrally. The basal part of the subcostal area is opaque reddish brown, and the costal area is opaque. The rest of the wing is hyaline with colorless veins.
.... *Americabaetis robacki* (Lugo-Ortiz and McCafferty, 1994) (Uruguay). Syn: *Acerpenna robacki* Lugo-Ortiz and McCafferty, 1994.
- The length is at least about 4 mm, and the fore-wing is at least 5 mm long. 8
8. The venation of the wing is barely perceptible (**Fig. 2.232**). Body length: 4.5 to 5 mm. Wing length: 5 to 6 mm.
....*Americabaetis jorgenseni* (Esben-Petersen, 1909) (Argentina). Syn: *Cloeon jorgenseni* Esben-Petersen, 1909, *Pseudocloeon jorgenseni* (Esben-Petersen, 1909).
- The venation of the wing is very obvious (**Fig. 2.233**). Length: of the female: 4 mm. Wing length of the female: 5.5 mm. The male has not been described.
....*Americabaetis weiseri* (Navás, 1922) (Argentina). Syn: *Pseudocloeon weiseri* Navás, 1922.

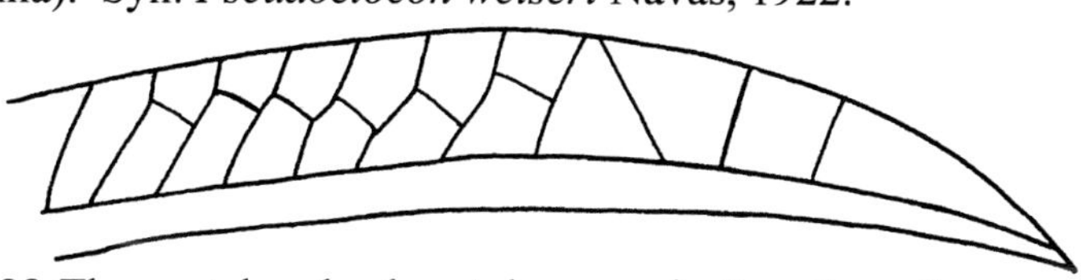

Fig. 2.233 The costal and subcostal spaces in the stigmatic area at the wing tip of *Americabaetis weiseri.*

Key to the species of *Americabaetis* larvae in South America

Information for the key is taken from Lugo-Ortiz and McCafferty (1996a) and Waltz and McCafferty (1999). Larvae of *Americabaetis bridarolli, A. jorgenseni, A. oldendorffi,* and *A. peterseni* remain undescribed.

1. A frontal keel is absent from the head. 2
- A frontal keel is present on the head (**Fig. 2.211**). 4
2. The second segment of the labial palp is much less than twice as long as wide at its apex.
.......................*Americabaetis robacki* (Lugo-Ortiz and McCafferty, 1994) (Peru). Syn: *Acerpenna robacki* Lugo-Ortiz and McCafferty, 1994.
- The second segment of the labial palp is at least twice as long as wide at its apex (**Fig. 2.234**). 3

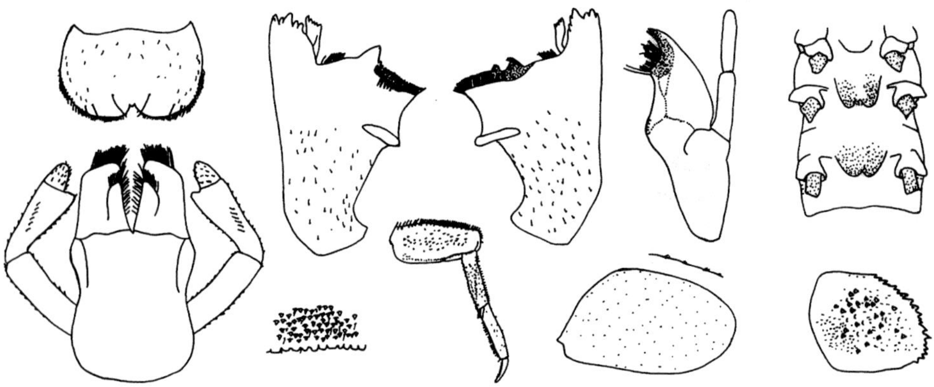

Fig. 2.234 *Americabaetis titthion* larva (upper row, left to right): labium, mandibles, maxilla, underside of the thorax showing the paired protuberances on the mesosternum and metasternum; (lower row, left to right): labium in dorsal view, pattern along the posterior margin of the fourth tergite, the leg, a gill plate from the fourth abdominal segment with an edge enlarged, and a paraproct.

3. There are paired protuberances covered with setae on the mesosternum and metasternum. The labrum is nearly twice as long as wide (**Fig. 2.234**). Length: 6.7 to 7.5 mm. Caudal filament length: 3.5 to 4.0 mm. Color: yellow brown with no distinct pattern, although the fourth abdominal segment is usually lighter in color than the others.
.........................*Americabaetis titthion* Lugo-Ortiz and McCafferty, 1996 (Paraná, Santa Catarina, Rio Grande do Sul).

- The mesosternum and metasternum lack paired setose protuberances. The length of the labrum is only about 1.2 times the width (**Fig. 2.235**). Length: 6.7 to 7.5 mm. Caudal filament length: 4.5 to 4.8 mm. Color: yellow brown, usually with no distinct pattern.
..........................*Americabaetis labiosus* Lugo-Ortiz and McCafferty, 1996 (Paraguay, Uruguay, Paraná, Santa Catarina, Rio Grande do Sul).

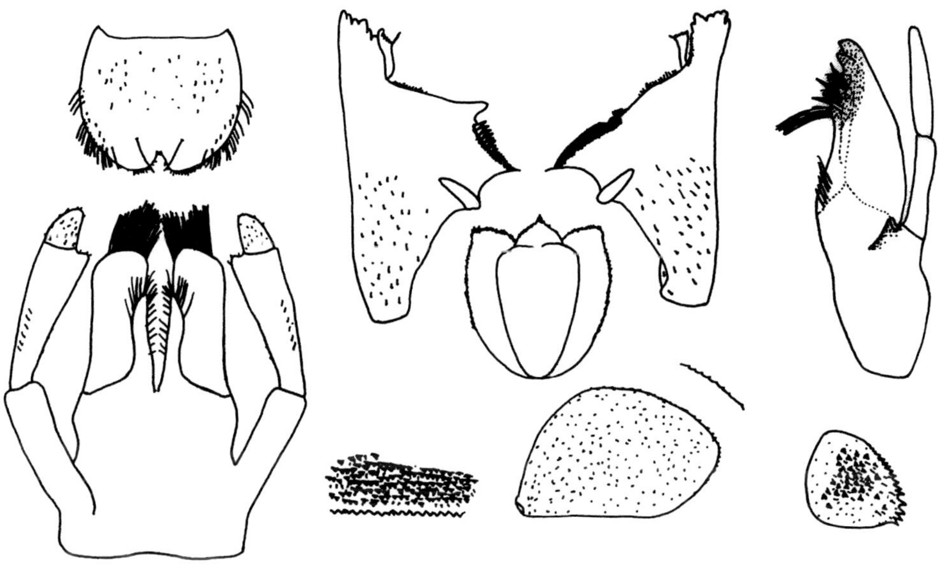

Fig. 2.235 *Americabaetis labiosus* larva (upper row, left to right): labium, mandibles, maxilla; (center, between mandibles): hypopharynx; (lower row, left to right): labium in dorsal view, pattern along the posterior margin of the fourth tergite, a gill plate from the fourth abdominal segment with an edge enlarged, and a paraproct.

4. The gills are 2.5 to 3 times as long as the tergite to which they are attached , and they bear few serrations along the margin (**Fig. 2.236**). Length: 3.9 to 4.1 mm. Caudal filament length: 1.8 to 2.0 mm. Color: pale yellow without any distinct pattern.

........................*Americabaetis maxifolium* Lugo-Ortiz and McCafferty, 1996 (Paraguay).
- The gills are only 1.5 to 2 times as long as the tergite to which they are attached, and the serrations on their margins are dense (**Fig. 2.237**). 5

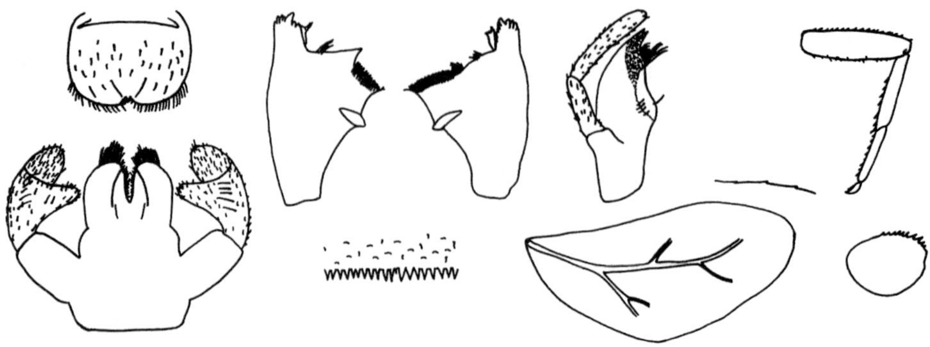

Fig. 2.236 *Americabaetis maxifolium* larva (upper row, left to right): labrum, mandibles, maxilla, fore-leg; (lower row, left to right): labium in dorsal view, pattern along the posterior margin of the fourth tergite, a gill plate from the fourth abdominal segment with an edge enlarged, and a paraproct.

5. The paraprocts each bear six or seven sharp, elongate spines. The males have distinctive white areas on the first, fourth, and seventh tergites, and the females have white lateral markings on the second through eighth tergites. (**Fig. 2.211**). Length: 4.2 to 5.8 mm. Caudal filament length: 2.7 to 3.0 mm. Color: head yellowish brown; thorax and abdomen mainly yellow brown to brown with lighter and darker markings.

............................*Americabaetis alphus* Lugo-Ortiz and McCafferty, 1996 (Paraguay, Bolivia, Chile, Argentina, Minas Gerais, Paraná, Santa Catarina, Rio Grande do Sul).

- The paraprocts each bear 11 to 13 slender spines, which increase in length distally. The males have distinctive white areas on the first, fourth, and seventh tergites, and the females have white lateral markings on the second through eighth tergites. There are 22 to 25 long, acute setae on each femur (**Fig. 2.237**). Length: 5.5 to 6.5 mm. Caudal filament length: 3.6 to 4.1 mm. Color: head yellow; thorax and abdomen mainly pale yellow brown.

.........................*Americabaetis longetron* Lugo-Ortiz and McCafferty, 1996 (Paraguay, Uruguay, Minas Gerais, Paraná, Santa Catarina). The Central American species, *Americabaetis lugoi* Waltz and McCafferty, 1999, is somewhat similar to this species but bears 8 to 10 irregular marginal spines on the paraprocts and 18 to 20 blunt bristles on each femur.

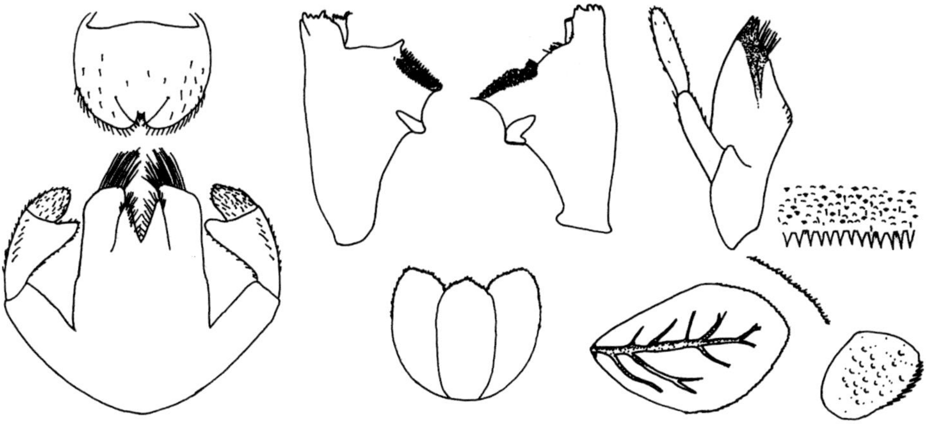

Fig. 2.237 *Americabaetis longetron* larva (upper row, left to right): labrum, mandibles, maxilla, pattern along the posterior margin of the fourth tergite; (lower row, left to right): labium in dorsal view, hypopharynx, gill plate from the fourth abdominal segment with an edge enlarged, and a paraproct.

Key to the species of adult *Baetodes* in South America

Information for the key was provided by Traver (1943). Most adults have not been described.

1. The head, thorax, and apex of the abdomen are light reddish brown. The thorax has yellow markings outlined in brown. The base of the male forceps bears a process directed inward which ends in an acute point and does not bear prominent setae (**Fig. 2.197**). Body length: 3.5 mm. Wing length: 3.5 mm. ....*Baetodes arawak* (Traver, 1943) (Venezuela). Syn: *Pseudocloeon arawak* Traver, 1943.

- The general color of the male adult is reddish brown with a purplish tinge. The anterior margin of the pronotum is covered by a wide black band. The female subadult differs from the male in having a yellow head. The process at the base of the male forceps is round and bears some short, spine-like setae (**Fig. 2.238**). Length of male: 4 mm. Fore-wing length of male: 4 mm. Length of female: 4.5 mm. Fore-wing length of female: 4.5 mm.

....*Baetodes spiniferum* Traver, 1943 (Venezuela). Syn: *Baetodes spinifer* (Traver, 1943).

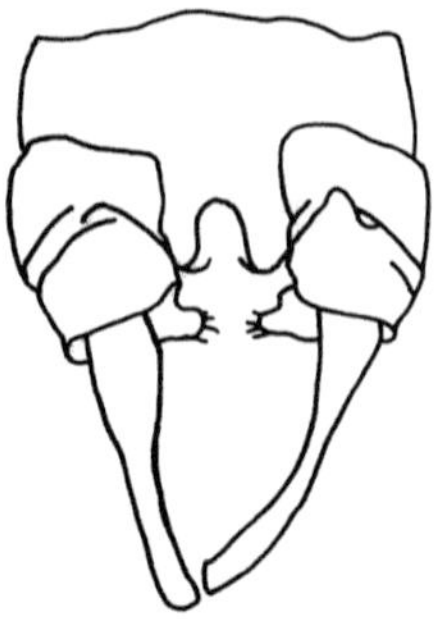

Fig. 2.238 The male genitalia of *Baetodes spiniferum.*

Key to the species of *Baetodes* larvae in South America

Information for the key was taken from Needham and Murphy (1924) and Mayo (1968b, 1972, 1973a). Descriptions of *Baetodes arawak* and *Baetodes spiniferum* are not available.

1. The head and mouthparts are elongated. The tarsi bear long spurs. Length of the female: 4 mm, cerci: 6 mm. Color: yellow with brown markings. ....*Baetodes longus* Mayo, 1973 (Mexico, not reported from South America).

- The head and mouthparts are not elongated. The maxillary and labial palpi are very weak, and the labial glossae are very reduced (**Fig. 2.239**). 2

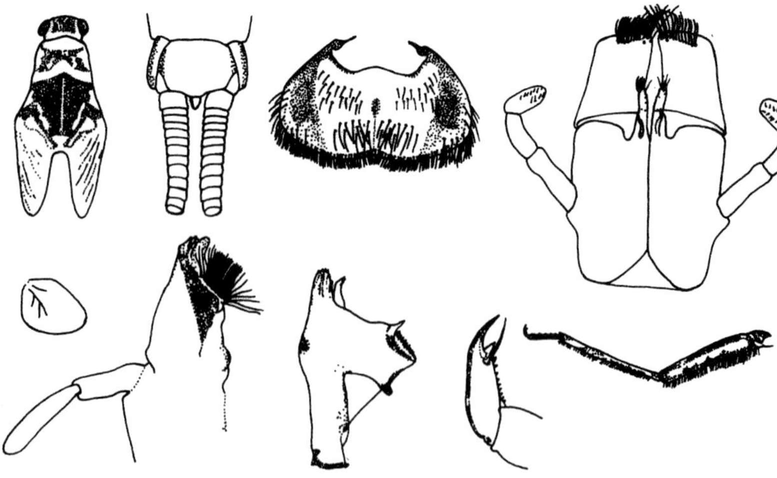

Fig. 2.239 *Baetodes traverae* larva (upper row, left to right): pattern on the head and thorax, apex of the abdomen with basal segments of the cerci, labrum, and labium; and (lower row, left to right): fourth gill, apex of a maxilla, left mandible, apex of a tarsal claw, and a middle leg.

2. There are only two cerci, the terminal filament being so short that it is not evident when the larva is viewed dorsally (**Fig. 2.239**). Length of male body: 5 mm, cerci: 5 mm; length of female body: 6 mm; cerci: 7 mm. Color: mainly dark brown with yellow markings.
....*Baetodes traverae* Mayo, 1972 (Peru).
- There are three cerci, although the caudal filament may be much shorter than the two lateral cerci (**Fig. 2.240**). 3

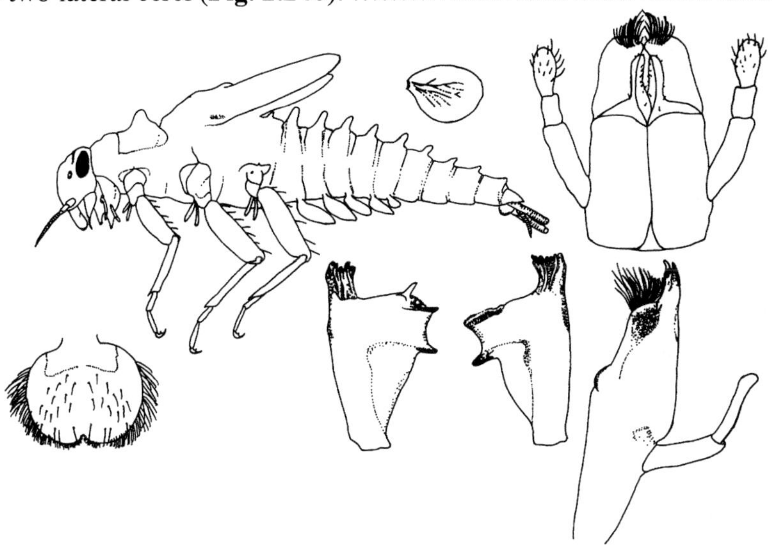

Fig. 2.240 *Baetodes spinae* larva: habitus without the apical portions of the cerci in lateral view showing the coxal gills and dorsal processes (upper left), the labrum (lower left), mandibles (lower center), maxilla (lower right), labium (upper right), and an abdominal gill (upper right center).

3. Dark brown spines encircle each segment of the cerci of the last instar, at least of the males (**Fig. 2.241**). 4
- There is no circle of dark spines on each segment of the cerci (**Fig. 2.240**). 5
4. There are prominent pale median tubercles on nine abdominal tergites. Each paraglossa bear fringed setae at the apex (**Fig. 2.241**). On less developed larvae, there is a species-specific round-tipped lobe at the rim of the sclerotized area around the apodeme on the second sternite. This structure disappears as the larvae become ready for metamorphosis. Length of last instar of male before metamorphosis, body 4 mm, cercal filaments: 6 mm. Body of female larva not ready for metamorphosis: 6 mm. Color: mainly dark brown with yellow markings.
....*Baetodes proiectus* Mayo, 1973 (Bolivia).

- There are no tubercles on the abdominal tergites. There is a pinnate seta at the apex of each glossa and a long, straight seta medially at the apex of each paraglossa (**Fig. 2.242**). Length of male body: 3 mm; length of female body: 4 mm; cerci: 4 mm. Color: mainly dark reddish brown with a yellow scutellum and other yellowish markings.
....*Baetodes sancticatarinae* Mayo, 1972 (Santa Catarina).

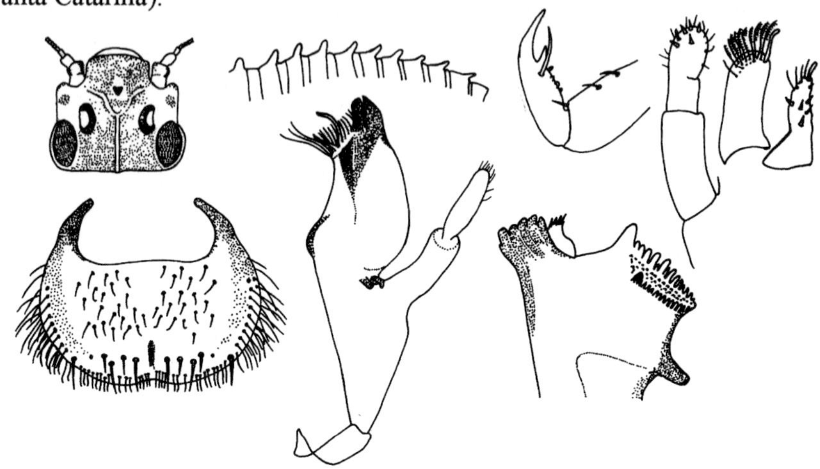

Fig. 2.241 *Baetodes proiectus* larva (upper row, left to right): the head in dorsal view, profile of the dorsal surfaces of the metanotum and abdomen showing the median tubercles, a tarsal claw, and the labial palp, paraglossa, and glossa of the labium; (lower row, left to right): the labrum, maxilla, and apex of a mandible.

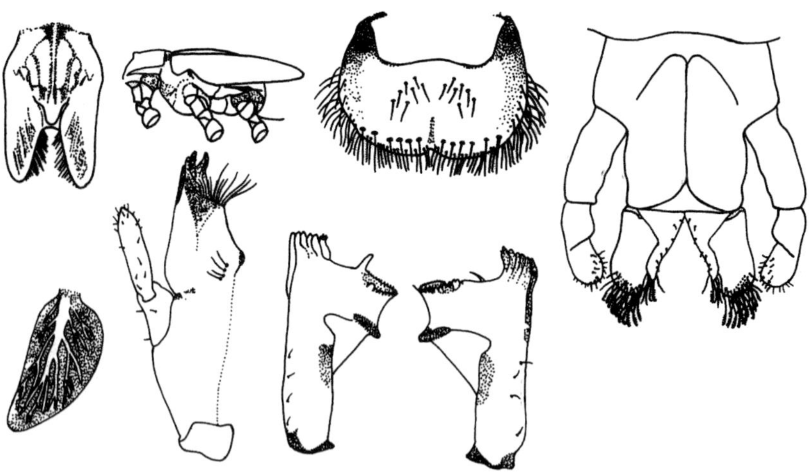

Fig. 2.242 *Baetodes sancticatarinae* larva (upper row, left to right): dorsal pattern on the thorax, thorax in lateral view, labrum, labium, and (lower row, left to right): the first gill, maxilla, and the two mandibles.

5. Ocelli are absent. The gills are inserted on the ventral sides of the abdominal segments (**Fig. 2.208**). Length: 6 mm.
....*Baetodes itatiayanus* Demoulin, 1955 (Rio de Janeiro).
- Ocelli are present (**Fig. 2.240**). 6
6. There are tubercles or hook-like processes on each of the first nine abdominal tergites (**Fig. 2.240**). 7
- The abdominal tergites lack tubercles or hook-like processes. 10
7. Each coxa bears two finger-like gills (**Fig. 2.240**). 8
- Coxal gills are absent (**Fig. 2.243**). 9

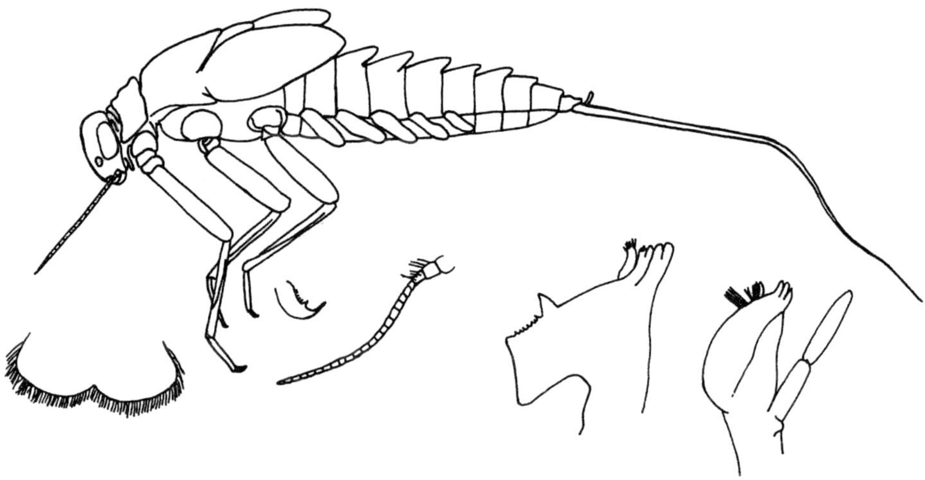

Fig. 2.243 *Baetodes serratus* larva: habitus in lateral view (above) with enlargements of the labrum, a claw, an antenna, a mandible, and a maxilla (below, left to right).

8. The pronotum and mesonotum bear large, blunt, thorn-like protuberances, which, like those on the abdominal sternites, are erect (**Fig. 2.240**). Length of body: 3 to 4 mm. Color: generally pale amber.
....*Baetodes spinae* Mayo, 1968 (Ecuador).
- The pronotum and mesonotum lack thorn-like protuberances, while those on the abdominal sternites, are directed posteriad and not too conspicuous (**Fig. 2.244**). Length of body: 3 to 4 mm. Color: generally pale reddish brown with yellow between the segments.
....*Baetodes levis* Mayo, 1968 (Ecuador).

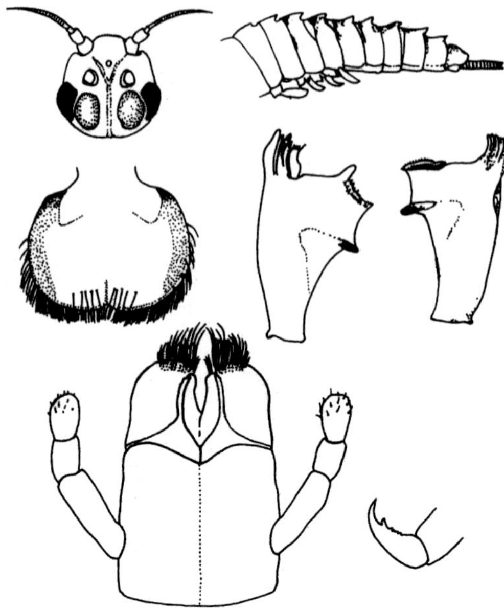

Fig. 2.244 *Baetodes levis* larva: head in dorsal view (upper left), profile of the abdominal segments (upper center), labrum (middle left), mandibles (center), maxilla (right), labium (lower left), and a tarsal claw (lower center).

9. The glossae are very much reduced and lack pinnate setae at the tips. The abdomen is generally a uniform greenish brown with white gills. Two ocelli are present. The gill lamellae are inserted on the lateral edges of the abdominal segments (**Fig. 2.243**). Length: 7 to 8 mm. Lateral caudal filament length: 10 mm; middle filament length: 1 mm. Color: mainly greenish brown.
....*Baetodes serratus* Needham and Murphy, 1924 (Rio de Janeiro).

- There is a blade-like pinnate seta at the tip of each glossa and a long, straight seta medially at the apex of each paraglossa (**Fig. 2.245**). The first three and last four abdominal tergites are brown, while the fourth through sixth are yellow dorsally and brown laterally. The gills are white. Body length: 5 to 6 mm. Coloration: mainly dark brown with yellow markings.
....*Baetodes peniculus* Mayo, 1973 (Venezuela).

10. Coxal gills are absent. The thoracic sternum is dark brown. The maxillary palp is barely as long as the width of the maxilla (**Fig. 2.246**). Length of female body: about 3 mm; the cerci are somewhat longer. Color: mainly yellowish brown with darker markings.
....*Baetodes solus* Mayo, 1972 (Peru).

- Coxal gills are present. The thoracic sternum is pale yellow with brown sclerites at the apodemes. The maxillary palp is much longer than the width of the maxilla (**Fig. 2.247**). 11

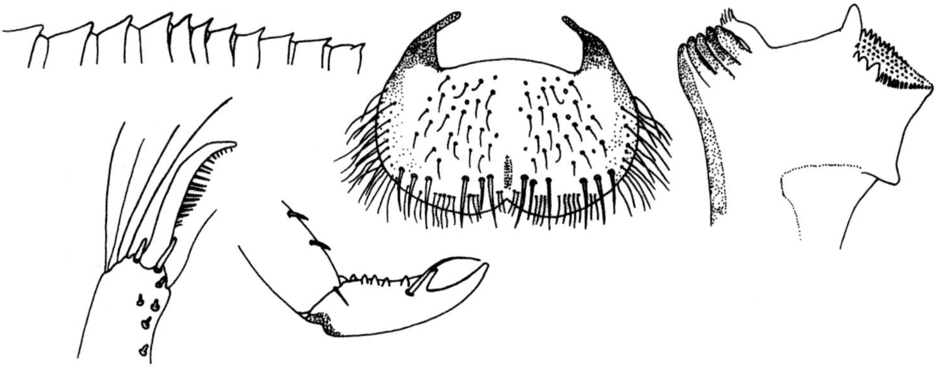

Fig. 2.245 *Baetodes peniculus* larva (upper row, left to right): dorsal profile of the metanotum and abdominal segments, labrum, and apical part of a mandible; the apex of a glossa with the blade-like pinnate seta (lower left), and a tarsal claw (lower center).

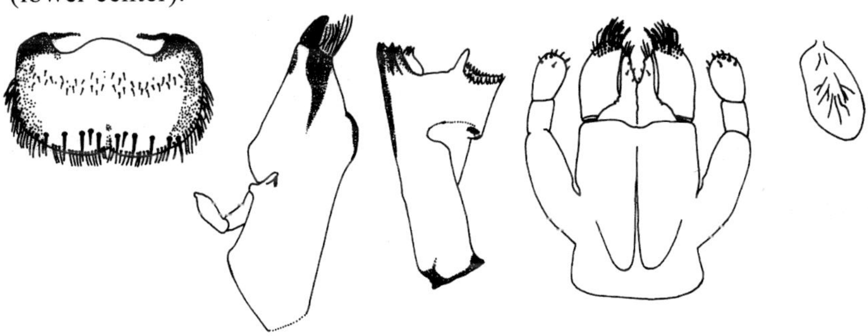

Fig. 2.246 *Baetodes solus* larva (left to right): labrum, maxilla, left mandible, labium, and a gill.

Fig. 2.247 *Baetodes chilloni* larva (left to right): labrum above a tarsal claw, maxilla, right mandible, and labium.

11. The first abdominal sternite is pale brown; the second through ninth are very light brown with darker markings on either side of the mid-line. All but two of the incisors on the mandible are not grouped into a single lobe (**Fig. 2.247**). Length of female body: about 4 mm; the cerci are somewhat shorter. Color: mainly brown dorsally with yellow markings. The abdomen is uniform brown dorsally, somewhat darker at the base of each tergite. Gills yellow. ....*Baetodes chilloni* Mayo, 1972 (Peru).

- The first abdominal sternum is pale yellow; the second through sixth are brown on median third of their width from the anterior to posterior borders and yellow with a dark marking on the anterior half of the lateral areas. The incisors on the inner margin of the mandible have two small lobes, while the rest are grouped into one single lobe (**Fig. 2.248**). Length of male body: 4.5 mm; the cerci are longer. Color: mainly dark brown dorsally with many yellow and reddish brown markings. ....*Baetodes andamagensis* Mayo, 1972 (Peru).

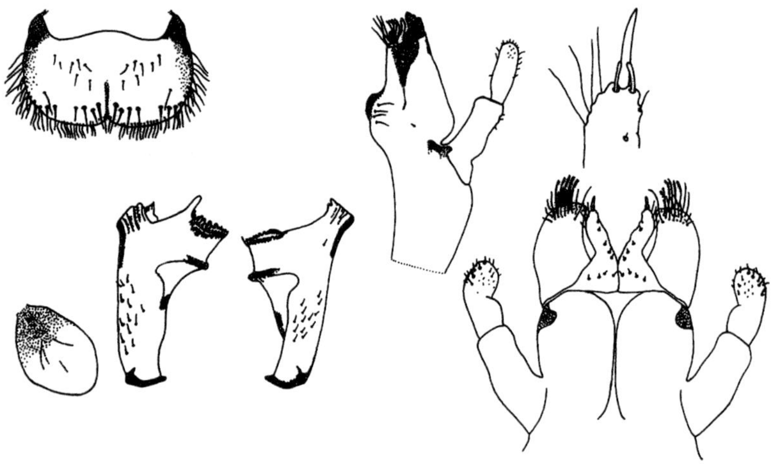

Fig. 2.248 *Baetodes andamagensis* larva (upper row, left to right): labrum, maxilla, and an enlargement of the apex of a glossa, and (lower row, left to right): the third abdominal gill, mandibles, and labium.

Key to the species of adult *Callibaetis* in South America

Information for the key is taken from Navás (1912, 1930c, d, 1934a), Needham and Murphy (1924), and Gillies (1990). No information is available about the larvae of the South American species.

1. The wings are completely hyaline without any trace of shading on the membrane, except at the base (**Fig. 2.249**). 2

- There is some darkening of the membrane of the female's wings beyond the base and usually other dark and light markings. If the wing of the male is completely hyaline, then there is a maroon median line on the second through eighth abdominal sternites (**Fig. 2.250**). 4

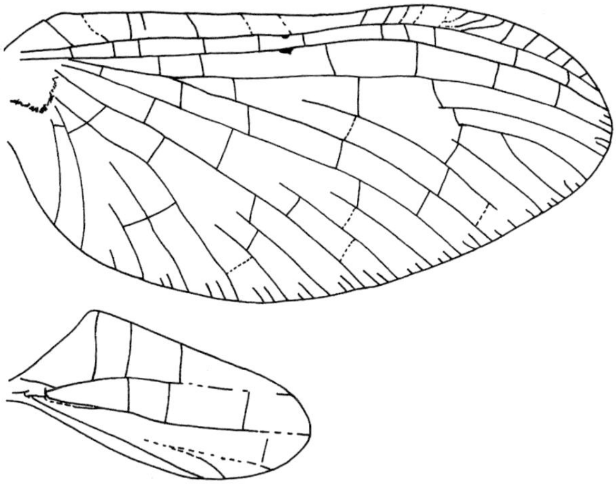

Fig. 2.249 The fore and hind wings of *Callibaetis abundans*.

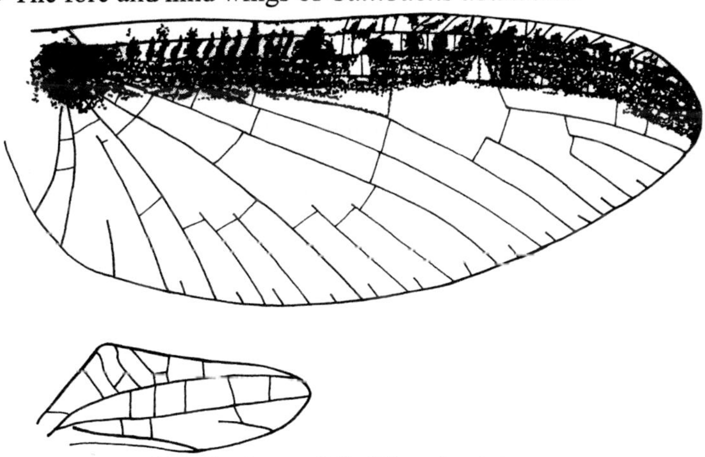

Fig. 2.250 The fore and hind wings of *Callibaetis viviparus*.

2. The hind wing has three long veins and one unattached intercalary vein between the second and third. There is a small darkened spot between the first and second long veins at the hind wing base (**Fig. 2.251**). Length: 5.5 mm. Fore-wing length: 6.9 mm. ....*Callibaetis gregarius* Navás, 1930 (Brazil).

- The hind wing lacks a distinct dark spot between the second and third veins at the wing base (**Fig. 2.249**). 3

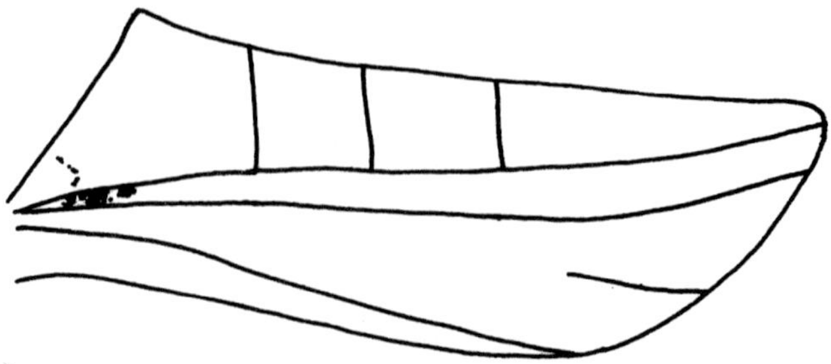

Fig. 2.251 The hind wing of *Callibaetis gregarius.*

3. The hind wings lack short intercalary veins between the second and third long veins, although the proximal portions of the third cross vein may be weak (**Fig. 2.249**).
....*Callibaetis abundans* Navás, 1912 (Paraguay). Syn: *Baetis abundans* Navás, 1912.
- The hind wings have short intercalary veins between the second and third long veins (**Fig. 2.198**). Body length: 5.5 to 6 mm. Fore-wing length: 5.5 to 5.7 mm. Caudal filament length: 10 mm. Thorax pale yellow.
....*Callibaetis pollens* Needham and Murphy, 1924 (Mato Grosso do Sul).
4. The wings have no distinct dark markings. The dark tinge on the membrane is prominent in the apical part of the costal field, while it is faint elsewhere. The veins are all dark brown or black. The color is dark brown with few markings other than light lateral areas on the apical abdominal segments. There are two moderately long intercalary veins in each space along the wing margin. In the costal area, there are about 10 to 12 cross veins. The expanse was reported as 24 mm, suggesting that the fore-wing should slightly exceed 10 mm. It is not clear whether the single specimen described was a male or female. The description of this species is exceptionally poor.
....*Callibaetis nigrivenosus* Banks, 1918 (Ecuador).
- The wings have distinct dark bands or other markings, or the wing veins are not all dark brown or black, or there is only a single intercalary vein in most of the spaces along the wing margin, or the body has a light ground color with darker markings. In case of doubt, the fore-wing is less than 10 mm in length.
....5
5. Vein Sc in the hind wing joins C just distal to the costal prominence. Two long veins in the hind wing end near the apex, and the second of these is joined to the first by at least four cross veins. On the opposite side, it forks to form three shorter veins in the apical half of the wing. These veins run from the posterior side of the of the long vein almost parallel as they curve to the wing margin; the first and second of these veins may be connected by a cross vein (**Fig. 2.252**). Length 5.5 mm; fore-wing length: 5 mm. The head is brownish

yellow with blackish eyes. The abdomen is yellow with dark markings. The male has not been described.
....*Callibaetis camposi* Navás, 1930 (Ecuador).
- Vein Sc in the hind wing does not reach the costal prominence, or the second long vein does not give rise to three shorter veins that curve nearly parallel to the wing margin (**Fig. 2.250**). 6

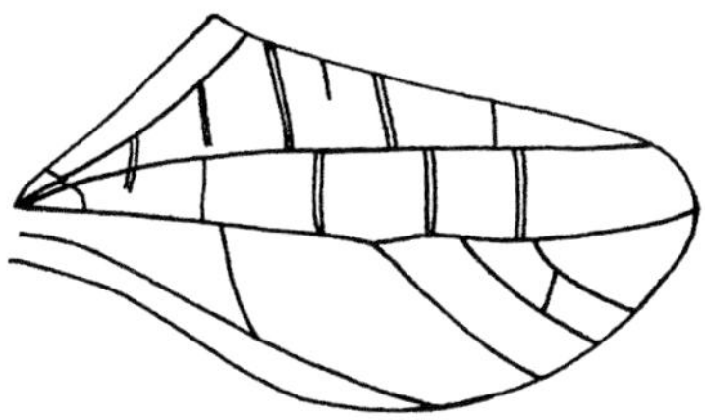

Fig. 2.252 The hind wing of *Callibaetis camposi.*

6. The short marginal intercalary veins are single in all but one or two spaces near the base or apex of the wing (**Fig. 2.250**). 7
- Most of the spaces contain two paired intercalary veins (**Fig. 2.253**). 9

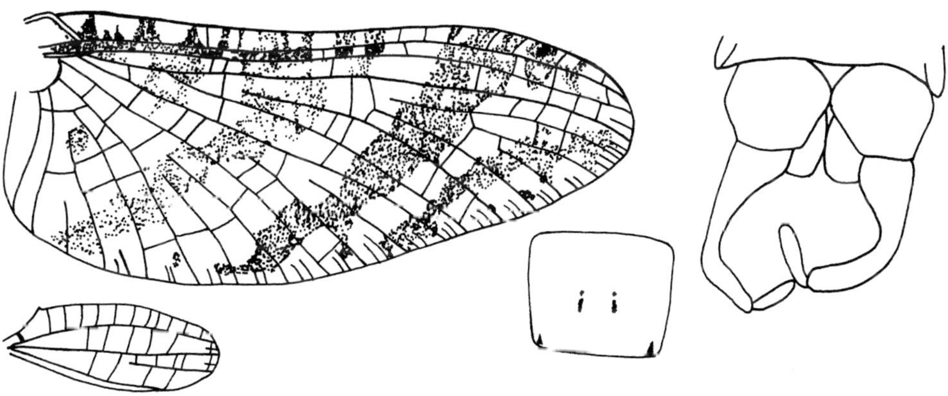

Fig. 2.253 *Callibaetis fasciatus:* the fore and hind wing (left), male genitalia (right), and the distinctive pattern on the first eight sternites with the posterior margin of the sternite at top (lower right center).

7. The dark markings on the wing are confined mainly to the costal part, covering only the costal, subcostal, and R_1 veins (**Fig. 2.250**). Length of female: 8.5 mm. Length of female's fore wing: 7.5 mm. Color: yellowish with spots of purplish brown on the thorax and abdomen. Males have not been found.
....*Callibaetis viviparus* Needham and Murphy, 1924 (Mato Grosso do Sul).

- The brown on the wings forms oblique cross bands, covering various parts of the wing (**Fig. 2.254**). 8

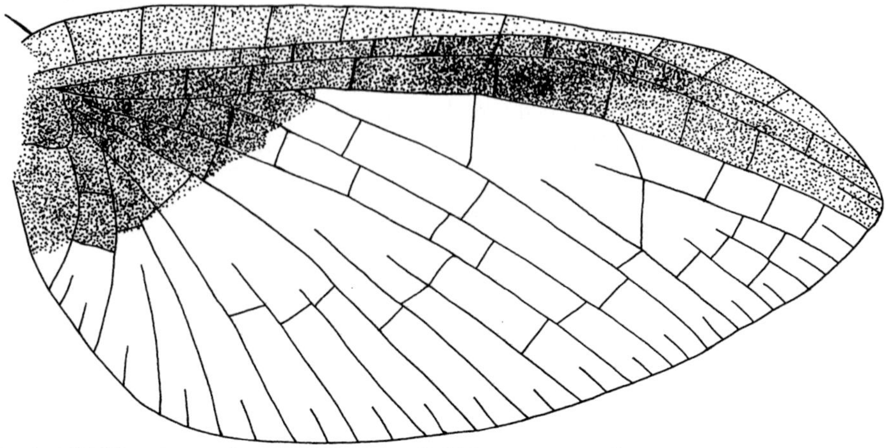

Fig. 2.254 The fore-wing of a female *Callibaetis gonzalezi.*

8. The vittae of the female do not have hyaline windows around the cross veins. Pigment extends broadly across the base of these wings to their inner margins. There are 13 to 15 cross veins in the outer third of the wing field posterior to vein R_2. The basal third to half of the hind wing is tinted brown (**Fig. 2.254**).
....*Callibaetis gonzalezi* (Navás, 1934) (Paraguay, Argentina). Syn: *Cloeon gonzalezi* Navás, 1934, so named because the type of Navás had apparently lost its hind wings (Gillies, 1990).
- There are numerous hyaline "windows" in the vittae, and the pigment overlaps the bases of the main veins but does not extend to the inner margin. The outer wing field has seven to nine cross veins. The hind wings are hyaline (**Fig. 2.255**).
....*Callibaetis radiatus* Navás, 1920 (Argentina, Paraguay). Syn: *Callibaetis venulosus* Navás, 1932.

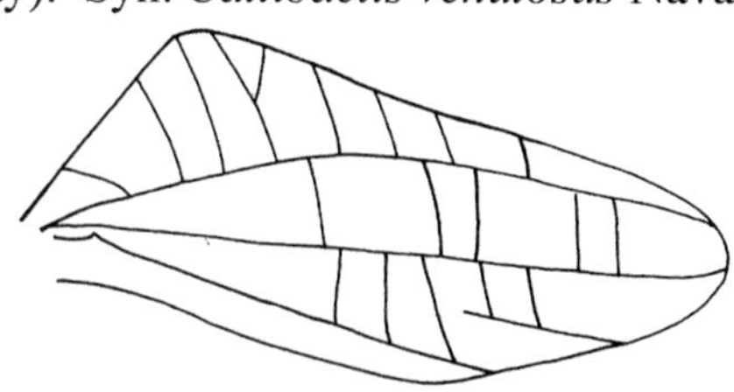

Fig. 2.255 The hind wing of *Callibaetis radiatus.*

9. The fore-wings are marked with conspicuous oblique or transverse pigmented bands separated by broad clear areas (**Fig. 2.256**). 10

- The wings lack clearly defined bands, although they may have lines of narrow hyaline areas across a lightly pigmented membrane (**Fig. 2.257**). 11

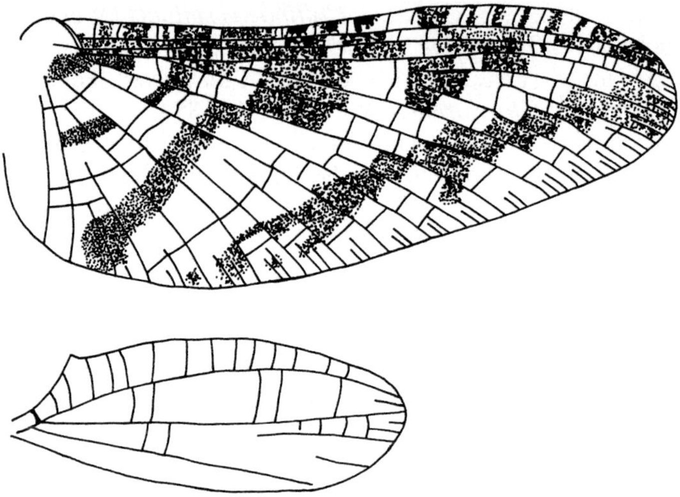

Fig. 2.256 The fore and hind wing of *Callibaetis sellacki.*

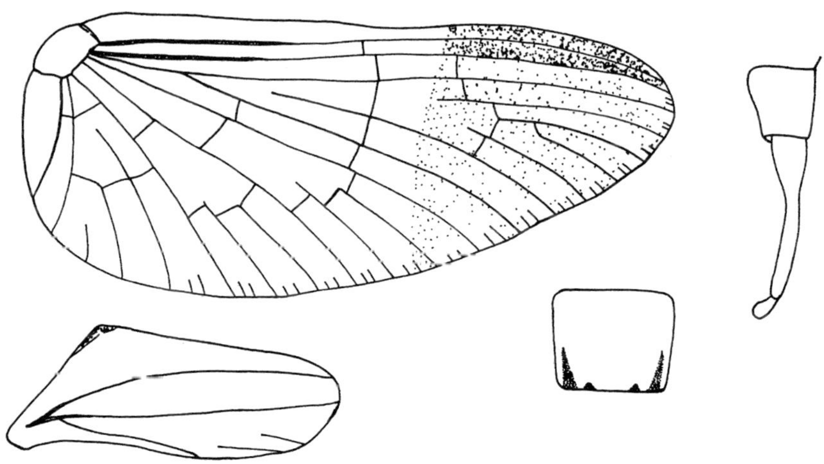

Fig. 2.257 The fore and hind wing of *Callibaetis dominguezi* (left), one limb of the forceps of a male (upper right), and the color pattern on a sternite of a male (lower right center).

10. The fore-wings have three or more or less well defined bands (**Fig. 2.253**). ....*Callibaetis fasciatus* (Pictet, 1843) (Chile, Argentina, Brazil). Syn: *Cloe fasciata* Pictet, 1843 *Cloeon fasciata* (Pictet, 1843), *Callibaetis trifasciatus* Esben-Petersen, 1912 partim; *Baetis gloriosus* Navás, 1923.

- There are two main bands of the fore-wing; the posterior one is parallel to the hind margin of the wing (**Fig. 2.256**).

....*Callibaetis sellacki* (Weyenbergh, 1883) (Chile, Argentina). Syn: *Cloe sellacki, Callibaetis lineatus* Navás, 1933; *Callibaetis fasciatus* (Pictet, 1843) partim; *Callibaetis trifasciatus* Esben-Petersen, 1912 partim; *Cloe lorentzii* Weyenbergh, 1883.

11. There are no cross veins in the hind wings. Only the costal and subcostal cells are darkened in the fore-wing of the female, and there are no contrasting hyaline spots. The outer third of the fore-wing of the male has a uniform brownish tint (**Fig. 2.257**). Length of male: 8 mm; of female: 9 mm. Wing length of male: 8 mm; of female: 9.5 mm. Color: generally dark brown.

....*Callibaetis dominguezi* Gillies, 1990 (Chile, Argentina).

- The hind wings have 6 to 20 cross veins. The darkened part of the fore-wing of the female extends at least to some of the membrane posterior to vein R_1, and numerous hyaline spots are evident. The fore-wing of the male is hyaline except for a small spot at the stigma (**Fig. 2.258**). 12

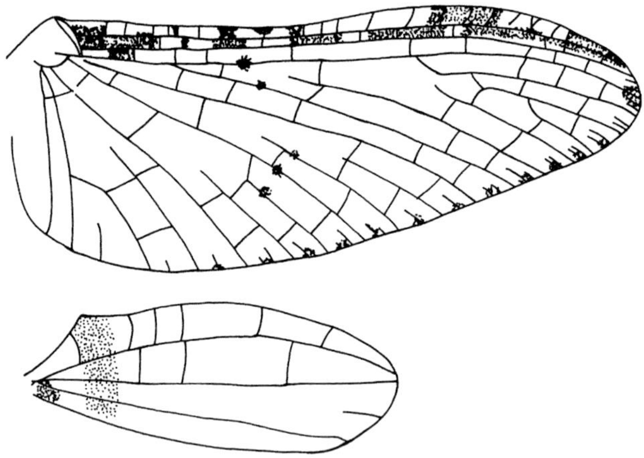

Fig. 2.258 The fore and hind wing of a female *Callibaetis guttatus*. The dark markings are variable and may be more extensive than shown here.

12. The tibiae may be dark at the knee or apex but there are never two discrete spots in the basal half. The abdominal sternites have paired, submedian markings of alternating dots and dashes. The fore-wings have dark anterolateral streaks and are otherwise lightly pigmented, sometimes with lines of narrow hyaline markings around the cross veins (**Fig. 2.259**). Length of female: 6.2 mm. Wing length of female: 5.5 mm.

....*Callibaetis jocosus* Navás, 1912 (Argentina, Chile, Paraguay, Mato Grosso do Sul). Syn: *Callibaetis jaffueli* Navás, 1918; *Callibaetis rimatus* Navás, 1932; *Callibaetis spegazzinus* Navás, 1920; *Callibaetis stictogaster* Navás, 1915.

- There are two dark spots on the basal half of the middle and hind tibiae as well as a subapical spot. The abdominal sternites have dark anterolateral streaks but lack paired submedian markings. The entire membrane of the fore-wing of the female is not pigmented, except on the distinct dark spots, and the fore-wing of the male is not pigmented at all, except on the stigma (**Fig. 2.258**). 13

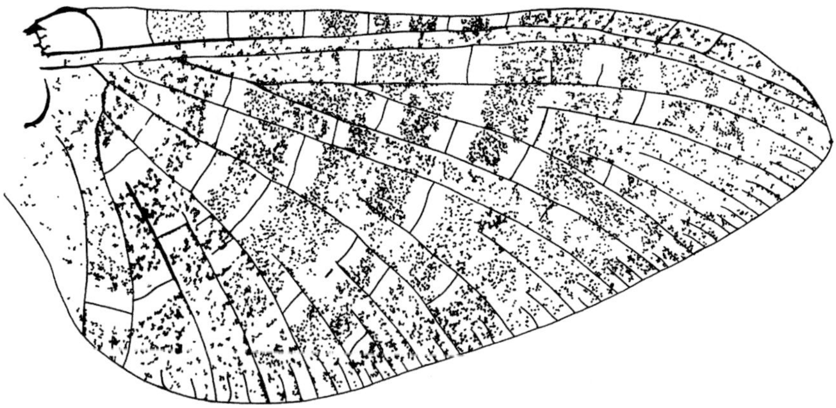

Fig. 2.259 The fore-wing of *Callibaetis jocosus*.

13. The fore-wings are either extensively spotted on the membrane, or the spots may be limited to a series along the posterior margin (**Fig. 2.258**). ....*Callibaetis guttatus* Navás, 1930 (Argentina). Syn: *Callibaetis apicatus* Navás, 1917; *Callibaetis bruchius* Navás, 1920; *Callibaetis zonatus* Navás, 1929.

- The pigment on the fore-wing of the female is confined to a broad line along the anterior margin, while the wings of the male are hyaline, except sometimes for pigment on the stigma (**Fig. 2.260**). 14

14. The dark anterior band on the fore-wing of the female overlaps less than half of the length of vein R_1 between with wing base and the origin of vein R_2. There is often a patch of smoky pigment on the stigma of the male wing. The hind wing is long and narrow (**Fig. 2.260**). There are no median markings on the abdominal sternites.
....*Callibaetis zonalis* Navás, 1915 (Argentina, Paraguay, Uruguay, Brazil). Syn: *Callibaetis amoenus* Navás, 1930; *Callibaetis apertus* Navás, 1917; *Callibaetis depressus* Navás, 1922; *Callibaetis sobrius* Navás, 1916; *Callibaetis vitreus* Navás, 1915; *Baetis opacus* Navás, 1915; *Baetis virellus* Navás, 1915.

- The dark anterior band on the fore-wing of the female overlaps half or more of the length of vein R_1 between with wing base and the origin of vein R_2 (**Fig. 2.261**). There is no trace of pigment on the stigma of the male wing, which is entirely hyaline. There is a maroon median line on the second through eighth abdominal sternites. Length of male: 8 mm; of female: 9 mm. Length of fore

wing of male: 7 mm; of female: 9 mm. Color yellowish with many small reddish spots on the abdomen.
....*Callibaetis willineri* Navás, 1932 (Argentina, Rio Grande do Sul). Syn: *Callibaetis alegre* Traver, 1944.

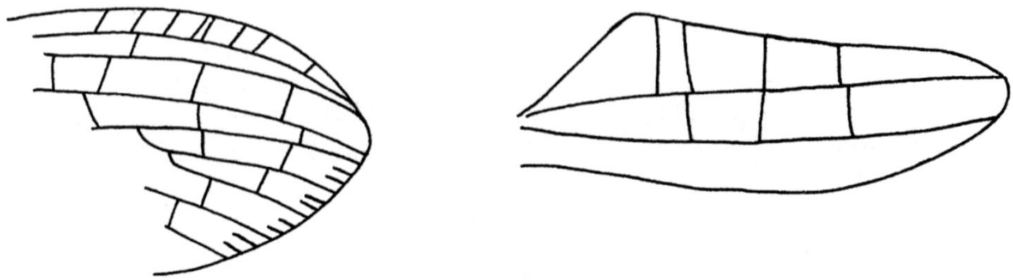

Fig. 2.260 The apical portion of the fore-wing and the hind wing of *Callibaetis zonalis.*

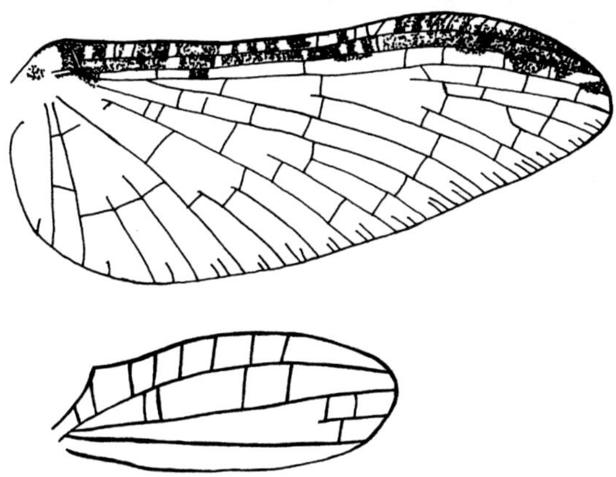

Fig. 2.261 The fore and hind wings of a female *Callibaetis willineri*.

Key to the species of adult *Fallceon* in South America

Information for the key was taken from Navás (1912), Needham and Murphy (1924), Traver (1971), Hubbard (1974), and McCafferty (2000). The larvae of the South American species have not yet been described.

1\. A cross vein connects the first longitudinal vein in the hind wing to the wing margin near the base (**Fig. 2.262**). The abdomen is light reddish brown and marked with wedge-shaped, chestnut brown lateral patches, which are most distinct on the basal segments but which extend to the seventh tergite. Length: 7 mm; fore-wing length: 7 mm.
....*Fallceon yaro* (Traver, 1971) (Uruguay). Syn: *Baetis yaro* Traver, 1971.

- The abdomen is not so strikingly marked. No cross veins connects the first longitudinal vein in the hind wing to the margin near the base or the veins to each other (**Fig. 2.199**). 2

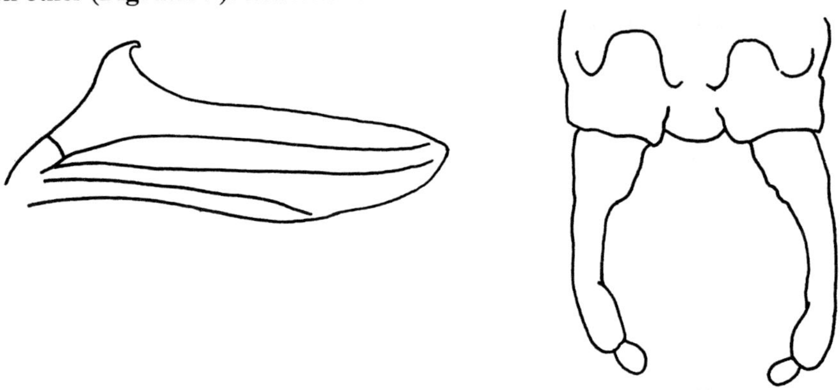

Fig. 2.262 The hind wing (left) and male genitalia (right) of *Fallceon yaro.*

2. The stigmatic area of the costal space in the fore-wing has a network that includes incomplete cross veins or cross veins with weak sections (**Fig. 2.263**). The longitudinal veins in the long, narrow hind wing are nearly parallel. Length: 4 mm. Fore-wing length: 4.5 mm. The thorax is mainly dark reddish brown dorsally with the midline of the mesonotum and a curved mark near the wing base yellowish. The underside of the thorax is dark yellowish brown. The abdomen is rust red dorsally with dark bands across the posterior margins of the tergites; it is light yellowish brown on the sternites. The wings are hyaline with whitish or yellowish veins. The description of this species is poor, and the type is in bad condition.
....*Fallceon inops* (Navás, 1912) (Argentina, Paraguay). Syn: *Baetis inops* Navás, 1912.
- The cross veins in the stigmatic area of the fore-wing are straight and not interconnected or incomplete (**Fig. 2.199**). Length: 5.5 mm; fore-wing length: 4.5 mm. Color: generally orange yellow dorsally and paler yellow ventrally. There are a few faint broken brown lines on the thorax near the wing root and on the carina. The upper parts of the eyes are amber. The segments in the middle of the abdomen of the male are translucent white with black pigment scattered on the tracheae. The legs and cerci are pale, and the wings are hyaline.
....*Fallceon murphyae* (Hubbard, 1974) (Peru, Bolivia, Chile, Argentina). Syn: *Baetis murphyae* Hubbard, 1974; *Baetis melleus* Needham and Murphy, 1924.

Fig. 2.263 The stigmatic area of the fore-wing of *Fallceon inops.*

Key to the species of adult *Moribaetis* in South America

Information for the key was taken from Navás (1912), Traver (1971), and McCafferty (2000). The larvae have not been described.

1. The second longitudinal vein in the hind wing is not forked. There is a cross vein and one or two intercalary veins in the apical ¾ of the hind wing (**Fig. 2.200**). Body length: 4 to 5 mm; fore-wing length: 3.5 to 4 mm. ....*Moribaetis aneto* (Traver, 1971) (Uruguay).

- In the hind wing, the second longitudinal vein is forked, and there are no cross veins (**Fig. 2.264**). 2

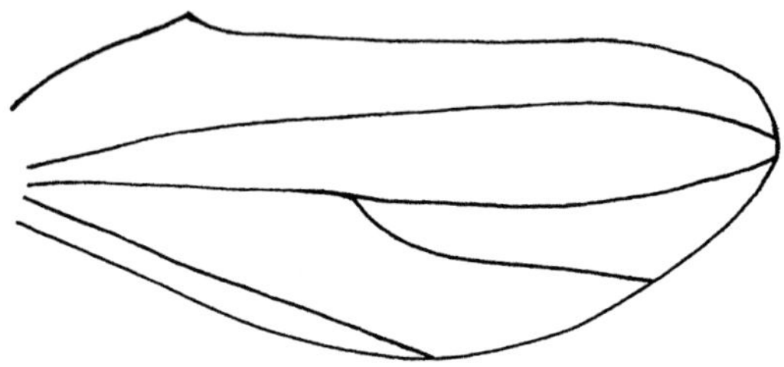

Fig. 2.264 The hind wing of *Moribaetis comes.*

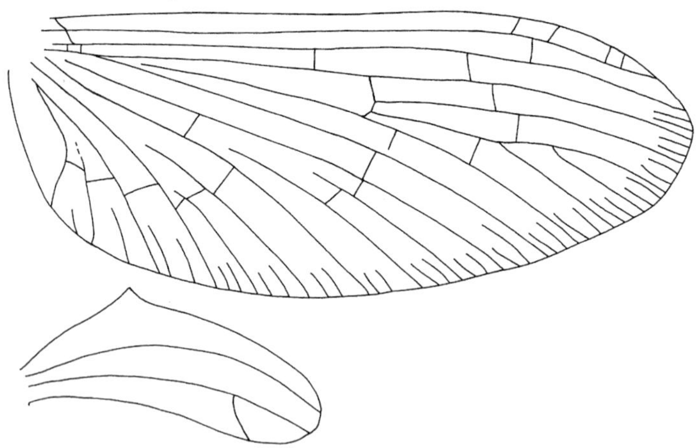

Fig. 2.265 The fore and hind wing of *Moribaetis socius* with the hind wing much enlarged.

2. The fork in the second vein of the hind wing is located near the midlength (**Fig. 2.264**). The head is rust color with black eyes. The thorax is dark green. The abdomen is uniformly dark reddish with cerci that are white with black rings around the joints. Length: 5 mm. Fore-wing length: 6.4 mm. ....*Moribaetis comes* (Navás, 1912) (São Paulo). Syn: *Baetis comes* Navás, 1912.

- The second vein in the hind wing is forked in the distal half (**Fig. 2.265**). The general color is pale brown with yellow on the head, thorax, tenth abdominal segment, legs, and cerci. There are dark brown markings on the thorax. Length of female: 8 mm; fore-wing length of female: 8 mm. The male has not been described.
....*Moribaetis socius* (Needham and Murphy, 1924) (Peru). Syn: *Baetis socius* Needham and Murphy, 1924.

Key to the species of adult *Aturbina* in South America

Information for this key was taken from Lugo-Ortiz and McCafferty (1996) and Gillies (2001).

1. The forceps of the male consists of three segments. There is no small dark spot near the wing base at the junction of the costal brace and the subcosta (**Fig. 2.201**). Length: 3.8 to 4.2 mm. Fore-wing length: 3.6 to 3.8 mm. Color: generally pale; cream with purple dorsal markings on the abdomen.
.... *Aturbina georgei* Lugo-Ortiz and McCafferty, 1996 (French Guiana, Paraguay, Amazonas, Pará, Mato Grosso, Acre Territory).
- The forceps of the male consists of four segments, the fourth being small and nearly spherical. There is a small, conspicuous dark spot near the wing base at the junction of the costal brace and the subcosta (**Fig. 2.266**). Length: 4.2 to 5.0 mm. Fore-wing length: 4.2 to 4.5 mm. Color: generally pale; cream with purple dorsal markings on the abdomen.
.... *Aturbina beatrixae* Gilles, 2001 (Uruguay).

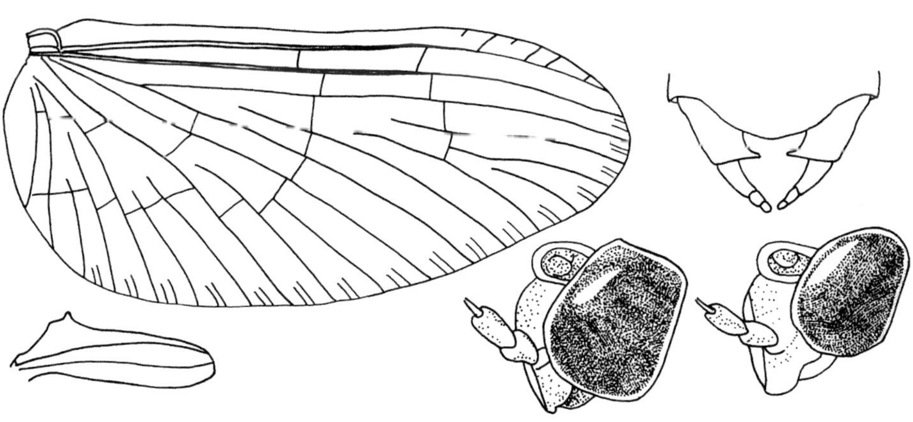

Fig. 2.266 *Aturbina beatrixae:* fore and enlarged hind wing (left), the forceps of a male (upper right), and lateral views of the heads of a male (lower right center) and a female (lower right).

Key to the species of larval *Aturbina* in South America

Information for this key was taken from Lugo-Ortiz and McCafferty (1996b) and Gillies (2001).

1. The long seta arising from the prostheca of the right mandible is obviously branched (**Fig. 2.218**). Length of last larval instar: 3.6 to 4.4 mm. Color: pale with a cream abdomen marked with purple.
.... *Aturbina georgei* Lugo-Ortiz and McCafferty, 1996 (French Guiana, Paraguay, Amazonas, Pará, Mato Grosso, Acre Territory).
- The long seta arising from the prostheca of the right mandible may have a fine inconspicuous division at the apex, but it is not obviously branched (**Fig. 2.267**). Length of last larval instar: 3.6 to 4.4 mm. Color: pale with a cream abdomen marked with purple.
.... *Aturbina beatrixae* Gilles, 2001 (Uruguay).

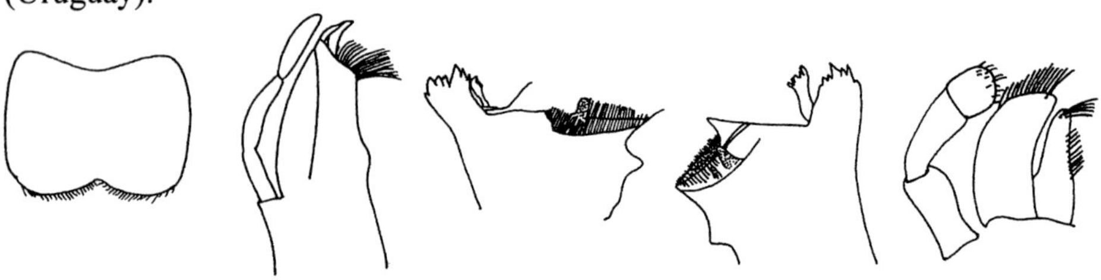

Fig. 2.267 The mouthparts of *Aturbina beatrixae* (left to right): labrum, maxilla, apices of the right and left mandible, and the palp, paraglossa, and glossa of the labium.

Key to the species of adult *Camelobaetidius* in South America

Information for the key was taken from Needham and Murphy (1924), Traver (1943, 1971), and Traver & Edmunds (1968). The larvae of most of these species have never been described, and the adults of all species described from larvae, except for *Camelobaetidius penai,* are unknown.

1. The abdomen has distinct bands across the posterior and sometimes anterior margins of the tergites. The long veins in the hind wing appear to fade out before reaching the margin (**Fig. 2.268**). 2
- The abdomen is pale and lacks brown bands on the tergites. The long veins in the hind wing reach the wing margin (**Fig. 2.269**). 4
2. The anterior margins of the abdominal tergites are narrowly darkened, more so laterally and more prominently in females. The tergites are dark olive brown with a reddish tinge and reddish posterior margins, especially dorsally. The second through sixth abdominal segments of the male are almost hyaline. The

thorax is reddish brown, darker in the male and sometimes blackish brown, with blackish and cream markings and yellowish mottling. The cross vein connecting the first longitudinal vein with the wing margin arises distal to the fork (**Fig. 2.203**). Length: 4.5 to 6.5 mm. Fore-wing length: 5.0 to 6.5 mm. .... *Camelobaetidius penai* (Traver & Edmunds, 1968) (Argentina). Syn: *Dactylobaetis penai* Traver & Edmunds, 1968.

- There are narrow dark brown bands across the posterior margins of the second through tenth abdominal tergites. 3

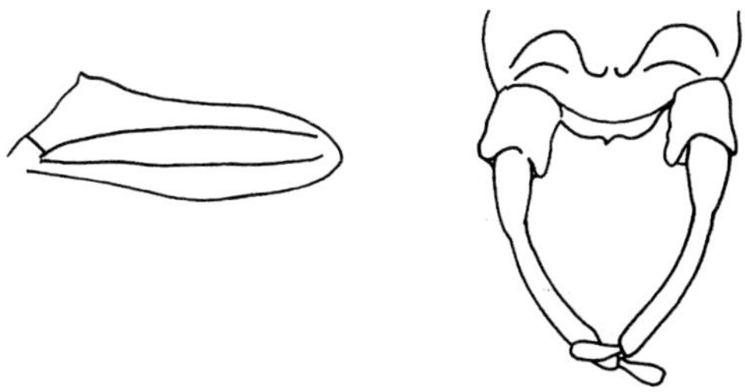

Fig. 2.268 The hind wing and male genitalia of *Camelobaetidius coveloae.*

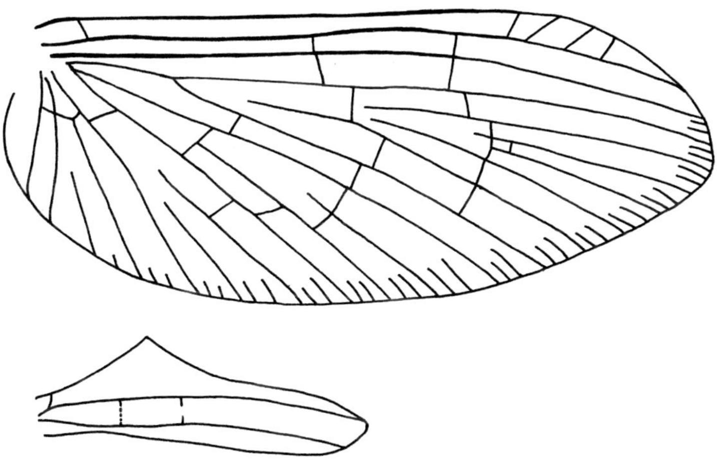

Fig. 2.269 The fore and hind wing of *Camelobaetidius tantillus* with the hind wing enlarged.

3. The thorax is dark reddish brown dorsally and ventrally without prominent dark brown markings. The posterior margins of the apical abdominal tergite and sternite are convex between the bases of the forceps (**Fig. 2.268**). Body length: 5 mm; fore-wing length: 4.4 to 5.25 mm. Color: the head is reddish brown with a dark brown scape of the antenna; and the abdomen is a fawn color, pale reddish brown, slightly paler ventrally and with narrowly dark posterior margins of the segments. The cerci are pale yellowish white. ....*Camelobaetidius coveloae* (Traver, 1971) (Uruguay). Syn: *Baetis coveloae* Traver, 1971.

- The thorax is obviously darker dorsally than ventrally, and dark and light markings are present. The longitudinal vein in the hind wing has two long branches and is connected to the base of the costa by a cross vein nearer to the base of the wing than to the fork (**Fig. 2.270**). The thorax is reddish dorsally and pale creamy white ventrally. The lateral edges of the pronotum and the sutures on the mesonotum and metanotum are dark brown. There is a pale yellow square surrounding the base of the fore-wing. The body and fore-wing lengths of the males is about 4 mm, and those of the females are about 5 mm. ....*Camelobaetidius alcyoneus* (Traver, 1943) (Venezuela). Syn: *Baetis alcyoneus* Traver, 1943.

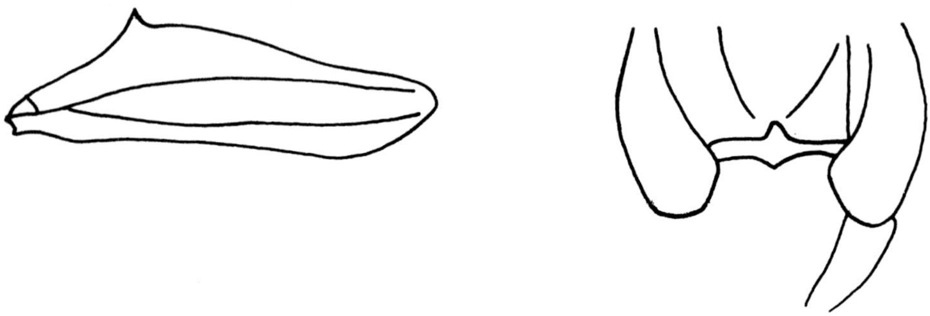

Fig. 2.270 *Camelobaetidius alcyoneus:* the hind wing (left) and male genitalia with one of the forceps removed and the other broken just beyond the base (right).

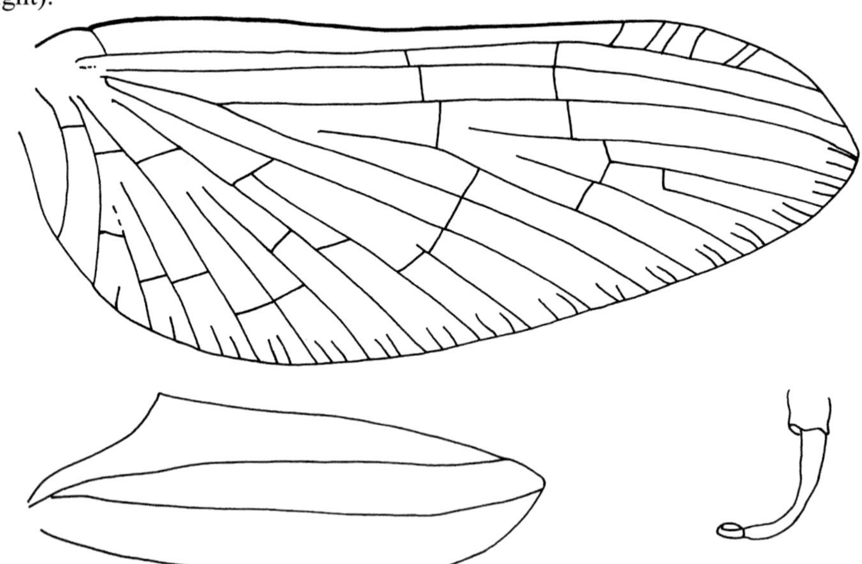

Fig. 2.271 *Camelobaetidius dryops:* fore and hind wings with the hind wing much enlarged and the forceps segments of a male (lower right).

4. There are traces of two cross veins between the longitudinal veins of the hind wing (**Fig. 2.269**). Length: 4 mm; fore-wing length: 3 mm.
..............................*Camelobaetidius tantillus* (Needham and Murphy, 1924) (Peru). Syn: *Baetis tantillus* Needham and Murphy, 1924.
- There are no traces of cross veins in the hind wing, and its costal angulation is not very high (**Fig. 2.271**). Length: 4.5 to 5 mm; fore-wing length: 5 to 5.5 mm.
...............................*Camelobaetidius dryops* (Needham and Murphy, 1924) (Peru). Syn: *Baetis dryops* Needham and Murphy, 1924.

Key to the larvae of *Camelobaetidius* in South America

Information for the key was taken from Traver and Edmunds (1968), Edmunds (1968), and Lugo-Ortiz and McCafferty (1999b). The larvae of all South American species known as adults, except for *Camelobaetidius penai,* have never been described.

1. There is a pair of conical, blunt tubercles along the midline on the posterior part of the pronotum and a single such tubercle at the apex of the metanotum. The median caudal filament is about 1.8 mm long, and the cerci are about 3.0 mm. The apical half of the basal segment of the maxillary palp is expanded on its inner margin. The spatulate tarsal claws bear about 15 to 17 denticles. Between the prostheca and mola of each mandible, there are a few fine, simple setae (**Fig. 2.272**). Length of the final larval instar: about 4.4 mm. Color: yellow brown head with a cream color on the frons, genae, and clypeus; cream or brown on the pronotum; femora cream with brown proximal and distal ends and a fine brown anterodorsal line; tarsi brown; the first five abdominal tergites are cream, the first suffused with black, the second through fifth brownish anteriorly, and the third and fourth also with a suffusion of black near the middle and laterally; the sixth through tenth tergites are yellowish brown; the abdominal sternites are light yellowish brown or cream.
...........................*Camelobaetidius tuberosus* Lugo-Ortiz and McCafferty, 1999 (Paraguay).
- There are no conical, blunt tubercles on the pronotum or at the apex of the metanotum (**Fig. 2.207**). 2
2. The median caudal filament is about as short as or shorter than the tenth abdominal segment. The fore-femur has a prominent blunt projection on the anterior margin near the base (**Fig. 2.207**). 3
- The median caudal filament is at least 75% as long as the cerci (**Fig. 2.273**).
....5
3. The claws are lined with 9 or 10 denticles. None of the tracheae in the gills are pigmented (**Fig. 2.207**).
....*Camelobaetidius leentvaari* Demoulin, 1967 (Surinam).
- The claws are lined with at least 25 denticles (**Fig. 2.274**). 4

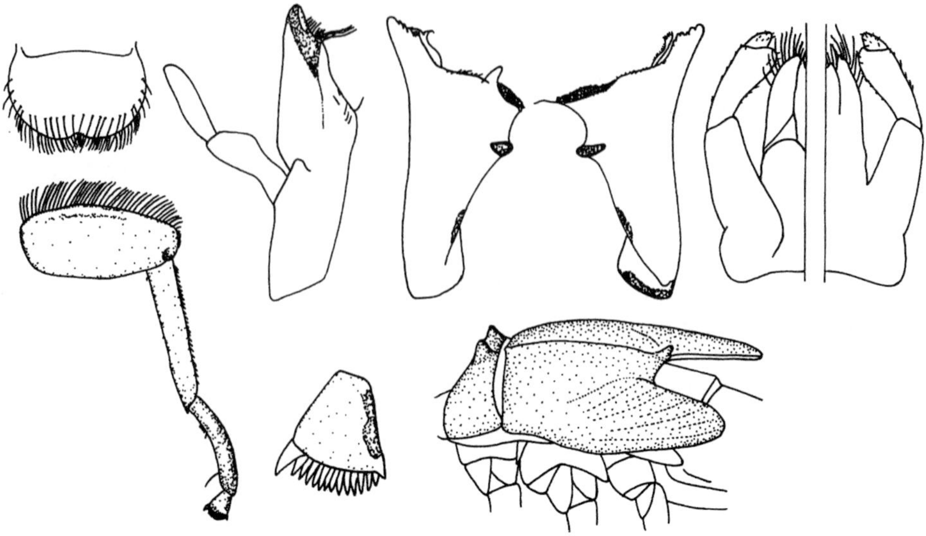

Fig. 2.272 *Camelobaetidius tuberosus* larva (upper row, left to right): labrum, maxilla, mandibles, half of the labium in dorsal and ventral view, and (lower row, left to right): fore-leg, enlargement of the tarsal claw, and the thorax in dorsolateral view showing the tubercle-like prominences.

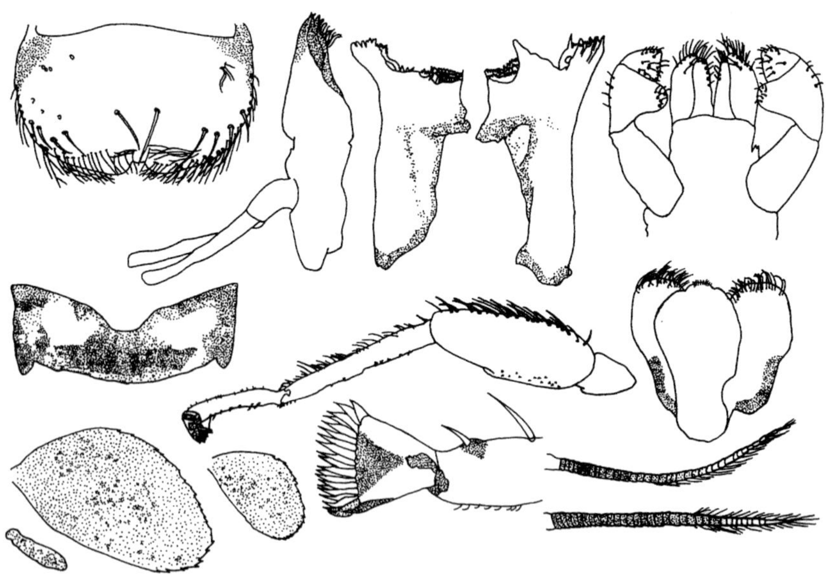

Fig. 2.273 *Camelobaetidius janae* larva (upper row, left to right): labrum, maxilla, mandibles, labium; (middle row, left to right): metanotum, fore leg, hypopharynx; (lower row, left to right): first, fourth, and seventh gills, tarsal claw in dorsal view, a cercus (above) and median caudal filament (below).

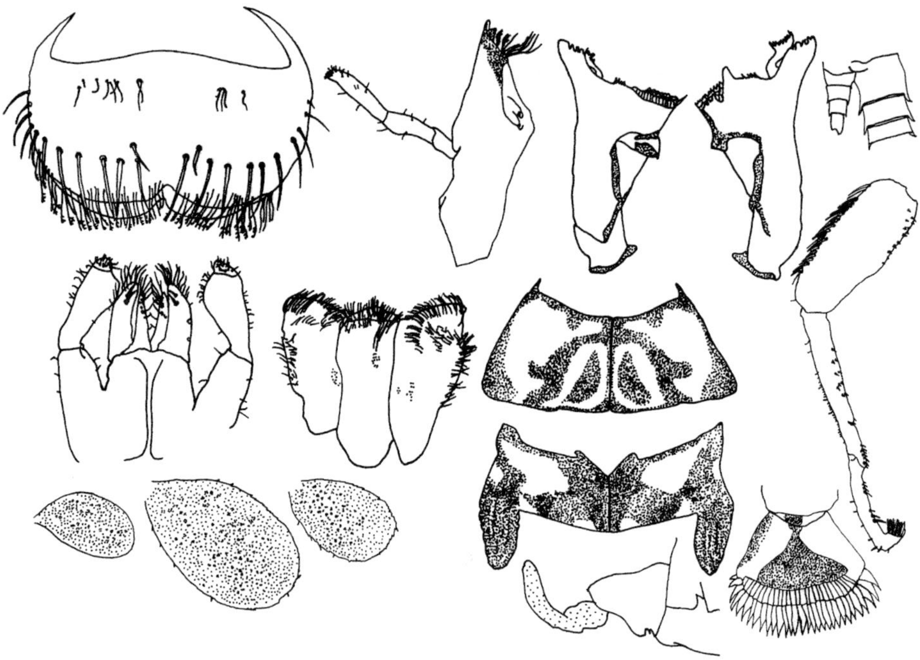

Fig. 2.274 *Camelobaetidius billi* larva (upper row, left to right): labrum, maxilla, mandibles, and the base of a cercus and the median caudal filament (above) and a fore-leg (below); the labium and hypopharynx (middle left, left and right, respectively); the pronotum and metanotum with the hind wing buds (right center, above and below, respectively); a tarsal claw in dorsal view (lower right); the base of a fore-leg showing the small thoracic gill (lower right center); and the first, fourth, and seventh gills (lower left, left to right).

4. The claws are lined with about 31 denticles. The tracheae in the gills are unpigmented. There is no proximal anterior protuberance on the fore-femur. There is a vestigial thoracic gill at the base of the fore-leg (**Fig. 2.274**). Length of the male final larval instar: c. 3.1 mm; female larval instar: c. 3.8 mm. Color: uniform brown with a contrasting light and dark pattern on the thorax. ....*Camelobaetidius billi* Thomas and Dominique, 2000 (French Guiana).

- The claws are lined with about 25 denticles. At least the middle trachea in each gill is pigmented. There is a thick process bearing a patch of short setae at the apex of the fore-tibia (**Fig. 2.275**). The description is based on only one specimen, a larval instar prior to the final one, which was 3 mm long. Color: pale yellow. ....*Camelobaetidius mantis* Traver and Edmunds, 1968 (Amazonas).

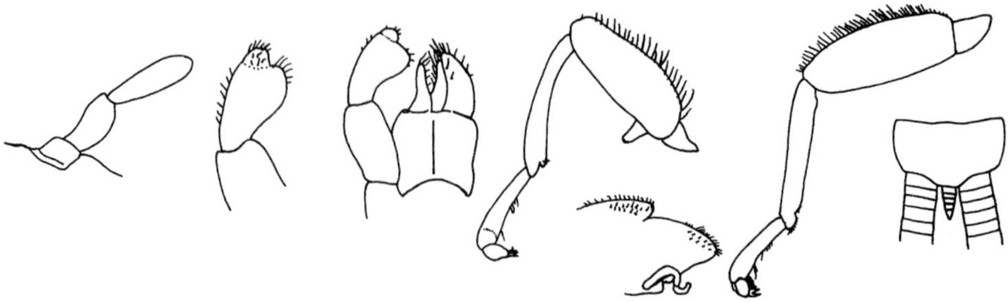

Fig. 2.275 *Camelobaetidius mantis* larva (left to right): maxillary palp, apex of the labial palp, labium with appendages on alternate sides removed, fore-leg with an enlargement of the apex of the fore-tibia below, the hind leg, and the apex of the abdomen with the median caudal filament and the bases of the cerci.

5. The maxillary palp is bifid. There are 21 denticles on the tarsal claws (**Fig. 2.273**). There are no thoracic gills. Length of final instar male larva: about 2.5 mm; final instar female larva: about 3.5 mm. Color: uniform brown except for light reddish brown patches on the third and sixth abdominal tergites.
..................................*Camelobaetidius janae* Dominique and Thomas, 2000 (French Guiana).
- The maxillary palp is not bifid. There is no projection on the fore-femur (**Fig. 2.276**). 6

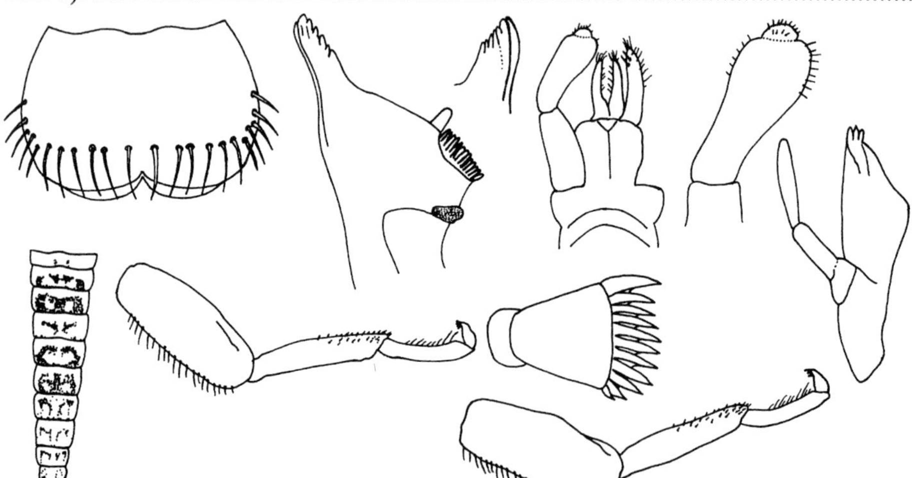

Fig. 2.276: *Camelobaetidius phaedrus* larva: labrum (upper left), pattern on the abdominal tergites (lower left), apical portion of the left mandible and the incisor of the right mandible (upper left center), labium with parts removed and an enlargement of the apex of its palp (upper right center), maxilla (right), fore-leg (lower left center), enlargement of a typical spatulate claw in dorsal view above a hind leg (lower right center).

6. Claws with 7 to 11 denticles. There is a gill at the base of each fore-leg. Dilation at inner apical margin of second segment of labial palp rounded or with a false point (**Fig. 2.276**). Length: 4.5 to 5 mm. Color: dark brown and reddish brown with black markings.
.................................*Camelobaetidius phaedrus* (Traver & Edmunds, 1968) (Rio Grande do Sul). Syn: *Dactylobaetis phaedrus* Traver & Edmunds, 1968.
- Claws with 15 or more denticles. 7
7. Claws with 15 to 17 denticles. There is no gill at the base of each fore-leg. Length: 3.5 mm. Color: pale olive brown with yellow markings and a dark brown metanotum. Abdomen yellow ventrally.
.................................*Camelobaetidius cayumba* (Traver & Edmunds, 1968) (Peru). Syn: *Dactylobaetis cayumba* Traver & Edmunds, 1968.
- Claws with 20 to 40 denticles. Dilation of inner apical margin of the second segment of labial palp with a sharp point. 8
8. There is no thoracic gill present near the base of the fore-leg. The three lateral spines on the labrum are set on a ridge or fold and may therefore appear to be connected basally. The denticles on the mandibular incisors are fused. The claws have 29 to 30 denticles. Length: 6 to 7 mm. Color: grayish olive to reddish brown dorsally with a pattern of two darker transverse markings on each side of the midline on the second through eighth tergites and one pair of transverse markings on the remaining tergites. The markings almost reach the anterior margins but end well before the posterior margins of the tergites.
.... *Camelobaetidius penai* (Traver & Edmunds, 1968) (Argentina). Syn: *Dactylobaetis penai* Traver & Edmunds, 1968.
- There is a thoracic gill present near the base of the fore-leg. The lateral spines on the labrum are not set on a ridge or fold and therefore are clearly separate. The six to eight denticles on the mandibular incisors are separate. The second segment of the maxillary palp is enlarged toward the apex (**Fig. 2.277**). 9

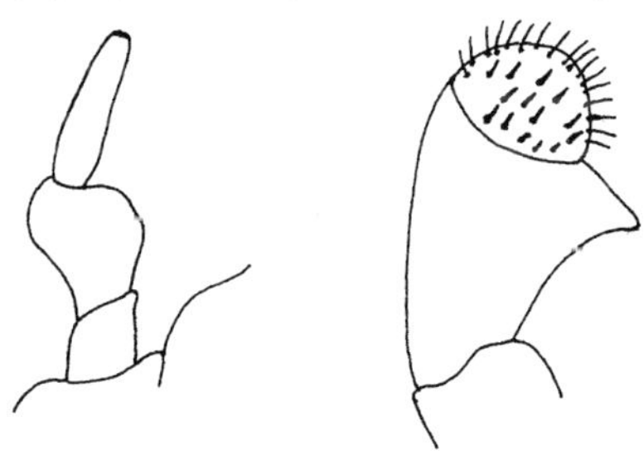

Fig. 2.277 The maxillary palp (left) and apex of the labial palp (right) of a larval *Camelobaetidius anubis.*

9. Claws with 20 to 27 denticles. Length: 6.5 to 10.0 mm. Color: dark reddish brown dorsally.
.... *Camelobaetidius anubis* (Traver & Edmunds, 1968) (Paraná). Syn: *Dactylobaetis anubis* Traver & Edmunds, 1968.

- Claws with 30 to 40 denticles. A denticle on the mandibular incisor is inserted out of the plane of the others. Length: 4.25 to 5.0 mm. Color yellowish to pale reddish brown with a yellow tinge and a pattern of darker markings (**Fig. 2.278**). Length: 4.5 to 5 mm.
....*Camelobaetidius serapis* (Traver & Edmunds, 1968) (Paraná). Syn: *Dactylobaetis serapis* Traver & Edmunds, 1968.

Fig. 2.278 *Camelobaetidius serapis* larva: apex of the left mandible showing the arrangement of the denticles (left) and the color pattern on the abdominal tergites (right).

Key to the species of larval *Paracloeodes* in South America

Information for this key was taken from Lugo-Ortiz and McCafferty (1996c). The adult female of only one South American species has been described: *Paracloeodes eurybranchus* (**Fig. 2.203**). Male adults of South American species are unknown.

1. The length of the final larval instar does not exceed 3.5 mm. The gills are wider in the distal half (**Fig. 2.213**). 2
- The length of the final larval instar is not less than about 3.9 mm. A color pattern is never distinct on the body. Hind wing buds are always present. Tubercles are never present between the fore-wing pads. On the labrum, a pair of long setae is always present on either side of the midline (**Fig. 2.279**). 3

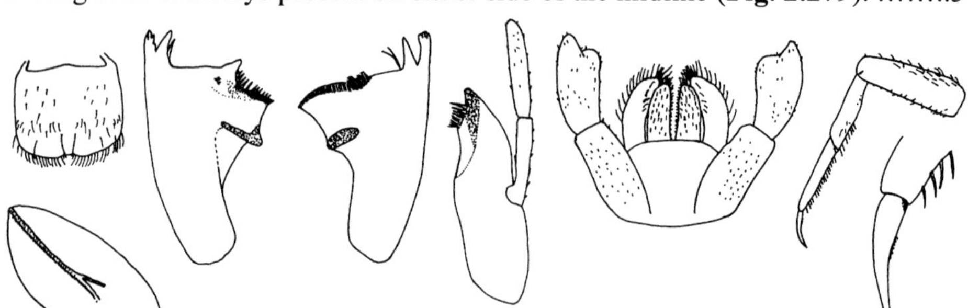

Fig. 2.279 *Paracloeodes eurybranchus* larva (left to right): labrum above the fourth gill, mandibles, maxilla, labium, and fore-leg with an enlargement of the tarsal claw.

2. The insect is yellowish brown without any distinct color pattern. Hind wing buds are absent. Two small tubercles are present between the fore-wing buds. On the labrum, there is no pair of long setae on either side of the midline. The pectinate prostheca on the right mandible is not bifid, and there is a deep fossa delimiting the mola. The distomedial process on the second segment of the labial palp is large and lacks a row of setae on the dorsal side. The gills lack any trace of a tracheal system (**Fig. 2.213**). Length: 2.7 to 3.4 mm. Color: yellowish brown without any pattern evident.
…………………….. *Paracloeodes binodulus* Lugo-Ortiz and McCafferty, 1996 (Pará).
- There is a distinct dark brown median stripe on the nine anterior segments of the abdomen. Hind wing buds are present. There are no tubercles present between the fore-wing buds. On the labrum, there is a pair of long setae on either side of the midline and about 15 other pairs of long setae (**Fig. 2.280**). The pectinate prostheca on the right mandible is bifid, and there is no deep fossa delimiting the mola. The distomedial process on the second segment of the labial palp had a row of five or six long setae on the dorsal side. There is a weakly developed tracheal system evident in the gills. Length: about 3 mm. Color: brown with darker brown markings on the abdomen and femora.
…...……………………. *Paracloeodes ibicui* Lugo-Ortiz and McCafferty, 1996 (Paraguay).

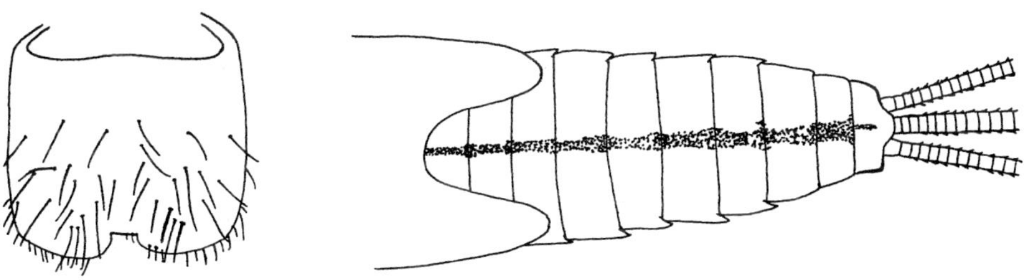

Fig. 2.280 The labrum (left) and the abdominal segments in dorsal view (right) of *Paracloeodes ibicui.*

3. The middle portion of the gills is widest and not serrate on the margins (**Fig. 2.279**). Length: 4.7 to 5.5 mm. Color: yellow brown with cream colored legs and not pattern evident.
…...……………….*Paracloeodes eurybranchus* Lugo-Ortiz and McCafferty, 1996 (Argentina, Rio Grande do Sul).
- The gills are narrow, tending to be widest in the distal half, and serrate on the margins (**Fig. 2.281**). Length: 3.9 to 4.1 mm. Color: yellow brown with no color pattern on the body but slightly marker markings on the femora.
…...…...……….*Paracloeodes leptobranchus* Lugo-Ortiz and McCafferty, 1996 (Paraguay, Paraná, Santa Catarina, Rio Grande do Sul).

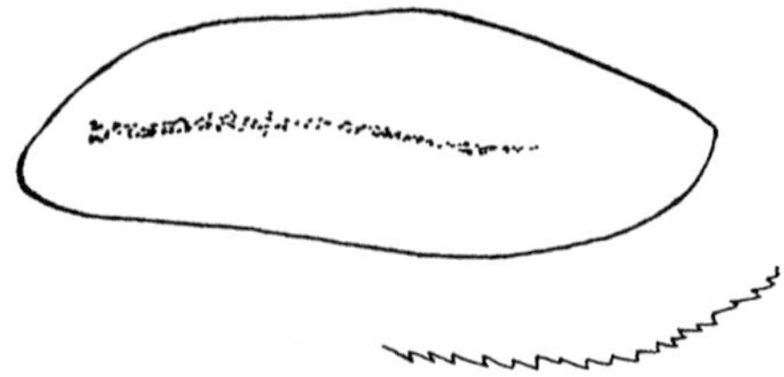

Fig. 2.281 The fourth gill of *Paracloeodes leptobranchus* with an enlargement of its margin.

Key to the South American species of adult *Cloeodes*

Information for the key was taken from Weyenbergh (1883), Navás (1923b), Traver (1943), Waltz and McCafferty (1987a, b), and McCafferty and Lugo-Ortiz (1995). Only the males of most species can be identified, and the adults of *Cloeodes irvingi* and *C. redactus* have not been described.

1. The whole body is a pearly gray color becoming almost steel blue on the dorsum. There is a dark longitudinal stripe running down the dorsal midline of the entire thorax and dark markings at the insertions of the wings. Some males have completely dark thoraces. The wings are darkened along the costal space, and the membrane has dark markings forming bands. Length of the male: 5.5 mm; wingspan of the male: 12 mm. The legs and antennae of the female are paler than those of the male.
....*Cloeodes stelzneri* (Weyenbergh, 1883) (Argentina). Syn: *Cloe stelzneri* Weyenbergh, 1883; *Baetis stelzneri* (Weyenbergh, 1883). This species has been very poorly described.
- The color is not pearly gray becoming almost steel blue on the dorsum, and there is no dark longitudinal stripe running down the dorsal midline of the thorax. 2
2. The distal part of the male's eye is cylindrical and parallel to the other eye (**Fig. 2.282**). Body length of the male: 5 mm. Fore-wing length of the male: 4.5 mm. Caudal filament length of the male: 9 mm.
....*Cloeodes binocularis* (Needham and Murphy, 1924) (Peru). Syn: *Pseudocloeon binocularis* Needham and Murphy, 1924.
- The distal part of the male's eye is not cylindrical, but it may be turbinate, that is, mushroom shaped, and as high as the head (**Fig. 2.195**). 3
3. At least the middle segments of the abdomen are translucent white. The head is mainly yellow, sometimes with blackish eyes or darker marks at the bases of the antenna, and the thorax is mainly yellow or light brownish, sometimes with blackish markings. 4
- The middle segments of the abdomen are not translucent white, or the head and thorax are not yellow. 6

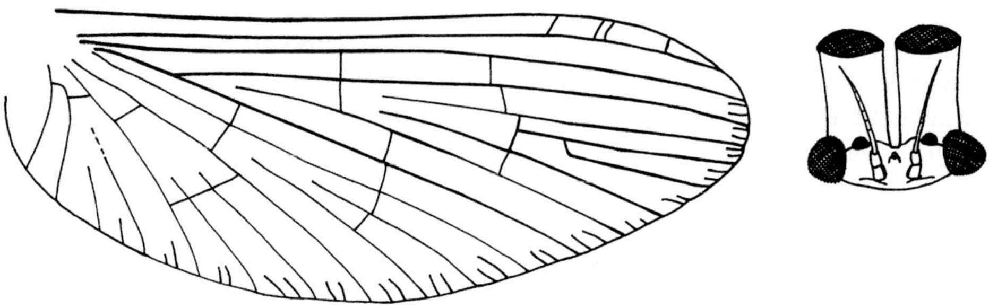

Fig. 2.282 The wing and head of a male *Cloeodes binocularis* in frontal view.

4. The length of the male and female is 2.75 mm, and their wing length is 2.5 mm. The head is yellow with reddish brown around the eyes, which are black at the bases. The thorax and legs are mainly reddish brown with some black markings. There is a short median spine at the apex of the subgenital plate (**Fig. 2.283**).
....*Cloeodes venezuelensis* (Traver, 1943) (Venezuela). Syn: *Pseudocloeon venezuelensis* Traver, 1943.
- The length is at least 4 mm. 5

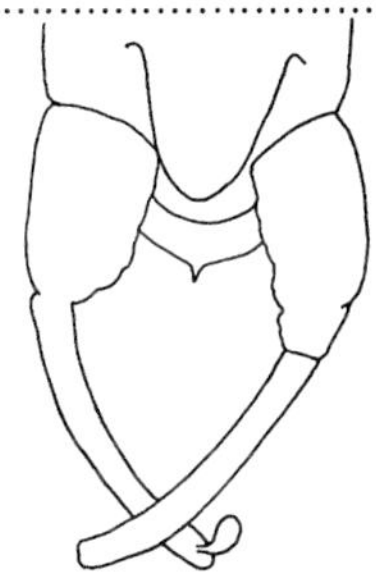

Fig. 2.283 The male genitalia of *Cloeodes venezuelensis* with the apical segment of a forceps arm broken off.

5. The costal angulation of the hind wing is strongly hooked. There are only two longitudinal veins in the hind wing. The basal third of the male forceps is thickened. The thickening in the basal third of the forceps limb is not obviously delimited as a segment (**Fig. 2.284**). Length: 5 mm; fore-wing length: 4.5 mm. Color: head and antenna yellowish; thorax yellowish or pale reddish brown with inconspicuous black submedian streaks running from the anterior margin of the pronotum; the anterior abdominal segments of the male are whitish; the middle segments are translucent, and the posterior segments are pale reddish brown. There are black stigmatic dots on the basal and middle segments.
....*Cloeodes aymara* (Traver, 1971) (Uruguay). Syn: *Baetis aymara* Traver, 1971.

- The costal angulation on the hind wing is fairly straight or obtusely rounded and not strongly hooked. The head and thorax are generally yellowish, and the abdomen is mainly translucent white. The basal third of the preapical segment of the male forceps is enlarged (**Fig. 2.285**). Body length of male: 4 mm. Wing length of male: 4 mm.
....*Cloeodes anduzei* (Traver, 1943) (Venezuela). Syn: *Pseudocloeon anduzei* Traver, 1943.

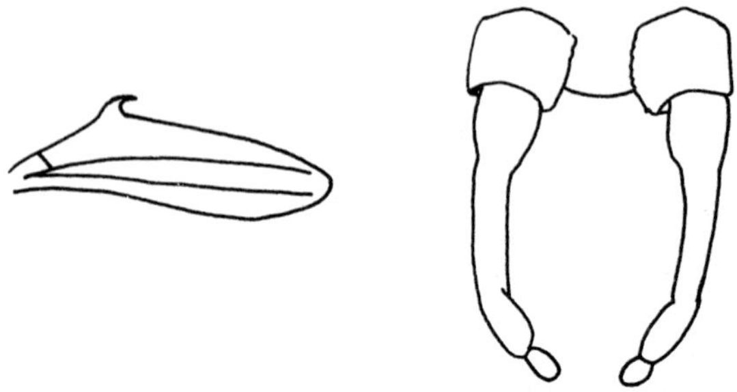

Fig. 2.284 The hind wing and male genitalia of *Cloeodes aymara.*

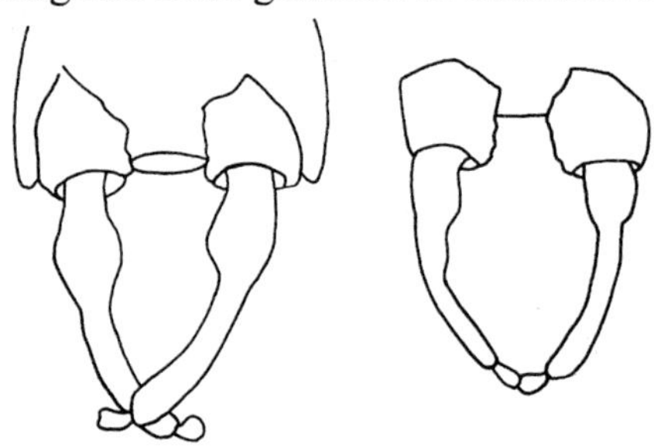

Fig. 2.285 The male genitalia of two specimens of *Cloeodes anduzei.*

6. Length: 3 to 3.5 mm. Wing length: 2.25 mm. The forceps arms of the male are not notably enlarged at the base (**Fig. 2.286**).
....*Cloeodes turbinops* (Needham and Murphy, 1924) (Guyana). Syn: *Pseudocloeon turbinops* Needham and Murphy, 1924.
- The length and length of the fore-wing are at least 3.8 mm. 7
7. The costal angulation of the hind wing is strongly hooked. The thickening in the basal third of the forceps limb is obviously delimited as an additional segment (**Fig. 2.195**). Length: 3.9 to 4.2 mm; fore-wing length: 3.8 to 4.0 mm. Hind wing length: 0.4 to 0.6 mm. Color: generally cream color with darker markings. The abdominal segments are cream with a faint darker pattern. This species is ecologically distinctive in its tolerance to short periods of habitat desiccation (Nolte *et al.*, 1996).
................................*Cloeodes hydation* McCafferty and Lugo-Ortiz, 1995 (Mato Grosso).

- The costal angulation on the hind wing is fairly straight or obtusely rounded but not strongly hooked (**Fig. 2.16**). In case of doubt, the length is at least 8 mm. 8

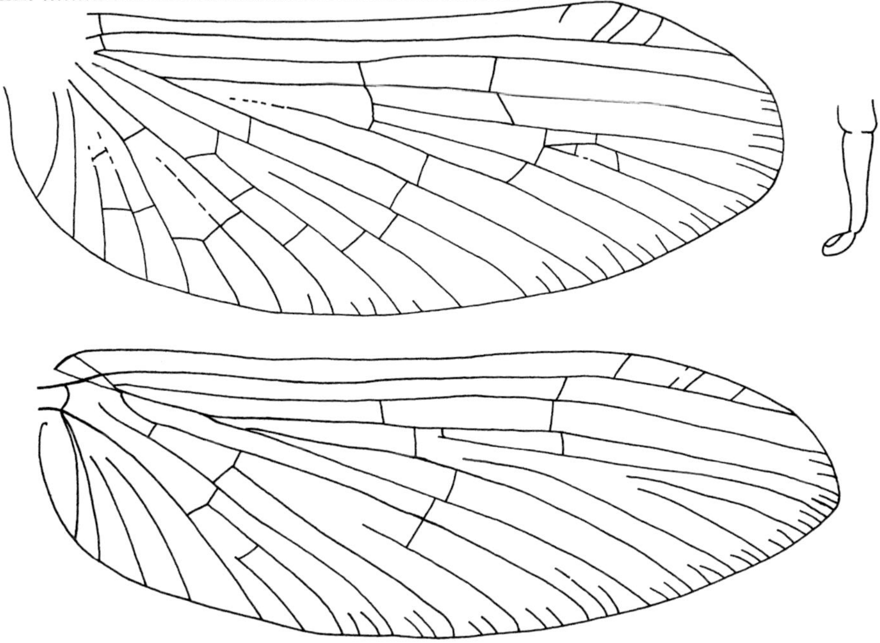

Fig. 2.286 *Cloeodes turbinops:* the wing of a female (above) and a male (below) and the forceps of a male (upper right).

8. The head and thorax are dark with pale marks. The abdomen is paler than the thorax with distinctive brown marks on each segment. The costal prominence on the hind wing is somewhat hooked with a broad base. The basal quarter of the basal segment of the male forceps is enlarged (**Fig. 2.16**). Length of male: 8 mm; fore-wing length of male: 7 to 8 mm; hind wing length of male: 0.5 to 0.8 mm. Length of female: 9 mm; fore-wing length of female: 9 mm; hind wing length of the female: 0.25 to 0.5 mm. .... *Cloeodes penai* (Morihara and Edmonds, 1980) (Argentina). Syn: *Notobaetis penai* Morihara and Edmonds, 1980.

- The head is yellowish brown with blackish eyes. The thorax is deep yellow with light yellow longitudinal stripes. The abdomen is yellowish brown with darkened articulations between the tergites. The hind wing has an obtusely rounded costal angulation and two longitudinal veins. It is three times as long as wide and a slightly concave posterior margin in the apical third of the wing. The legs are whitish yellow with gray anterior femora. Length of male: 3.8 mm. Fore-wing length: 5 mm. ....*Cloeodes nocturnus* (Navás, 1922) (Argentina). Syn: *Baetis nocturnus* Navás, 1922.

Key to the South American species of larval *Cloeodes*

Information for the key was taken from Waltz and McCafferty (1987a, b) and McCafferty and Lugo-Ortiz (1995). Larvae of most species remain undescribed.

1. Hind wing pads are absent. The maxillary palp apparently consists of two segments. The hind margin of the third abdominal tergite has 15 to 18 spines on each side of the midline. The paraprocts bear 7 to 12 marginal spines. Length: 4 mm. The head is brown. The thorax is brown mottled with white. The abdomen is brown with lighter markings dorsally (**Fig. 2.229**).
.................................*Cloeodes redactus* (Waltz and McCafferty, 1987) (Peru). If the adult shows no trace of hind wings, the species will probably have to be placed in another genus.
- Hind wing pads are present. The maxillary palp apparently consists of three segments (**Fig. 2.287**). 2

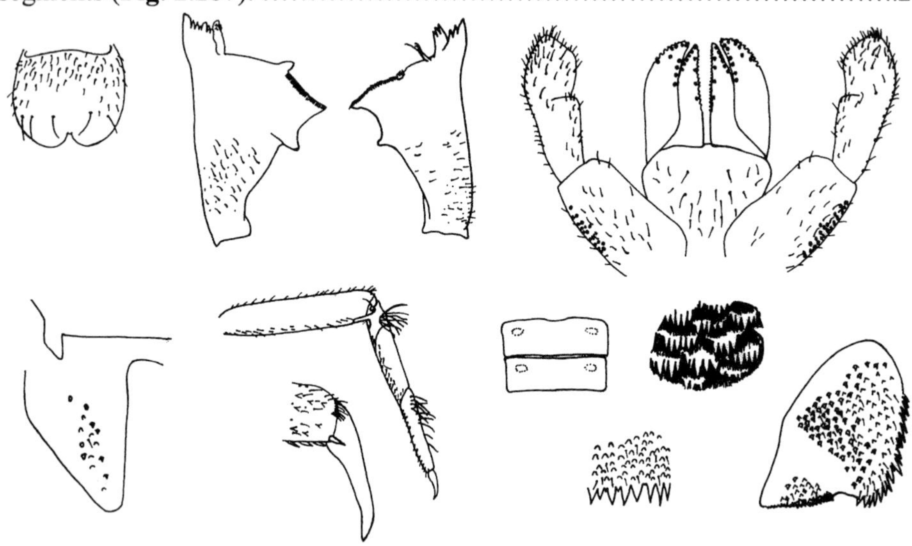

Fig. 2.287 *Cloeodes penai* larva (upper row, left to right): labrum, mandibles, labium; (lower row, left to right) the posterolateral corner of the metathorax showing the hind wing bud and the bud enlarged below, the hind leg with an enlargement of the claw (below), the sixth and seventh sternites with the location of the spiny area shown enlarged at right, the posterior margin of the sixth tergite (below the sternites), and a paraproct.

2. The dorsal color pattern on the abdomen consists of three or fewer light spots on any tergite. Some large, strong setae on the femora are serrate (**Fig. 2.287**). Length: 7 to 10 mm. Color: The head is light, and the thorax, uniformly tan.
....*Cloeodes penai* (Morihara and Edmunds, 1980) (Argentina). Syn: *Notobaetis penai* Morihara and Edmunds, 1980.

- The anterior tergites have more than three light spots (**Fig. 2.288**). None of the large, strong setae on the femora are serrate. 3

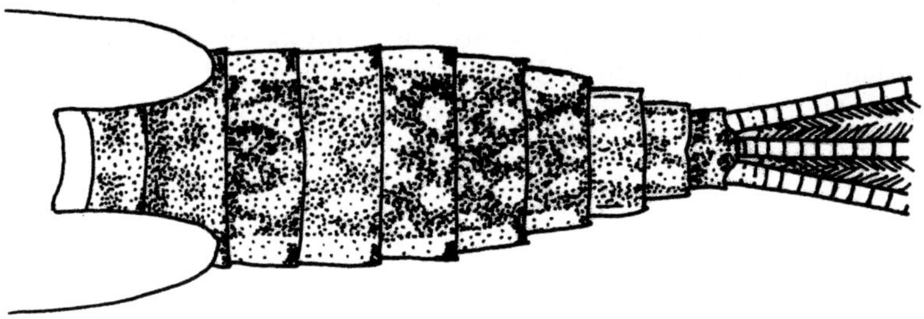

Fig. 2.288 The pattern on the abdominal tergites of a larval *Cloeodes irvingi.*

3. Each antenna is about 1.5 times as long as the frontal length of the lead capsule. Length: 5.5 mm. Color: the head is light brown with lighter markings. The thorax is uniformly tan. The abdomen is brown with lighter and darker markings dorsally (**Fig. 2.288**) and pale ventrally.
.....................................*Cloeodes irvingi* (Waltz and McCafferty, 1987) (Paraguay).
- Each antenna is about twice as long as the frontal length of the head capsule. Length: 3.4 to 4.7 mm. Color: the head is brown with lighter markings. The thorax is not uniform in color and has a pattern of light and darker brown markings (**Fig. 2.209**). This species is ecologically distinctive in the tolerance of the larvae to short periods of habitat desiccation (Nolte *et al.*, 1996).
................................*Cloeodes hydation* McCafferty and Lugo-Ortiz, 1995 (Mato Grosso).

Key to the species of *Rivudiva* larvae in South America

Information for the key was provided by Lugo-Ortiz and McCafferty (1998). The adults have not been described.

1. The scape and pedicel, or first two segments of the antenna, lack setae. There is a deep median emargination on the anterior margin of the labrum, and this margin is notably wider than the posterior one. The lingua, or central plate, of the hypopharynx bears a thick tuft of strong, simple setae (**Fig. 2.219**). Hind wing buds are evident, and there is no color pattern on the abdomen. Length: 5.9 to 6.4 mm. Length of caudal filament: 1.9 to 2.0 mm. Coloration: mainly light yellowish brown with no color pattern evident.
..............................*Rivudiva minantenna* Lugo-Ortiz and McCafferty, 1998 (Rio Grande do Sul, Santa Catarina).
- The scape and pedicel, or first two segments, of the antennae have a coating of long, simple setae on one side. The median emargination on the anterior margin

of the labrum is shallow, and the anterior margin is only slightly wider than the posterior one. The lingua, or central plate, of the hypopharynx does not bear a thick tuft of strong, simple setae (**Fig. 2.289**). Hind wing buds are not evident, and there are brown markings along the midline of the third through sixth abdominal tergites. Length: 5.5 to 5.5 mm. Length of caudal filament: 1.5 to 1.7 mm. Coloration: mainly light yellowish brown with a brown color pattern on the third through sixth abdominal tergites.
..............................*Rivudiva trichobasis* Lugo-Ortiz and McCafferty, 1998 (Rio Grande do Sul).

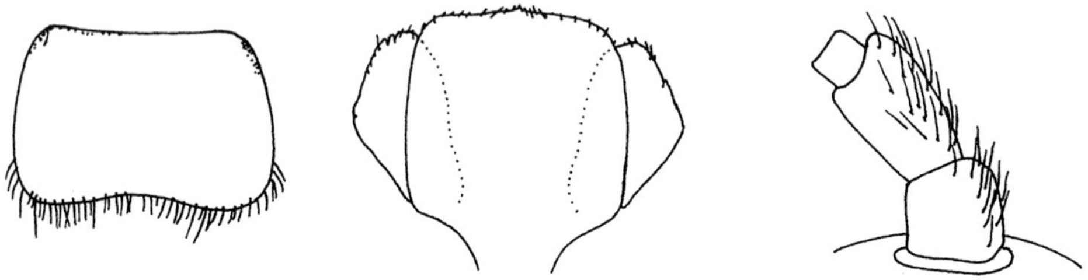

Fig. 2.289 *Rivudiva trichobasis* (left to right): labrum, hypopharynx, and the scape and pedicel of the antenna.

Key to the South American species of *Deceptiviosa* adults

Information for the key was taken from Navás (1933a), Lugo-Ortiz and McCafferty (1999a), and McCafferty (2000). The adult *Deceptiviosa ardua* has not been described.

1. There are three long veins in the apical part of the hind wing, the first and third of which arise from a common fork, while the second is a long intercalary vein. In the costal space of the stigmatic area, some of the cross veins are joined to each other by other short veins (**Fig. 2.290**). Length of subadult: 4 mm. Fore-wing length: 6.5 mm; hind wing: 2 mm. Color of subadult: reddish brown and black with gray pubescence. Adults have not been described.
....*Deceptiviosa angolina* (Navás, 1933) (Chile). Syn: *Baetis angolinus* Navás, 1933.
- In the hind wing, there are five longitudinal veins, the second and fourth of which arise from a common fork (**Fig. 2.206**). Length: 4.0 to 5.1 mm. Fore-wing length: 5.0 to 5.5 mm.
..............................*Deceptiviosa torrens* Lugo-Ortiz and McCafferty, 1999 (Chile, Argentina).

Fig. 2.290 The stigmatic area of the fore-wing and the hind wing of *Deceptiviosa angolina.*

Key to the South American species of larvae in the genus *Deceptiviosa*

Information for the key was taken from Lugo-Ortiz and McCafferty (1999a). The larva of *Deceptiviosa angolina* has not been described.

1. The median caudal filament is 50% to 60% as long as a cercus. The head capsule is longer than wide. The labrum has a submarginal row of 10 to 12 long, fine setae. The margin of the gills is weakly serrated and bears many fine, minute setae (**Fig. 2.291**). Length: 6.5 to 8.2 mm; length of the cerci: 5 to 5.5 mm. Color: head, brown or pale yellowish brown; thorax, dark yellow or brown with darker markings; abdomen, yellowish brown with dark brown markings. ....*Deceptiviosa ardua* Lugo-Ortiz and McCafferty, 1999 (Chile).

- The median caudal filament is 15% to 20% as long as a cercus. The head capsule is wider than long. The labrum has a submarginal row of four to six long, fine setae. The margin of the gills is not serrated and bears fine, minute setae arranged in groups of three and four (**Fig. 2.220**). Length: 6 to 7 mm; length of the cerci: 4.5 to 5.5 mm. Color: brown or cream with darker markings. ....*Deceptiviosa torrens* Lugo-Ortiz and McCafferty, 1999 (Chile, Argentina).

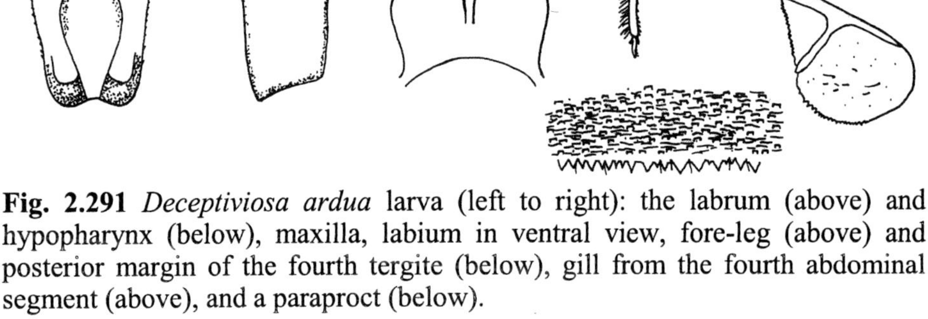

Fig. 2.291 *Deceptiviosa ardua* larva (left to right): the labrum (above) and hypopharynx (below), maxilla, labium in ventral view, fore-leg (above) and posterior margin of the fourth tergite (below), gill from the fourth abdominal segment (above), and a paraproct (below).

Leptophlebiidae

Key to the genera of adults in South America

Information was provided by Ulmer (1920a, b), Navás (1934b), Traver (1946, 1947a, 1959b), Demoulin (1955a, b, d, f, 1966a), Peters (1969), Peters and Edmunds (1972), Savage and Peters (1978), Flowers (1980, 1987), Pescador and Peters (1982), Savage (1983a, 1987a), Domínguez (1988b), Savage and Domínguez (1992), and Domínguez *et al.* (1992, 1997). Some genera can be identified positively only from adult males. Recent authors have tended to create a new genus for almost every new species, although morphological differences among many of them are minimal. Hence, it is more difficult to identify many of the genera in this family than to identify species of many other insect taxa. Adult *Bessierus*, *Dactylophlebia, Gonserellus, Magallanella,* and *Perissophlebiodes* are unknown; these genera can still only be identified from larvae. Adult *Leentvaaria* have not been described, but the arrangement of the wing veins was determined from wings dissected by Demoulin (1966a) from larval wing buds **(Fig. 2.292)**. Adult characteristics of South American *Atalophlebioides* Peters, 1930, were determined for species that have since been transferred to other genera. Hence, the genus can presently be identified reliably only from larval characteristics. Several South American species were assigned to the genera *Deleatidium* Eaton, 1899, and *Atalophlebia* Eaton, 1881, by the authors describing them, but all have since been transferred to other genera.

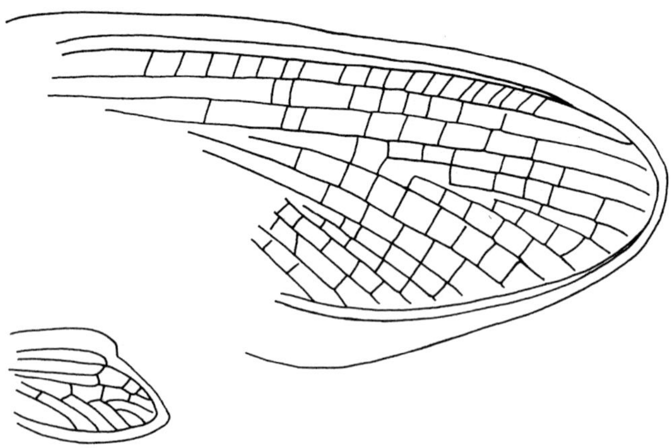

Fig. 2.292 *Leentvaaria palpalis:* the undeveloped fore and hind wings of a subadult dissected from the last instar larva.

1. Hind wings are absent. The female has an ovipositor (**Fig. 2.293**). 2
 - Hind wings are present, but may be very small (**Fig. 2.294**). An ovipositor is lacking, except in *Hapsiphlebia* Peters and Edmunds (1972); the West Indian *Borinquena* Traver, 1938; and *Hagenulus* Eaton, 1882. ………..…………...........3

2. Vein ICu is not attached to vein CuA at its base, and the fork in vein MA is symmetrical. The penis lobes of the male bear sub-apical ventral spines (**Fig. 2.295**). The ovipositor is very long.
....*Hagenulopsis* Ulmer, 1920..p. 281
- Vein ICu is attached to vein CuA at its base. The fork in vein MA is asymmetrical. The penis lobes of the male do not bear sub-apical ventral spines. The ovipositor reaches only to the posterior end of the ninth abdominal segment. The claws in each pair are dissimilar: one is hooked and the other, pad-like (**Fig. 2.293**).
....*Askola* Peters, 1969

Askola froehlichi Peters, 1969, found in São Paulo, Paraná, and Santa Catarina, is the only known *Askola* species. The male is 6.4 mm long with 7.0 mm wings. Its general coloration is brownish with lighter and darker markings.

3. The claws are similar; all are acutely pointed (**Fig. 2.294**). 4
- The tarsal claws are dissimilar, one is hooked apically, and the other is widened and pad-like or blunt (**Fig. 2.296**). 13
4. The basal 1/2 to 2/3 of the penis lobes are fused (**Fig. 2.297**). 5
- The penis lobes are tubular, separate for their whole lengths, and bear appendages (**Fig. 2.294**). 8

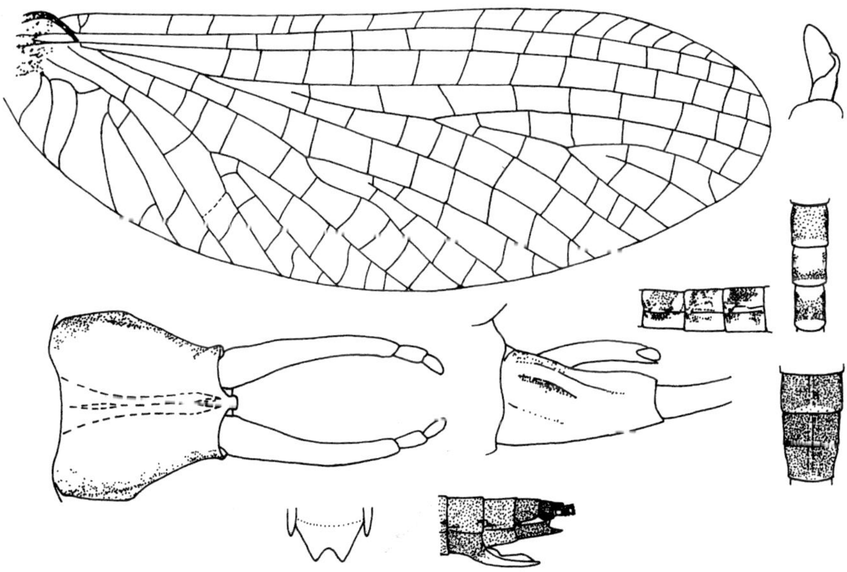

Fig. 2.293 *Askola froehlichi:* wing (upper left), the tarsal claws (upper right), the apex of the abdomen of a male in ventral (lower left) and lateral (middle) view, the fifth through seventh abdominal segments of an adult male in lateral and dorsal view (middle right center and right), the fifth through seventh segments of an adult female in dorsal view (lower right), the ninth sternite of an adult female (lower left center), and the apex of a female abdomen in lateral view showing the ovipositor (lower right center).

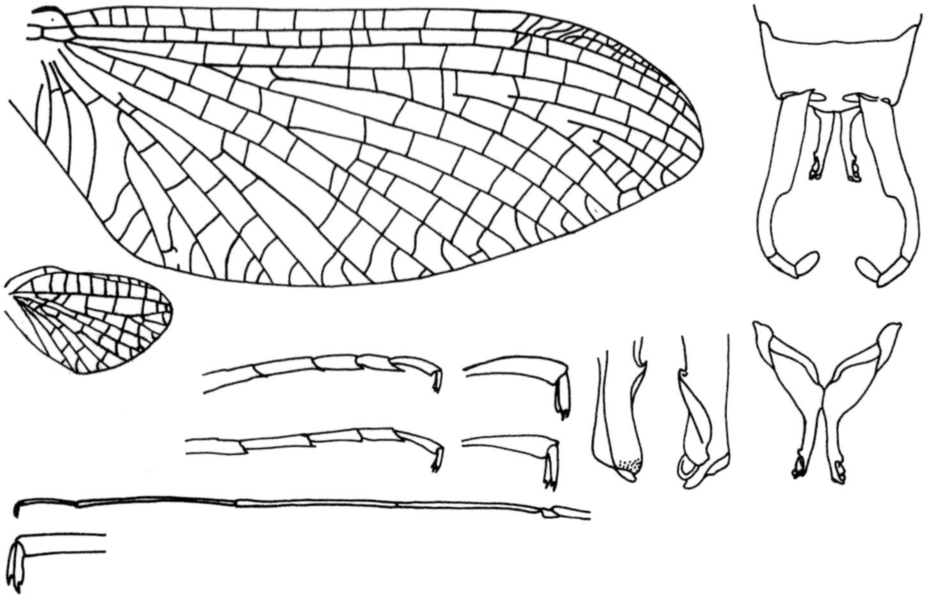

Fig. 2.294 *Hapsiphlebia anastomosis* adult male: fore and hind wing (upper left); genitalia in ventral view (upper right); penis in ventral view (lower right) beside enlarged dorsal and ventral views of the apex of a penis lobe (lower right center); and the hind, middle, and fore-tarsus beside enlargements of the tarsal claws of each (lower left center and left, top to bottom).

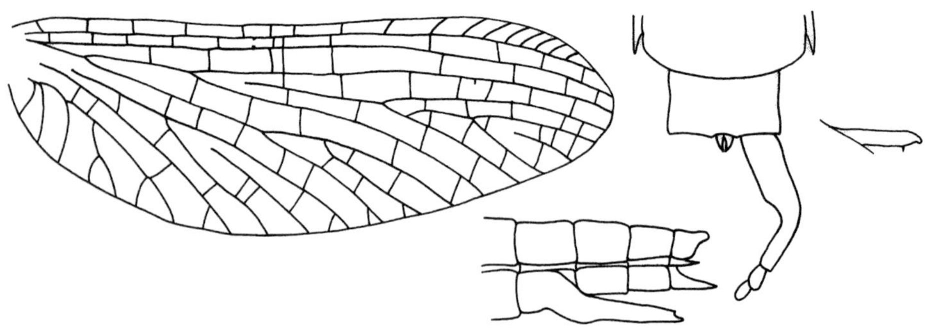

Fig. 2.295 *Hagenulopsis diptera* male: the wing (upper left), male genitalia in ventral view with one of the forceps limbs removed (upper right center), penis in lateral view showing subapical spine (right), and the female genitalia in lateral view showing the ovipositor (lower center).

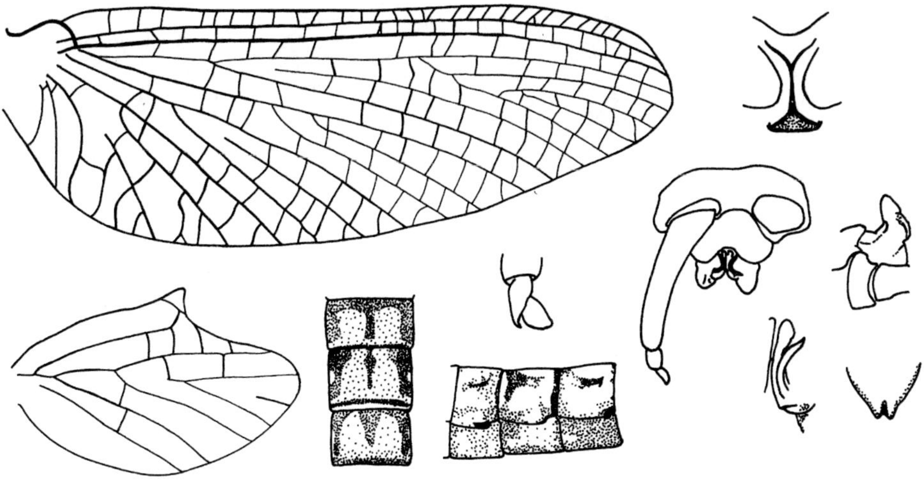

Fig. 2.296 *Hylister plaumanni:* the fore and hind wing with the hind wing drawn larger than scale (left), the prosternum (upper right), the fifth through seventh abdominal segments in dorsal (lower left center) and lateral view (lower center), the tarsal claws of a male (above abdomen), the male genitalia in ventral view with the left forceps segments removed above an enlargement of the ventral process on the penis and a lateral view of the genitalia (middle right), and the apex of the ninth sternite of a female (lower right).

5. In for fore-wing, vein MP_2 has a moderate sigmoid curve and is attached at the base to MP_1 by a cross vein more than 1/3 of the distance from the base to the margin. The proximal half to two thirds of the penis lobes are fused and apical third of each is tubular, separate from the other, and bears a short sclerotized process on the inner margin. The subanal plate of the female is excavated apically (**Fig. 2.298**).
....*Nousia* Navás, 1918..p. 281
A common junior synonym of this genus is *Atalonella* Needham and Murphy, 1924.
- The apical portions of the penis lobes remain in contact and bear no processes or only the basal half of the penis lobes are fused, and a large subapical spine is present (**Fig. 2.297**). 6

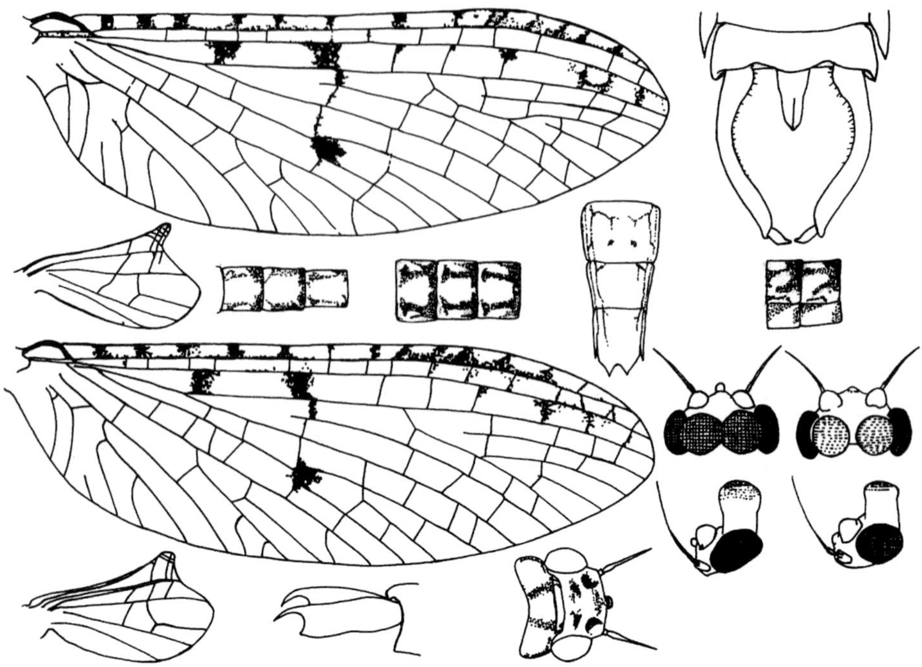

Fig. 2.297 *Fittkaulus maculatus:* fore-and hind wings of a male (upper left) and a female (lower left) drawn with the hind wings enlarged to about double their natural size relative to the fore-wing, male genitalia in ventral view (upper right), heads (lower right) of a male imago (left) and subimago (right) in dorsal and lateral view, head and pronotum of a female in dorsal view (lower right center), claw on the fore-leg of a male (lower left center), and (middle row, left to right from hind wing): color pattern on the fourth to sixth tergites of a male and a female, seventh to ninth abdominal segments of a female in ventral view, and third and fourth abdominal segments of a female in lateral view.

6. The basal half of the penis lobes are fused, and a large subapical spine is present. The fork in vein MA of the fore-wing is symmetrical. The claws on each tarsus are similar, with one acute point and a smaller, opposing point (**Fig. 2.299**). The female has not been described.
....*Archethraulodes* Pescador and Peters, 1982

The only species in this genus yet described is *Archethraulodes spatulus* Pescador and Peters, 1982, from Argentina and Chile. The body of the male is 7 to 8 mm long, with a 7 to 8.5 mm fore-wing. It is mainly yellow with dark markings.

- The basal 2/3 of the penis lobes are fused, and they bear no sclerotized processes. The fork in vein MA of the fore-wing is asymmetrical (**Fig. 2.297**).
....7

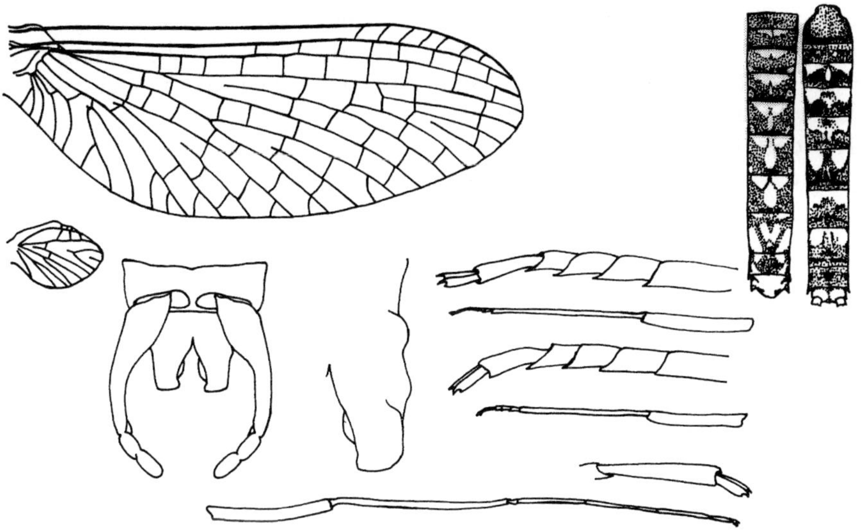

Fig. 2.298 *Nousia minor,* male adult: wings (upper left), pattern on the dorsal (right center) and ventral (far right) surfaces of the abdomen, genitalia with an enlargement of the right penis lobe (lower left center), and (right center, top to bottom) the hind, middle and fore-legs, each with an enlargement of its tarsi or its apical tarsal segment with claws.

7. The costal projection on the hind wing is very large, extending the anterior margin considerably. The eyes are turbinate, set on stalks (**Fig. 2.297**). Fore-wing length of male: 4.8 to 5.2 mm; of female: 5.3 to 6.0. Color: yellowish with darker brownish markings.
....*Fittkaulus* Savage and Peters, 1978..p. 287
- The costal projection on the hind wing is rounded and only moderately alters the profile of the wing (**Fig. 2.300**).
....*Terpides* Demoulin, 1966

The only species in this genus known from South America is *Terpides guyanensis* Demoulin, 1966, from Surinam.

8. The costal prominence in the hind wing is very well developed, and vein Sc ends in the angulation apical to this costal prominence or in a cross vein at this level; the penis lobes are more or less conical, each bearing a slender projection, and the forceps base is not divided (**Fig. 2.301**).
....*Thraulus* Eaton, 1881

Traver (1959b) did not believe that this genus occurs in the Neotropical zone, so the species assigned to it would have to be reassigned to still undescribed genera. There are no definitive ways to delimit its species from those of *Choroterpes* Eaton, 1892, and some of the other genera, but this is true of many other nominal genera in this family, as well. Because of its poor description,

Thraulus costalis Navás, 1934, could not be transferred to another genus by Traver (1960) and remains tentatively in this genus. However, the costal prominence of the hind wing (**Fig. 2.301**) does not fit the description of *Thraulus* in this couplet. Its true taxonomic status will require examination of additional specimens to determine.

- The costal prominence in the hind wing is rounded and sometimes only slightly developed, and vein Sc ends well beyond it (**Fig. 2.302**). 9

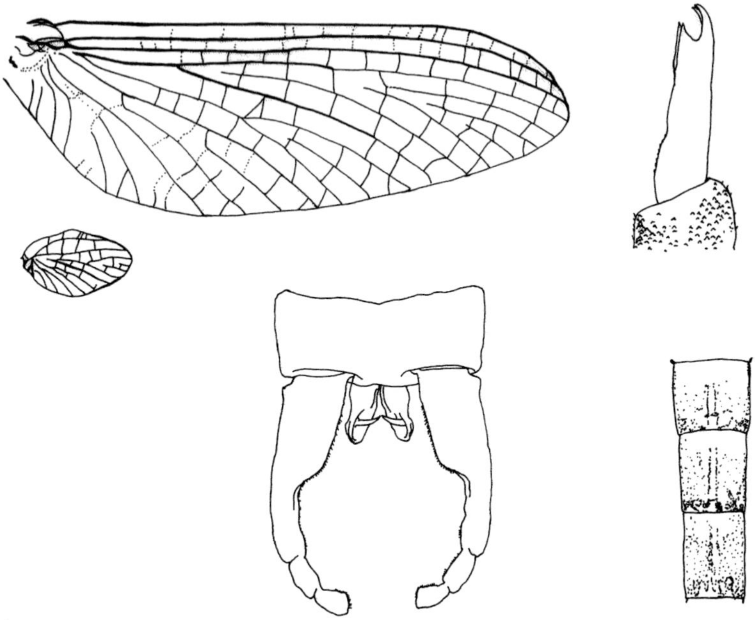

Fig. 2.299 *Archethraulodes spatulus:* fore and hind wings (upper left), claws (upper right), pattern on abdominal tergites (lower right), and genitalia of a male (lower center).

9. The transverse veins in the apical third of the costal area of the fore-wing form an anastomosing network. The female has a well-developed ovipositor that extends just beyond the anterior margin of the eighth abdominal segment (**Figs. 2.294**).
....*Hapsiphlebia* Peters and Edmunds, 1972

The only species of this genus yet described is *Hapsiphlebia anastomosis* (Demoulin, 1955) from Chile, formerly known as *Atalophlebia anastomosis* Demoulin, 1955. The length is 13 to 14 mm. The length of the fore-wing is 14 to 16 mm, and the hind wing is 4 to 4.5 mm. The general color is brown with yellowish markings.

- The transverse veins in the costal area of the fore-wing do not form an anastomosing network. The female lacks an ovipositor (**Fig. 2.302**). 10

10. Vein MP_2 in the fore-wing shows a strong sigmoid curve. The subanal plate of the female is excavated apically **(Fig. 2.302)**. 11
- Vein MP_2 in the fore-wing is only moderately curved **(Fig. 2.303)**. 12

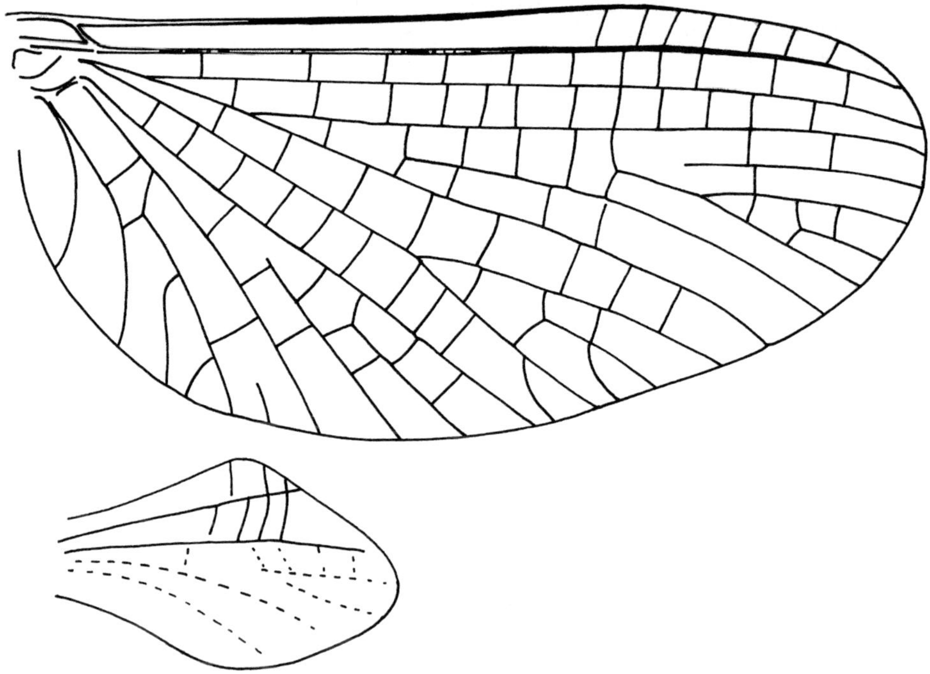

Fig. 2.300 The fore-wing and enlarged hind wing of *Terpides guyanensis.*

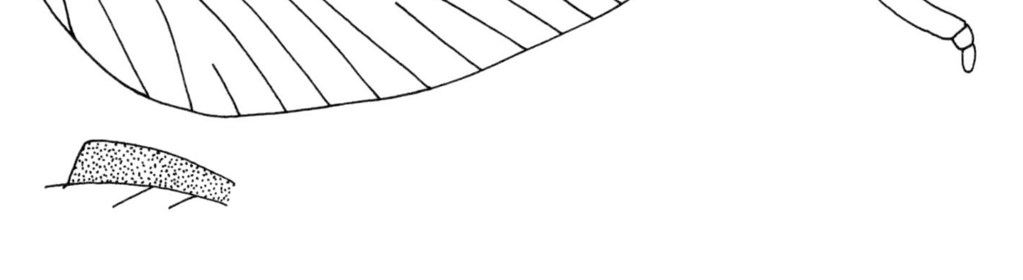

Fig. 2.301 The fore-wing with many cross veins omitted, costal area of the hind wing, forceps arm, and apex of the penis of *Thraulus costalis* (left to right).

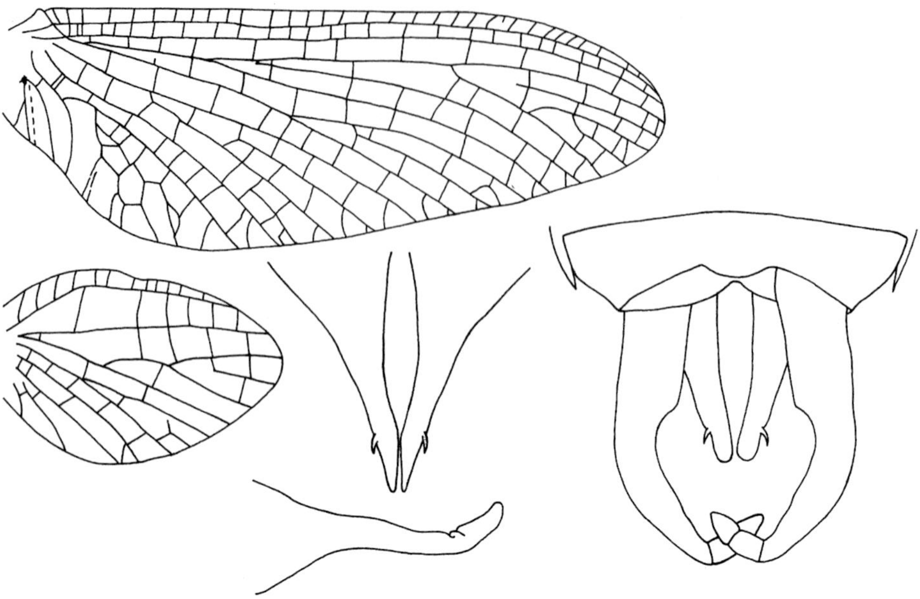

Fig. 2.302 *Massartella alegrettae* male: fore and hind wings (left) not showing the dark clouds and brownish yellow toning of the wing membrane and with the hind wing shown larger than scale, the male genitalia in ventral view (lower right), the penis lobes in posteroventral (center) and lateral view (lower center).

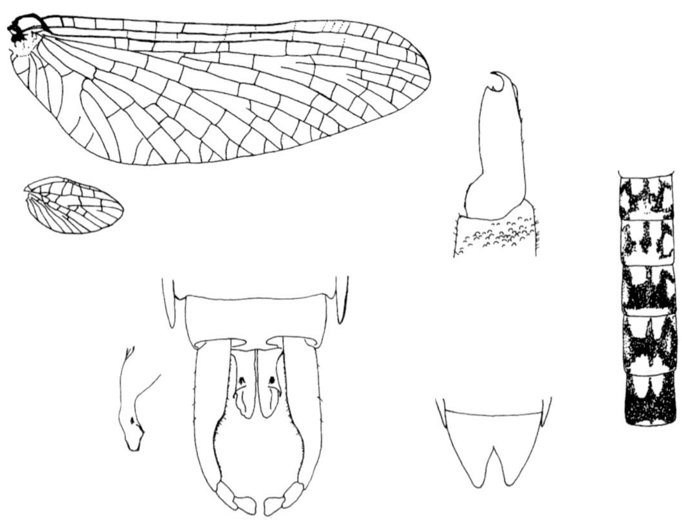

Fig. 2.303 *Rhigotopus andinensis*: fore and hind wings (upper left), claw (upper right center), color pattern on middle abdominal tergites (right), genitalia of a male with one penis in lateral view to the left (lower left), and the subanal plate of a female (lower right center).

11. Vein ICu_1 in the fore-wing joins vein CuA at the base. The distal section of vein MA is not bowed toward the posterior margin (**Fig. 2.17**).
....*Penaphlebia* Peters and Edmunds, 1972..p. 288
- Vein ICu_1 in the fore-wing is joined to vein CuA+CuP by a cross vein. Cu_2 in the forewing is forked, giving rise to the first cubital intercalary, which is attached by a cross vein to CuA and CuP. The distal section of vein MA is bowed toward the posterior margin (**Fig. 2.302**). The separated tubular penis lobes each bear one or two lateral, subapical spines or projections on the outer margin.
....*Massartella* Lestage, 1930..p. 299

12. The subanal plate of the female is excavated apically. Each of the penis lobes of the male has an outward and a dorsal projection from 1/2 to 2/3 of the distance from the base to the apex (**Fig. 2.303**). The larvae are undescribed.
....*Rhigotopus* Pescador and Peters, 1982

The only species yet described in this genus is *Rhigotopus andinensis* Pescador and Peters, 1982, from Argentina and Chile. Its body is 8 to 9.5 mm long, and its fore-wing, 9 to 10.5 mm. Coloration: brown and yellow with a few black markings.

- The subanal plate of the female is not excavated apically. The penis lobes are separated and tubular, each with a dorsal, apical projection (**Fig. 2.304**).
....*Massartellopsis* Demoulin, 1955

The only species in this genus yet described is *Massartellopsis irarrazavali* Demoulin, 1955, from Argentina, Chile, and Bolivia. Its body is 8.5 to 9 mm, and its fore-wing is 10 to 11 mm.

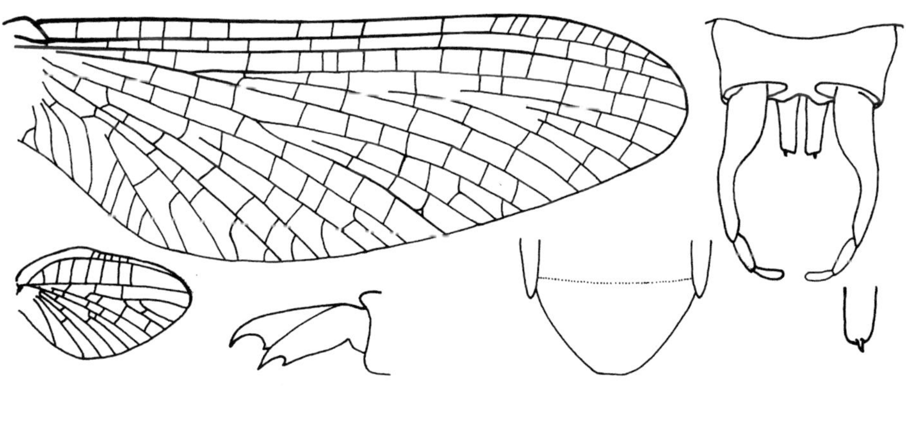

Fig. 2.304 *Massartellopsis irarrazavali:* fore and an enlarged hind wing (left), tarsal claws (lower left center), male genitalia with an enlarged apex of a penis (right), and apex of a female abdomen in ventral view (lower right center).

13. The fork in vein MA of the fore-wing is symmetrical (**Fig. 2.296**). 14
- The fork in vein MA of the fore-wing is asymmetrical (**Fig. 2.305**). 24
14. The basal 1/3 to 1/2 of the penis lobes are fused, and the apical parts bear some kind of appendage (**Fig. 2.296**). 15
- The penis lobes are not fused basally, and they lack appendages (**Fig. 2.306**). 20
15. The body and legs of adults and sub-adults are yellow with black or dark brown species-specific markings on the tibiae and apical segments of the abdomen. The apical 2/3 of the penis lobes are narrow and twisted ventrally and laterally near the apex. The upper portions of the eyes are on short stalks. The costal projection on the hind wing is 5/6 of the distance from the base to the apex, and the cross veins are concentrated in the apical third. Vein MP_2 in the fore-wing is connected to vein IMP by a cross vein. The costal cross veins in the apical part of the fore-wing are strongly curved and show some anastomosis (**Fig. 2.307**).
....*Atopophlebia* Flowers, 1980..p. 300
- The adults and subadults are not yellow with dark markings on the tibiae and apical abdominal segments. The penis lobes are not twisted ventrally and laterally at the apex but bear more differentiated appendages. If in doubt, one of the other characteristics is absent (**Fig. 2.296**). 16

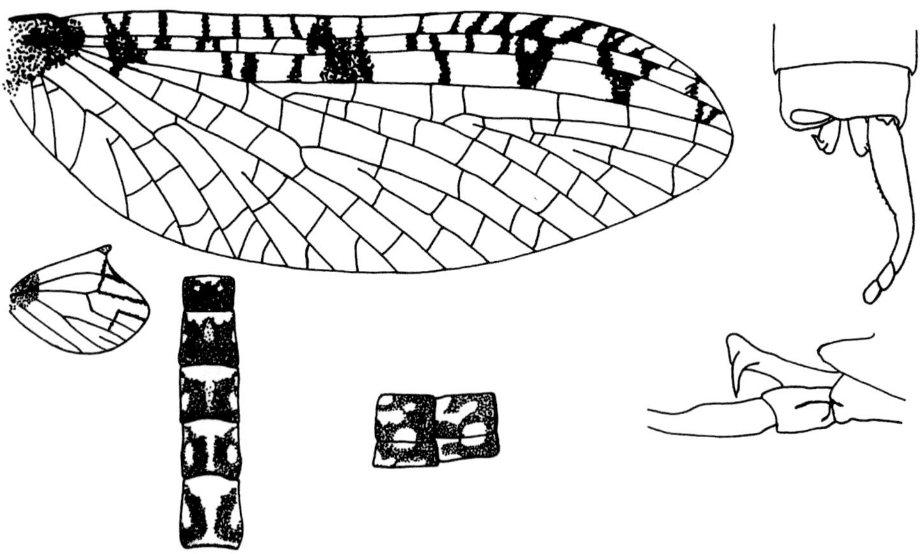

Fig. 2.305 *Ecuaphlebia rumignaui* male: the fore and hind wings (left), the dorsal color pattern of the second through sixth abdominal tergites (lower left center), the lateral color pattern on the fifth and sixth abdominal segments (lower center), the genitalia in ventral view with one of the forceps removed to reveal the appendage on the penis lobule (upper right), and a penis and base of a forceps in lateral view (lower right).

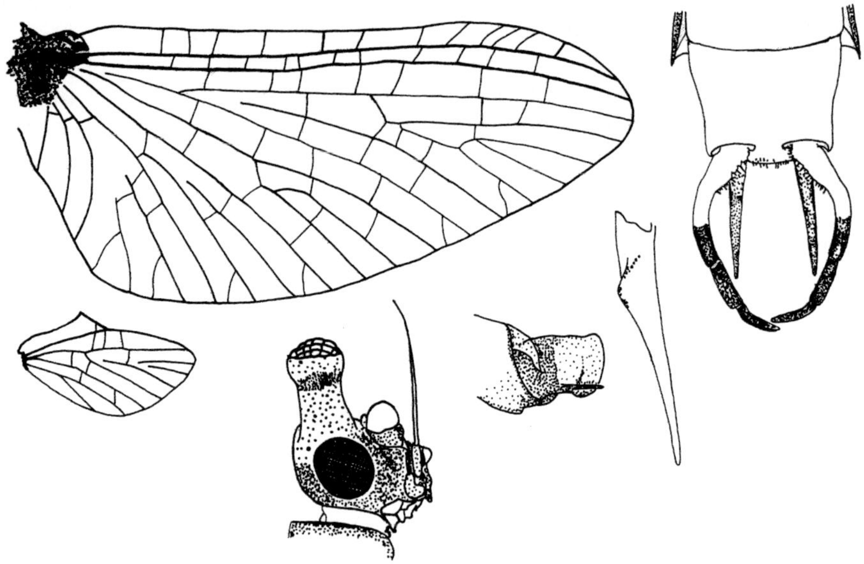

Fig. 2.306 *Miroculis rossi* male: fore and hind wing (left), the head in lateral view (lower left center), the tarsal claw on the fore-leg (lower right center), the male genitalia in ventral view (right), and an enlarged penis lobe in ventral view.

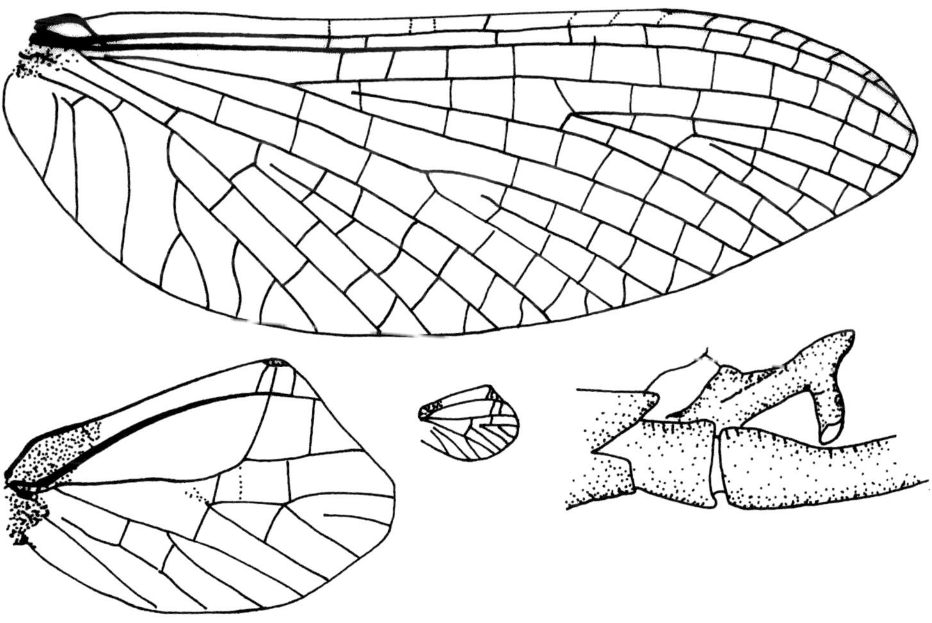

Fig. 2.307 *Atopophlebia obrienorum:* fore-wing (above), the hind wing greatly enlarged (lower left) and another to the same scale as the fore-wing (lower center), and the male genitalia in lateral view (lower right).

16. The fork in vein MP of the fore-wing is symmetrical. Each of the penis lobes bears a ventral appendage that is directed laterad (**Fig. 2.308**).
....*Farrodes* Peters, 1971..p. 302
- The fork in vein MP of the fore-wing is asymmetrical. Each of the penis lobes bears a ventral appendage that is directed toward the base of the penis (**Fig. 2.296**). 17

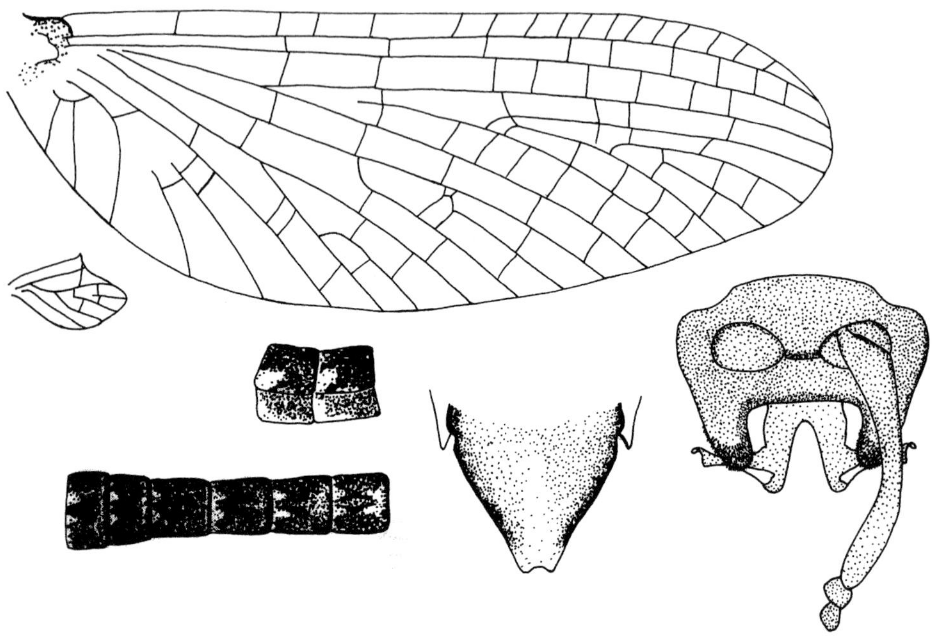

Fig. 2.308 *Farrodes yungaensis:* the fore and hind wing (upper and middle left), the fifth and sixth abdominal segments in lateral view above the second through seventh segments in dorsal view (lower left), the male genitalia in ventral view with the left forceps segments removed (lower right), and the ninth sternite of the female in ventral view (lower right center).

17. The subgenital plate of the male lacks spine-like projections. The apical fourth of the penis lobes are divergent, and their ventral processes are in the form of apically divergent spines (**Fig. 2.296**).
....*Hylister* Domínguez and Flowers, 1989

The only known South American species in this genus is *Hylister plaumanni* Domínguez and Flowers, 1989, known from Minas Gerais, Paraná, Rio de Janeiro, and Santa Catarina. The length of the males and females is 8.1 to 9.2 mm, and their fore-wings are 8.2 to 10 mm long.
- The subgenital plate has projecting processes (**Fig. 2.309**). 18

Fig. 2.309 *Traverella montium:* the fore and hind wings (left) and the genitalia of a male in ventral (lower right) and lateral (lower center) view.

18. The processes on the penis lobes of the male are nearly straight spines, usually directed toward the edge of the body. The subanal plate of the female is slightly excavated apically. The base of the forceps of the male sometimes bears two slender, rod-like projections (**Fig. 2.309**).
....*Traverella* Edmunds, 1948..p. 313
- The processes on the penis lobes are curved spines or lobes (**Fig.2.310**). 19

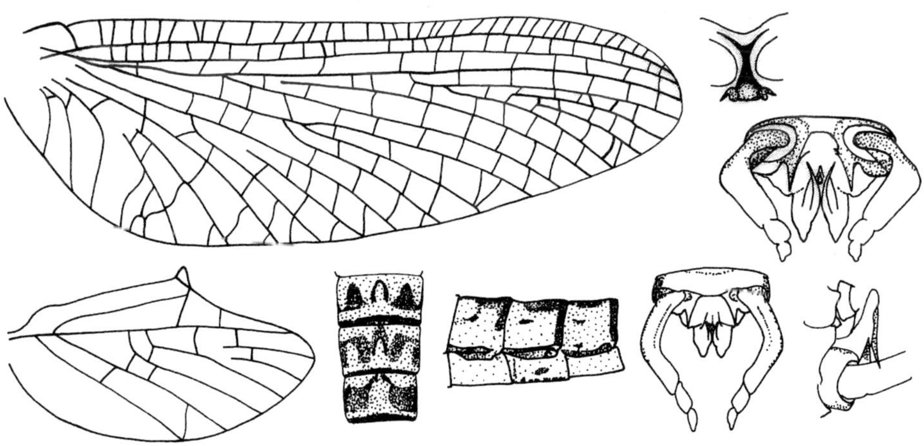

Fig. 2.310 *Needhamella ehrhardti:* the fore and hind wing with the hind wing drawn larger than scale (left), the prosternum (upper right), the fifth through seventh abdominal segments in dorsal (lower left center) and lateral view (lower center), and the male genitalia in ventral (lower right center), apical (middle right), and lateral view (lower right).

19. The ventral processes on the penis lobes are narrow, apically divergent spines. Those on the subgenital plate are acute spines around the base of the forceps (**Fig. 2.310**).
....*Needhamella* Domínguez and Flowers, 1989

The only species in this genus is *Needhamella ehrhardti* (Ulmer, 1920) from Argentina, Uruguay, Santa Catarina, and probably Paraguay, as well. It is 8 to 9 mm long; the length of the fore-wing of the male is 8 to 9 mm, and that of the female is 11 mm. It has also been called *Thraulus ehrhardti* Ulmer, 1920, and *Traverella ehrhardti* (Ulmer, 1920). The bases of both wings and the costal margins of the fore-wings are distinctly tinged with yellow or orange brown; the rest of the wings are almost hyaline. Each femora has two dark bands.

- The ventral processes of the penis lobes are small lobes or short curved spines. The subgenital plate bears processes at the bases of the penis lobes. There are only about 12 cross veins between the costa and subcosta of the fore-wing (**Fig. 2.311**).
....................................*Hermanella* Needham and Murphy, 1924..p. 316

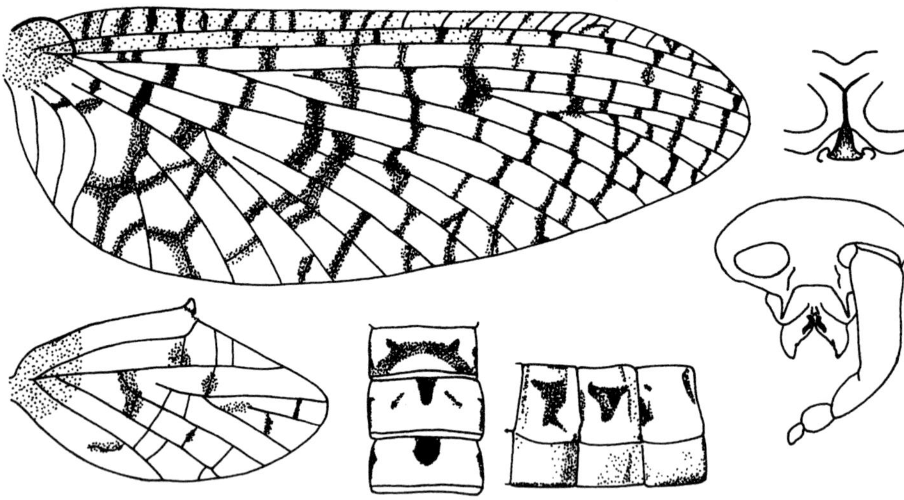

Fig. 2.311 *Hermanella maculipennis:* the fore and hind wing with the hind wing drawn larger than scale (left), the prosternum (upper right), the fifth through seventh abdominal segments in dorsal (lower left center) and lateral view (lower right center), and the male genitalia in ventral view with the right segments of the forceps removed (lower right).

20. The claws on the fore-tarsi of the male are different in shape than the claws on the middle and hind legs. The eyes are turbinate, set on stalks (**Fig. 2.306**).
....*Miroculis* Edmunds, 1963..p. 320

- The corresponding claws on all tarsi are almost the same shape. 21

21. The costal projection on the hind wing is only moderately elevated but pointed, and its posterior side forms an obtuse angle with the rest of the costal margin. Vein Sc terminates beyond the costal projection (**Fig. 2.312**). The differences between this genus and the two in the next couplet are minimal.
.... *Miroculitus* Savage and Peters, 1982

The only South American species in this genus is *Miroculitus emersoni* (Needham and Murphy, 1924) from Guyana, formerly known of as *Choroterpes emersoni* Needham and Murphy, 1924. It lacks a distinct color pattern on the head and thorax, which are reddish brown darkening to black on the head. It is 4 to 4.5 mm long, and its fore-wing length is 5 mm.

- The costal projection on the hind wing is well developed and projects considerably beyond the line of the costal margin, its posterior side forming approximately a right angle with the rest of the costal margin. Vein Sc terminates in the costal margin on the posterior side of the projection (**Fig. 2.313**). 22

22. The costal projection on the hind wing is acutely pointed at the tip. The lobes of the penis terminate in acute points (**Fig. 2.313**).
.... *Hermanellopsis* Demoulin, 1955..p. 331

- The costal projection on the hind wing is rounded at the apex. The apices of the lobes of the penis are narrow and rounded (**Fig. 2.314**). 23

23. Each forceps arm of the male consists of four segments (**Fig. 2.314**). The stem of MA in the fore-wing does not curve in an arc.
.... *Choroterpes* Eaton, 1892

The only South American species presently assigned to this genus is *Choroterpes bilineata* Needham and Murphy, 1924, from Peru and Surinam. Its thorax is yellow with two black stripes. Its body length is 6.5 to 7 mm, and its wing is 7 to 9 mm long.

- Each forceps arm of the male consists of three segments (**Fig. 2.315**).
.... *Microphlebia* Savage and Peters, 1982..p. 332

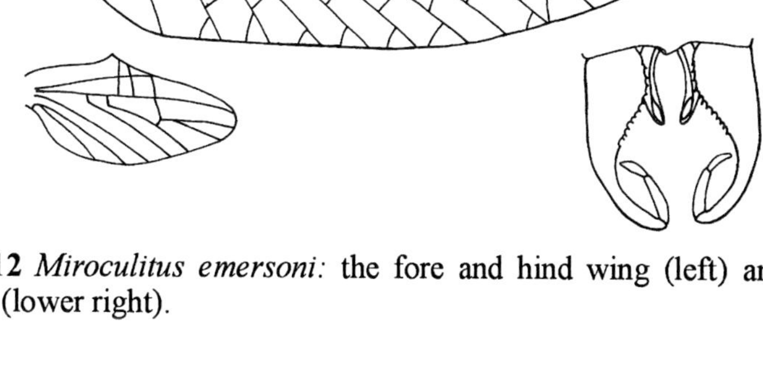

Fig. 2.312 *Miroculitus emersoni:* the fore and hind wing (left) and the male genitalia (lower right).

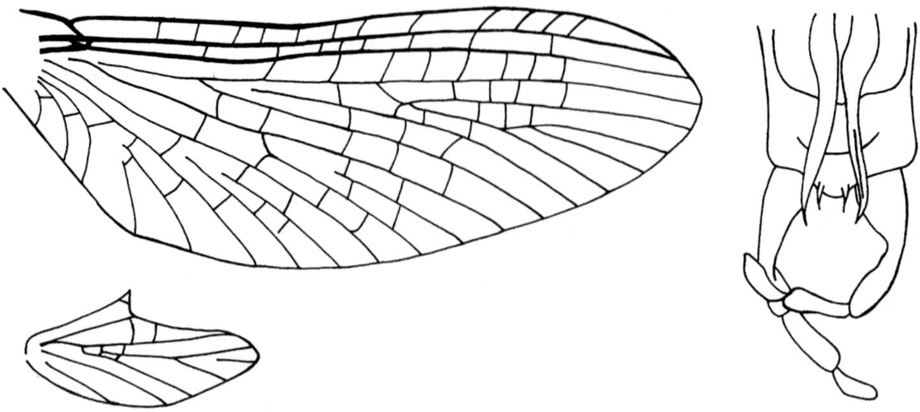

Fig. 2.313 *Hermanellopsis incertans:* the fore and hind wings with the hind wing enlarged (left) and the genitalia of the male (right).

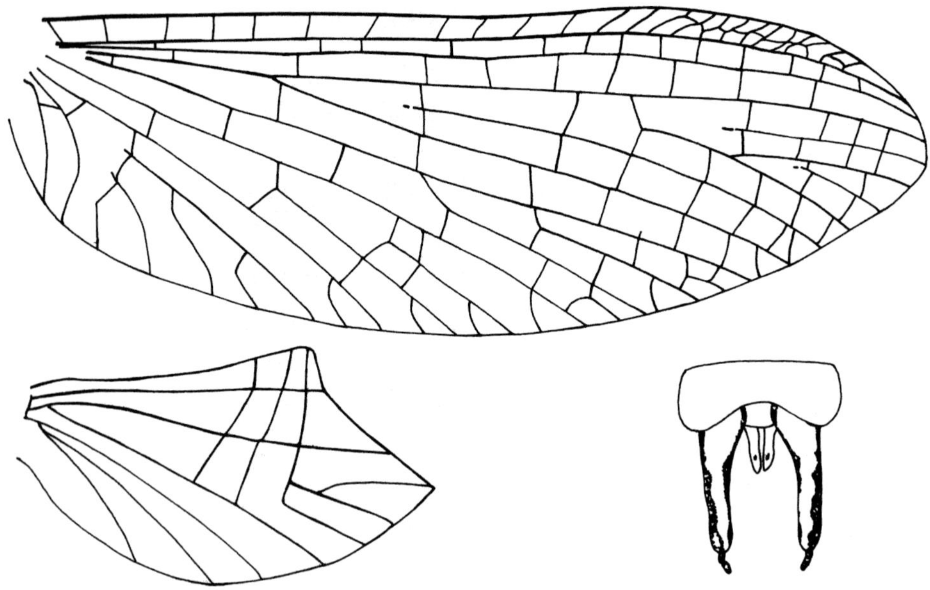

Fig. 2.314 The fore and enlarged hind wing (left) and the genitalia of a male subadult *Choroterpes bilineata* (lower right).

24. The penis lobes are fused except at the apical fissure, and sometimes, one large or many small apical spines are present (**Fig. 2.316**).
....*Meridialaris* Peters and Edmunds, 1972..p. 333

- The penis lobes are separated for their whole lengths or fused only in the basal half. Vein MP_2 of the fore-wing is moderately curved; its base may sometimes be attached to MP_1 by a cross vein 1/5 of the distance from the base to the margin. The distal section of vein MA is bowed toward the posterior edge of the wing (**Fig. 2.305**). 25

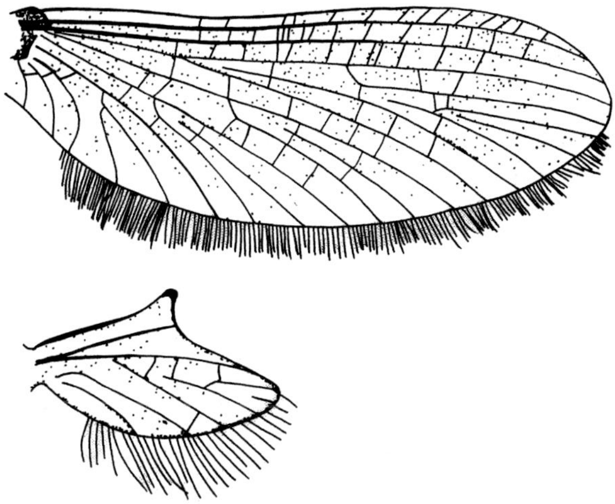

Fig. 2.315 The fore and hind wing of *Microphlebia surinamensis.*

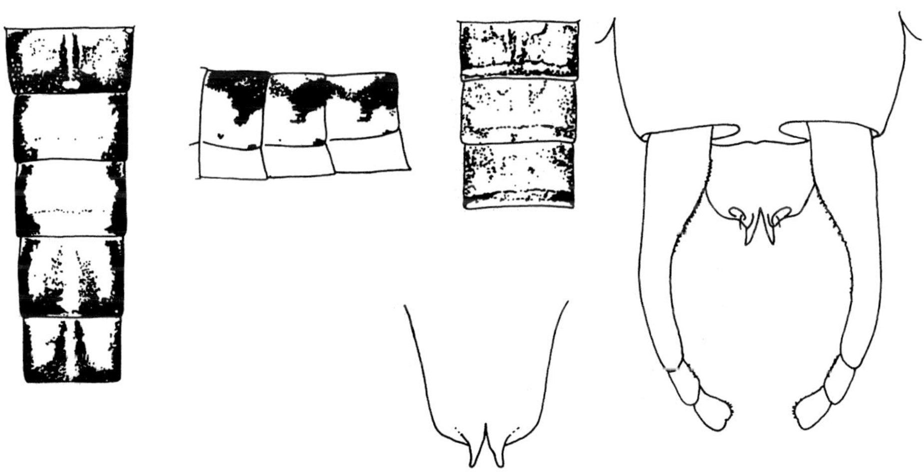

Fig. 2.316 *Meridialaris tintinnabula:* the third through seventh abdominal segments of a male in dorsal view (left), third through fifth segments in lateral view (upper left center), fourth through sixth segments reportedly of a female in dorsal view (upper right center), the male genitalia in ventral view (right), and the penis in dorsal view (lower center).

25. The penis lobes are separated for their whole length (**Fig. 2.317**). 26
- The lobes of the penis are fused in their basal halves (**Fig. 2.318**). 28

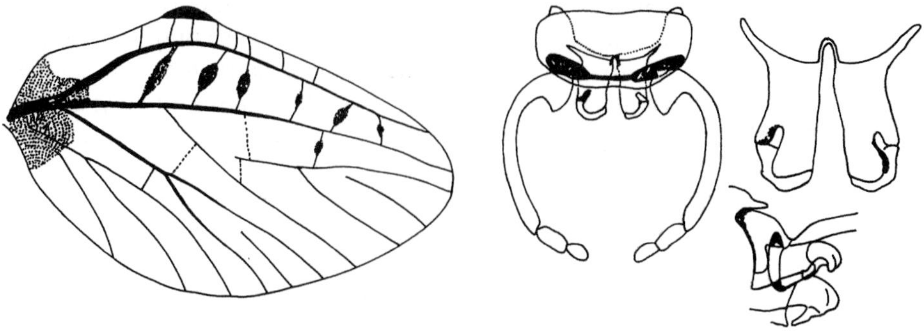

Fig. 2.317 *Ulmeritus saopaulensis:* hind wing (left), male genitalia cleared showing the concealed parts as dotted lines (center), penis (upper right), and lateral view of genitalia (lower right).

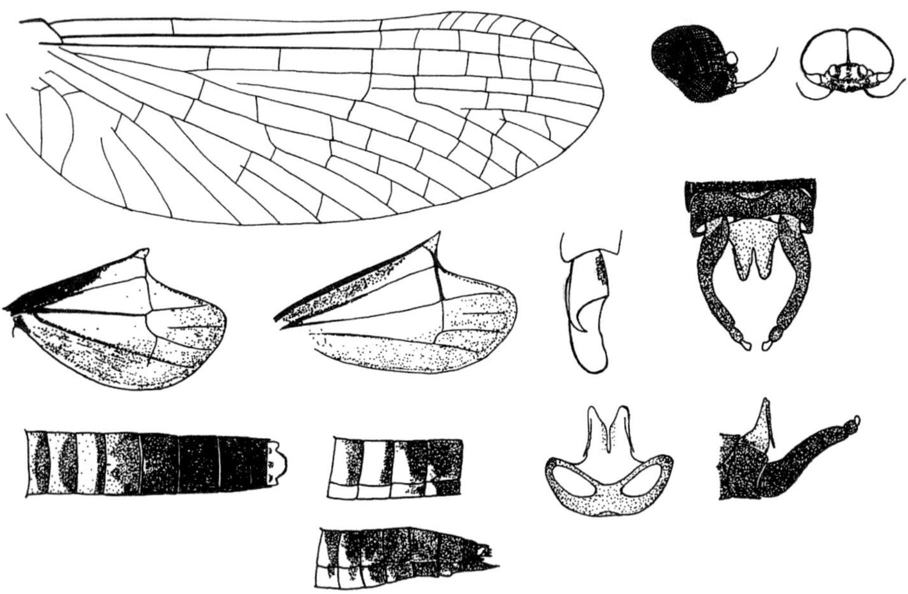

Fig. 2.318 *Simothraulopsis demerara:* fore and hind wing of a male adult with the hind wing drawn larger than scale (left), the hind wing of the subadult (left center), the head of an adult male in lateral (upper right center) and frontal view (upper right), the color pattern on the second through tenth abdominal tergites of the male adult (lower left), the third through sixth abdominal segments of the male adult in lateral view (below subadult hind wing), the third through tenth abdominal segments of the female in lateral view (lower left center), the tarsal claws on the fore-leg of a subadult (right of the subadult hind wing), the apex of the male abdomen in ventral (middle right) and lateral (lower right) view, and the male genitalia in posterior view with the forceps removed (lower right center).

26. The forceps bases are well separated; the basal segment of each arm is long, slender, and inflated over the basal 1/6 of its length. It forms an angle with the internal margin. The second long cubital intercalary vein in the fore-wing joins CuP with a shorter free intercalary between the two long ones. Intercalaries in Rs of the fore-wing form two triangles between R_{2+3} and the bisector of the fork. The subanal plate of the female is excavated apically (**Fig. 2.317**).
....*Ulmeritus* Traver, 1956..p. 340
This includes the subgenus *Ulmeritoides* Traver, 1959, which has been treated by some recent authors as a separate genus (Domínguez *et al*., 1992).
- The bases of the forceps are situated side be side; the basal segment is either not very long and slender or its basal part is not inflated (**Fig. 2.306**). 27

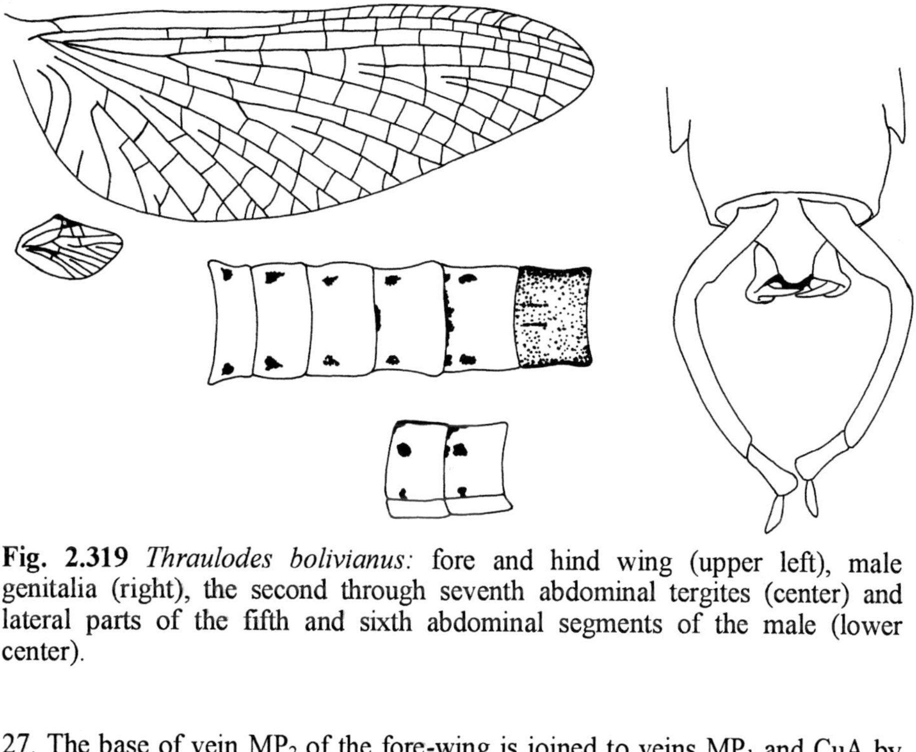

Fig. 2.319 *Thraulodes bolivianus:* fore and hind wing (upper left), male genitalia (right), the second through seventh abdominal tergites (center) and lateral parts of the fifth and sixth abdominal segments of the male (lower center).

27. The base of vein MP_2 of the fore-wing is joined to veins MP_1 and CuA by cross veins. There is a strong bend at the base of vein MP_2. Each lobule on the penis lobes has a spiniform subapical projection directed ventrally and outward. Each forceps arm has a separate cavity (**Fig. 2.305**).
....*Ecuaphlebia* Domínguez, 1988
There is a single described species in this genus: *Ecuaphlebia rumignaui* Domínguez, 1988, known only from Ecuador. A second species, described only from a subadult, was referred to as "*Ecuaphlebia* sp. A" by Domínguez (1988b). This form has no taxonomic status and remains undescribed until someone decides to do so and give it a name.

- The base of vein MP_2 of the fore-wing is not joined to veins MP_1 and CuA by cross veins, or there is no strong bend at the base of vein MP_2. The subcosta of the hind wing does not approach the tip but reaches only as far as a costal prominence, which may be well developed; the costal space is short and wide in most species. The first long cubital intercalary of the fore-wing runs into CuA (**Fig. 2.319**). The subanal plate of the female is excavated apically. The lobules on each penis lobe have a spiniform apical or subapical projection directed toward the midline. The cavities at the bases of the forceps arms are united to form a single cavity.
....*Thraulodes* Ulmer, 1920..p. 348

28. Vein MP in the hind wing is not forked. The base of the forceps of the male bears no rod-like projections (**Fig. 2.318**). 29
- Vein MP in the hind wing is forked (**Fig. 2.320**). 32

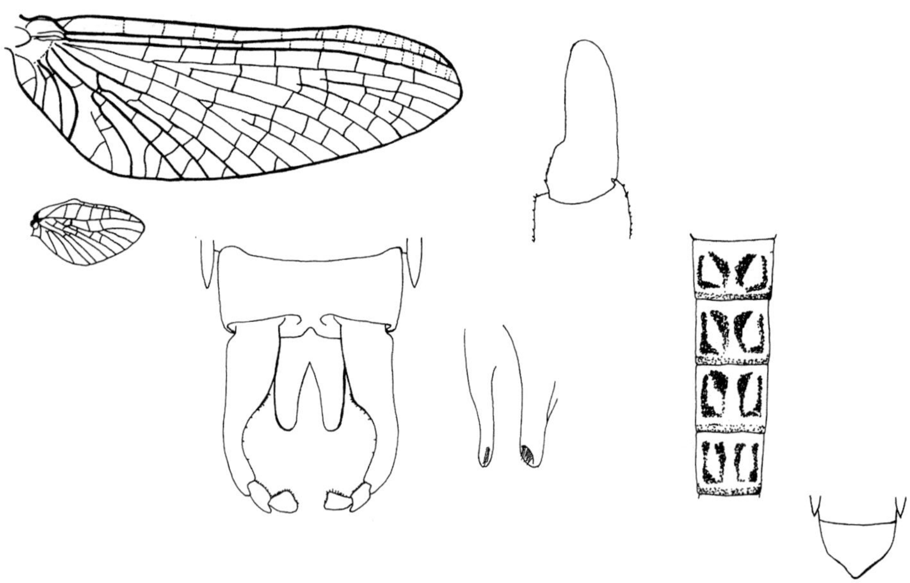

Fig. 2.320 *Demoulinellus coloratus:* fore and hind wings (upper left), the claw on the fore-leg (upper center), pattern on the fourth through seventh abdominal tergites of a male (middle right), male genitalia with enlargements of two aspects of a penis lobe (lower center), and the ninth sternite of a female (lower right).

29. Each lobe of the penis bears a spine-like appendage near the middle that points anteromesally. There is a median ridge on the styliger plate that curves dorsad between the two penis lobes (**Fig. 2.321**). Length of fore-wing: 8.25 to

8.33 mm. Length of the hind wing: 1.5 to 1.7 mm. Color: mainly various shades of brown. Only the adult male is known.

....*Paramaka* Savage and Domínguez, 1992

The only species in this genus is *Paramaka convexa* (Spieth, 1943), known from Surinam and Pará.

- Each lobe of the penis each has a ventral subapical projection directed toward the base (**Fig. 2.318**). 30

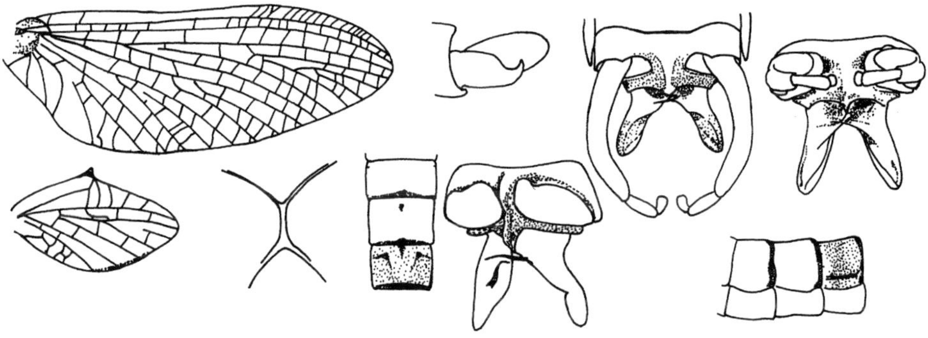

Fig. 2.321 *Paramaka convexa* adult male: fore and hind wing (left), tarsal claw on the fore-leg (upper left center), genitalia in ventral view (upper right center), rotated to a posterior view of an inverted specimen (upper right), and a ventrolateral view with the forceps removed (lower right center), the middle carina in the prosternum (lower left center), and the fifth through seventh abdominal segments in dorsal (lower center) and lateral view (lower right).

30. The costal projection of the hind wing is not well developed. The subanal plate of the female is not excavated at all apically (**Fig. 2.322**).

....*Homothraulus* Demoulin, 1955..p. 365

A genus similar to this one is known only from Puerto Rico: *Neohagenulus* Traver, 1938, distinguished by the very high angle on the hind wing. It has not been encountered in South America.

- The costal projection of the hind wing is well developed and ends about 4/7 to about 2/3 of the distance from the wing base to the wing tip (**Fig. 2.317**). 31

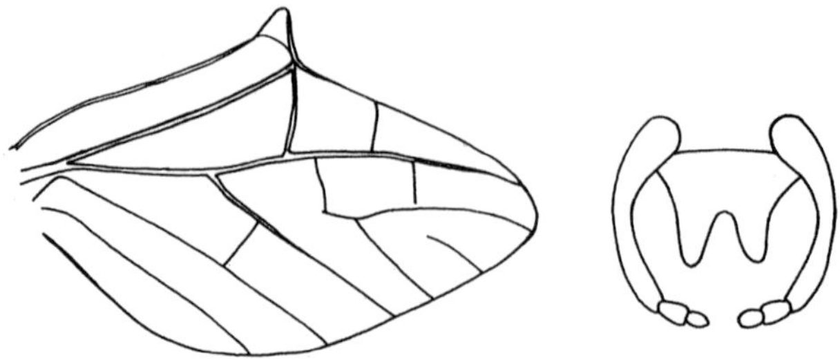

Fig. 2.322 *Homothraulus misionensis:* hind wing (left) and a diagrammatic illustration of the male genitalia (right).

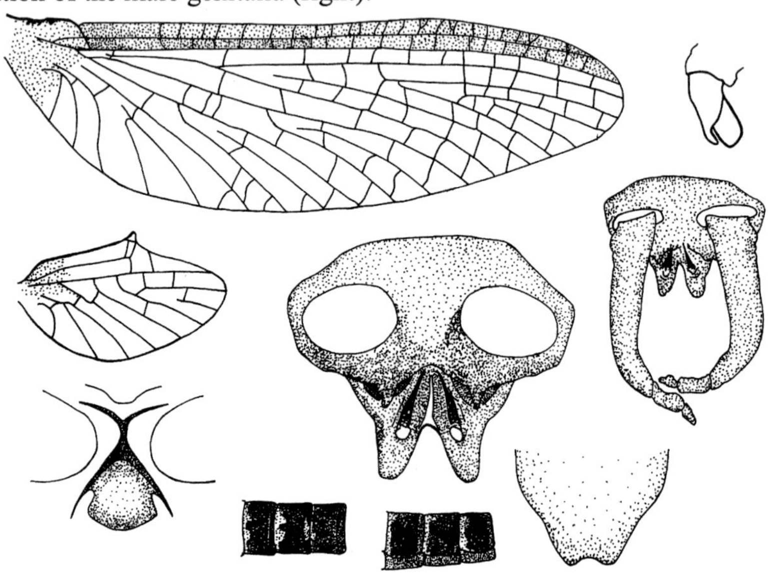

Fig. 2.323 *Hydrosmilodon saltensis:* fore and hind wing (upper and middle left), male tarsal claws (upper right), prosternum (lower left), color pattern on the fifth through seventh abdominal segments in dorsal (lower left center) and lateral (lower right center) view, male genitalia in ventral (middle right) and apical (center below fore-wing) view, and ninth sternite of a female (lower right).

31. The knees are yellow. The fore-femur is yellow with purplish brown subapical band and lines along the lower and apical margins. The basal half of the fore-tibia is purplish brown, and the apical half is white. The middle femur has a complete purplish subapical band, and the middle tibia is purplish brown except at its apex. The middle of the hind femur is broadly purplish brown, while its ends are pale yellow. The hind tibia is purplish brown except at its apex. The hind wing is darkened posteriorly (**Fig. 2.318**). Body length of male: 4.5 mm; of female: 5 mm. Fore-wing length of male: 5 mm; of female: 5.5 mm. Color: dark reddish brown on the head; thorax reddish brown dorsally

and yellow with black markings laterally; the first five abdominal segments are yellowish, translucent in the male and opaque in the female; the apical segments have darker markings.

....*Simothraulopsis* Demoulin, 1966

The only known species in this genus is *Simothraulopsis demerara* (Traver, 1947) from Guyana and Surinam. Synonyms include *Thraulus demerara* Traver, 1947; *Homothraulus demerara* (Traver, 1947); and *Simothraulopsis surinamensis* Demoulin, 1966. Domínguez *et al.* (1997) speculated that this genus may actually be indentical with *Homothraulus*. From all evidence, this seems likely.

- The knees are brown washed with black, the same as the rest of the femora and tibiae. The middle tibia and tarsi are lighter than those on the fore and hind legs. The subgenital plate bears short, acute projections near the base of the forceps (**Fig. 2.323**).

...................................*Hydrosmilodon* Flowers and Domínguez, 1992..p.366

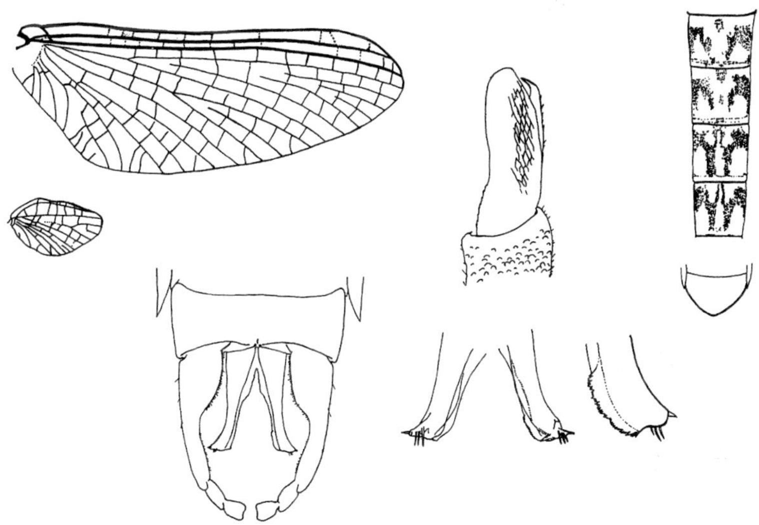

Fig. 2.324 *Secochela illiesi:* fore and hind wings (upper left), the claw on the fore-leg (upper center), pattern on the fourth through seventh abdominal tergites of a male (upper right), male genitalia with an enlargement of the penis lobes and details of the apex on one lobe (lower row, left to right), and the ninth sternite of a female (middle right).

32. The apex of each penis lobe is rounded and lacks spines but has a semicircle of subapical setae (**Fig. 2.320**).

....*Demoulinellus* Pescador and Peters, 1982

The only species of this genus yet described is *Demoulinellus coloratus* Pescador and Peters, 1982, from Chile. The body length of the male is 6 to 7.2 mm, fore-wing length: 6 to 7 mm. Color: pale yellow with darker markings.

- There are short spines and setae at the apex of each penis lobe (**Fig. 2.324**). ....*Secochela* Pescador and Peters, 1982

Only one species in this genus has been described: *Secochela illiesi* Pescador and Peters, 1982, from Chile. The body length of the male is 6.5 to 8 mm, fore-wing length: 6.5 to 8.5 mm. Color: mainly yellow and brown, with a few darker markings on the head.

Key to the genera of larvae of Leptophlebiidae in South America

Information provided by Ulmer (1904, 1920c), Demoulin (1955a), Edmunds (1963), Peters (1969), Savage and Peters (1982), Flowers (1980, 1987), Domínguez (1995a, b, 1997), Domínguez *et al.* (1992), and Thomas *et al.* (2000). Descriptions of larval *Archethraulodes, Choroterpes, Miroculitus, Paramaka,* and *Rhigotopus* are not available. The larva of the questionable *Thraulus costalis* is also unknown.

1. Posterior wing pads are absent. 2
- Posterior wing pads are present, although they may be small and partially concealed by the anterior wing pads (**Fig. 2.325**). 4

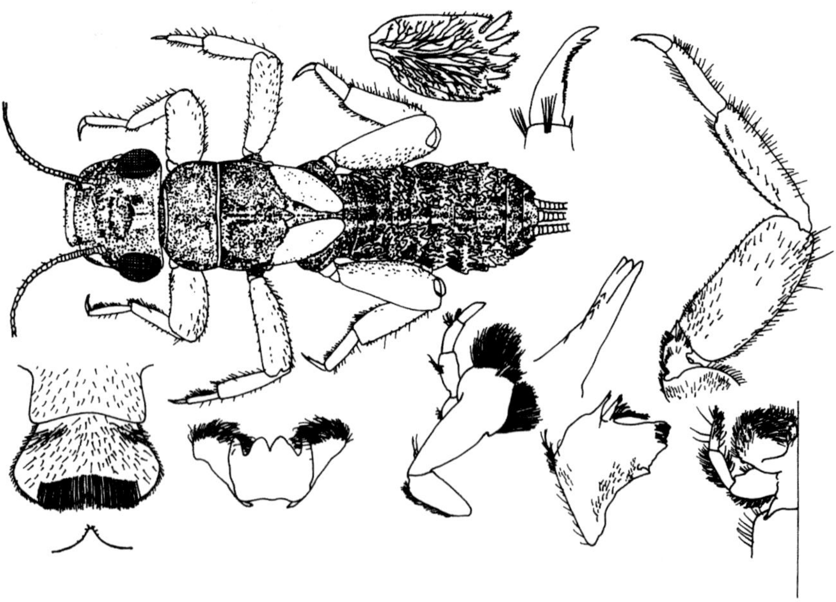

Fig. 2.325 *Dactylophlebia carnulenta* larva: habitus without cerci (upper left), fourth gill (upper center), tarsal claw from the fore-leg, (upper right center), fore-leg (upper right), and (below, left to right): labrum with an enlargement of its median emargination, hypopharynx, maxilla, mandible with an enlargement of its outer incisor above it, and half of the labium in dorsal view.

2. The gills on all abdominal segments are the same ovoid shape; their lamellae have a fringe of narrow processes along the apical half. In the median emargination of the labrum there are two denticles. On the tarsal claws, the denticles become progressively larger from the base to the apex (**Fig. 2.326**). There are posterolateral spines of the abdominal segments beginning with the fifth or sixth and continuing through the ninth. The color of the only known South American species is brown dorsally and paler ventrally. The abdomen is usually somewhat darker than the head and thorax, and a variable pattern may be observed.
....*Askola* Peters, 1969

Askola froehlichi Peters, 1969, found in São Paulo, Paraná, and Santa Catarina, is the only species in this genus.

- The gills are narrow and gradually become thinner toward the apex (**Fig. 2.327**). 3

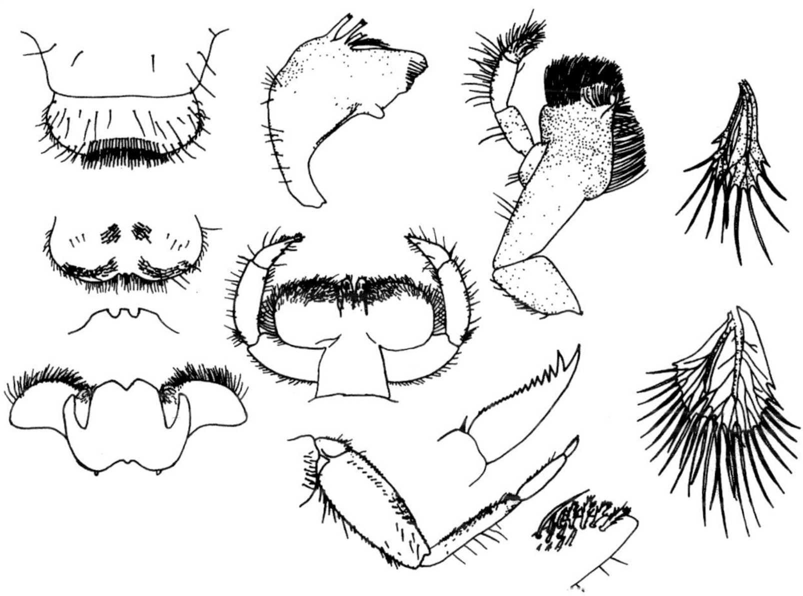

Fig. 2.326 *Askola froehlichi* larva (left, top to bottom): outside and inside of the labrum with an enlargement of its anterior margin, hypopharynx, left mandible (upper left center), right maxilla (upper right center), labium (center), fore-leg with enlargements of its claw and apex of its tibia (lower center), and the gills on the first (upper right) and fourth (lower right) abdominal segments.

3. There are flattened denticles in the median emargination of the labrum. The fore and middle femora are not fringed with long setae. There is a single row of denticles on the tarsal claws, and large posterolateral spines are present on the seventh through ninth abdominal segments.
....*Hagenulopsis* Ulmer, 1920..p. 281

- No flattened denticles are present in the median emargination of the labrum. The fore and middle femora are fringed with long, fine setae. There are two rows of denticles on the tarsal claws. There are large posterolateral spines only on the eighth and ninth abdominal segments (**Fig. 2.327**). The final larval instar of the only known species is 3.2 to 3.3 mm long with cerci 3.1 to 3.4 mm. It is light yellow with a prominent violet band across the first abdominal segment.
....*Bessierus* Thomas and Orth, 2000

The only known species in this genus is *Bessierus doloris* Thomas and Orth, 2000, from French Guiana.

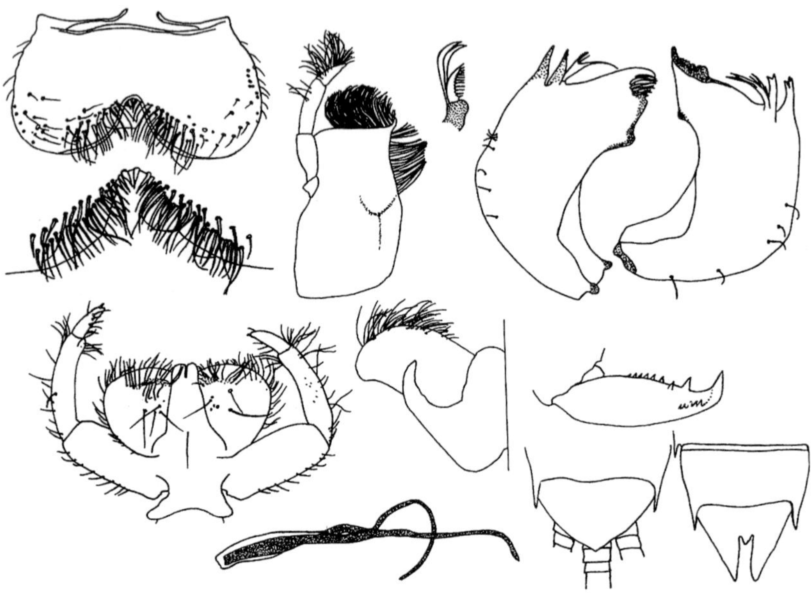

Fig. 2.327 *Bessierus doloris* larva (upper row, left to right): labrum with an enlarged outline of its anterior margin, maxilla in ventral view with an enlargement of the inner apex, mandibles, and (below, left to right): the labium, the fourth abdominal gill (below the labium and hypopharynx), half of the hypopharynx, and a tarsal claw above the subgenital plates of a female and a male (lower right, left and right, respectively).

4. The abdominal gills vary in their basic form on the different segments, or the dorsal and ventral laminae on a single segment are different (**Fig. 2.328**). 5

- The abdominal gills are of the same basic form on all abdominal segments, although some may be smaller on some than others (**Fig. 2.325**). 10

5. Gills are present only on the first six abdominal segments. The dorsal lamellae of the gills are almost as broad near the base as at the broadest point and end in short filaments (**Fig. 2.329**).
....*Massartella* Lestage, 1930..p. 299

- Gills are present on the first seven abdominal segments (**Figs. 2.32, 2.328**). ...6

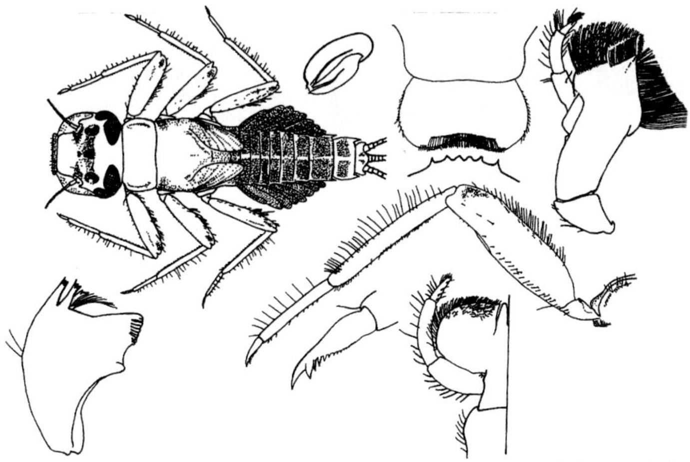

Fig. 2.328 *Microphlebia surinamensis* larva (upper row, left to right): habitus with apical portions of the antennae and cerci removed, an enlarged gill from the fourth abdominal segment, the labrum above an enlarged outline of the margin of the median excavation, and the maxilla; and the mandible (lower right), a leg with an enlargement of the tarsal claw (middle right), and half of the labium in dorsal view (lower right).

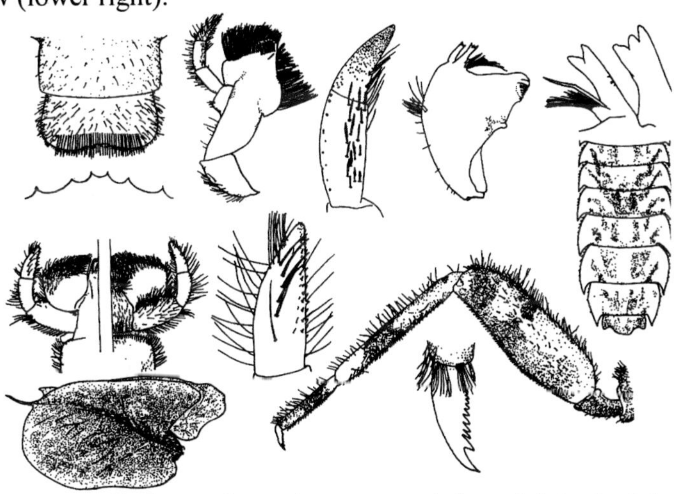

Fig. 2.329 *Massartella brieni* larva (upper row, left to right): the labrum with an outline of the margin of the excavation on its apical margin, the maxilla, an enlargement of the two apical segments of the maxillary palp, the mandible, and enlargements of the teeth on the mandible; (middle row, left to right): two views of half of the labium with the dorsal at left and the ventral at right, an enlargement of the apical segment of the labial palp, the fore-leg, and the fourth through tenth abdominal segments in dorsal view; (lower row, left and right): the gill on the fourth abdominal segment and an enlargement of the tarsal claw on the fore-leg.

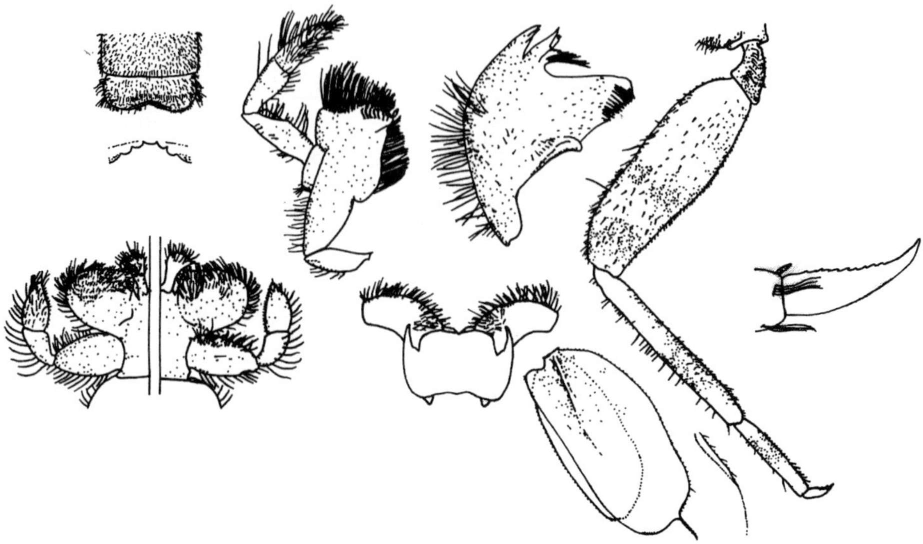

Fig. 2.330 *Hapsiphlebia anastomosis* larva: the labrum with an outline of the margin of its median excavation (upper left), the maxilla (upper left center), mandible (upper right center), fore-leg with an enlarged tarsal claw beside it (right), half of the labium in dorsal and ventral view (lower left, left and right, respectively), the hypopharynx (lower center), the laminar gills from the third abdominal segment and filamentous gill from the seventh segment (lower right center).

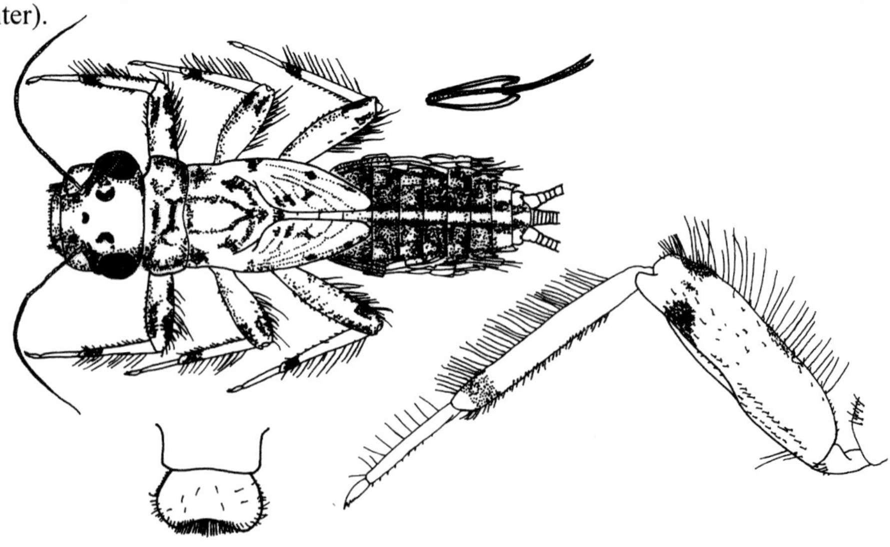

Fig. 2.331 *Miroculis mourei* larva: habitus with much of the cerci and caudal filament removed (upper left), an enlarged gill from the fourth abdominal segment (above center), a leg (lower right), and the labrum (lower left).

6. The denticles on the tarsal claws become progressively larger toward the apex, and the subapical tooth is much larger than the others (**Fig. 2.328**). 7
- The denticles on the tarsal claws are almost equal in size (**Fig. 2.330**). 9
7. The dorsal part of the gills is 75% as long and 50% as wide as the ventral part (**Fig. 2.328**).
....*Microphlebia* Savage and Peters, 1982..p. 332
- The dorsal and ventral parts of the gills are almost the same size (**Fig. 2.331**).
....8
8. Gills on the first six abdominal segments are fringed on the posterior margin, while those on the seventh segment are reduced and filamentous. The outer margin of the mandible is curved and fringed with long setae, most dense at the sharpest part of the curvature. The labrum has distinct broad-based denticles. There are posterolateral spines on the second through ninth abdominal segments (**Fig. 2.332**).
....*Atopophlebia* Flowers, 1980..p. 300
- The gills on the first seven abdominal segments are similar with their dorsal and ventral portions approximately equal in length. Each has a long median projection at the apex and two posterolateral projections, which are short. The dorsal portion is not symmetrical, which the ventral portion is, or nearly so. The outer border of the mandible is curved with scattered hair-like setae near the middle of the curve. There are four or five small, rounded denticles along the margin of the median emargination. Posterolateral spines may be present on the second through ninth or third through ninth segments (**Fig. 2.331**).
....*Miroculis* Edmunds, 1963..p. 320

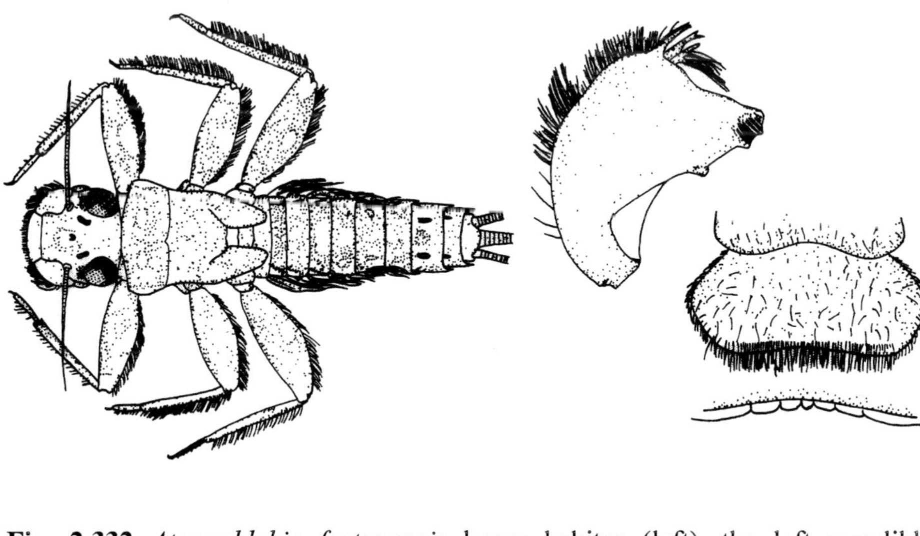

Fig. 2.332 *Atopophlebia fortunensis* larva: habitus (left), the left mandible (center), and the labrum with an enlarged outline of the anterior margin showing the broad-based denticles (lower right).

9. The gills on the first five abdominal segments are laminar in form and similar, but that on the fifth segment is somewhat smaller. The dorsal part terminates in a narrow process. The gills on the sixth and seventh segments consist of a narrow filament bearing setae (**Fig. 2.32** and **2.330**).
....*Hapsiphlebia* Peters and Edmunds, 1972

The only known South American species in this genus is *Hapsiphlebia anastomosis* (Demoulin, 1955) from Chile, formerly known as *Atalophlebia anastomosis* Demoulin, 1955.

- The gills on the first seven abdominal segments are narrow becoming finer toward the apex. The dorsal part of each gill on the second through sixth segments is large with narrowed bases and a truncate apex bearing a median apical projection. The ventral portions of these gills are elliptical (**Fig. 2.333**).
....*Fittkaulus* Savage and Peters, 1978..p. 287

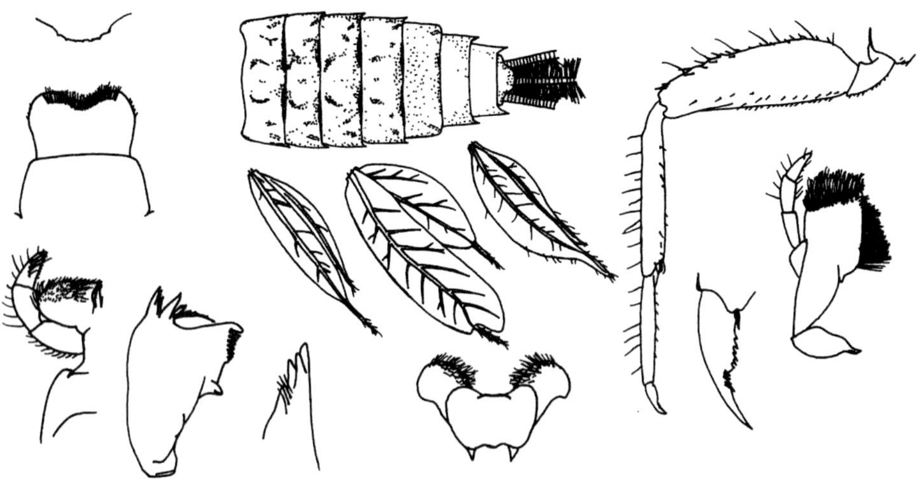

Fig. 2.333 *Fittkaulus maculatus:* the clypeus and labrum with an enlargement of the tooth pattern on the anterior edge of the labrum (upper left); the abdomen in dorsal view above the gills (left to right) on the first, fourth, and seventh abdominal segments (upper center); the fore-leg with an enlargement of its claw (upper right and right center); the right maxilla (right); and (lower row, left to right center) half of the labium in dorsal view, the left mandible and the incisor of the right mandible, and the hypopharynx.

10. One or more denticles in the middle part of the tarsal claw stand out as being much larger than the others (**Fig. 2.334**). 11
- The denticles in the middle part of the tarsal claw are nearly equal in size or they form a row with the size increasing regularly from the base toward the apex; the subapical denticle may be much larger than the others. The head is prognathous (**Fig. 2.335**). 12

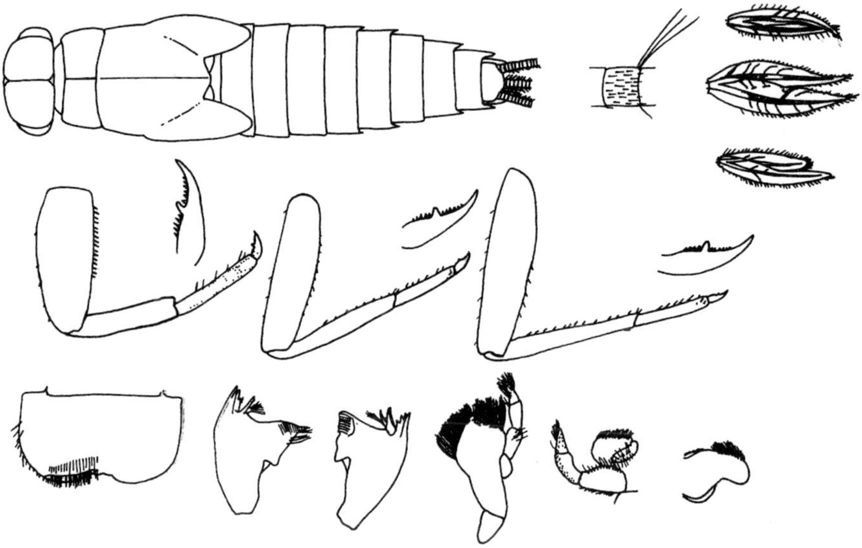

Fig. 2.334 *Terpides guyanensis* larva: outlines of the head, thorax, and abdomen in dorsal view (upper left); an enlarged segment of the caudal appendage (upper right center); the first, fourth, and seventh abdominal gills (upper right, top to bottom); the fore, middle, and hind legs with the enlarged tarsal claws of each (middle row, left to right); and (lower row, left to right) the labrum with the setae removed on one side, the two mandibles, the maxilla, one side of the labium, and one side of the hypopharynx.

11. There is one large denticle in the middle of the inner margin of the tarsal claw, with denticles becoming progressively larger basal to it and a row of very small denticles apical to it. The head is hypognathous (**Fig. 2.334**).
....*Terpides* Demoulin, 1966
The only species in this genus known from South America is *Terpides guyanensis* Demoulin, 1966, from Surinam.
- There are three large denticles in the middle of the inner margin of the tarsal claw, with a subapical row of denticles also on the interior surface (**Fig. 2.336**).
....*Perissophlebiodes* Savage, 1983
The only species in this genus known from South America is *Perissophlebiodes flinti* (Savage, 1982), described as *Perissophlebia flinti* Savage, 1982, from southern Brazil.

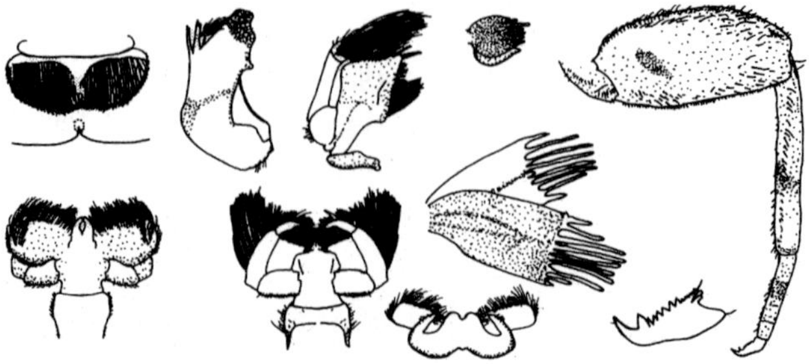

Fig. 2.335 *Hylister plaumanni* larva (upper row, left to right): labrum above the outline of its enlarged center emargination, left mandible, maxilla, anteriormost abdominal gill above the gill on the sixth abdominal segment, fore-leg; (lower row, left to right) labium in dorsal and ventral view, hypopharynx, and enlarged tarsal claw.

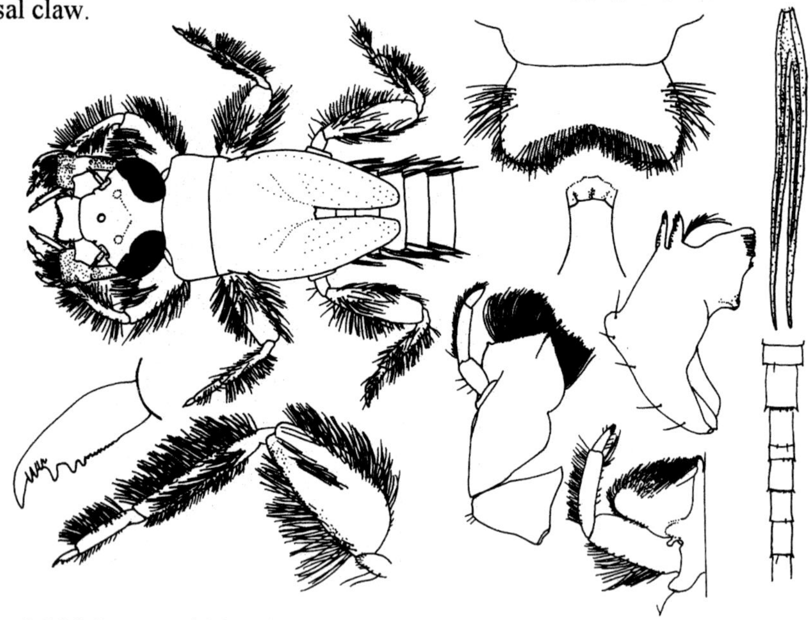

Fig. 2.336 *Perissophlebiodes flinti* larva: anterior portion of habitus (upper left), labrum above its enlarged anterior border and median excavation (upper right center), mandible (middle right center), maxilla (lower center), half the labium in dorsal view (lower right center), fore-leg below an enlarged tarsal claw (lower left), gill from the seventh abdominal segment (upper right), and basal portion of a cercus (lower right).

12. The denticles on the tarsal claws are nearly equal or they become progressively larger from the base to the apex without an exceptionally large subapical denticle (**Fig. 2.337**). 13

- The denticles of the tarsal claws are nearly equal or become progressively larger toward the apex, but the subapical denticle is clearly larger than the others (**Fig. 2.335**). 22

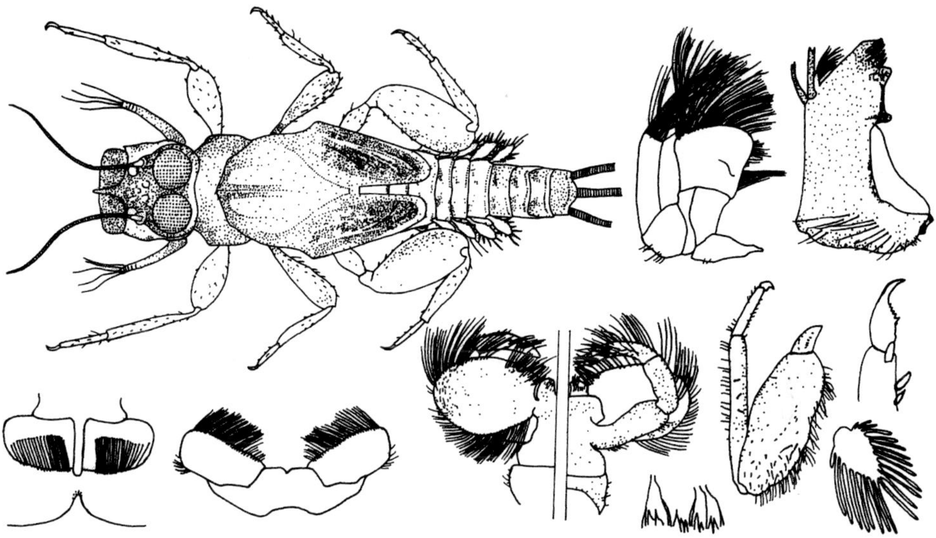

Fig. 2.337 *Traverella calingastensis* larva: habitus without apical sections of the cerci (upper left), maxilla and mandible (upper right, left and right, respectively), and (lower row, left to right), labrum above an enlarged outline of the anterior margin, hypopharynx, half of labium in dorsal and in ventral view, spines located along the posterior margin of the fifth abdominal tergite, fore-leg, and a tarsal claw above the gill from the second abdominal segment.

13. The gills are fringed with fine setae (**Fig. 2.337**). 14
- The gills are not fringed, but sometimes a narrow apical filament is present (**Fig. 2.338**). 16
14. The gills on the first through fifth abdominal segments are much larger than those on the sixth and seventh, and that on the seventh segment may be lacking. The labrum is narrower than the head in dorsal view (**Fig. 2.337**).
....*Traverella* Edmunds, 1948..p. 313
- The gills on the first through seventh abdominal segments are all similar. The labrum is the same width as the clypeus (**Fig. 2.339**). 15
15. The long dorsal setae on the labrum are arranged in two rows. There are five flattened denticles on the anteromedial emargination of the labrum (**Fig. 2.339**).
....*Ulmeritus (Ulmeritus)* Traver, 1956..p. 340

- The long dorsal setae on the labrum are arranged in a single continuous, sinuous row. The middle denticle of the five on the anteromedial emargination of the labrum is well developed (**Fig. 2.340**).
.... *Ulmeritus (Ulmeritoides)* Traver, 1959..p. 340

This subgenus has been treated by some recent authors as a separate genus (Domínguez *et al.*, 1992), but because there is no definitive way of distinguishing the adults of these two taxa from one another, they are incorporated in the same keys as subgenera. The great similarity of their features was confirmed by cladistic analysis (Flowers and Domínguez, 1991).

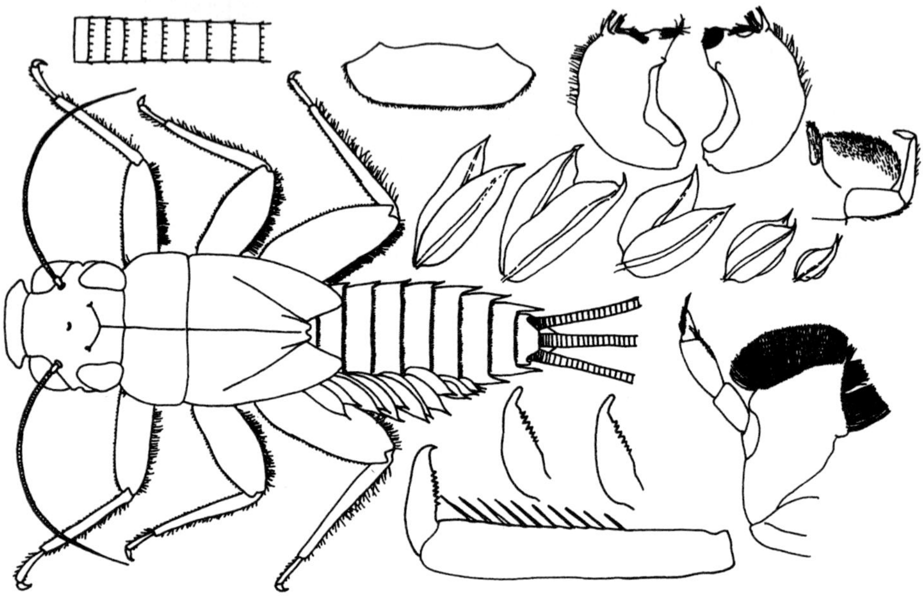

Fig. 2.338 *Thraulodes* sp. larva: habitus (left); the fore-tarsus and claw and the claws from the middle and hind legs (lower right center, left to right); first, third, fifth, sixth, and seventh abdominal gills (upper middle right, left to right); section of a cercus (upper left); labrum, mandibles, and one side of the labium (upper center to right, left to right); and maxilla (lower right).

16. Posterolateral spines are present on the second or third through the ninth abdominal segments (**Fig. 2.338**). 17
- Posterolateral spines are present only on the sixth, seventh, or eighth through the ninth abdominal segments (**Fig. 2.341**). 20
17. The lateral edges of the labrum are acutely pointed. There are posterolateral spines present on the third through ninth abdominal segments (**Fig. 2.338**).
....*Thraulodes* Ulmer, 1920..p. 348
- The lateral edges of the labrum are rounded (**Fig. 2.342**). The posterolateral spines are present on the second through ninth abdominal segments. 18

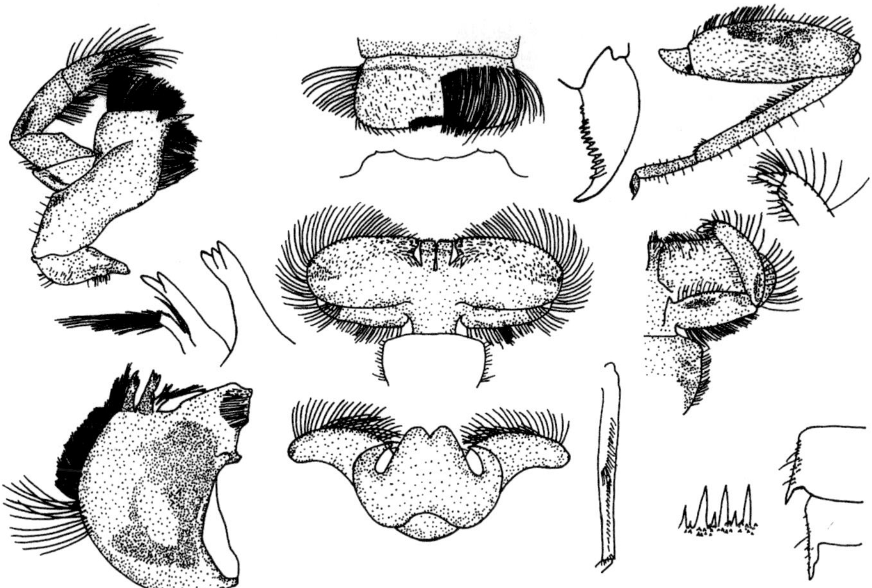

Fig. 2.339 *Ulmeritus (Ulmeritus) carbonelli* (larva): left maxilla (upper left), left mandible (lower left) and enlarged incisors and prostheca of the right mandible (middle left), labrum with the long setae on the right side removed and an outline of the anteromedian emargination (upper center), labium in dorsal view (center) and its left half in ventral view (middle right) with the apex of its palp enlarged to the upper right, hypopharynx (lower left center), fore-leg with an enlargement of its tarsal claw (upper right), arrangement of setae on the hind tibia (lower right center), posterior border of the fifth tergite (right of tibia), and outline of the posterolateral processes on the eighth and ninth abdominal segments (lower right).

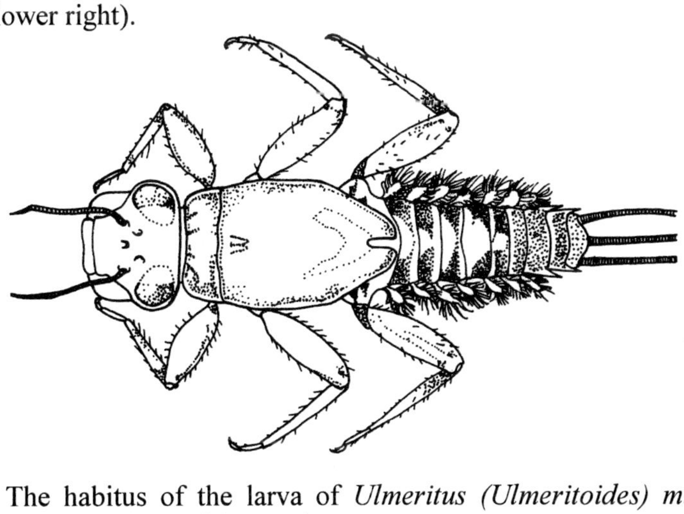

Fig. 2.340 The habitus of the larva of *Ulmeritus (Ulmeritoides) misionensis* without the apical portions of the cerci.

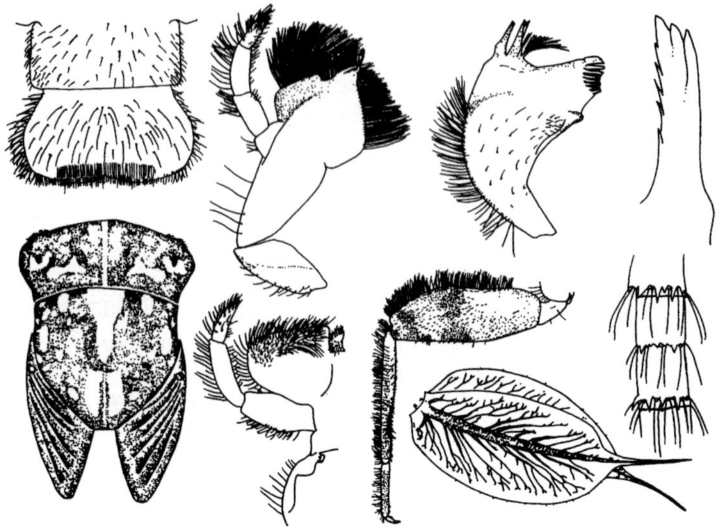

Fig. 2.341 *Penaphlebia barriai* larva (upper row, left to right): the labrum, maxilla, mandible, and the outer tooth on the mandible enlarged; and (lower row, left to right): the color pattern on the dorsal surface of the thorax, the dorsal surface of the left half of the labium, fore-leg, gill from the fourth abdominal segment, and details of segments from the caudal filament.

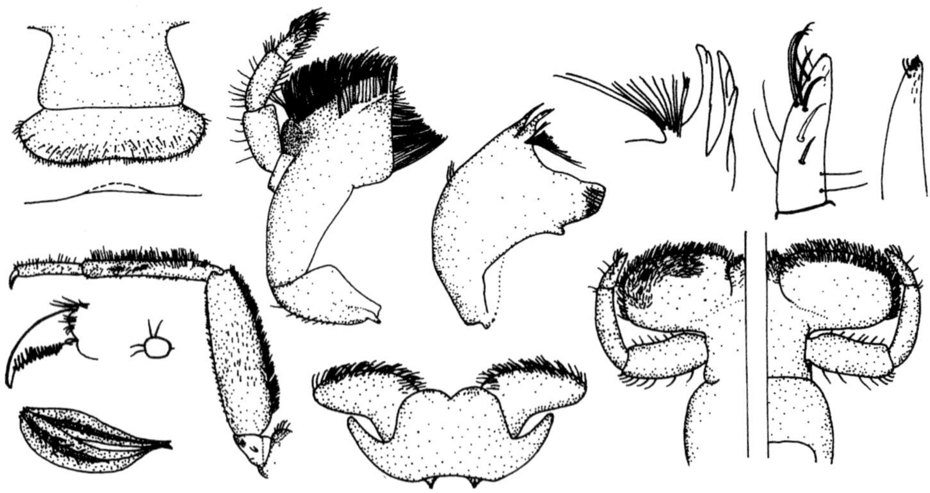

Fig. 2.342 *Secochela illiesi* larva (upper row, left to right): the labrum with an outline of the median emargination along its anterior border, maxilla, left mandible, enlargement of the incisor on the right mandible, labial palp in dorsal and ventral view, and the hypopharynx (lower center), half of the labium in dorsal and ventral view (lower right, left and right, respectively), the fore-leg with enlargements of its tarsal claw and a cross-section of its tibia (middle left), and the fourth gill (lower left).

18. The tarsal claws have basal denticles that are unequal in size followed by a row of denticles that gradually increases in size toward the apex of the claw. There is a tuft of a few setae near the middle of the outer margin of the mandible, which is otherwise free of setae. There is a row of spatulate setae along each tibia, and there are spoon-shaped setae on the femora (**Fig. 2.342**). Each seta bears a tiny papilla at the apex.
....*Secochela* Pescador and Peters, 1982

Only one species in this genus has been described: *Secochela illiesi* Pescador and Peters, 1982, from Chile. The body length of the final instar larva is 6 to 7 mm. Color: various shades of yellow with brown and black wash.

- The tarsal claws have do not have basal denticles that are unequal in size followed by a row of denticles that gradually increases in size toward the apex of the claw, or the margins of the mandible are not free of setae with the exception of a small tuft near the middle of the outer margin. A row of spatulate setae bearing tubercles at the apex is not found along each tibia, and the setae on the femora are not spoon-shaped (**Fig. 2.343**). 19

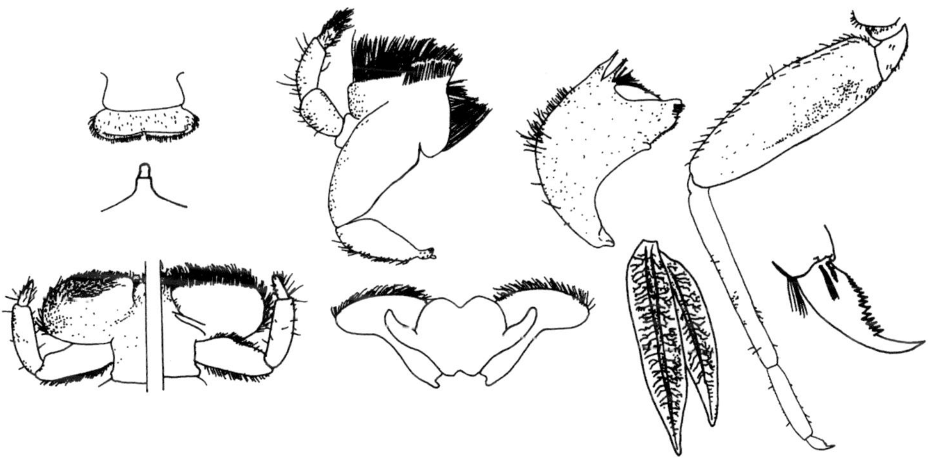

Fig. 2.343 *Massartellopsis irarrazavali* larva: labrum with an enlargement of its anterior margin below (upper left), maxilla in ventral view (upper left center), left mandible (upper right center), fore-leg with an enlargement of its tarsal claw (right), half of the labium in dorsal and in ventral view (lower left, left and right, respectively), hypopharynx (lower center), and the gills on the fourth abdominal segment (lower right center).

19. The main trachea of each gill and its branches are well pigmented (**Fig. 2.343**).
....*Massartellopsis* Demouliin, 1955

The only species in this genus yet described is *Massartellopsis irarrazavali* Demoulin, 1955, from Argentina, Chile, and Bolivia.

- The main trachea of each gill is pigmented, while there is much less pigment on its branches (**Fig. 2.344**).
....*Meridialaris* Peters and Edmunds, 1972..p. 333

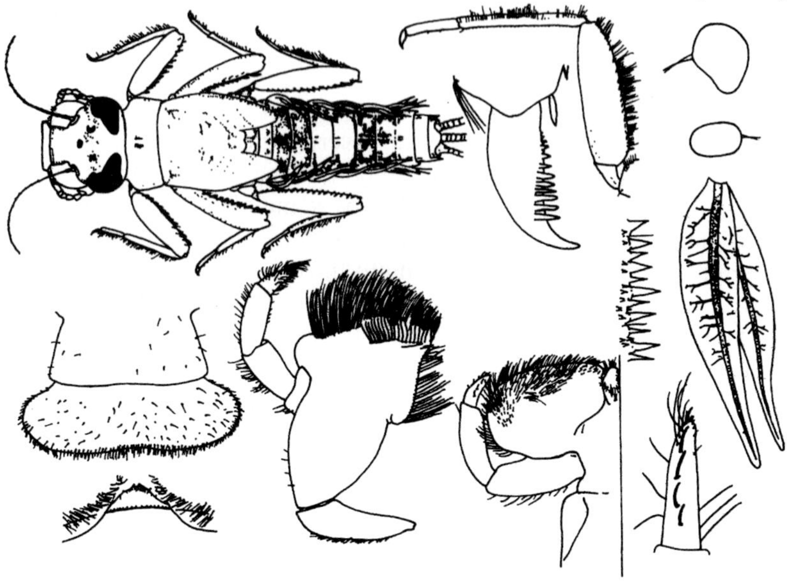

Fig. 2.344 *Meridialaris laminata* larva: habitus without the apical portions of the cerci and caudal filament (upper left), the fore-leg with an enlargement of the tarsal claw (upper right center), cross-sections of its tibia and tarsus (upper right, top and bottom, respectively), the labrum with an enlargement of the anterior margin of the median excavation (lower left), the maxilla (lower left center), half of the labium in dorsal view with the apical segment of the labial palp enlarged (lower right), the gill on the fourth abdominal segment (middle right), and the posterior margin of a tergite (left of the gill).

20. The gills are laminar and end in a narrow process (**Fig. 2.341**).
....*Penaphlebia* Peters and Edmunds, 1972..p. 288
- The gills are narrow and become thinner continually toward the apex, at least in the apical 2/3 of their length (**Fig. 2.345**). 21
21. At the base of each tarsal claw, there is a smaller, claw-like structure. The anterior margin of the labrum is nearly straight with a small, shallow indentation at the midline (**Fig. 2.345**). The final larval instars of the only South American species are about 9 mm long.
....*Atalophlebioides* Phillips, 1930

The only South American species in this genus is *Atalophlebioides lestagei* Ulmer, 1938, from Chile. It was referred to as *Atalophlebia fulvipes* by Lestage (1931), although this name was applied by Needham and Murphy (1924) to a different species.

- There is no small, claw-like structure at the base of each tarsal claw. The anterior margin of the labrum is broadly emarginated and bears incisor-like processes in the middle of the emargination (**Fig. 2.346**).
....*Demoulinellus* Pescador and Peters, 1982

The only species of this genus yet described is *Demoulinellus coloratus* Pescador and Peters, 1982, from Chile.

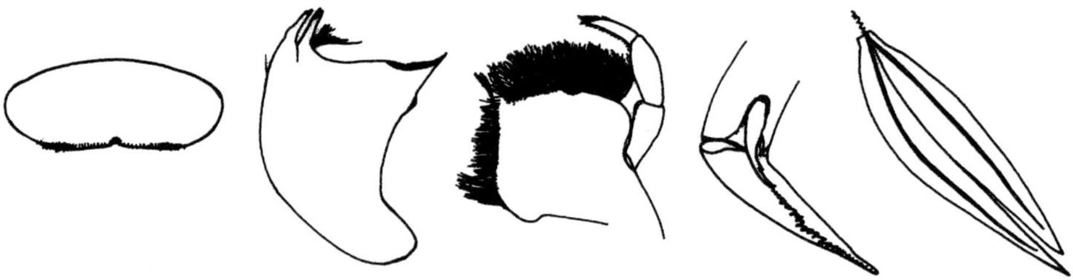

Fig. 2.345 *Atalophlebioides lestagei* larva (left to right): labrum, mandible, apex of the maxilla, tarsal claw, and a gill.

22. The labrum is about as wide or wider than the head (**Fig. 2.335**). 23
- The labrum is much narrower than the head (**Fig. 2.347**). 28

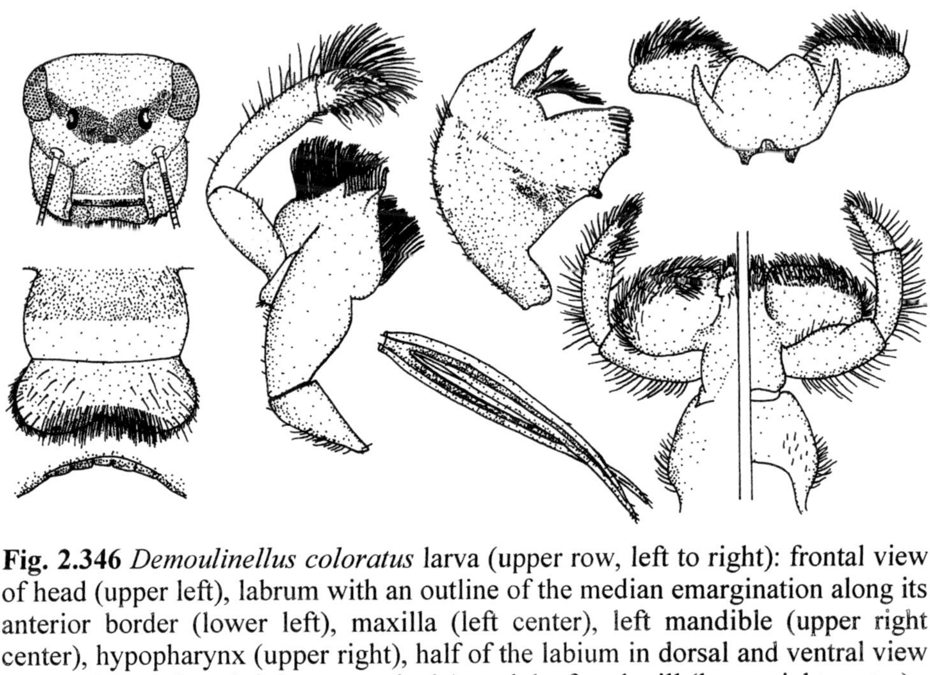

Fig. 2.346 *Demoulinellus coloratus* larva (upper row, left to right): frontal view of head (upper left), labrum with an outline of the median emargination along its anterior border (lower left), maxilla (left center), left mandible (upper right center), hypopharynx (upper right), half of the labium in dorsal and ventral view (lower right, left and right, respectively), and the fourth gill (lower right center).

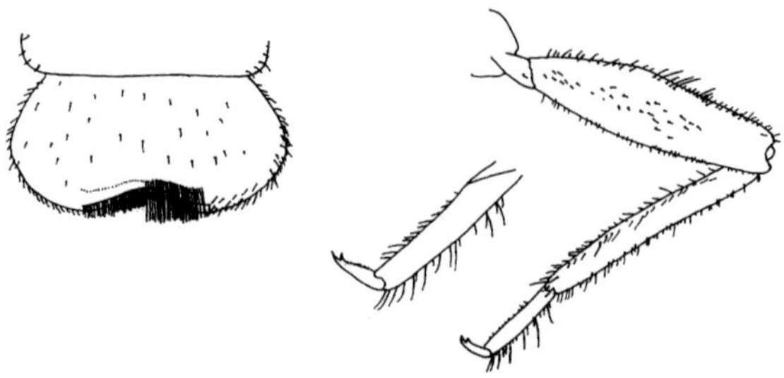

Fig. 2.347 *Farrodes caribbianus* larva (left to right): the labrum with the long setae on one side of the anterior margin removed, hind leg, and an enlargement of its tarsus and claw.

23. The gills are narrow and become continually thinner toward the apex (**Fig. 2.348**).
....*Leentvaaria* Demoulin, 1966
The only known South American species in this genus is *Leentvaaria palpalis* Demoulin, 1966, from Surinam, French Guiana, and Amazonas.
- The gills are thin but do not narrow continually toward the apex (**Fig. 2.335**).
....24

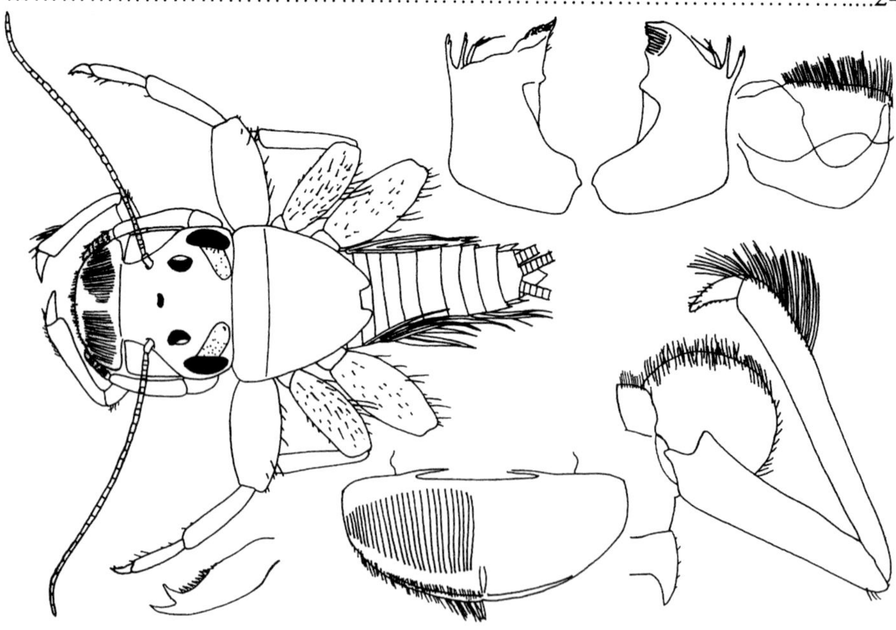

Fig. 2.348 *Leentvaaria palpalis* larva: habitus with the cerci broken off (left), the fore-tarsal claw enlarged (lower left), the mandibles (upper center), and half of the labium and hypopharynx (lower right), labrum (lower center), and the apex of a maxilla (upper right).

24. There are posterolateral spines on the seventh through ninth abdominal segments. The gills on the first through sixth abdominal segments are truncate apically with a fringe of 8 to 10 serrae (**Fig. 2.335**).
.... *Hylister* Domínguez and Flowers, 1989

The only known South American species in this genus is *Hylister plaumanni* Domínguez and Flowers, 1989, from Minas Gerais, Paraná, Rio de Janeiro, and Santa Catarina. The mature larva is 10 to 14 mm long.

- There are posterolateral spines only on the eighth and ninth abdominal segments (**Fig. 2.349**). 25

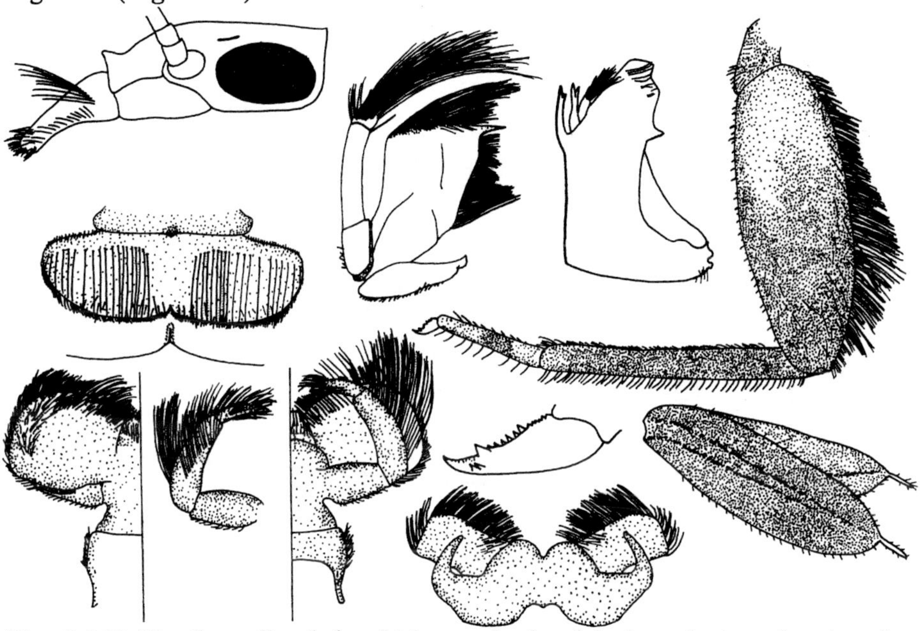

Fig. 2.349 *Needhamella ehrhardti* larva: the head in lateral view showing the prominence along the anterior margin of the clypeus (upper left), anterior margin of the clypeus and labrum in dorsal view above an outline of the anterior margin (middle left), maxilla (upper left center), left mandible (upper right center), fore-leg (upper right), enlarged tarsal claw (below tarsus of the fore-leg), half of the labium without the palp and the labial palp in dorsal view (lower left, left and right), a ventral view of half of the labium with the palp (lower left center), the hypopharynx (lower right center), and the seventh gill (lower right).

25. The gills end in a single finger-like process. There is a small medial projection on the anterior margin of the clypeus (**Fig. 2.349**). 26

- The gills end in two lateral lobes and one medial digitiform projection. The clypeus lacks a medial projection (**Fig. 2.350**). 27

26. The spines along the inner margin of the fore-tibia are not pectinate. The anterior margin of the maxilla forms a narrow tooth on its inner corner (**Fig. 2.349**).

....*Needhamella* Domínguez and Flowers, 1989

The only known South American species in this genus is *Needhamella ehrhardti* (Ulmer, 1920), reported from Argentina, Uruguay, and Santa Catarina, and probably Paraguay, as well. It has also been called *Thraulus ehrhardti* Ulmer, 1920, and *Traverella ehrhardti* (Ulmer, 1920). It is 7.9 to 8.8 mm long.

- The spines along the inner margin of the fore-tibia are pectinate. A blade-like tusk is formed along the anterior margin of the maxilla (**Fig. 2.351**).

................................*Hydrosmilodon* Flowers and Domínguez, 1992..p.366

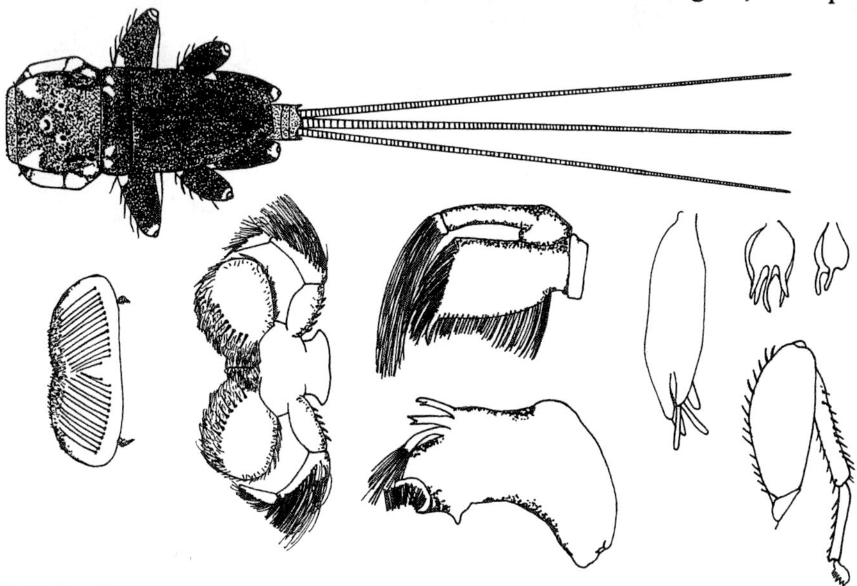

Fig. 2.350 *Hermanella thelma* larva: habitus (above); labrum (lower left); labium (lower left center); maxilla (center); left mandible (lower center); gills on the fifth, sixth, and seventh abdominal segments (middle right, left to right); and the hind leg (lower right).

27. The head is voluminous, its length being more than half the length of the rest of the body. The abdomen is only slightly longer than the thorax. The cerci are much longer than the total length of the rest of the insect (**Fig. 2.350**).

....*Hermanella* Needham and Murphy, 1924..p. 316

- The head is much shorter than half of the rest of the body. The abdomen is elongated and much longer than the thorax. The cerci are only about as long as the abdomen (**Fig. 2.352**). The larva of only one unidentified species has been described (Demoulin, 1955a).

....*Hermanellopsis* Demoulin, 1955..p. 331

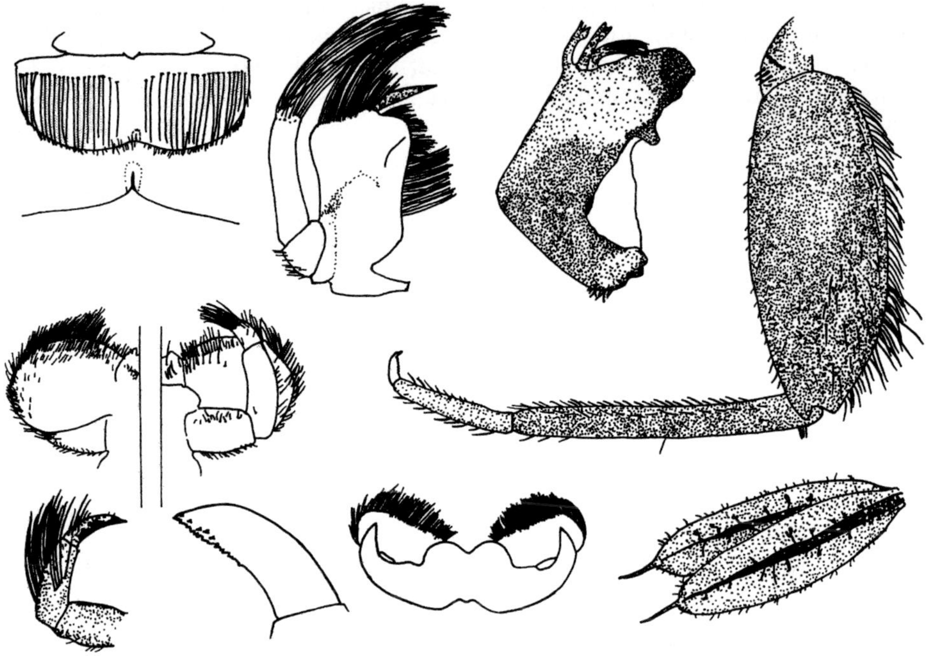

Fig. 2.351 *Hydrosmilodon primanus* larva (upper row, left to right): the labrum with an enlargement of the apical margin below, maxilla, left mandible, and fore-leg; an outline of the labium with half in dorsal and half in ventral view (middle left, left and right, respectively); and (lower row, left to right): the left palp in dorsal view, its third segment without setae enlarged, the hypopharynx, and the gills from the fourth abdominal segment.

28. Posterolateral spines are present only on the eighth and ninth abdominal segments. 29
- Posterolateral spines are present on the sixth through the ninth abdominal segments (**Fig. 2.353**). 31
29. There is a V-shaped border to the anterolateral angle of the maxilla (**Fig. 2.347**).
....*Farrodes* Peters, 1971..p. 302
- There is no border of this kind on the anterolateral angle of the maxilla (**Fig. 2.354**). 30
30. Each gill is at least six times as long as wide, and it bears no long processes along its margin. There is a greatly enlarged apical denticle on each claw. The widest part of the labrum is about 2/3 of the way from the base, and the lateral margins are rounded. A mandibular tuft is present. The paraglossae of the labium are located ventral to the glossae. There is a row of three thick setae on the margin of the third labial palp segment (**Fig. 2.354**). The only known species has a yellowish head with a brown wash on the frons between the ocelli, a brown thorax with black on the lateral areas of the nota and the on the carina,

black pleurae and black on parts of the sternum. The eight anterior abdominal tergites are yellow with a blackish band across the posterior margins. The legs are yellowish with brownish areas at about midlength and at the apex of the femora and forming a subapical band on the tibiae and a basal band on the tarsi.
....*Simothraulopsis* Demoulin, 1966

The only known species in this genus is *Simothraulopsis demerara* (Traver, 1947) from Guyana and Surinam. Synonyms include *Thraulus demerara* Traver, 1947; *Homothraulus demerara* (Traver, 1947); and *Simothraulopsis surinamensis* Demoulin, 1966. Domínguez *et al.* (1997) speculated that this genus may actually be indentical with *Homothraulus*. From all evidence, this seems likely.

- Each gill is considerably less than six times as long as wide, and the margins of the gills usually bear long processes. There is no greatly enlarged apical denticle on each claws. A detailed illustration by Demoulin (1955a) of a larva designated *Homothraulus* sp. apparently does not belong to that genus.
....*Homothraulus*, Demoulin, 1955..p. 365

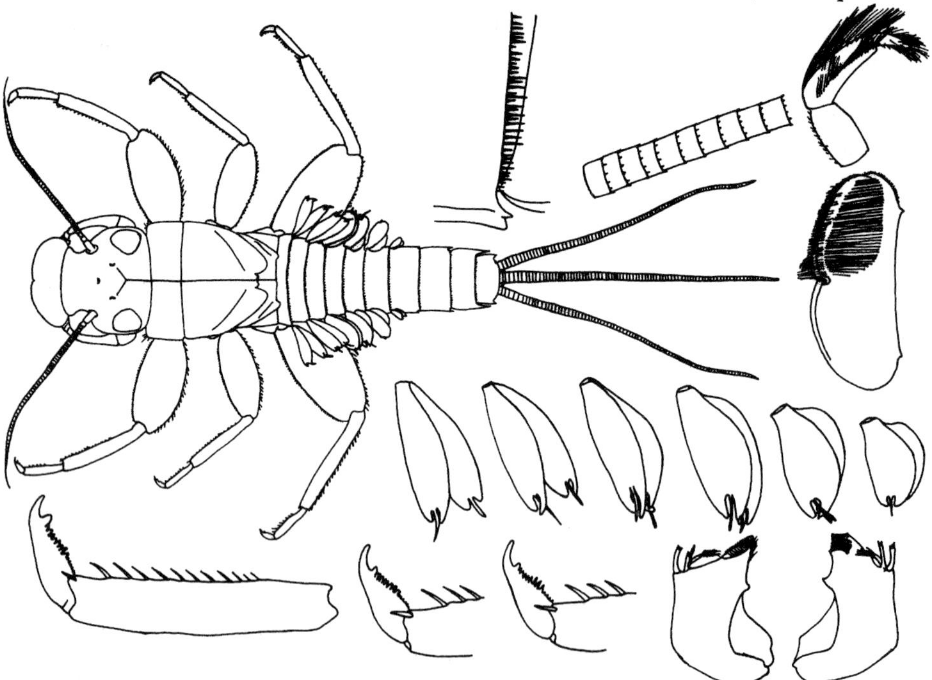

Fig. 2.352 *Hermanellopsis* sp. larva: habitus (upper left), the tarsus of the foreleg and the tarsal claws of the middle and hind legs (lower left to center, left to right), the gills on the first through sixth abdominal segments (lower middle, left to right), details of the posterior margin of the fourth abdominal tergite (upper center), a section of a cercus (upper right center), the labrum with the setae removed from one side (right of habitus), the mandibles (lower right), and the right labial palp (upper right).

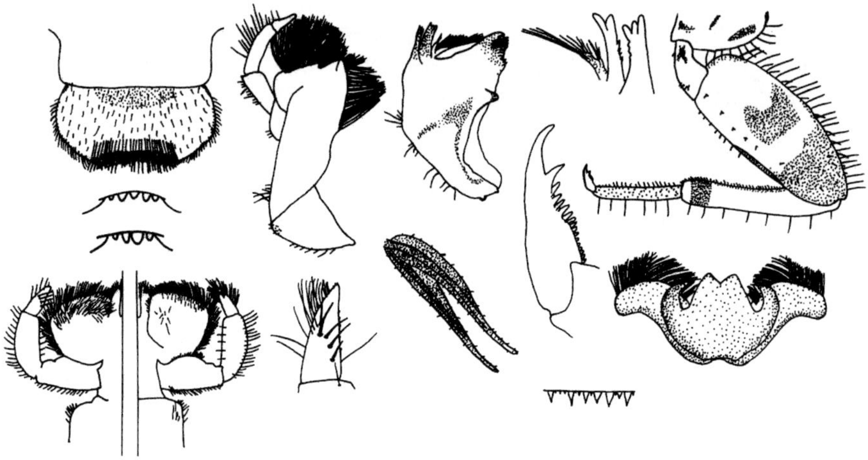

Fig. 2.353 *Ecuaphlebia* sp. larva: (upper row, left to right): labrum above enlargements of the edge of the anterior emargination in two varying specimens, maxilla, left mandible, details of the prostheca and incisors on the right mandible, fore-leg with and enlargement of the tarsal claw to its left, and (lower row, left to right), half of the labium in dorsal and in ventral view, the apical segment of the labial palp enlarged, the second gill, the margin of the fifth abdominal tergite, and the hypopharynx.

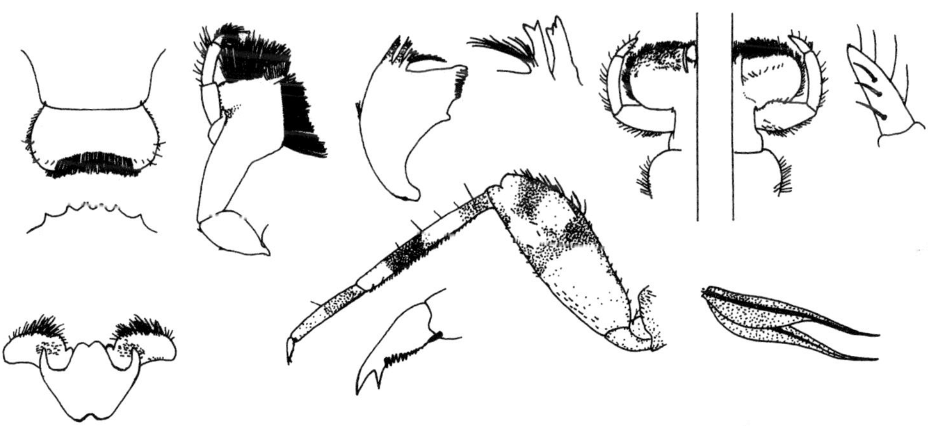

Fig. 2.354 *Simothraulopsis demerara* larva (upper row, left to right): the clypeus and labrum above an enlargement of the margin of the median emargination of the labrum with setae removed, maxilla, mandible, enlargement of its incisors, half of the labium in dorsal and half in ventral view, apical segment of the labial palp, and (lower row, left to right): hypopharynx, fore-leg above an enlargement of its tarsal claw, and a fourth abdominal gill.

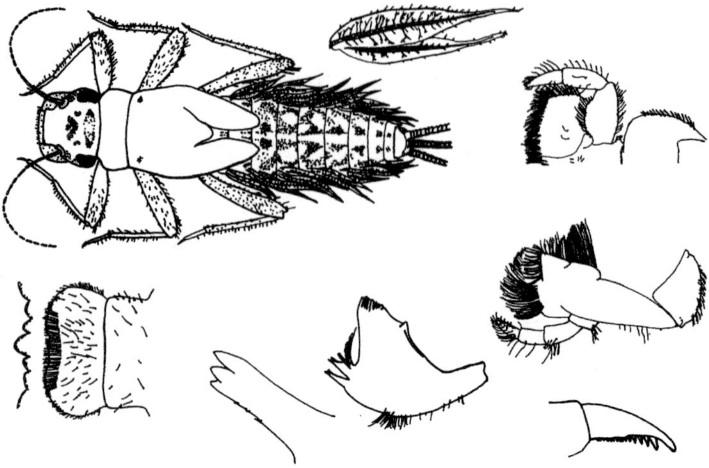

Fig. 2.355 *Nousia delicata* larva: habitus without the apical portions of the cerci and caudal filament (upper left), enlargement of a gill (upper center), the labrum with an enlargement of the emargination in its anterior margin (lower left), the left mandible with an enlargement of the outer incisor of the right mandible (lower center), one side of the labium in ventral view (upper right), right maxilla (right middle), and a claw (lower right).

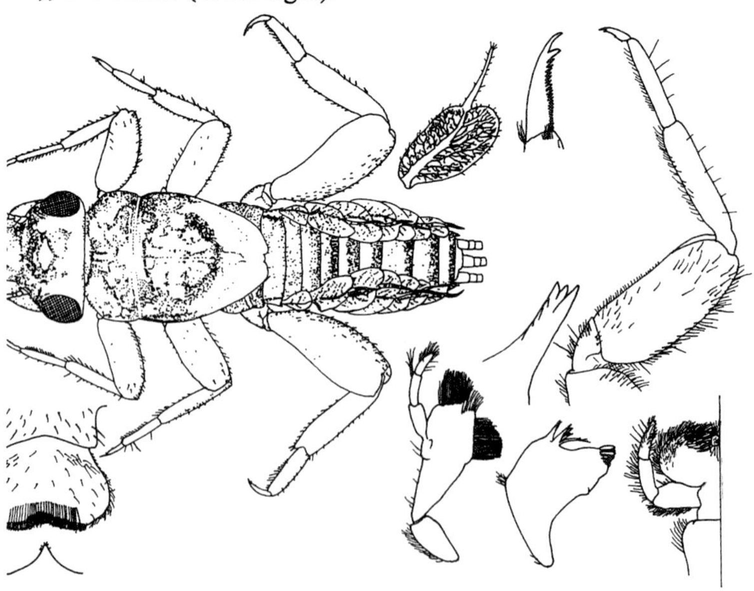

Fig. 2.356 *Magallanella flinti* larva: habitus without cerci (upper left), the fourth gill (upper center), the tarsal claw from the fore-leg (upper right center), the fore leg (upper right), and (lower row, left to right): the labrum with an enlargement of its median emargination, maxilla, mandible with an enlargement of its outer incisor above it, and half of the labium in dorsal view.

31. The gills are lanceolate, narrowing gradually toward the apex (**Fig. 2.355**). 32

- The gills are not notably narrow and do not become gradually thinner toward the apex (**Fig. 2.325**). 34

32. The antenna is 2.5 times as long as the head. The lateral margins of the pronotum are armed with small spines (**Fig. 2.353**). .... *Ecuaphlebia* Domínguez, 1988

There is a single described species in this genus: *Ecuaphlebia rumignaui* Domínguez, 1988, known only from Ecuador. A second species from Ecuador, described only from a subadult and larvae, was referred to as "*Ecuaphlebia* sp. A" by Domínguez (1988b). This form has no taxonomic status and remains undescribed until someone decides to do so and give it a name.

- The antenna is 1.5 times as long as the head. The lateral margins of the pronotum are smooth or there are short spines anterolaterally (**Fig. 2.355**). 33

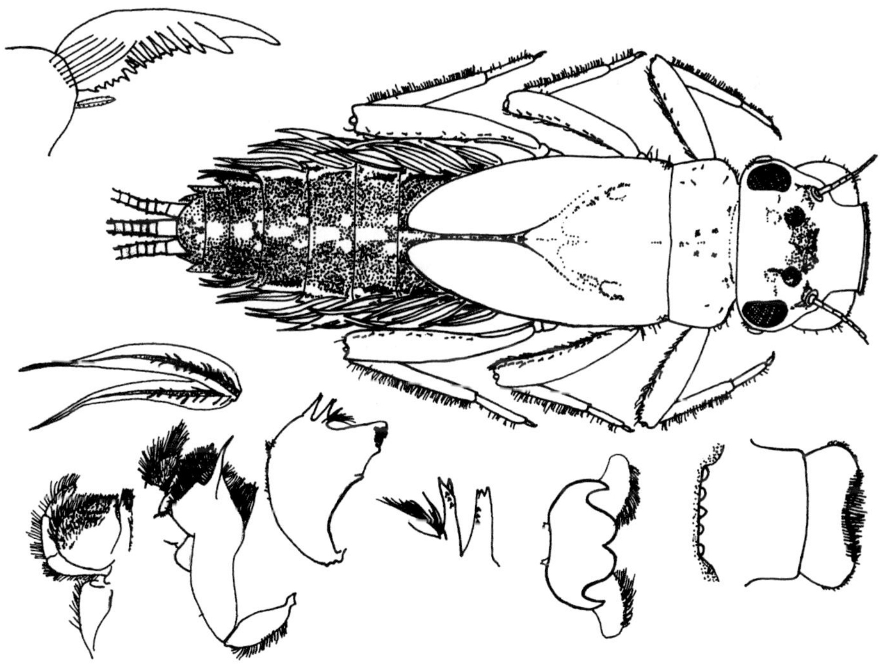

Fig. 2.357 *Gonserellus atopus* larva: habitus without apices of cerci and caudal filament (upper right), an enlargement of a claw (upper left), an enlarged abdominal gill from the fourth segment (middle left), and (bottom row, left to right) a dorsal view of one side of the labium, ventral view of the right maxilla, left mandible, enlarged incisor on the right mandible, hypopharynx, enlargement of the margin of the anteromedial emargination of the labrum, and the clypeus and labrum.

33. There is a well-developed subapical tusk on the galealacinia of the maxilla (**Fig. 2.357**) and bipectinate setae on the tibiae. The only species described is 6.0 to 7.5 mm long. It is yellowish with darker brownish markings and black eyes.

....*Gonserellus* Pescador, 1997

The only species described in the genus is *Gonserellus atopus* Pescador, 1997, from south central Chile.

- The galealacinia of the maxilla lacks a well-developed tusk (**Fig. 2.355**), and the tibiae lack bipectinate setae.

....*Nousia* Navás, 1918..p. 281

34. The gills are ovoid and terminate abruptly in a filamentous process. Only a single lamina is present (**Fig. 2.356**). The adults and subadults have not yet been described.

....*Magallanella* Pescador and Peters, 1980

The only species in this genus is *Magallanella flinti* Pescador and Peters, 1980, from Chile.

- The gills are quadrangular, terminating in various kinds of digitiform projections. Both a dorsal and ventral lamina are present (**Fig. 2.325**). The adults and subadults have not yet been described.

....*Dactylophlebia* Pescador and Peters, 1980

The only species yet described in this genus is *Dactylophlebia carnulenta* Pescador and Peters, 1980, from southern Argentina and Chile.

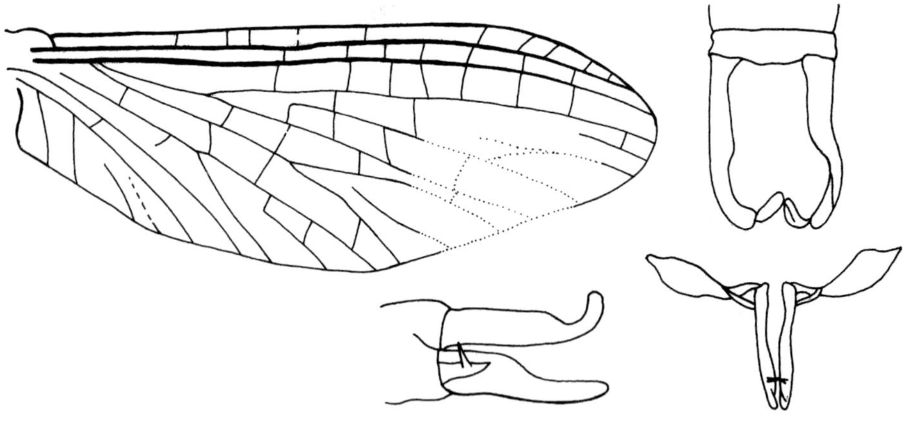

Fig. 2.358 *Hagenulopsis minuta* male: the wing (upper left) with the posterior portion near the apex restored to show the probable locations of the veins with dotted lines, forceps of the male (upper right), penis in ventral view (lower right), and the genitalia in ventrolateral view (lower center).

Key to the species of adult *Hagenulopsis* in South America

Information for the key was taken from Ulmer (1920a), Spieth (1943), and Traver (1946). Reliable descriptions of the larvae are not available.

1. Fewer than 12 cross veins are in costal space of the fore-wing (**Fig. 2.295**). ....*Hagenulopsis diptera* Ulmer, 1920 (Santa Catarina).
- More than 16 cross veins are in the costal space of the fore-wing (**Fig. 2.358**). ....*Hagenulopsis minuta* Spieth, 1943 (Guyana, Surinam).

Key to the species of adult *Nousia* in South America

The key is taken mainly from that of Pescador and Peters (1985). According to Hubbard *et al.* (1992) and other authors, the genus *Atalonella* Needham and Murphy, 1924 is a junior synonym for *Nousia*, and it is treated as such here.

1. The markings on the abdominal tergites are very small or absent entirely. The depth of the median emargination in the posterior border of the ninth sternite of the female is about 1/7 the maximum width of the sternite (**Fig. 2.359**). Body length: 7.5 to 10 mm; fore-wing length: 9 to 12.5 mm; hind wing length: 2.5 mm. Color: mainly yellow to brownish yellow, sometimes with small dark brown markings. ....*Nousia grandis* (Demoulin, 1955) (Argentina, Chile). Syn: *Atalonella grandis* Demoulin, 1955.
- The abdominal tergites have prominent dark markings. The emargination in the ninth sternite of the female is about 1/3 of the maximum width of the sternite (**Fig. 2.360**). 2
2. The body length is 5 to 6.5 mm, and the fore-wing ranges from 5 to 7 mm. The middle abdominal tergites are mainly dark with light midlines and anterior or posterior margins and small lateral spots. The sternites also have extensive dark markings (**Fig. 2.298**). Hind wing length: 1.0 to 1.3 mm. Color: brown to light maroon on the head and thorax and light ochre with brown and light brown markings on the abdomen. ....*Nousia minor* (Demoulin, 1955) (Argentina, Chile). Syn: *Atalonella minor* Demoulin, 1955.
- The body length is greater than 6.5 mm. The dark markings on the tergites and sternites are usually much less extensive (**Fig. 2.260**). 3
3. The compound eyes of the male are not in contact along the midline of the head. The dark markings on the middle tergites of the female are well separated from the anterior margin. The dark markings on the tergites of the male are relatively faint. At least the first seven abdominal sternites are entirely yellow in color except for a small central marking (**Fig. 2.360**). Body length: 7 to 9 mm;

fore-wing length: 7 to 10 mm; hind wing length: 1.8 to 2.0 mm. Color: mainly brown to light maroon on the head and thorax and light to dark brown on the abdomen.
....*Nousia delicata* Navás, 1918 (Argentina, Chile). Syn: *Atalonella ophis* Needham and Murphy, 1924.
- The compound eyes of the male are in contact along the midline of the head. The dark markings on the abdominal tergites of the female come close or are in contact with the anterior margin. There are usually more extensive dark markings on the abdominal sternites than a small dark spot in the center (**Fig. 2.361**). 4

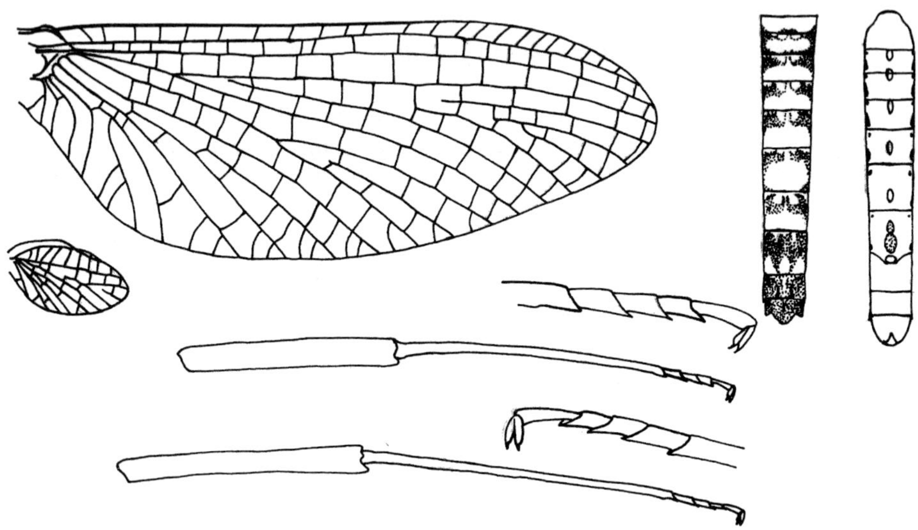

Fig. 2.359 *Nousia grandis* female adult: wings (upper left), pattern on the dorsal (right center) and ventral (far right) surfaces of the abdomen, and the middle leg above the hind leg, each with an enlargement of the tarsi.

4. The markings on the abdominal tergites of the female are dark brown. The penis lobes diverge and are not in contact along the midline except at their bases (**Fig. 2.361**). Body length: 7 to 9 mm; fore-wing length: 7 to 9 mm; hind wing length: 2 to 2.2 mm. Color: the head and thorax are mainly dark yellow, brown, or grayish.
....*Nousia maculata* (Demoulin, 1955) (Argentina, Chile). Syn: *Atalonella maculata* Demoulin, 1955.
- The markings on the abdominal tergites of the female are black. Processes on the inner side of the penis lobes are in contact or nearly so along the midline (**Fig. 2.362**). 5

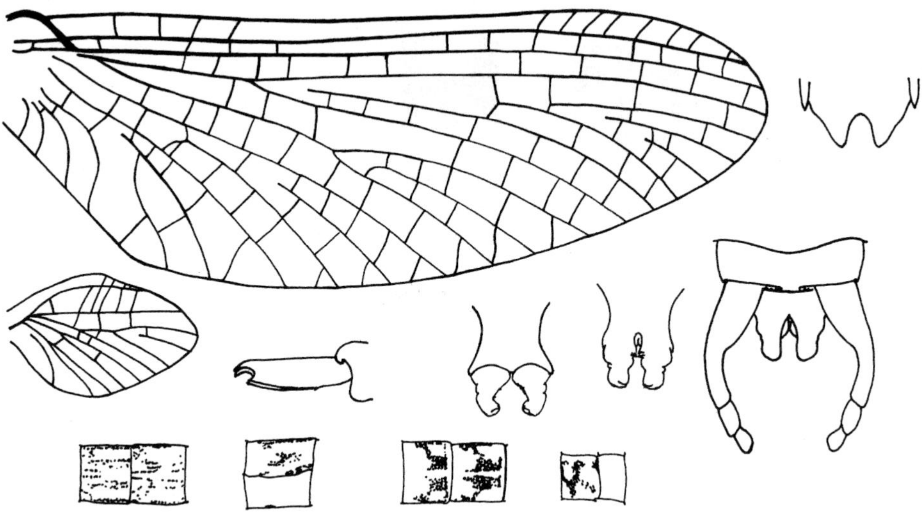

Fig. 2.360 *Nousia delicata:* the fore and hind wing (above left), emargination in the ninth sternite of the female (upper right), the claw of the fore-leg (right of hind wing), the penis in dorsal and ventral view and the entire male genitalia (middle row right, left to right), and the color pattern (lower row, left to right) on the fifth and sixth abdominal tergite in dorsal view and the sixth in lateral view of a male and the same segments of a female in dorsal and lateral view.

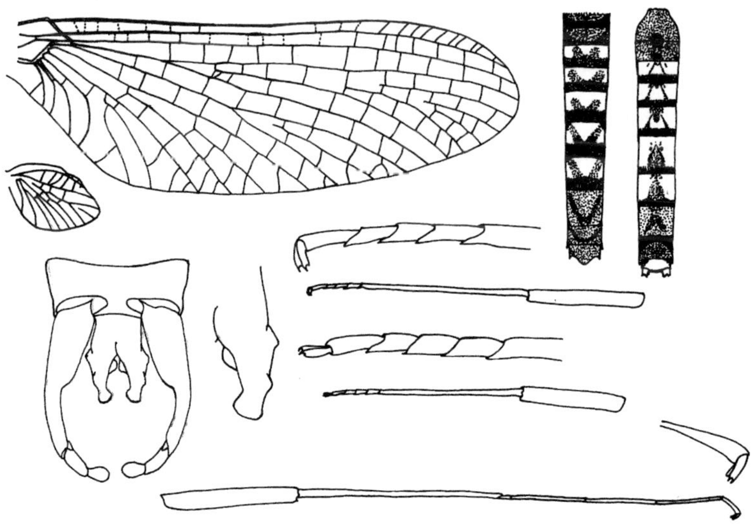

Fig. 2.361 *Nousia maculata,* male adult: wings (upper left), pattern on the dorsal (right center) and ventral (far right) surfaces of the abdomen, genitalia with an enlargement of the right penis lobe (lower left center), and (right center, top to bottom) the hind, middle and fore-legs, each with an enlargement of its tarsi or its apical tarsal segment with claws.

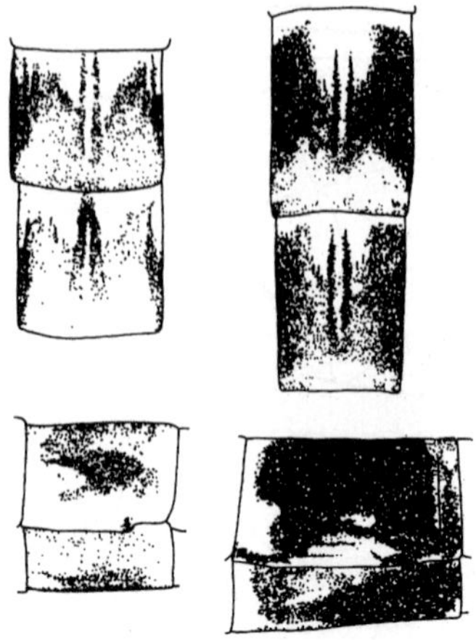

Fig. 2.362 *Nousia bella:* the color pattern on the fifth and sixth abdominal tergites of two males in dorsal view (upper left and center) and on the sixth segment of the males in lateral view (lower left and center) and the penis in ventral view (right).

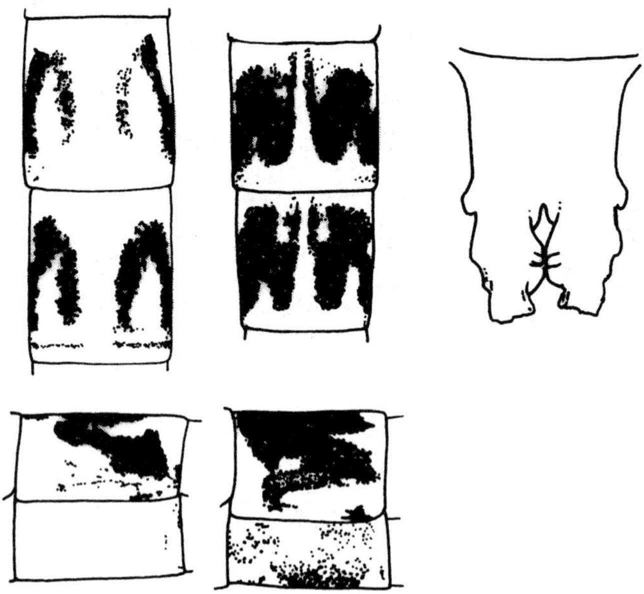

Fig. 2.363 *Nousia crena:* the color pattern on the fifth and sixth abdominal tergites of the male (upper left) and female (upper center) in dorsal view and on the sixth segment of the male (lower left) and female (lower center) in lateral view and the penis in ventral view (right).

5. The light markings along the midline of the abdominal tergites are narrow and often do not reach from the anterior to posterior margins without interruption by

dark pigment. The markings of both males and females are black. The processes of the penis lobes that are in contact along the midline are broad (**Fig. 2.362**). Body length: 6.5 to 9 mm; fore-wing length: 6.5 to 9.5 mm. Color: Yellow with dark brown and black markings.
....*Nousia bella* Pescador and Peters, 1985 (Argentina, Chile).
- The light markings along the midline of the abdominal tergites are broad and usually extend from the anterior to posterior margins without interruption by dark pigment. The dark pigment on the female abdomen is black, while that of the male is dark brown on a translucent yellow background. The processes of the penis lobes that are in contact along the midline are narrow (**Fig. 2.363**). Body length: 7 to 9.5 mm; fore-wing length: 7 to 10 mm. Color: Reddish black and reddish brown on the head and thorax, respectively.
....*Nousia crena* Pescador and Peters, 1985 (Argentina, Chile). Aside from color, the two species in this couplet are similar, and the argument that they are distinct is not fully convincing.

Key to the species of *Nousia* larvae in South America

Information for the key was taken from Pescador and Peters (1985).

1. The gills are about eight times as long as their width at the widest point (**Fig. 2.364**). There is a brown band apical to the midpoint of each tibia. Body length of last instar larva: 5 to 6 mm. Color: yellow with dark brown markings.
....*Nousia minor* (Demoulin, 1955) (Argentina, Chile). Syn: *Atalonella minor* Demoulin, 1955.
- The gills are not longer than five times their width at the widest point. The tibiae lack bands apical to the midpoint (**Fig. 2.365**). 2

Fig. 2.364 The fourth gill from a larva of *Nousia minor*.

Fig. 2.365 The tarsal claw on the fore-leg (left) and the fourth abdominal gill (right) from the larva of *Nousia maculata*.

2. The abdomen is yellowish or light brown with darker markings confined to the posterior edge of the tergites and, sometimes, small lateral marks. Each lateral margin of the submentum bears fewer than 10 spines. Body length of last instar larva: 7 to 8.5 mm. Color: mainly dark yellow with a yellow or light brown abdomen with few darker markings.
....*Nousia grandis* (Demoulin, 1955) (Argentina, Chile). Syn: *Atalonella grandis* Demoulin, 1955.
- There are extensive dark markings on the abdominal tergites, and each lateral margin of the submentum bears more than 10 spines. 3
3. The gills on the abdomen are black or dark gray. The abdominal sternites are dark brown with pale yellow lateral margins. Body length of last instar larva: 6 to 8 mm. Color: mainly brown with blackish markings dorsally.
....*Nousia bella* Pescador and Peters, 1985 (Argentina, Chile).
- The gills on the abdomen are smoky white, and the abdominal sternites are yellow. 4
4. The length of the hind femur is more than three times the maximum width. The main trunk of each abdominal gill bears a few weakly developed branches (**Fig. 2.365**). Body length of last instar larva: 7 to 8.5 mm. Color: head mainly dark yellow with blackish ocelli and compound eyes; thorax mainly yellow or light brown with pale yellow spots; anterior abdominal tergites brownish with dark brown markings, and the eighth through tenth tergites, pale yellow.
....*Nousia maculata* (Demoulin, 1955) (Argentina, Chile). Syn: *Atalonella maculata* Demoulin, 1955.
- The length of the hind femur is less than three times the maximum width. The main trunk of each abdominal gill bears many well-developed branches (**Fig. 2.354**). 5

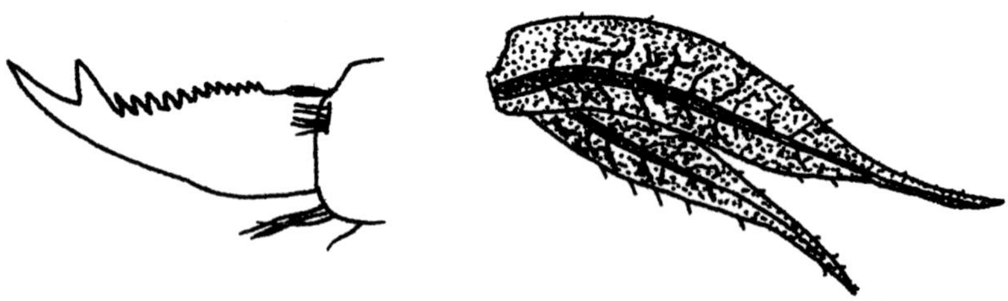

Fig. 2.366 The tarsal claw on the fore-leg (left) and the fourth abdominal gill (right) from the larva of *Nousia crena*.

5. The tarsi bear more than ten spines. The subapical denticle on each claw is distinctly narrower at the base than the width of the claw just apical to that denticle (**Fig. 2.354**). Length of the last instar larva: 7 to 9 mm. Color: mainly yellow or light brown with dark brown markings dorsally.
....*Nousia delicata* Navás, 1918 (Argentina, Chile). Syn: *Atalonella ophis* Needham and Murphy, 1924.

- The subapical denticle on each claw is wider at the base than the width of the claw just apical to the base (**Fig. 2.366**). The tarsi bear fewer than ten spines. Body length of last instar larva: 6 to 9 mm. Color: mainly dark yellow to light brown with dark brown and black markings dorsally.
....*Nousia crena* Pescador and Peters, 1985 (Argentina, Chile).

Key to the species of adult *Fittkaulus* in South America

Information for this key was provided by Savage (1986). The key is based on descriptions of the females. The larva of *Fittkaulus maculatus* is the only one described (**Fig. 2.333**).

1. The fork of vein MA is not covered by a dark marking, although such markings may cover adjacent cross veins (**Fig. 2.367**). The pleura of the female do not bear stripes running the entire length of the thorax. The episternites of the mesopleuron and metapleuron are brownish black. The first and second abdominal sternites have small brownish black posteromedian marks. Length: 5.8 to 6.1 mm. Fore-wing length of female: 5.9 mm. Color: yellowish with darker brownish markings. Only the female has been described.
....*Fittkaulus cururuensis* Savage, 1986 (Pará).
- There is a dark marking covering the fork in vein MA of the fore-wing (**Fig. 2.368**). There is a brownish black median stripe running the full length of the thorax on the pleura of the female. The first and second abdominal sternites do not have small brownish black markings at the center of the posterior margins.
....2

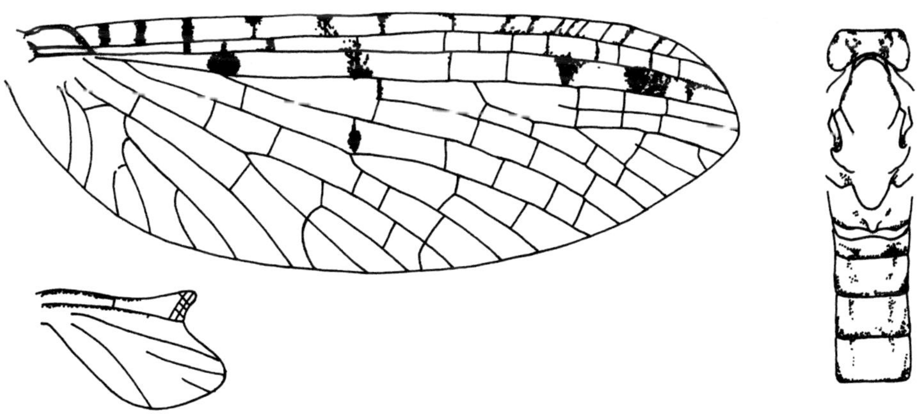

Fig. 2.367 *Fittkaulus cururuensis:* the fore and hind wing of a female (left) and the head, thorax, and first four abdominal segments of an adult female.

2. The hind wing is obtuse and broadly rounded at the apex, and the costal projection is well developed and situated at or beyond 7/10 of the distance from the base to the apex of the wing. The fork in vein MP of the fore-wing is asymmetrical (**Fig. 2.297**). Length of male: 4.8 to 6.0 mm; of female 5.5 to 6.8 mm. Fore-wing length of male: 4.8 to 5.3 mm; of female: 5.3 to 6.0. Color: head pale yellow with black markings; thorax light and yellowish brown with a brownish black pronotum, median longitudinal mark, carinae, and peripheral margins. The abdomen is pale yellow with brownish black posterior borders to most tergites and rectangular posterolateral markings.
.... *Fittkaulus maculatus* Savage and Peters, 1978 (Surinam, Pará).

- The hind wing is acutely rounded at the apex, and the costal projection is well developed and situated less than 7/10 of the distance from the base to the apex of the wing. The fork in vein MP of the fore-wing is symmetrical (**Fig. 2.368**). Length of female: 5.3 to 6.5 mm. Fore-wing length of female: 5.6 to 6.8 mm. Color: yellowish with darker brownish markings. Only the female has been described.
....*Fittkaulus cuiabae* Savage, 1986 (Mato Grosso).

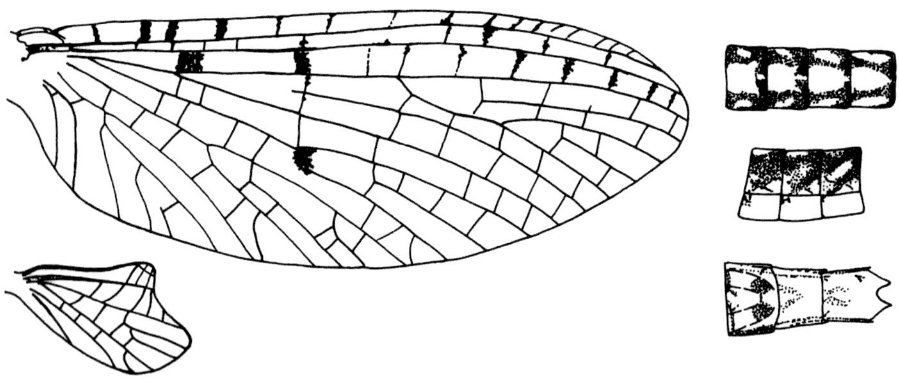

Fig. 2.368 *Fittkaulus cuiabae:* the fore and hind wing of a female (left), the color pattern on the abdominal tergites (right, top to bottom), the fourth through seventh in dorsal view, the third through fifth in lateral view, and the seventh through ninth in ventral view.

Key to the adult *Penaphlebia* species in South America

Information for the key was provided by Navás (1928, 1930e, 1934a, 1936a), Demoulin, (1955d), Thew (1960), Domínguez and Pescador (1983), and Pescador and Peters (1991).

1. The wing base is yellowish, and the base of the costal space is dark brown as far as the humeral vein. In females, there is a strong angle in the subcostal vein in the stigmatic area (**Fig. 2.369**). Length of the male: 8.5 mm. Length of the fore-wing of the male: 10.3 mm; of the female: 13.4 mm. Hind wing length of the male: 3.4 mm; of the female: 4.5 mm. The head and thorax are dark with a pair of longitudinal yellow stripes on the mesonotum. The abdomen is yellow with a pair of dark longitudinal stripes on some of the tergites. The legs are mainly dark with darker bands on the femora near the middle and at the apex and yellow apices of the tibiae.
....*Penaphlebia rofucensis* (Navás, 1934) (Chile). Syn: *Atalophlebia rofucensis* Navás, 1934. From the description, it is not certain that this species belongs to the genus *Penaphlebia*, and Domínguez *et al*. (2001) regard its name as a *nomen dubium*.
- Either the costal and subcostal spaces are almost entirely infuscated, at least around the cross veins, or they are entirely hyaline (**Fig. 2.370**). 2

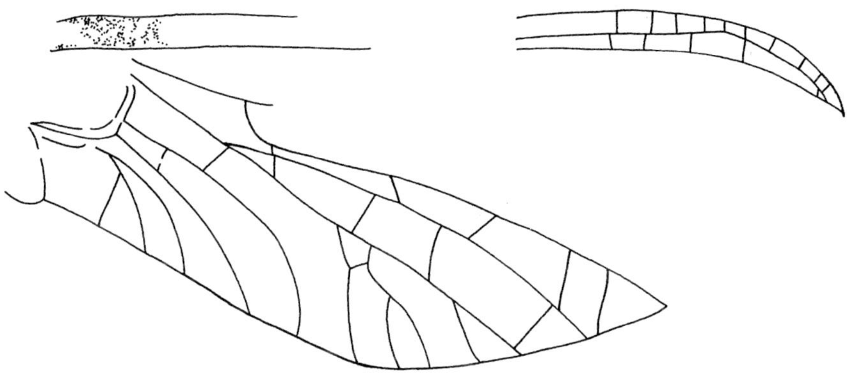

Fig. 2.369 *Penaphlebia rofucensis:* the veins in the posterior area of the wing near the base (left) and in the stigmatic area (right).

2. The cross veins in the costal and subcostal spaces of both males and females are strong and heavily infuscated (**Fig. 2.370**). The length of the male is 8 to 12 mm, and the female is 9 to 14 mm. The fore-wing length is 9 to 14 mm. 3
- The cross veins in the costal and subcostal spaces of the male are weak and not infuscated. Those of the female may be lightly infuscated (**Fig. 2.371**). 6
3. The eight anterior abdominal tergites are yellowish brown with irregular dark markings arranged to form two or three longitudinal stripes: either one median stripe with a lateral stripe on each side or only two lateral stripes. The ninth tergite is almost entirely dark. The tenth tergite is slightly produced in the center. Length of the male: 10.5 mm; female: 8.4 mm. Fore-wing length of the male: 12 mm; of the female: 10.2 mm. Hind wing length of the male: 3.4 mm; of the female: 2.9 mm. The cerci are from 19.5 to 21 mm long; they are light with dark rings. The head is yellow with dark eyes and antenna bases. The

thorax is shiny yellowish brown, and that of the male is darkened dorsally. The legs are dark brown with dark rings on the fore-legs of the male.
....*Penaphlebia discolor* (Navás, 1928) (Chile). Syn: *Atalophlebia discolor* Navás, 1928. The description of this species is not adequate to confirm that it belongs to the genus *Penaphlebia,* and Domínguez *et al*. (2001) regard it as a *nomen dubium*
- The dark markings on the eight anterior abdominal tergites do not form two or three longitudinal stripes (**Fig. 2.370**). 4

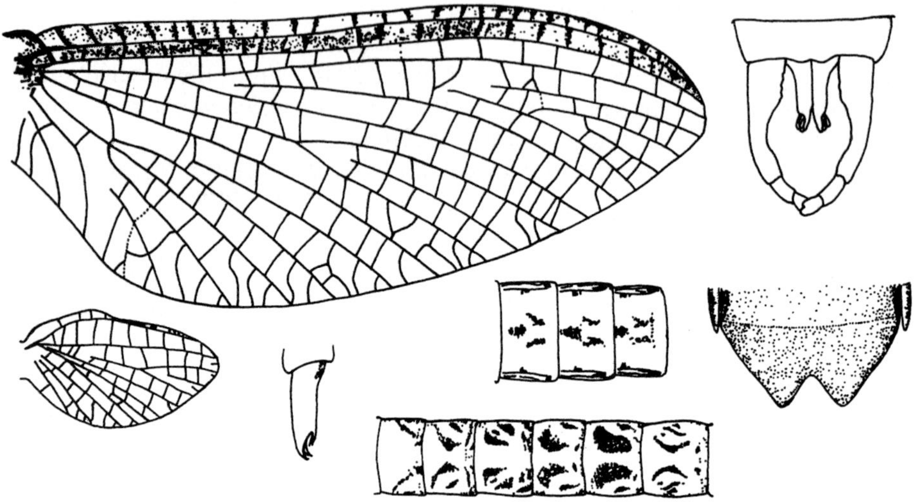

Fig. 2.370 *Penaphlebia barriai:* fore and hind wings of a female imago (left), genitalia of an adult male (upper right), the apex of the abdomen of a female in ventral view (lower right), the tarsal claw from the fore-leg of a male (lower left center), the fifth through seventh abdominal sternites (center), and the third through eighth abdominal tergites (lower right center) oriented with the head of the larva toward the left.

4. The first abdominal tergite lacks extensive dark markings similar to those on the third, sixth, seventh, and ninth tergites. The costal membrane of the female wing is hyaline, while the subcostal membrane is yellowish (**Fig. 2.370**). Length of male: 9 to 12 mm; of female: 10 to 14 mm. Fore-wing length of male: 10 to 12 mm; of female: 10 to 14 mm. Color: generally yellowish with brown or black markings, including the ventral part of the compound eyes. Light yellow with distinct dark brown or black markings.
....*Penaphlebia barriai* Pescador and Peters, 1991 (Chile). That this species is distinct from the two in couplet 3 is very doubtful.
- The first abdominal tergite has extensive black or dark brown markings, similar to or more extensive than those on the third, sixth, seventh, and ninth tergites (**Fig. 2.372**). 5

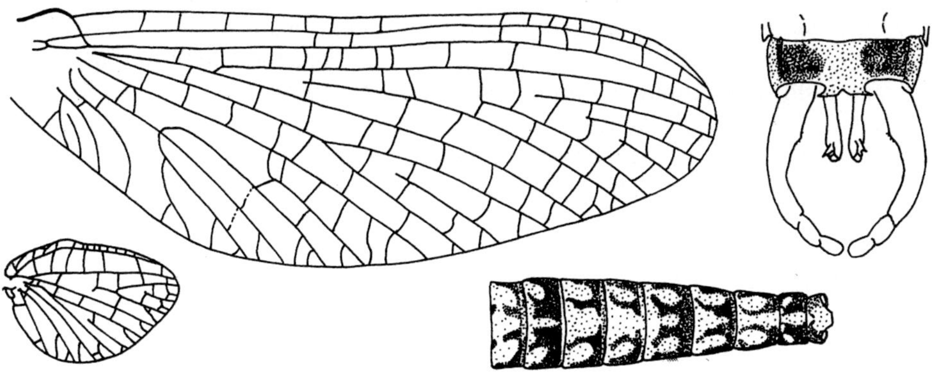

Fig. 2.371 *Penaphlebia exigua:* the fore and hind wings (left), male genitalia in ventral view (upper right) and the markings on the first through ninth abdominal tergites of a male (lower right).

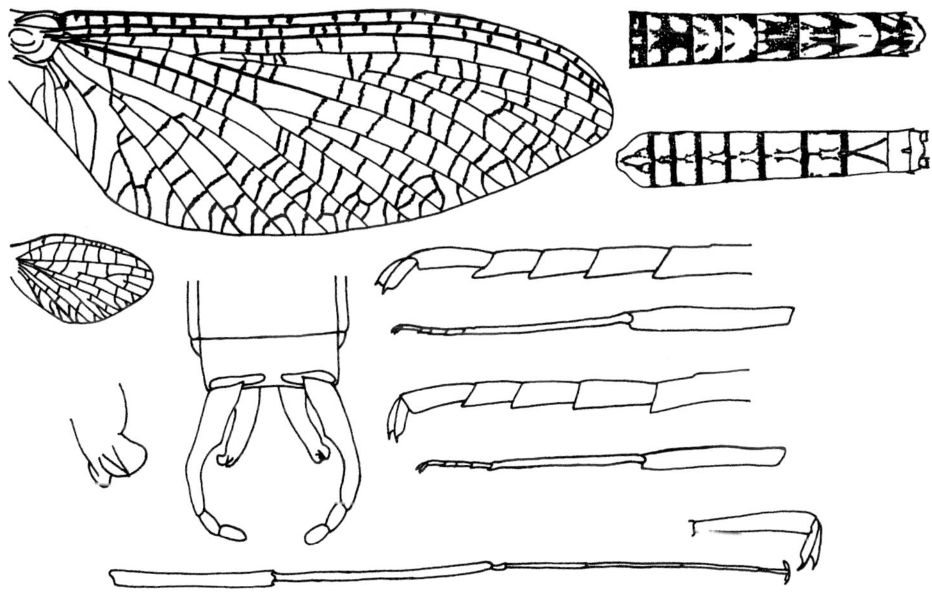

Fig. 2.372 *Penaphlebia chilensis:* fore and hind wings of a female sub-imago (above); genitalia of an adult male with the enlarged left penis lobe to the left (lower left); the hind, middle, and fore legs below enlargements of their respective tarsi (lower right, top to bottom); and the pattern on the dorsal (upper right) and ventral (middle right) abdominal surfaces of the male.

5. The first abdominal tergite has dark markings more extensive than those on the third, sixth, seventh, and ninth tergites. The cross veins in the hind wing are narrowly darkened. Length of the male: 8.2 mm. Fore-wing length of the male: 10 mm; hind wing length: 2.9 mm. Color: mainly yellow with dark markings. ....*Penaphlebia fenestrata* (Navás, 1930) (Chile). Syn: *Atalophlebia fenestrata* Navás, 1930. This species is very poorly described and is regarded by Domínguez *et al.* (2001) to be a *nomen dubium.* It would not be surprising if all three species with infuscated costal cross veins turned out to belong to a single species.

- The costal and subcostal membrane of the female wing is dark brownish. There is a long stretch of the costal space distal to R_{2+3} that lacks cross veins. The cross veins in the costal space are always thickened. The second segment of the fore-tarsus is 7/8 the length of the third (**Fig. 2.372**). Length: 8 to 11 mm. Fore-wing length: 9 to 14 mm. Color: generally yellowish with black markings on the head and blackish longitudinal stripes on the thorax. The abdomen is translucent yellowish with blackish markings. ....*Penaphlebia chilensis* (Eaton, 1884) (Argentina, Chile). Syn: *Atalophlebia chilensis* Eaton, 1884. This name would have priority if one or both of the other nominal species in couplets two and three turn out to be conspecific with this one.

6. The length of the male is not greater than about 8.2 mm, and the fore-wing length is not greater than 9.2 mm. 7

- The length is at least 9 mm, and the fore-wing is at least 10 mm long. Any dark markings on the femora form bands or spots rather than longitudinal stripes (**Fig. 2.373**). 10

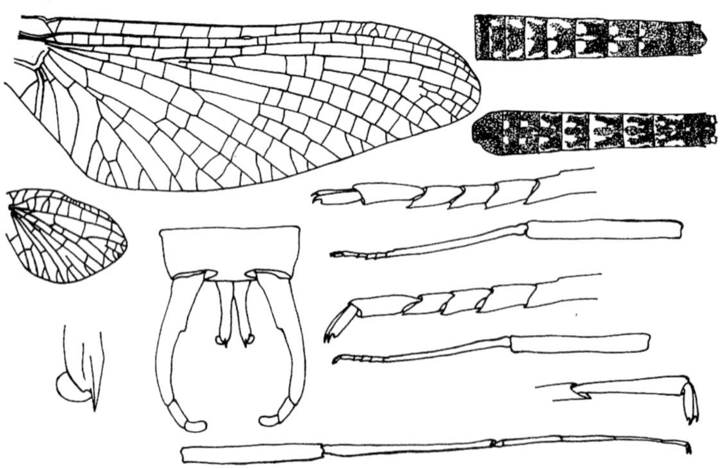

Fig. 2.373 *Penaphlebia vinosa,* male adult: fore and hind wings (above); genitalia with the enlarged left penis lobe to its left (lower left); the hind, middle, and fore legs below enlargements of their respective tarsi (lower right, top to bottom); and the pattern on the dorsal (upper right) and ventral (middle right) abdominal surfaces of the male.

7. Length of the male: 6.4 mm to 6.9 mm; male fore-wing length: 6.5 to 7.0 mm. Length of female: 7.7 mm; female fore-wing length: 8 mm. All apical tergites of the abdomen are yellow with dark lateral markings. The pleurae are brown without dark stripes. The thorax is mainly brown; the pronotum is lighter with black margins and two pairs of longitudinal black stripes on either side of the dorsal midline. The femora have faint black bands arranged transversely near the midlength. The abdomen is translucent yellow with blackish markings (**Fig. 2.371**).
................................*Penaphlebia exigua* Domínguez and Pescador, 1983 (Argentina).
- The length of the male is greater than 7 mm, and fore-wing of the male is longer than 8.5 mm. Dark dorsal stripes on the thorax, if present, are not black, or the head and thorax is mainly blackish with yellow humeral stripes. 8
8. The head and thorax are mainly blackish with a few yellow or whitish markings. The first seven abdominal segments of the male are whitish and translucent with dark markings. The apical segments are blackish. The female is mainly yellowish brown, and its abdomen is not translucent. Length of male: 7.5 mm; of female: 6 mm. Fore-wing length of male: 8.6 mm; of female: 9.2 mm. Hind wing length of male: 2.8 mm; of female: 3 mm.
....*Penaphlebia hyalina* (Navás, 1930) (Chile). Syn: *Atalophlebia hyalina* Navás, 1930. Domínguez *et al.* (2001) regard this as a *nomen dubium.*
- The head and thorax of the male are mainly yellow with dark markings, including dark dorsal stripes on the thorax. 9
9. The apical segments of the abdomen are mainly dark. The thorax is dark reddish brown with darker lateral stripes on the anterior part of the mesonotum, a dark median stripe on its posterior part, and dark stripes on the pleurae. There is a dark longitudinal stripe on each femur. The abdomen is mainly yellow with dark oblique lateral markings and dark apical segments. The posterior borders of the apical segments of the male abdomen are nearly straight or slightly concave without emarginations or processes (**Fig. 2.374**). Length of the male: 8.2 mm. Male fore-wing length: 9.2 mm. Male hind wing length: 4 mm.
....*Penaphlebia valdiviae* (Navás, 1936) (Chile). Syn: *Atalophlebia valdiviae* Navás, 1936. The description of this species is superficial, and Domínguez *et al.* (2001) consider it a *nomen dubium.*
- All abdominal tergites are mainly yellow with lighter and darker markings. The legs are dark brown with two blackish bands on the hind femur. The posterior margin of the apical abdominal sternite is truncate (**Fig. 2.375**). Length of the male: 7.5 mm. Fore-wing length of the male: 9.6 mm. The female has not been described.
....*Penaphlebia anastasii* (Navás, 1930) (Chile). Syn: *Atalophlebia anastasii* Navás, 1930. The description of this species is very superficial, and Domínguez *et al.* (2001) consider it a *nomen dubium.*

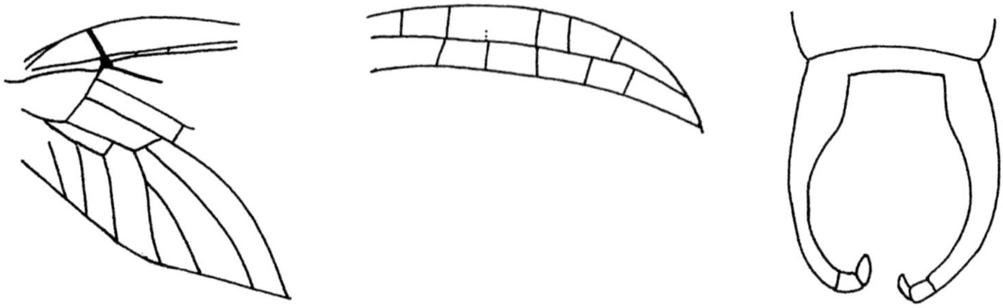

Fig. 2.374 *Penaphlebia valdiviae:* the basal portion of the fore-wing (left), the stigmatic area of the same wing (center), and the male genitalia (right).

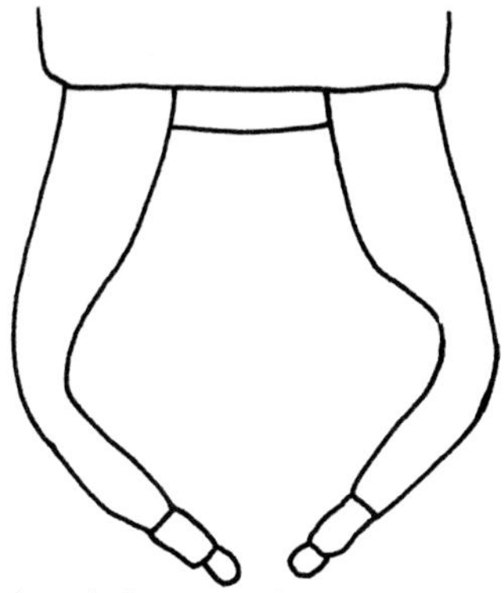

Fig. 2.375 The apex of the abdomen of a male *Penaphlebia anastasii,* redrawn from the illustration with the original description by Navás (1930e). Without descriptions or figures of the penis lobes, this species cannot be recognized again unequivocally.

10. The visible portions of the penis lobes appear generally parallel sided when viewed from above (**Fig. 2.376**). The head is yellow with reddish compound eyes. The thorax is yellow with two dark longitudinal stripes. The abdomen is yellow with dark markings on each tergite that generally form four longitudinal stripes, most notable on the second through fourth tergites. The basal ¼ of the costal area has an orange tinge. The legs are yellow with dark markings at the apices of the tibiae and the middle and hind femora. The tarsi are pale. Length of the male: 12 mm. Fore-wing length of the male: 13.6 mm; hind wing length: 3.4 mm. The female has not been described.
....*Penaphlebia athanasii* (Navás, 1930) (Chile). Syn: *Atalophlebia athanasii* Navás, 1930. The original description of this species, including the illustration of the genitalia, is so poor, that it is considered a *nomen dubium* (Domínguez *et al*., 2001). From the color pattern as described by Navás (1930e), there seems to be considerable similarity between this species and *P. flavidula* in couplet 7.
- The visible portions of the penis lobes taper or are enlarged toward the apex (**Fig. 2.373**). 11

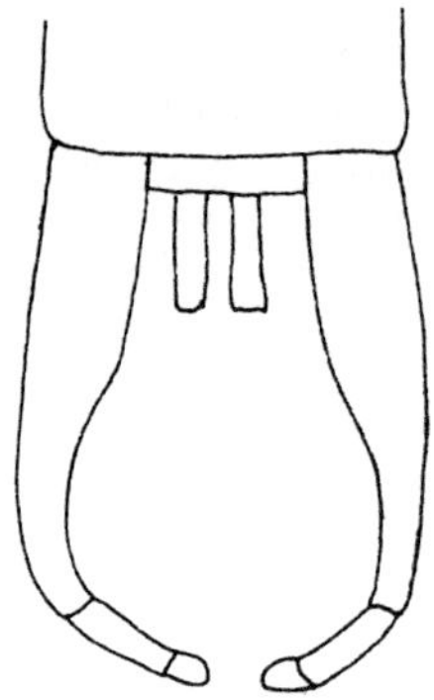

Fig. 2.376 The male genitalia of *Penaphlebia athanasii,* redrawn from the original illustration of Navás (1930e). Enough details are lacking to call the accuracy of the illustration into question.

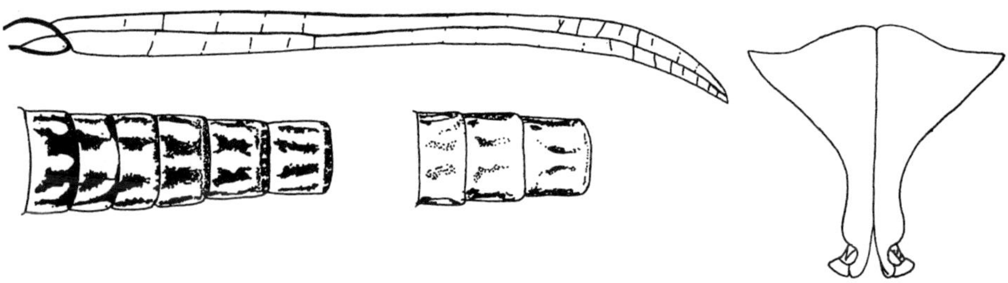

Fig. 2.377 *Penaphlebia flavidula:* the anterior edge of the fore-wing showing the costal and subcostal spaces (above left), penis (right), third through eighth abdominal tergites (lower left), and fifth through seventh abdominal sternites (lower center).

11. The third and fourth segments of the fore-tarsus of the male are nearly equal in length, and the second segment is longer than the third. Cross veins appear at more or less regular intervals with only fairly short stretches without such veins in the costal space distal to R_{2+3}, and they are not notably thickened. The hind wing is more ovate than elliptical. There is no median emargination along the posterior margin of the styliger plate. The penis lobes are enlarged considerable in their apical 1/3, and a slender subapical spine projects posteriad (**Fig. 2.373**). Length: 14 to 18 mm. Fore-wing length of male: 14 to 18 mm; of female: 16 to 20 mm. Hind wing length: 4 to 4.5 mm. Length of the cerci: 22 mm. Color: generally yellow or orange brown with dark brown markings. .... *Penaphlebia vinosa* (Demoulin, 1955) (Argentina, Chile). Syn: *Atalophlebia vinosa* Demoulin, 1955. This species has been placed in a separate subgenus, *Megalophlebia* Pescador and Peters (1991).

- The third tarsal segment of the fore-leg of the male is considerably longer than the fourth. The hind wings are approximately elliptical. There is a median emargination along the posterior border of the styliger plate. The enlargement of the apical third of the penis lobes is slight to moderate, and the subapical spine is directed posterolaterad **(Fig. 2.377)**. 12

12. The fourth, fifth, and eighth abdominal tergites are largely light in the center and are not as extensively marked as the others. Dark submedian markings occupy less than half the length of the eighth tergite (**Fig. 2.17**). Length: 10 to 13 mm. Fore-wing length: 10 to 14 mm. Length of the cerci: 13 mm. Color: brownish dorsally and yellowish ventrally with black on the ventral rim of the eyes and on the collar of the pronotum. The middle and hind legs are tawny yellow with a brownish ring just apical to the middle of the femora, which are also brown at their apices.
.................................*Penaphlebia fulvipes* (Needham and Murphy, 1924) (Chile, Argentina). Syn: *Atalophlebia fulvipes* Needham and Murphy, 1924; *Atalophlebia sepia* Thew, 1960.

- All abdominal tergites have extensive dark markings, and those on the submedian part of the eighth segment occupy more than half of the length. (**Fig. 2.377**). Length of male: 11 to 14 mm; of female: 11 to 15 mm. Fore-wing length of male: 11 to 15 mm; of female: 13 to 18 mm. Color: light yellow with distinct dark brown or black markings.
.....................................*Penaphlebia flavidula* Pescador and Peters, 1991 (Chile, Argentina).

Key to the species of *Penaphlebia* final instar larvae in South America

Information for the key was provided by Pescador and Peters (1991). The larvae of the six species described by Navás are unknown.

1. The sixth through ninth abdominal segments bear posterolateral spines. There is a blade-like process at the apex of each abdominal gill. The mandibles are deeply grooved on the dorsal side (**Fig. 2.378**). Length: 14 to 17 mm. Color: brown to dark brown dorsally with yellow markings.
....*Penaphlebia vinosa* (Demoulin, 1955) (Argentina, Chile). Syn: *Atalophlebia vinosa* Demoulin, 1955. This species has been placed in a separate subgenus, *Megalophlebia* Pescador and Peters (1991).

- Only the eighth and ninth abdominal segments bear posterolateral spines. There is a filament-like process at the apex of each abdominal gill. The mandibles are either not grooved on the dorsal side or have only superficial grooves (**Fig. 2.341**). 2

2. The apical denticles on the claws are smaller than the middle ones (**Fig. 2.379**). 3

- The apical denticles on the claws are larger than the middle ones (**Fig. 2.380**).
....5

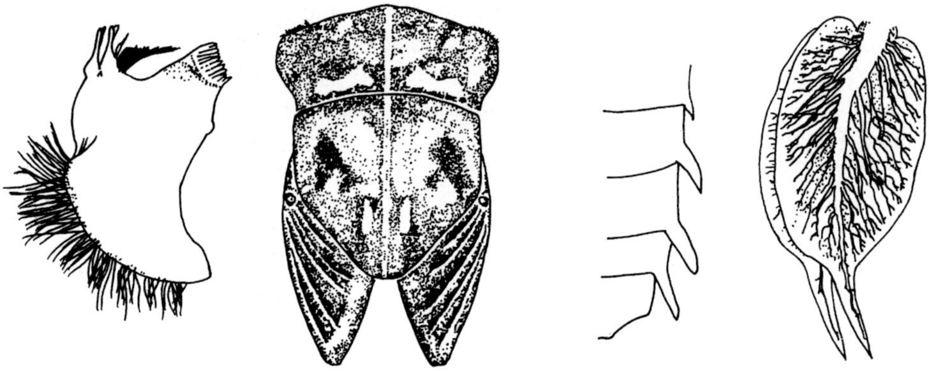

Fig. 2.378 *Penaphlebia vinosa* larva (left to right): mandible, color pattern on the dorsal surface of the thorax, the posterolateral spines on the apical segments of the abdomen, and a gill from the fourth abdominal segment.

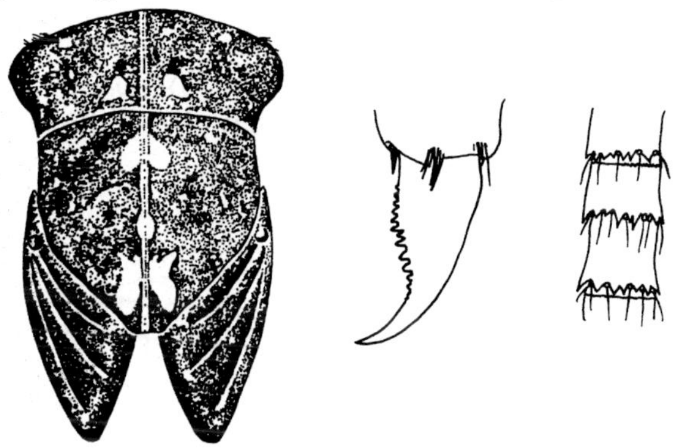

Fig. 2.379 *Penaphlebia flavidula* larva (left to right): the pattern on the thorax, the tarsal claw on the fore-leg, and details of the segments of a caudal filament.

3. The feathered setae are considerably longer than the spines along the posterior margins of the tergites. Some of the setae at the articulations along the cerci are longer than the segments (**Fig. 2.341**). The dark markings on the abdomen are greatly reduced or absent. Length: 8 to 12 mm.
....*Penaphlebia barriai* Pescador and Peters, 1991 (Chile).

- The feathered setae are equal to or slightly shorter than the spines along the posterior margins of the tergites. The setae at the articulations along the cerci are short (**Fig. 2.379**). The dark markings on the abdomen are distinct. 4

4. The posterolateral spines on the eighth and ninth abdominal segments are not well developed. The abdominal gills are whitish (**Fig. 2.381**). Length: 6.9 to 7.2 mm.
..................................*Penaphlebia exigua* Domínguez and Pescador, 1983 (Argentina).

- The posterolateral spines on the eighth and ninth abdominal segments are well developed. The abdominal gills are grayish black (**Fig. 2.379**). Length: 10 to 15 mm.
..................................*Penaphlebia flavidula* Pescador and Peters, 1991 (Chile, Argentina).

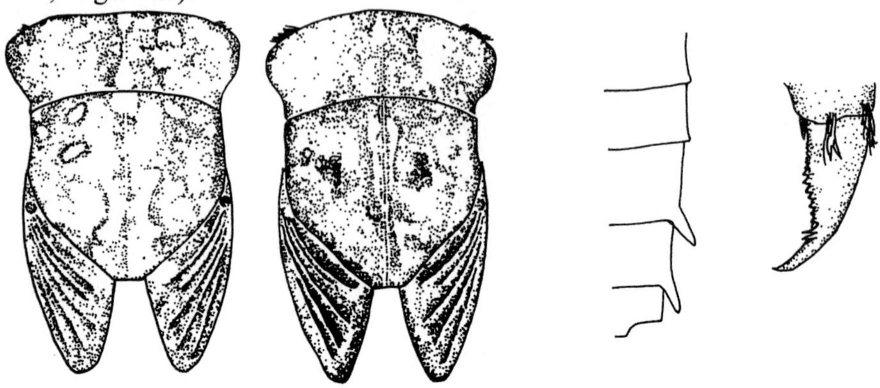

Fig. 2.380 *Penaphlebia chilensis* larva (left to right): two varying color patterns, the posterolateral spines on the apical segments of the abdomen, and the tarsal claw from a fore-leg.

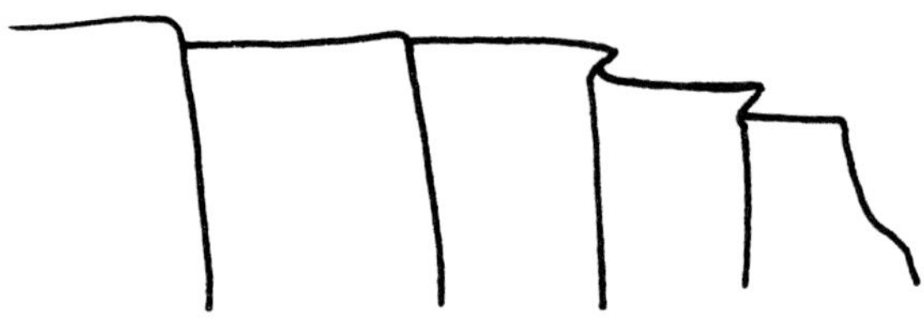

Fig. 2.381 The poorly developed posterolateral spines on the apical abdominal segments of *Penaphlebia exigua.*

5. The tarsal segments are brown with yellow basal and apical bands. The hair-like setae along the posterior margins of the tergites are twice as long as the spines (**Fig. 2.380**). Length: 9 to 11 mm.
....*Penaphlebia chilensis* (Eaton, 1884) (Argentina, Chile). Syn: *Atalophlebia chilensis* Eaton, 1884.
- The tarsal segments are uniformly dark brown. The length of the hair-like setae along the posterior margins of the tergites is equal to slightly greater than that of the spines. There are large light areas along the midline contrasting with the darker ground color on the dorsal surface of the thorax (**Fig. 2.382**). Length: 9 to 12 mm.
................................*Penaphlebia fulvipes* (Needham and Murphy, 1924) (Chile, Argentina). Syn: *Atalophlebia fulvipes* Needham and Murphy, 1924; *Atalophlebia sepia* Thew, 1960.

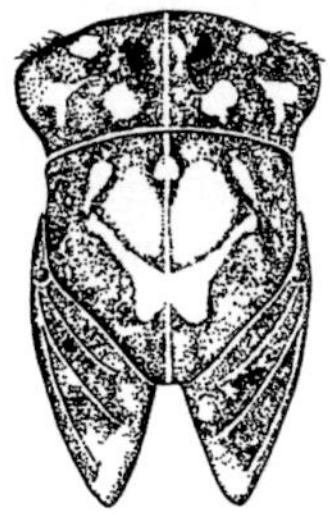

Fig. 2.382 The color pattern on the thorax of a *Penaphlebia fulvipes* larva.

Key to the adult *Massartella* species in South America

Information for the key was provided by Navás (1934b), Ulmer (1943), and Pescador and Peters (1990). A third species in this genus, *Massartella venezuelensis* Pescador and Peters, 1990, is known only from larvae.

1. The wing membrane is faintly toned with brown. Cross veins are clouded with dark brown. The Cu area in the fore-wing encloses a network of cross veins (**Fig. 2.302**).
....*Massartella alegrettae* Ulmer, 1943 (Rio Grande do Sul, Paraná).
- The wing membrane is hyaline except for the costal and subcostal membranes of the fore-wing, which are toned with brownish-yellow. Cross veins are not clouded. The Cu area in the fore-wing encloses little or no network formed by cross veins (**Fig. 2.3**).
....*Massartella brieni* (Lestage, 1924) (Minas Gerais, Paraná, São Paulo, Rio Grande do Sul). Syn: *Atalophlebia brieni* Lestage, 1924; *Atalophlebia axillata* Navás, 1934; *Massartella fruhstorfferi* Ulmer, 1943.

Key to the larvae of *Massartella* species in South America

Information for the key was provided by Pescador and Peters (1990). The larva of *Massartella alegrettae* is unknown.

1. Posterolateral spines on abdominal segments are less than 50% the median length of the segment. Abdominal sternites are covered by a coat of thick hairs. There are pectinate setae on the second segment of the maxillary palp. A thin median hair tuft on the mandible (**Fig. 2.329**).
....*Massartella brieni* (Lestage, 1924) (Minas Gerais, Paraná, São Paulo, Rio Grande do Sul). Syn: *Atalophlebia brieni* Lestage, 1924; *Atalophlebia axillata* Navás, 1934; *Massartella fruhstorfferi* Ulmer, 1943.

- Posterolateral spines on abdominal segments 75% to 100% of the median length of the segment. Sternites of abdomen bare. Spine-like setae on the second segment of the maxillary palp. There is a thick median hair tuft on the mandible (**Fig. 2.383**).
..................................*Massartella venezuelensis* Pescador and Peters, 1990 (Venezuela).

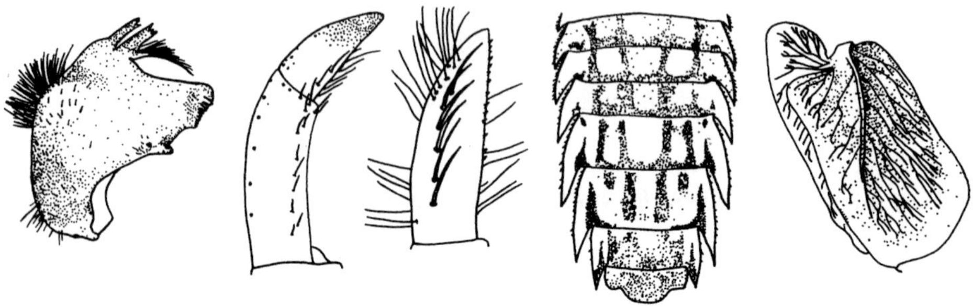

Fig. 2.383 *Massartella venezuelensis* larva (left to right): mandible, apical segments of the maxillary palp, apical segment of the labial palp, the fourth through tenth abdominal tergites, and a gill.

Key to the species of sub-adult and adult *Atopophlebia* in South America

Information for the key was provided by Flowers (1987) and Domínguez and Molineri (1996).

1. The costal projection in the hind wing is located near its midlength. Vein MP in the hind wing is forked. Vein ICu_1 in the fore-wing is joined to CuA and CuP near its base by cross veins. There is a row of small spines on each penis lobe (**Fig. 2.384**). Length of male adult: 8.5 to 8.7 mm; male subadult: 10.5 to 10.8 mm; female adult: 10.8 to 11.0 mm; female sub-adult: 9.3 to 10.8. Fore-wing length of male adult: 12.8 to 13 mm; of male sub-adult: 11.6 to 12.8 mm; of female adult: 13.7 to 14.0 mm; of female sub-adult: 12.0 to 13.8. Hind wing length of male adult: 1.8 to 1.9 mm; of male sub-adult: 1.5 to 2.0 mm; of female adult: 2.0 to 2.2; of female sub-adult: 1.2 to 2.0 mm.
...............................*Atopophlebia flowersi* Domínguez and Molineri, 1996 (Bolivia, Argentina).
- The costal projection is clearly in the apical half of the wing (**Fig. 2.307**). 2
2. The fore-tibiae and two apical abdominal segments are entirely dark brown. The base of the hind wing has a smoky brown wash. The section of the penis lobe that is twisted ventrad and laterad is not strongly hooked and has a spur at the base that continues posteriad (**Fig. 2.307**). Length: 7.5 mm. Fore-wing length: 9 mm. Color: yellow with dark brown markings.
....*Atopophlebia obrienorum* Flowers, 1987 (Ecuador).

- All tibiae and the two apical abdominal segments are yellow with dark markings. The section of the penis lobe that is twisted ventrad and laterad is strongly hooked and has a spur at the base or it is not hooked but lacks any trace of a spur at its base (**Figs. 2.385 and 2.386**). 3

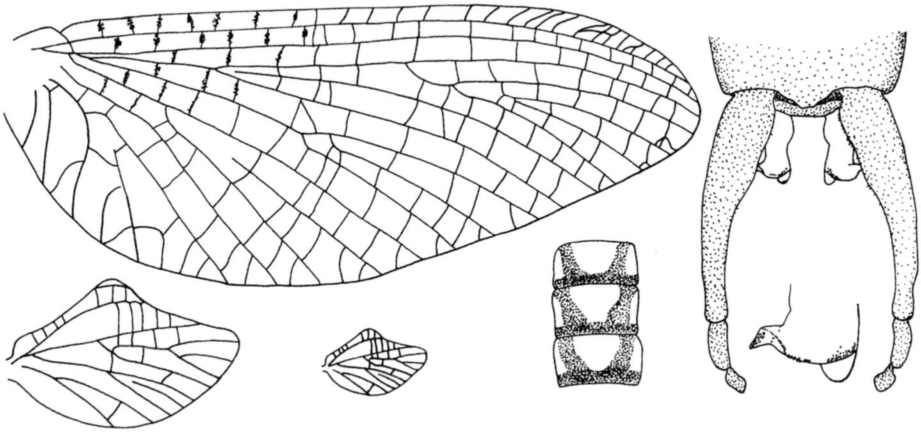

Fig. 2.384 *Atopophlebia flowersi:* fore-wing (upper left); the hind wing greatly enlarged (lower left) and another to the same scale as the fore-wing (lower left center); the second, third, and fourth abdominal tergites of the male (lower right center); the male genitalia in ventral view with the apex of one enlarged penis lobe shown between the forceps (right).

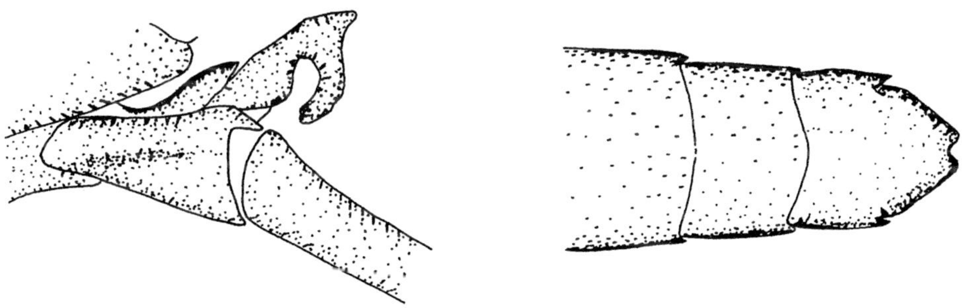

Fig. 2.385 The genitalia of a male *Atopophlebia yarinacocha* in lateral view (left) and the apex of the abdomen of a female sub-adult in ventral view (right).

3. The first abdominal tergite has black anterolateral markings (**Fig. 2.385**). Length: 7.1 mm. Fore-wing length: 9.5 mm. Color: yellow with dark brown markings.
....*Atopophlebia yarinacocha* Flowers, 1987 (Peru).
- The first abdominal tergite is brown only along the posterior margin of the segment (**Fig. 2.386**). Length of male: 9.1 to 9.7 mm. Fore-wing length of

male: 9.3 to 9.9 mm. Length of female: 10.4 mm. Fore-wing of female: 10.6 mm. Color: yellow with dark brown markings.
....*Atopophlebia fortunensis* Flowers, 1980 (Costa Rica, Panama, Colombia).

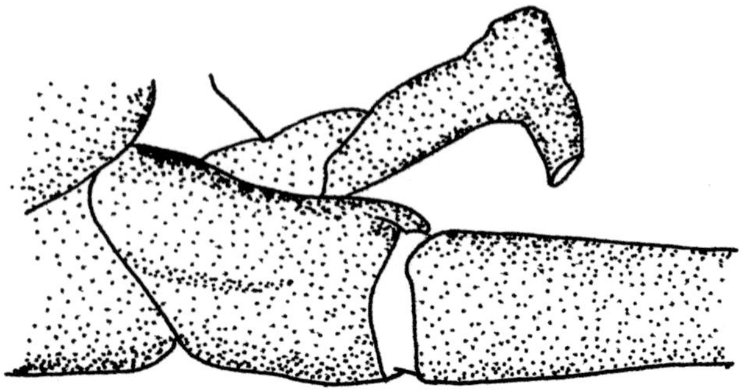

Fig. 2.386 The genitalia of a male *Atopophlebia fortunensis* in lateral view.

Key to the species of known larval *Atopophlebia* in South America

Information for the key was provided by Flowers (1987) and Domínguez and Molineri (1996). The larvae of *Atopophlebia yarinacocha* and *A. obrienorum* have not been described.

1. The abdomen is orange brown with a pattern of blackish posterior margins of the tergites and broad oblique bands running to the anterior margin of each tergite on either side of the lighter median area, as in the adult (**Fig. 2.384**). There are also paired blackish spots on the posterior margins of the first through eighth tergite. Length of last female larval instar: 10.5 to 11 mm. General coloration: yellowish brown.
..................................*Atopophlebia flowersi* Domínguez and Molineri, 1996 (Bolivia, Argentina).
- The first seven abdominal tergites are brown with large oval yellow areas in the middle of the fourth and fifth tergites. The eighth and ninth tergites are yellow with paired black spots (**Fig. 2.332**). The length was not reported, but based on the size of the adults, the larvae should be approximately the same size as those of *A. flowersi*.
....*Atopophlebia fortunensis* Flowers, 1980 (Costa Rica, Panama, Colombia).

Key to the species of adult *Farrodes* in South America

Information for the key was taken from Traver (1943), Domínguez *et al.* (1996), Domínguez and Savage (1987) and Domínguez (1999). Only the males of all species in the key can be identified. Unfortunately, Domínguez (1999) does not state how he measured the height of the costal projection and width of the hind wing, which is a major feature for distinguishing his species groups. In addition, using this measurement, several species of his *bimaculatus* group fall into the *caribbianus* group. To calculate the relative size, as determined from personal measurements of the wings in the illustrations, the costal projection should be measured from its apex along the outer edge to the midpoint of the curve along its costal margin turning outward toward the wing tip. The hind wing width is then measured from the apex of the projection to the posterior costal margin at its widest point.

1. On the hind wings, the costal projection is no more than 0.16 times the total width of the wing. The projections of the penis are long and cylindrical (**Fig. 2.387**). 2
- The costal projections of the hind wing are greater than 0.19 times the total width of the wing. The projections of the penis are usually funnel-shaped or conical and curved toward the midline, or they curve dorsad at the apex (**Fig. 2.388**). 4

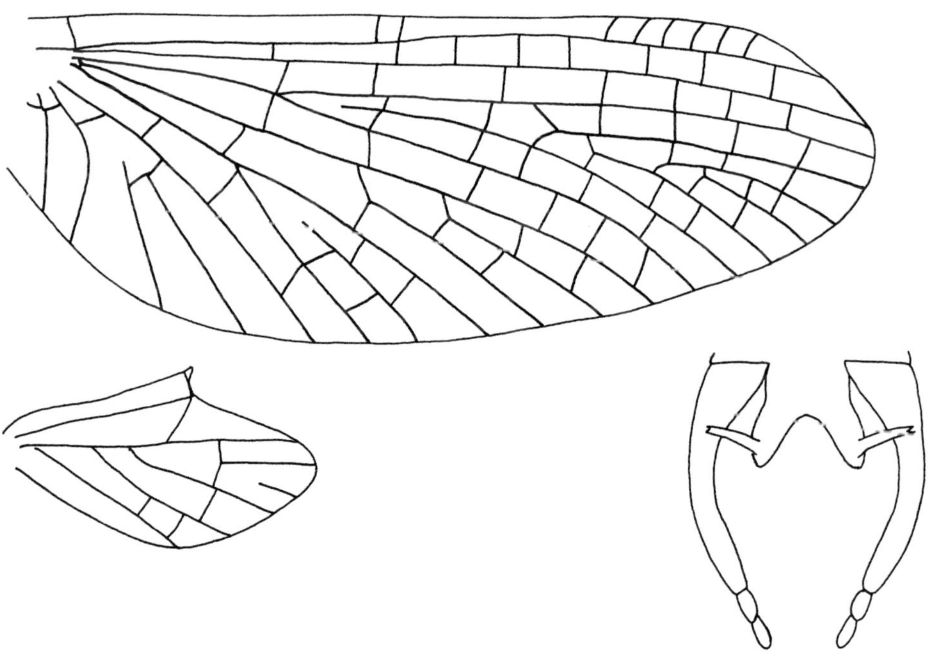

Fig. 2.387 The fore and enlarged hind wing (left) and the male genitalia (lower right) of *Farrodes maculatus*.

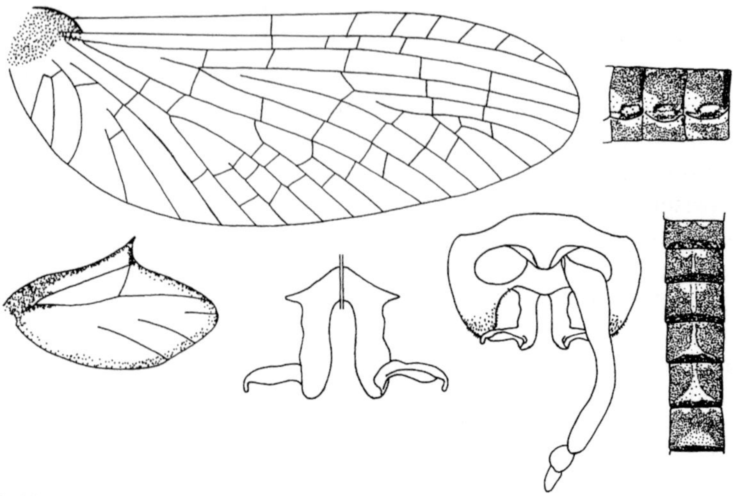

Fig. 2.388 *Farrodes carioca*: the fore-wing (upper left) and hind wing drawn larger than scale (lower left), second through seventh abdominal tergites of a male in dorsal view (lower right), third through fifth abdominal segments in lateral view (upper right), male genitalia in ventral view with the right forceps segments removed (lower right center), and half of the partially divided penis in dorsal view (left half, lower left center) and half in ventral view (right half).

2. The first segment of the forceps is entirely brown of grayish brown, while the second and third segments are lighter in color. A considerable area of the abdominal segments is white. The legs are whitish with a faint brownish tint and obscure subapical rings on the femora (**Fig. 2.387**). Body length of male: 5.5 mm. Fore-wing length: 5.5 mm. The female has not been described. ....*Farrodes maculatus* (Needham and Murphy, 1924) (Argentina). Syn: *Thraulus maculatus* Needham and Murphy, 1924; *Homothraulus maculatus* (Needham and Murphy, 1924).

- The basal half of the first segment of the forceps is yellow or yellowish brown, and the apical half and the second and third segments are yellowish white. The whitish marks on the abdominal segments are small (**Fig. 2.389**). ...3

3. The basal half of the first segment of the forceps is brown with a brownish black wash; the rest of the forceps is yellowish white. The styliger plate is brown with a brownish black wash. The abdominal segments of the male are predominantly dark with pairs of light markings along the anterior margins and light areas in the middle along the posterior margins (**Fig. 2.308**). The heads of both males and females are brownish yellow with heavy brownish black wash. Length of male: 5.5 to 6.5 mm; of female: 5.0 to 6.2 mm; fore-wing length of male: 6.4 to 8.1 mm; of female: 6.2 to 7.0 mm. ....*Farrodes yungaensis* Domínguez and Savage, 1987 (Argentina)

- The basal half of the first segment of the forceps is yellow, and the rest is yellowish white. The styliger plate is yellow. The abdominal segments of the male have dark bands across the middle and are broadly light along the anterior and posterior margins (**Fig. 2.389**). The heads of the females are yellowish white with a black ring around each antennal socket. The head of the male is mainly brownish yellow. Length of male: 5.1 to 5.2 mm; of female: 4.9 to 5.2 mm; fore-wing length of male: 5.2 to 5.3 mm; of female: 6.5 to 6.6 mm.

...................................*Farrodes iguazuanus* Domínguez and Savage, 1987 (Argentina). The male genitalia do not seem distinct enough to prevent interbreeding between these two forms. It is therefore questionable whether the two should be considered distinct species.

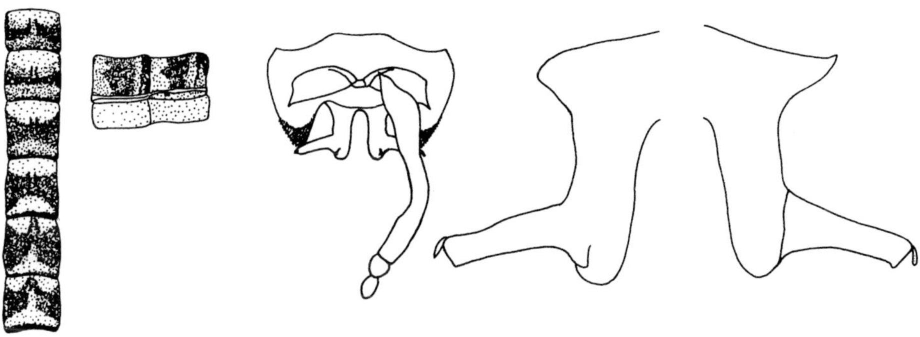

Fig. 2.389 *Farrodes iguazuanus* (left to right): the second through seventh abdominal tergites of a male in dorsal view, the fifth and sixth abdominal segments in lateral view, the male genitalia in ventral view with the right forceps segments removed, and one side of the partially divided penis in ventral and the other in dorsal view.

4. The posterolateral projections of the styliger plate are short and wide. The penis lobes are fused only in the basal ¼ of their length. The projections from the penis lobes have terminal flaps that are curved dorsad (**Fig. 2.388**). Length of the male: 4.9 to 5.1 mm; of female: 4.2 to 4.3 mm. Fore-wing length of male: 5.6 to 5.7 mm; of female: 5.1 to 5.3 mm. Hind wing length of male: 0.7 to 0.8 mm; of female: 0.5 to 0.6 mm. Color: mainly yellowish and orange brown washed with black in places and with brown and black markings.
.............................*Farrodes carioca* Domínguez, Molineri, and Peters, 1996 (Rio de Janeiro).
- The posterolateral projections of the syliger plate are long and narrow (**Fig. 2.390**). 5

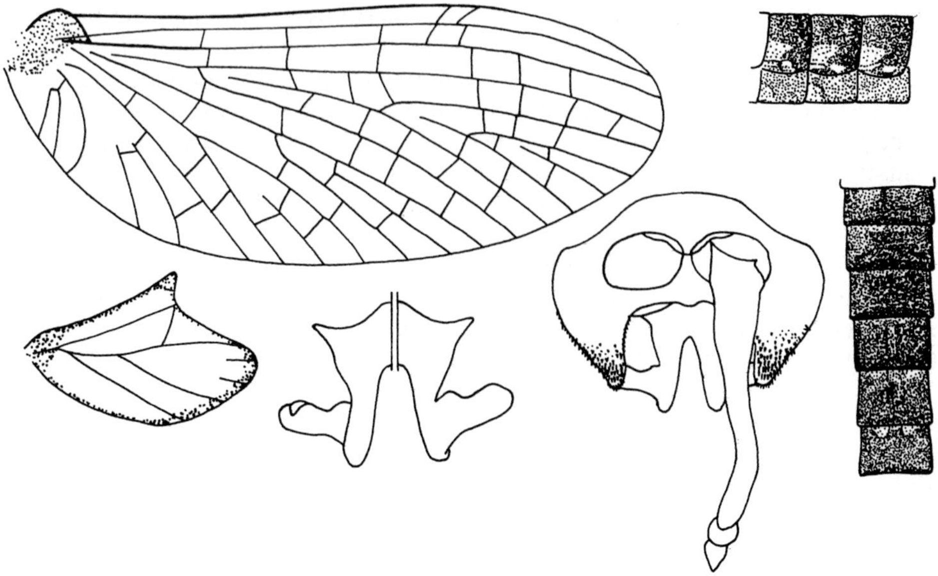

Fig. 2.390 *Farrodes ochraceous*: the fore-wing (upper left) and hind wing drawn larger than scale (lower left), second through seventh abdominal tergites of a male in dorsal view (lower right), third through fifth abdominal segments in lateral view (upper right), male genitalia in ventral view with the right forceps segments removed (lower right center), and half of the partially divided penis in dorsal view (left half, lower left center) and half in ventral view (right half).

5. The abdomen is yellow washed with orange brown dorsally except for small patches along the anterior border; the color is darker along the posterior margins of the tergites. The styliger plate has posterolateral projections that are long and narrow with acute apices. The basal 2/5 of the penis lobes are fused, and they are rounded apically and bear long and wide cylindrical projections laterally bearing flaps that are short and folded ventrad (**Fig. 2.390**). Length: 3.9 to 4.1 mm. Fore-wing length: 4.1 to 4.3 mm. Hind wing length: 0.5 to 0.6 mm. Color: mainly orange brown. Only the male has been described.
.......................*Farrodes ochraceous* Domínguez, Molineri, and Peters, 1996 (Amazonas).
- The abdominal tergites have extensive darker and lighter markings. The apices of the lateral projections of the penis lobes are not rounded and do not bear short flaps (**Fig. 2.391**). 6
6. Dark wash on the second through sixth abdominal tergites covers less than half of the dorsal surface but includes the posterior margins of most segments; the remainder of the area is relatively light in color (**Fig. 2.391**). 7
- More than half the dorsal surface of the second through sixth abdominal tergites is covered by a dark or blackish wash (**Fig. 2.392**). 8

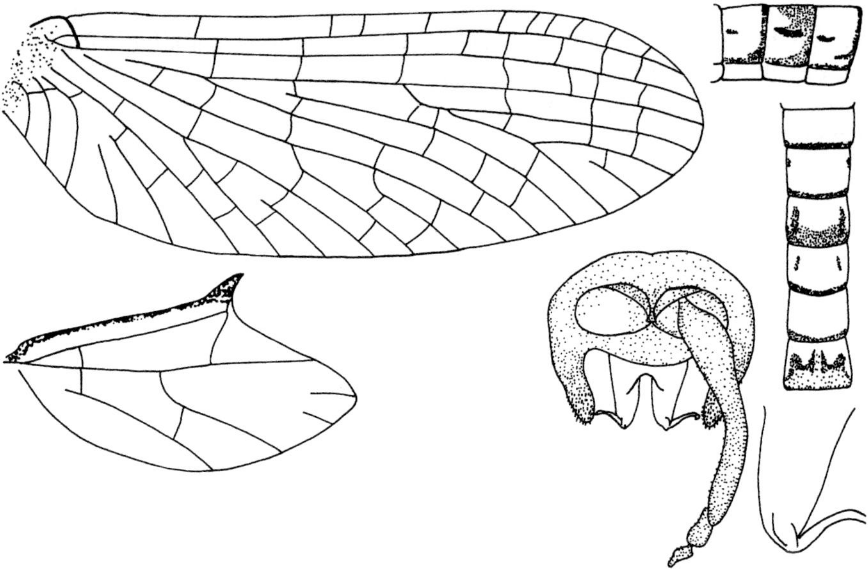

Fig. 2.391 *Farrodes savagei:* the fore and a greatly enlarged hind wing (left), the third through fifth abdominal segments in lateral view (upper right) and second through seventh in dorsal view (middle right), and the male genitalia in ventral view with an enlarged penis lobe (lower right).

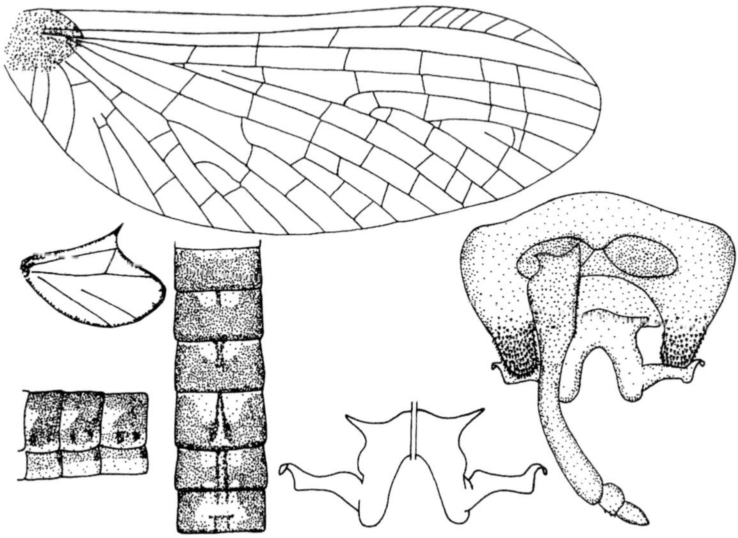

Fig. 2.392 *Farrodes xingu*: the fore-wing (upper left) and hind wing drawn larger than scale (middle left), second through seventh abdominal tergites of a male in dorsal view (lower left center), third through fifth abdominal segments in lateral view (lower left), male genitalia in ventral view with the left forceps segments removed (lower right), and half of the partially divided penis in dorsal view (left half, lower right center) and half in ventral view (right half).

7. The color is variable, but the blackish wash usually extends dorsolaterally from the darkened posterior margins of the third and fourth abdominal tergites to form two darkened areas. There is less extensive darkening in small dorsolateral patches on second, fifth, and sixth tergites. The seventh tergite is extensively washed in black except for a narrow light area behind the anterior margin. The lateral projections from the styliger plates are short and blunt with wide bases. The projections from the lateral surfaces of the penis lobes are conical and curved inward (**Fig. 2.393**).
....*Farrodes caribbianus* (Traver, 1943) (Central America, Venezuela, Ecuador). Syn: *Thraulus caribbianus* Traver, 1943.
- Aside from the blackish posterior margins, the blackish markings on the hyaline second, third, fifth, and sixth abdominal tergites are confined to small rectangular patches on the dorsolateral surfaces. The fourth tergite has the largest blackish markings and is covered by a blackish wash usually forming a band contiguous to the blackish posterior margin. The seventh tergite has an irregular pair of blackish markings in the middle. The apical 75% to 80% of the penis lobes are separated. The apices of the lateral projections on the penis lobes curve upward. The styliger plates bears straight, cylindrical lateral projections that are rounded apically (**Fig. 2.391**). Length of male: 4.2 to 4.5 mm; of female: 4.5 to 4.7 mm. Fore-wing length of male: 4.4 to 4.6 mm; of female: 5.0 to 5.2 mm. Hind wing length of male: 0.6 to 0.7 mm; of female: 0.7 to 0.8 mm.
....*Farrodes savagei* Domínguez, 1999 (Venezuela).
8. The entire first segment of the forceps is brown or grayish brown, and the two apical segments are lighter. The first tergite is entirely washed with black, while the blackish wash on the yellowish white second through eighth tergites is interrupted in semicircular areas along the anterior margins. There are blackish stripes paralleling the midlines of the posterior tergites. The styliger plates have long, narrow projections that are rounded at the apices. The penis lobes are fused approximately along the basal third and are rounded at their apices (**Fig. 2.392**). Length of male: 4.1 to 4.3 mm; of female: 3.9 to 4.0 mm. Fore-wing length of male: 4.5 to 4.6 mm; of female: 4.3 to 4.4 mm. Hind wing length of male: 0.6 to 0.7 mm; of female: 0.4 to 0.5 mm. Color: manly orange brown with blackish markings.
.................................*Farrodes xingu* Domínguez, Molineri, and Peters, 1996 (Pará).
- The basal half of the first segment of the forceps is obviously darker than the apical half (**Fig. 2.394**). 9
9. There is a long ventral spine near the base of the terminal flap on each lateral projection of the penis. The basal halves of the penis lobes. The second through sixth abdominal tergites are translucent yellowish with a blackish wash that is interrupted along the anterior margin to form a pair of light semi-circular markings. The seventh through tenth tergites are mainly yellowish brown with

diffuse blackish markings (**Fig. 2.394**). Length: 3.4 to 3.7 mm. Fore-wing length: 3.9 to 4.0 mm. Hind wing length: 0.4 to 0.5 mm. Color: mainly yellowish brown with blackish markings. Only the male has been described.
........................*Farrodes longispinus* Domínguez, Molineri, and Peters, 1996 (Venezuela).
- The lateral projections on the penis lobes do not bear long ventral spines (**Fig. 2.395**). 10

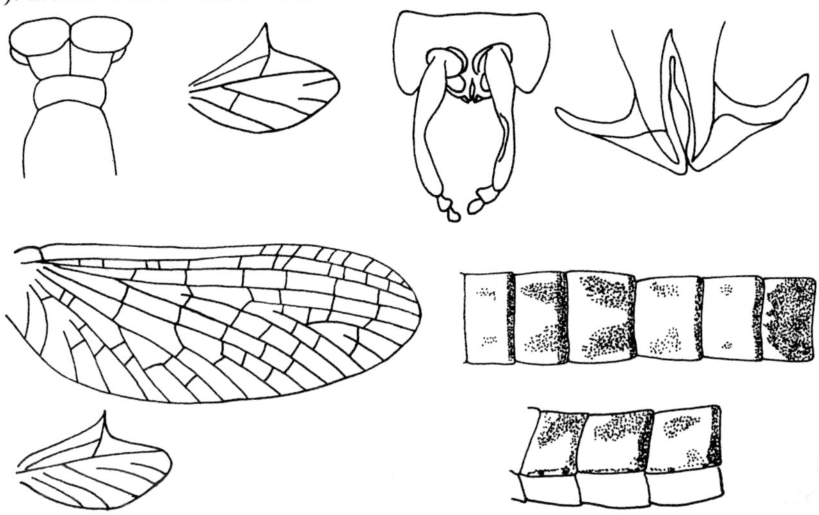

Fig. 2.393 *Farrodes caribbianus* male (upper row, left to right): head in dorsal view, hind wing as illustrated by Traver (1943), genitalia, and the penis enlarged; (middle and lower row, left to right): the fore and hind wing as illustrated by Domínguez (1999) and the color pattern on the second through seventh abdomen tergites (above) and on the lateral sides of the third through fifth segments (below) illustrated with the head of the insect to the left.

10. The apices of the posterolateral projections of the styliger plates are nearly acute apically. The basal half of the penis lobes are fused, and their apices are rounded (**Fig. 2.395**). Length: 4.4 to 4.6 mm. Fore-wing length: 4.5 to 4.7 mm. Hind wing length: 0.5 to 0.6 mm. Color: mainly yellowish and orange brown with blackish markings. Only the male has been described.
...........................*Farrodes pakitza* Domínguez, Molineri, and Peters, 1996 (Peru).
- The apices of the posterolateral projections of the long, narrow styliger plates are blunt. The basal 60% of the penis lobes are fused, and their apices are rounded (**Fig. 2.396**). Length of male: 3.8 to 4.2 mm Fore-wing length of male: 4.1 to 4.3 mm. Hind wing length of male: 0.5 to 0.6 mm. Color: light orange darker and lighter markings.
..............................*Farrodes tepui* Domínguez, Molineri, and Peters, 1996 (Venezuela).

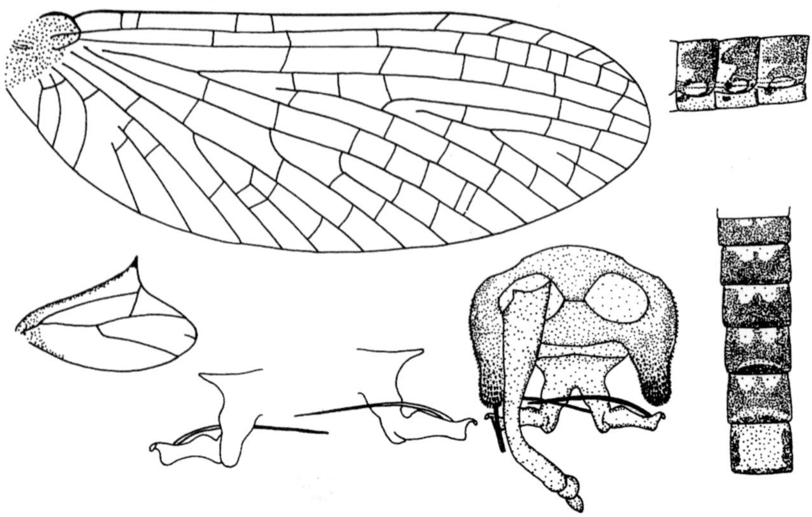

Fig. 2.394 *Farrodes longispinus*: the fore-wing (upper left) and hind wing drawn larger than scale (lower left), second through seventh abdominal tergites of a male in dorsal view (lower right), third through fifth abdominal segments in lateral view (upper right), male genitalia in ventral view with the left forceps segments removed (lower right center), and half of the partially divided penis in dorsal view (lower left center) and half in ventral view (lower center).

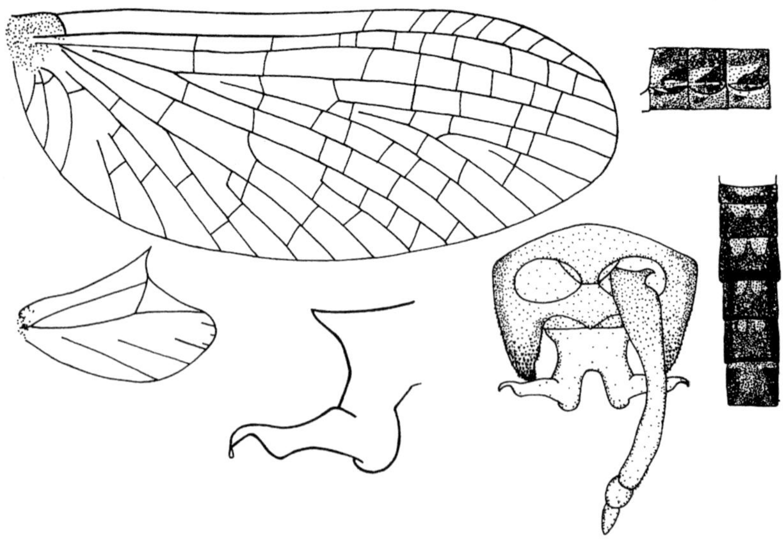

Fig. 2.395 *Farrodes pakitza*: the fore-wing (upper left) and hind wing drawn larger than scale (lower left), second through seventh abdominal tergites of a male in dorsal view (lower right), third through fifth abdominal segments in lateral view (upper right), male genitalia in ventral view with the right forceps segments removed (lower right center), and half of the partially divided penis (lower left center).

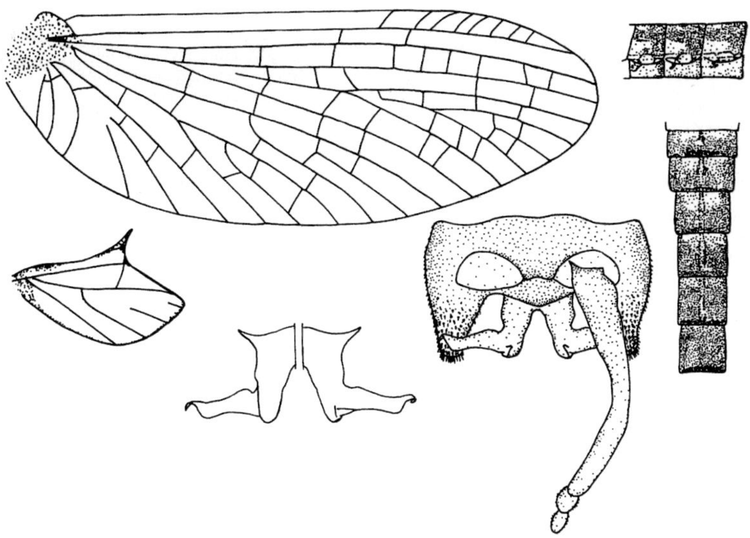

Fig. 2.396 *Farrodes tepui*: the fore-wing (upper left) and hind wing drawn larger than scale (lower left), second through seventh abdominal tergites of a male in dorsal view (lower right), third through fifth abdominal segments in lateral view (upper right), male genitalia in ventral view with the right forceps segments removed (lower right center), and half of the partially divided penis in dorsal view (left half, lower left center) and half in ventral view (right half).

Key to the species of known *Farrodes* larvae in South America

Information for the key was taken from Domínguez (1999). Descriptions of larvae of only three species are available, while eight are known only as adults.

1. The abdominal tergites, like those of the adults (**Fig. 2.308**), are mainly dark due to an extensive dark wash over the middle portion. On some specimens, the darker area is more extensive than that of the adults. Markings with heavy dark wash on the head include the genae, ring around the antenna sockets, a transverse area posterior to the median ocellus, and a U-shaped mark bordering a light area anterior to the median ocellus. There are black lateral, anterosubmedian, and posteromedian marks on the pronotum. There are 16 to 20 denticles on the tarsal claws. Dark markings on the legs include the subapical section of the femora, basal and subapical rings on the tibiae, and subbasal section of the tarsi; the markings on the fore-tarsi are more extensive than those on the other legs (**Fig. 2.397**). Length: 4.5 to 6.7 mm.
................................*Farrodes yungaensis* Domínguez and Savage, 1987 (Argentina)

- At least some of the abdominal tergites are mainly light in color (**Fig. 2.398**).
....3

Fig. 2.397 *Farrodes yungaensis* larva (left to right): the middle part of the posterior margin of the sixth tergite, the tarsal claw on the fore-leg, and the fourth gill.

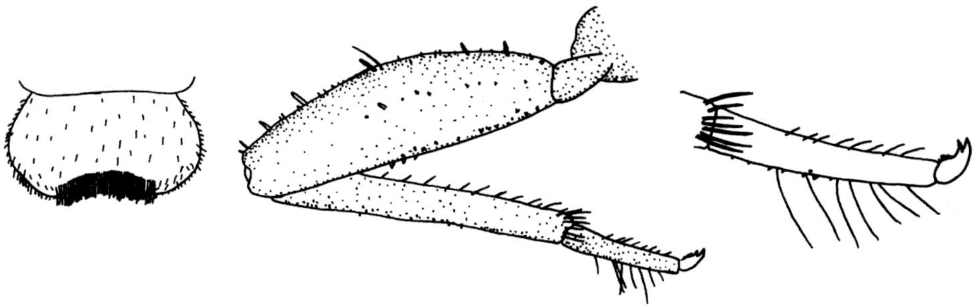

Fig. 2.398 *Farrodes savagei* larva (left to right): the labrum, hind leg, and an enlargement of its tarsus and claw.

2. The spines along the outer margin of the hind tarsus are long. The spines of the dorsal side of the femora are acutely pointed (**Fig. 2.347**). The color pattern on the abdomen, like that of the adult (**Fig. 2.393**), includes a pair of dark markings on the third and fourth tergites that reach from the posterior margin at least ¾ of the way to the anterior margin. Length: 3.8 to 4.6 mm.
....*Farrodes caribbianus* (Traver, 1943) (Central America, Venezuela, Ecuador). Syn: *Thraulus caribbianus* Traver, 1943.

- The spines along the outer margin of the hind tarsus are relatively short. The spines on the dorsal side of the hind femora are blunted (**Fig. 2.398**).
....*Farrodes savagei* Domínguez, 1999 (Venezuela).

Key to the species of adult *Traverella* in South America

Information for the key was taken from Needham and Murphy (1924), Ulmer (1943), Traver (1963), and Domínguez (1995a).

1. In the fore-wing, vein ICu_1 ends basally in a free spur. The subgenital plate of the male bears a pair of broad processes (**Fig. 2.399**). The wings are brownish at the base and otherwise hyaline, except for the clouded stigmatic area of the fore-wing. Length of the male: 8.2 to 9.2 mm; female: 5.9 to 7.4 mm. Fore-wing length of male: 8.2 to 9.1 mm; of female: 8.2 to 9.2 mm. Hind wing length of male: 1.4 to 1.6 mm; of female: 1.2 to 1.8 mm. Color: brown with grayish brown and orange brown markings. The first six segments of the abdomen of the male are translucent with grayish brown markings. The claws are whitish. The first seven tergites of the female are orange brown with the dark markings extending closer to the midline. The eighth through tenth tergites of the female are yellowish brown.
.... *Traverella (Zonda) calingastensis* Domínguez, 1995 (Argentina).
- In the fore-wing, vein ICu_1 joins vein CuA at its base. The paired processes on the subgenital plate of the male are long and narrow (**Fig. 2.400**).
....Subgenus *Traverella*..............2

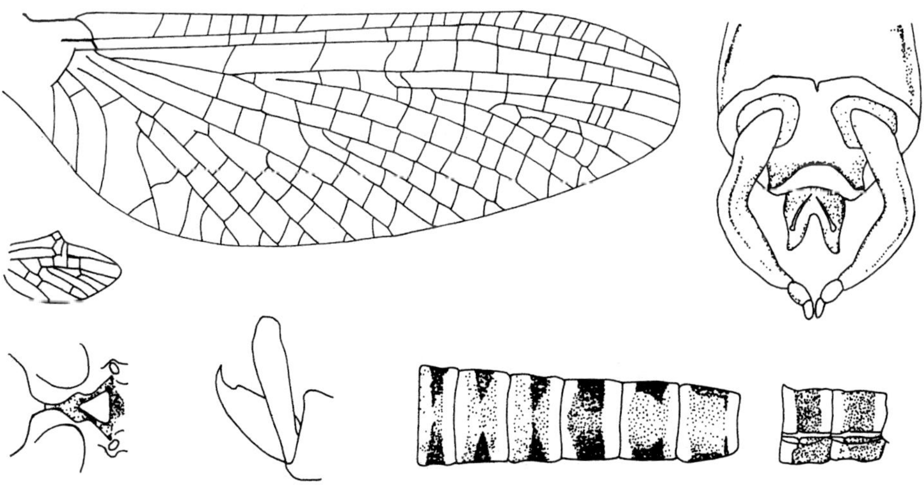

Fig. 2.399 *Traverella (Zonda) calingastensis* adult male: fore and hind wing (upper left), the prosternum (lower left), tarsal claws (lower left center), second through seventh abdominal tergites (lower right center), fifth and sixth abdominal segments in lateral view (lower right), and genitalia (upper right).

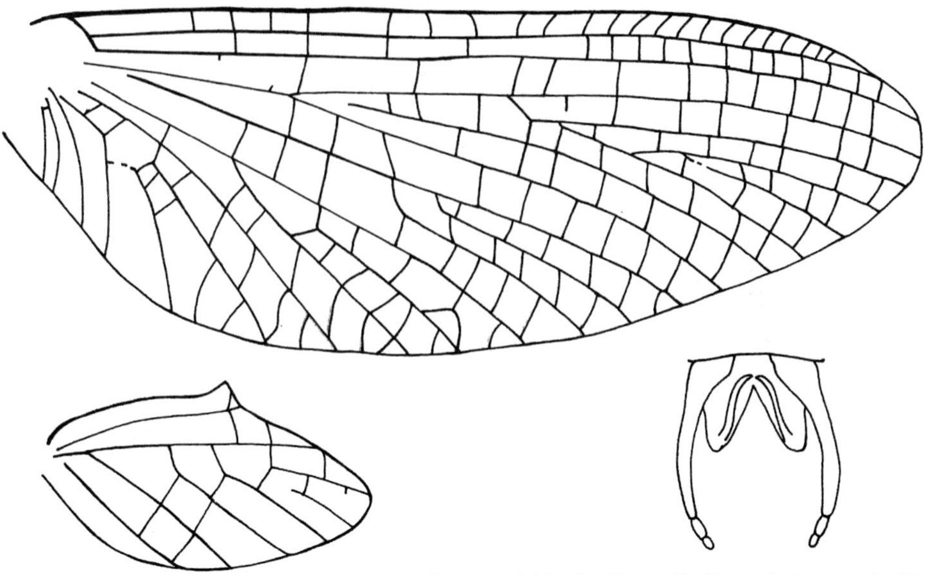

Fig. 2.400 *Traverella bradleyi:* The fore and hind wings (left) and the genitalia of a male (lower right).

2. The fore-wings are pale except for prominent arcuate brown areas at the base. The hind wings are similarly marked. The second through sixth abdominal tergites are whitish and translucent with brown posterior margins (**Fig. 2.400**). The subadult is similar to the adult, but its colors are fainter. Length of the male: 7 mm, female: 6.5 mm; fore-wing length of the male: 6.5 mm, female: 7 mm.
....*Traverella bradleyi* (Needham & Murphy, 1924) (Argentina, Brazil). Syn: *Thraulus bradleyi* Needham & Murphy, 1924.
- The wings have a distinct brownish tinge, at least in the costal and subcostal spaces. The second through sixth abdominal tergites are translucent only along the anterior margin and anterolaterally; the posterior portions are mainly dark grayish brown (**Fig. 2.401**). 3
3. All legs have extensive blackish markings. The wings have a brownish tinge in the costal and subcostal spaces. The wings have relatively few cross veins (**Fig. 2.309**). Length of the body: 7.5 mm; length of the fore-wing: 9 mm.
.... *Traverella montium* (Ulmer, 1943) (Peru). Syn: *Thraulus montium* Ulmer, 1943.
- The wings have and intense brown tinge, but the cross veins are not clouded. The costal and subcostal spaces are blackish brown. There are no dark bands on the femora (**Fig. 2.401**). Length of the body: 6 mm; length of the fore-wing: 7.5 mm.
....*Traverella valdemari* (Esben Petersen, 1912) (Argentina). Syn: *Thraulus valdemari* Esben-Petersen, 1912.

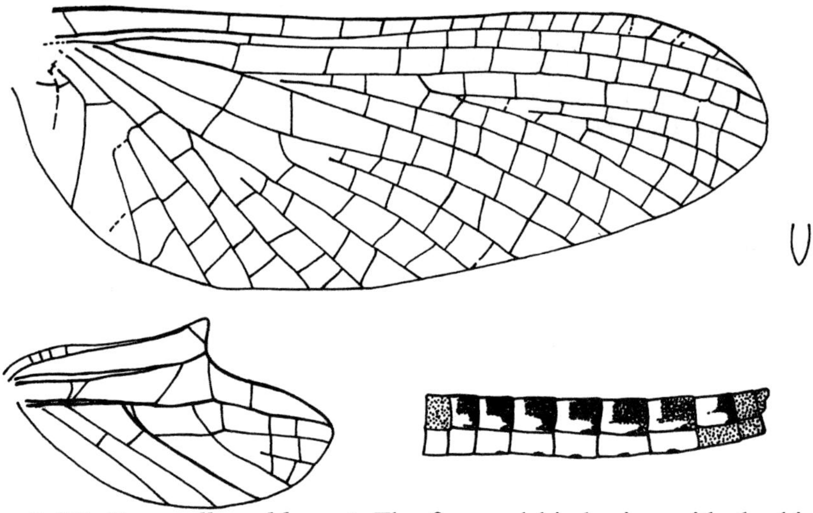

Fig. 2.401 *Traverella valdemari:* The fore and hind wing with the hind wing enlarged relative to the fore-wing (left), the abdomen of a male in lateral view (lower right), and the apex of the penis (middle right), which is mainly concealed within the genitalia.

Key to the species of known larval *Traverella* in South America

Information for the key was taken from Domínguez (1995a). The larvae of most South American species have not been described.

1. At the inner anterior angle of the maxilla, the tusk, if evident at all, is a small prominence visible only in the final larval instar. The second segment of the maxillary palp is less than twice as long as the first segment. The third segment of the labial palp is at least 1.2 times as long as the second. There is a row of spines on the dorsal surface of the fore-tibia. There is no gill on the seventh abdominal segment (**Fig. 2.337**). Length: 6.8 to 7.5 mm. Color: yellowish white with blackish markings.
.... *Traverella (Zonda) calingastensis* Domínguez, 1995 (Argentina).
- At the inner anterior angle of the maxilla, there is a well-developed tusk. The second segment of the maxillary palp is more than twice as long as the first segment. The third segment of the labial palp is 0.8 to 1.2 times as long as the second. There is only a row of setae on the dorsal surface of the fore-tibia. There is a small gill on the seventh abdominal segment.
..............................*Traverella (Traverella) bradleyi* (Needham & Murphy, 1924) (Argentina, Brazil). Syn: *Thraulus bradleyi* Needham & Murphy, 1924. This description should also match the other South American species in the subgenus *Traverella,* but a description of them is not available at this time.

Key to the species of adult *Hermanella* in South America

Information for the key was taken from Ulmer (1920a, b), Needham and Murphy (1924), Spieth (1943), Demoulin (1955a), Domínguez and Flowers (1989), and Ferreira and Domínguez (1992).

1. The projections on the subgenital plate at the base of the penis are slender. The ventral projections on the penis are curved and spine-like (**Fig. 2.402**). ....Subgenus *Guayakia* Domínguez and Flowers, 1989............2
- The projections on the subgenital plate at the base of the penis are broad (**Fig. 2.403**). ....Subgenus *Hermanella* Needham and Murphy, 1924. 4

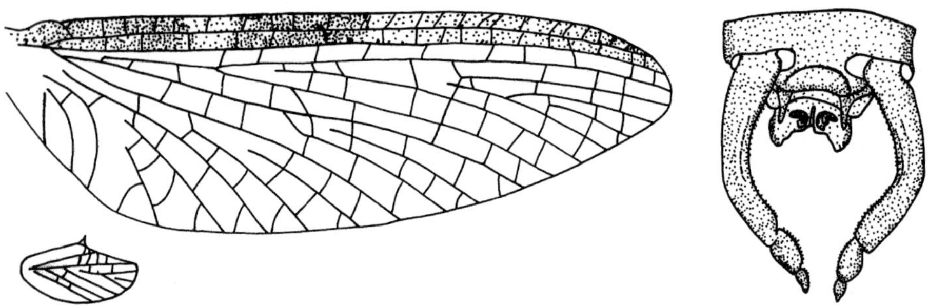

Fig. 2.402 The fore and hind wing (left) and male genitalia (right) of *Hermanella froehlichi*.

2. The wing membrane around many of the cross veins is clouded with brown. The costal margin of the wing has a yellow tinge. The second through sixth abdominal tergites are yellowish and translucent, tinged with russet brown and with black posterior margins (**Fig. 2.311**). Length of male and female: 7.5 to 8.8 mm. Fore-wing length of male: 8.5 to 9 mm; of female: 10.0 to 10.5 mm. Hind wing length of male and female: 1.1 to 1.5 mm. Color: orange brown washed with black to yellowish orange.
.................................*Hermanella (Guayakia) maculipennis* (Ulmer, 1920) (Argentina, Santa Catarina, Paraná). Syn: *Thraulus maculipennis* Ulmer, 1920; *Traverella maculipennis* (Ulmer, 1920).
- The membrane of the fore-wing is mainly hyaline, clouded with orange brown only in the costal and subcostal spaces (**Fig. 2.402**). 3

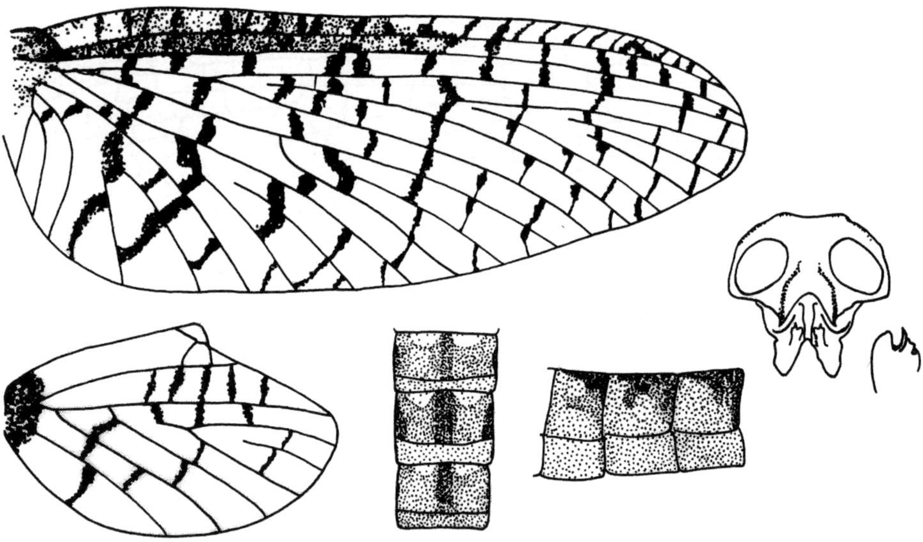

Fig. 2.403 *Hermanella guttata:* the fore and hind wing with the hind wing enlarged (left), the fifth through seventh abdominal segments in dorsal (lower center) and lateral view (lower right), and the male genitalia in ventral view with an enlargement of the apex of the ventral process on the penis (middle right).

3. There is a black median dorsal stripe on the light orange brown abdomen of the male. The abdomen of the female is brownish dorsally and yellowish ventrally with a similar pattern. The legs are yellowish orange with a dark spot on the basal third of the fore-femur, dark apices of the tibiae, and yellowish tarsi. The apex of the costal prominence of the hind wing is acute (**Fig. 2.402**). Length of male: 8.2 to 9.2 mm. Length of female: 6.4 to 8.4 mm. Fore-wing length of male: 8.2 to 8.7 mm; of female: 8.5 to 9.2 mm. Hind wing length of male: 1.4 to 1.5 mm; of female: 1.0 to 1.4 mm. Color: mainly light orange brown with a black wash and black and brown markings.
..................*Hermanella (Guayakia) froehlichi* Ferreira and Domínguez, 1992 (São Paulo).

- The abdomen is yellowish brown with a black wash, which is most prominent in the middle and lateral areas. The abdominal segments become lighter toward its apex. The legs are orange brown with black spots on the basal half and at the apex of all femora. The fore-tibia is blackish, and there is a black spot at the apex of the middle tibia. The apex of the costal prominence is broadly rounded and darkened on the membrane (**Fig. 2.404**). Length of female: 10.4 to 10.8 mm. Fore-wing length of female: 12.0 to 12.5 mm. Hind wing length of female: 1.6 to 1.8 mm. Color: yellowish to light brown, partially washed with black. The male has not been described.
....................*Hermanella (Guayakia) grandis* Domínguez and Flowers, 1989 (Argentina, Paraná, Santa Catarina).

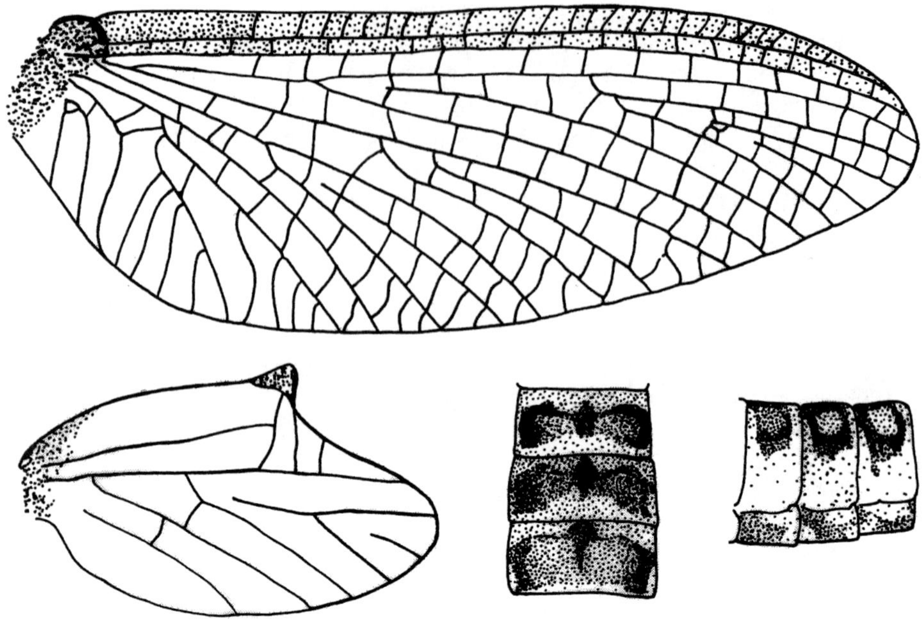

Fig. 2.404 *Hermanella grandis:* the fore and hind wing with the hind wing drawn larger than scale (left), and the fifth through seventh abdominal segments in dorsal (lower center) and lateral view (lower right).

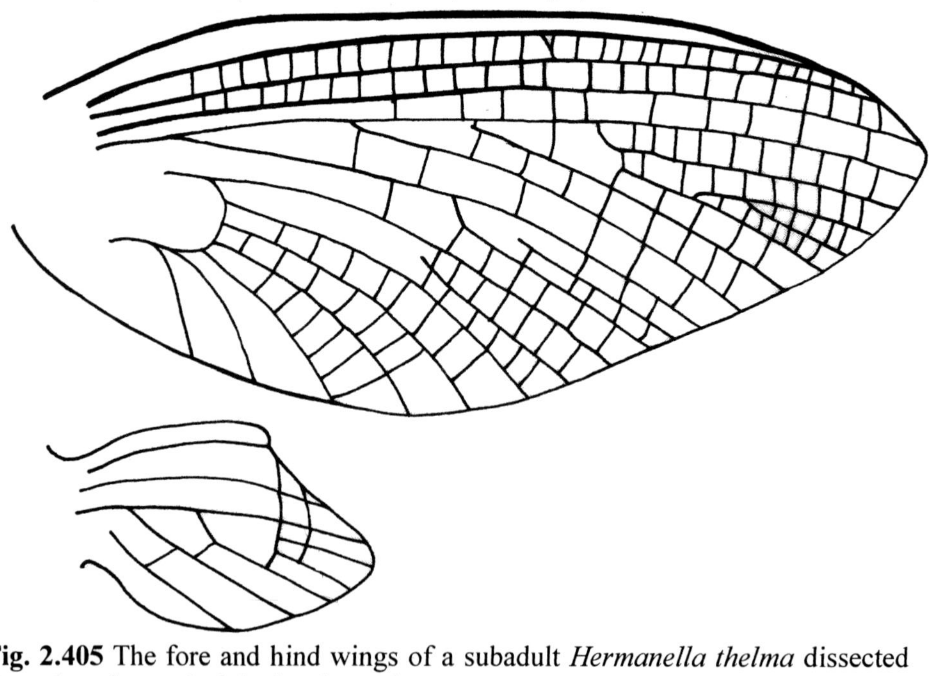

Fig. 2.405 The fore and hind wings of a subadult *Hermanella thelma* dissected from the wing pod of the last instar larva.

4. The wings of the subadult are unspotted. The abdomen is mainly blackish brown (**Fig. 2.405**). Length of male subadult: 5.4 to 5.8 mm. Fore-wing length of male subadult: 6.2 to 6.4 mm. Hind wing length of male subadult: 1.0 to 1.1 mm. Color: mainly blackish brown. The adults have not been described.
……………….*Hermanella (Hermanella) thelma* Needham and Murphy, 1924 (Argentina).
- The wings are spotted. The abdomen is mainly orange brown (**Fig. 2.403**). Length of male: 7.8 to 8.0 mm. Length of female: 6.7 to 6.9 mm. Fore-wing length of male: 8.3 to 8.5 mm; of female: 8.2 to 8.5 mm. Hind wing length of male: 1.2 to 1.3 mm; of female: 1.4 to 1.6 mm. Color: mainly orange brown with black, brown, and gray markings.
……………….*Hermanella (Hermanella) guttata* Domínguez and Flowers, 1989 (Argentina, Paraguay, Uruguay, Paraná).

Key to the species of mature *Hermanella* larvae in South America

Information for the key was taken from Needham and Murphy (1924), Domínguez and Flowers (1989), and Ferreira and Domínguez (1992).

1. The tibiae bear short dorsal setae and have glabrous outer margins. The length exceeds 9 mm.
……………………………………………………………………………2
- The tibiae bear rather long setae, and setae are present along their outer margins (**Fig. 2.406**). ……………………………………………………………3

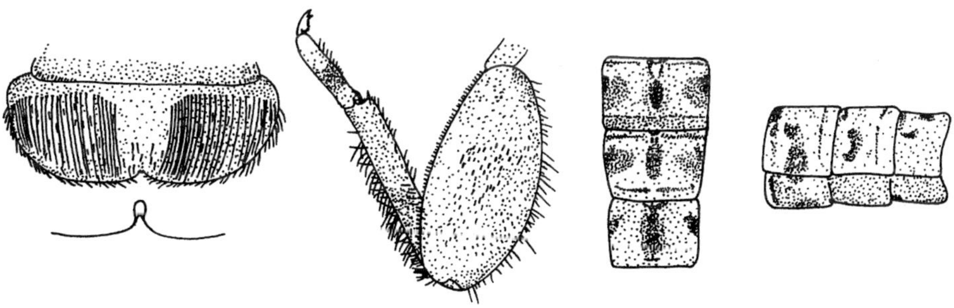

Fig. 2.406 *Hermanella froehlichi* larva (left to right): the labrum with an enlarged view of the median emargination, the fore-leg, and the color pattern on the fifth through seventh abdominal segments in dorsal and lateral view.

2. Spotted wings show through on the wing pads. Length: 9.5 to 10 mm. Color: mainly orange brown.
……………………………*Hermanella (Guayakia) maculipennis* (Ulmer, 1920) (Argentina, Paraná, Santa Catarina). Syn: *Thraulus maculipennis* Ulmer, 1920; *Traverella maculipennis* (Ulmer, 1920).

- The wing pads show no traces of spots and have dark outer margins. Length: 10 to 12.5 mm. Color: mainly orange brown.
.....................*Hermanella (Guayakia) grandis* Domínguez and Flowers, 1989 (Argentina, Paraná, Santa Catarina).
3. The long setae on the outer margins of the middle and hind tibiae are dense (**Fig. 2.406**). The gills are whitish at the base and apex and blackish in the middle. Length of fully grown larva: 4.0 to 5.6 mm. Color: mainly orange brown with brown markings and black wash.
...................*Hermanella (Guayakia) froehlichi* Ferreira and Domínguez, 1992 (São Paulo).
- Relatively long setae are only scattered along their outer margins of the tibiae (**Fig. 2.350**). The fully grown larvae do not exceed 8 mm in length.
.....................Subgenus *Hermanella* Needham and Murphy, 1924. 4
4. The abdominal tergites are almost entirely black without a pattern (**Fig. 2.350**). Length: 3.4 to 4.0 mm. Color: mainly blackish brown.
.....................*Hermanella (Hermanella) thelma* Needham and Murphy, 1924 (Argentina).
- The tergites of the abdomen are mainly orange brown with a blackish pattern. Length: 7 to 7.5 mm. Color: mainly orange brown with a dark pattern.
...................*Hermanella (Hermanella) guttata* Domínguez and Flowers, 1989 (Argentina, Paraguay, Uruguay, Paraná).

Key to the species of known adult *Miroculis* in South America

Information for the key was taken from Savage and Peters (1983) and Savage (1983, 1987b). The key applies mainly to adult males because few adult females have been described. The adults of *Miroculis (Atroari) colombiensis* and *Miroculis (Miroculis) brasiliaensis* have not yet been described.

1. The upper portions of the eyes of the male adult are not stalked, and they are separated by a distance equal to half to one width of an upper portion (**Fig. 2.407**). There are 27 to 40 facets in the longest row across the compound eyes. The eyes of the only female described form this group are separated by a distance equal to 3.2 to 3.3 times the maximum width of one eye. 2
- The upper portions of the eyes of adult males are set on stalks (**Fig. 2.408**). ...5
2. The upper portions of the eyes are separated by a distance equal to 0.5 to 0.6 of the maximum width of one upper portion. The fore-wing width is equal to 0.47 to 0.51 times the length. The posteromedian margin of the styliger plate is rounded (**Fig. 2.407**). ....Subgenus *Atroari* pars...........................3
- The upper portions of the eyes are separated by a distance approximately equal to the maximum width of one upper portion. The fore-wing width is equal to 0.41 to 0.43 times the length. The posteromedian margin of the styliger plate has a large U-shaped indentation (**Fig. 2.409**). ......Subgenus *Ommaethus*.........4

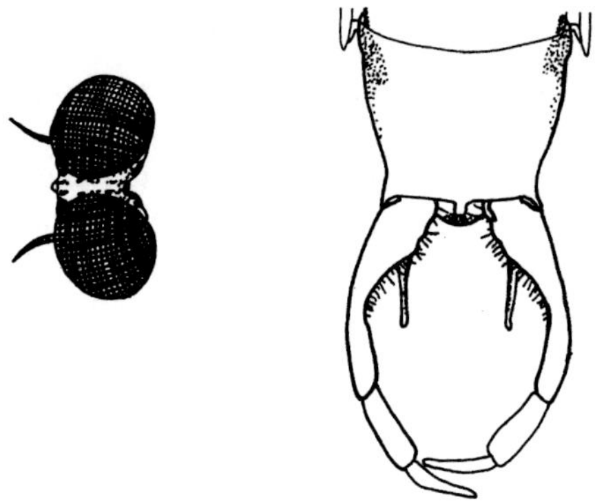

Fig. 2.407 *Miroculis duckensis:* the head of a male subadult in dorsal view (left), and the genitalia of a male adult in ventral view (right).

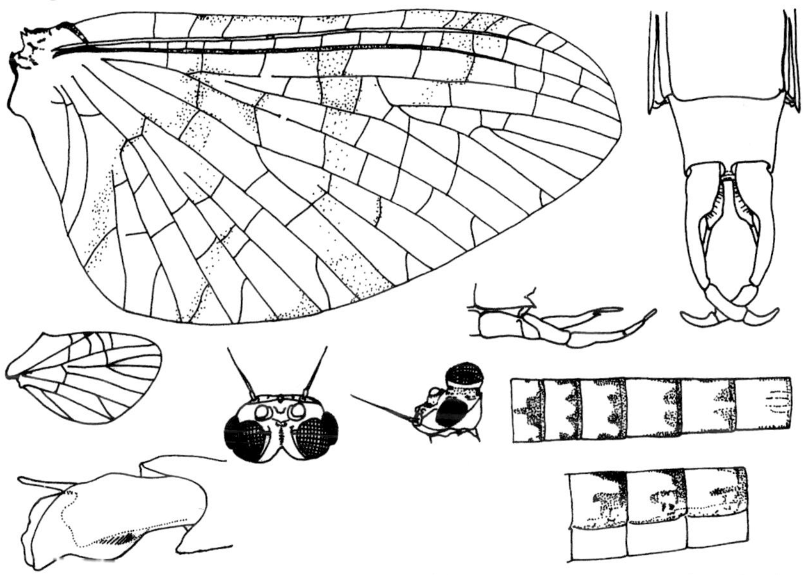

Fig. 2.408 *Miroculis nebulosus:* fore and hind wing (upper left), the head of a male in dorsal (lower left center) and lateral view (lower center), second through seventh abdominal tergites (middle right) and fourth through sixth abdominal segments in lateral view of a male (lower right), male genitalia in ventral (upper right) and lateral view (right center), and the tarsal claw on the fore-leg (lower left).

3. The upper portions of the eyes are separated by a distance equal to 0.5 of the maximum width of one upper portion. The stigmatic area of the fore-wing is very dark, and the apical part of the hind wing is mainly dark brown (**Fig. 2.410**). The length of the penis is equal to 1.7 to 1.9 the length of the first segment of the forceps. The second segment of the forceps is rectangular in

profile. Body length: 5.9 mm. Fore-wing length: 4.6 mm. Color: mainly yellowish with blackish brown markings.
.............................*Miroculis (Atroari) amazonicus* Savage and Peters, 1983 (Amazonas).

- The upper portions of the eyes are separated by a distance equal to 0.6 of the maximum width of one upper portion. The fore-wing has very light brown bands and blotches. The length of the penis is equal to 1.1 to 1.4 the length of the first segment of the forceps. The second segment of the forceps widens toward the apex (**Fig. 2.407**). Body length: 4.8 to 6.0 mm. Fore-wing length: 4.4 to 4.6 mm. Color: mainly yellowish with blackish brown markings.
.................................*Miroculis (Atroari) duckensis* Savage and Peters, 1983 (Amazonas).

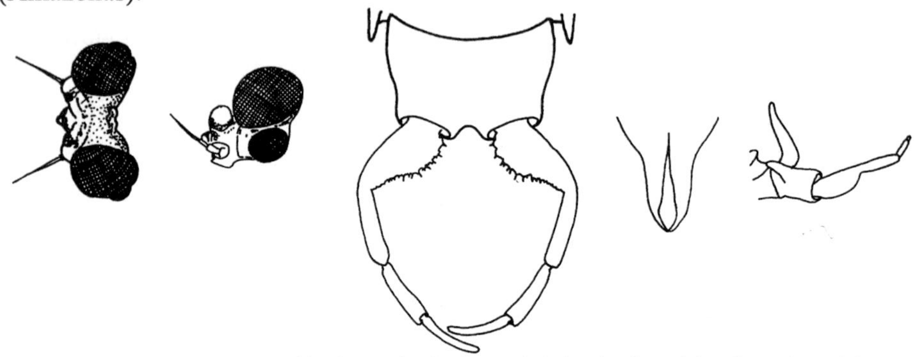

Fig. 2.409 *Miroculis froehlichi* male (left to right): the head in dorsal and lateral view, the male genitalia in ventral view, the penis in ventral view, and the genitalia in lateral view.

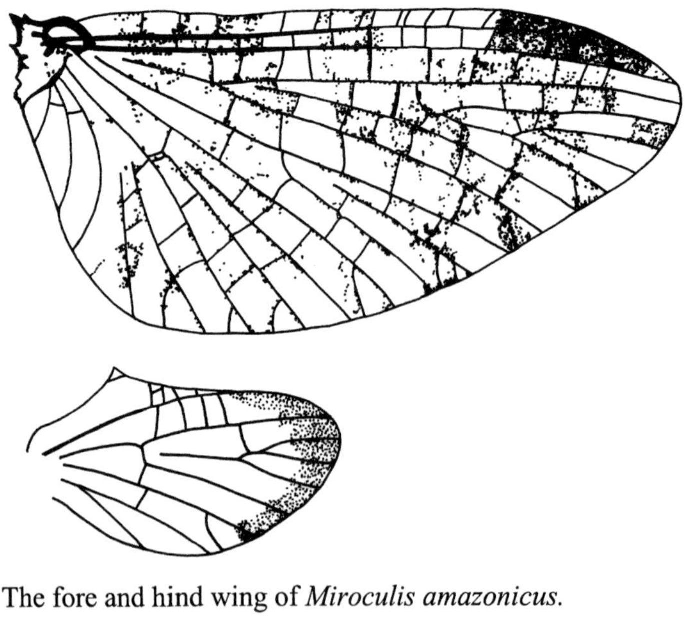

Fig. 2.410 The fore and hind wing of *Miroculis amazonicus*.

4. The fifth and sixth abdominal tergites are white and somewhat translucent with faint submedian bars on the sixth tergite. The forceps and caudal filaments are light to pale yellow (**Fig. 2.409**). Body length: 6.2 to 7.8 mm. Fore-wing length: 5.8 to 6.5 mm. Color: mainly yellowish with brownish black markings.
……………………….*Miroculis (Ommaethus) froehlichi* Savage and Peters, 1983 (São Paulo).
- The fifth and sixth abdominal tergites are brownish yellow with brownish black bars and not obviously translucent. The distal third of the first forceps segment and all of the second and third segments have a blackish tone. The caudal filaments are brownish yellow with brownish black bands on the proximal third to half. The membrane of the fore-wing has a light brown pigmentation, which somewhat obscures a darker pattern (**Fig. 2.411**). The female subimago has a somewhat similar coloration. Body length of male adult: 3.8 to 4.6 mm; of female subadult: 4.9 to 5.1 mm. Fore-wing length of adult male: 5.4 to 5.6 mm; of subadult female: 5.5 to 5.7 mm. Color: mainly brownish yellow with brownish black markings.
……………………….*Miroculis (Ommaethus) mourei* Savage and Peters, 1983 (Paraná).

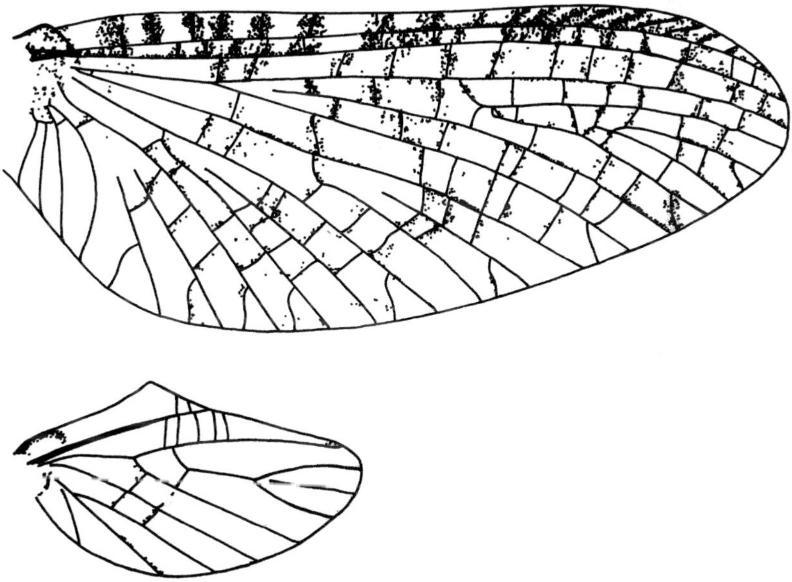

Fig. 2.411 The fore and hind wing of *Miroculis mourei* with the hind wing drawn larger than scale.

5. The stalks supporting the upper portion of the eyes on the heads of adult males are separated by a distance equal to 0.1 to 0.2 times the maximum width of one upper portion of an eye. The stalks are short and wide. There are 32 to 37 facets in the longest row across the compound eye. The membrane of the wings is very lightly tinged with brown and lacks a dark brown pattern. Body

length: 5 mm. Fore-wing length: 4.5 mm. Color: mainly brownish yellow with brownish black markings including dark bands across the posterior parts of the abdominal tergites (**Fig. 2.412**).
..................................*Miroculis (Yaruma) wandae* Savage and Peters, 1983 (Amazonas).
- The stalks supporting the upper portion of the eyes on the heads of adult males are separated by a distance equal to 0.5 to 1.5 times the diameter of one upper portion. The stalks are long and narrow. The longest row across a compound eye consists of 5 to 14 large facets (**Fig. 2.408**). The compound eyes of the known female subadults are separated by a distance equal to about three times the width of one eye. The sixth abdominal tergite of the female subadult is almost hyaline. 6

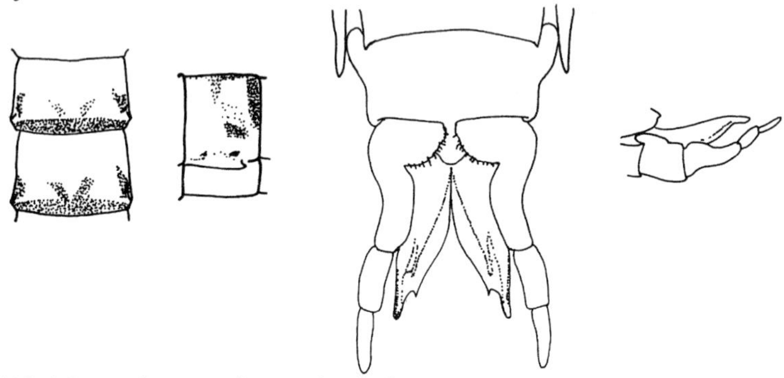

Fig. 2.412 *Miroculis wandae* male (left to right): the fifth and sixth abdominal tergites, the sixth abdominal segment in lateral view, the genitalia of a male in ventral and lateral view.

6. The longest row in the upper portions of each eye consists of 11 to 14 facets. The upper portions are separated by 0.5 to 1.0 times the diameter of one portion. The brown markings on the fore-wings form broken, irregular transverse bands and many blotches (**Fig. 2.408**). In the middle of the posterolateral margin of the styliger plate, there is a shallow notch. Length of male: 5.4 to 5.8 mm. Fore-wing length: 4.1 to 4.7 mm. Coloration: brownish yellow and light brown with a blackish wash and blackish brown markings. The first eight segments of the abdomen are somewhat translucent. The female has not been described.
....*Miroculis (Atroari) nebulosus* Savage, 1987 (Venezuela).
- The longest row in the upper portions of each eye consists of 5 to 10 facets. The upper portions are separated by 0.7 to 1.5 times the diameter of one portion (**Fig. 2.413**). 7
7. There are brown clouds across the cross veins of the fore and hind wings, which are otherwise hyaline or weakly tinged with brown. The inner margin of the first segment of the forceps is abruptly narrowed about 1/3 to 1/2 of the distance from the base. The posteromedian margin of the styliger plate is

gradually rounded and has a shallow indentation. The inner apical margin of each penis lobe is excavated. There is a V-shaped brownish black mark with the open part of the V facing posteriad in the middle of the sixth abdominal tergite of both males and females. The caudal filaments of the female are uniformly light brown. The fore-wings of the female have dark clouds across the cross veins, and the membrane of the female hind wings is dark toward the apex (**Fig. 2.413**). Body length: 4.9 to 6.2 mm. Fore-wing length: 4.5 to 5.2 mm. Color: mainly light brown and brownish yellow with brownish black markings.
................................*Miroculis (Miroculis) fittkaui* Savage and Peters, 1983 (Venezuela, Surinam, Pará).

- There are no brown clouds across any of the cross veins. Either the first segment of the male forceps narrows abruptly in the basal 1/8 or it narrows gradually in the basal half (**Fig. 2.414**). 8

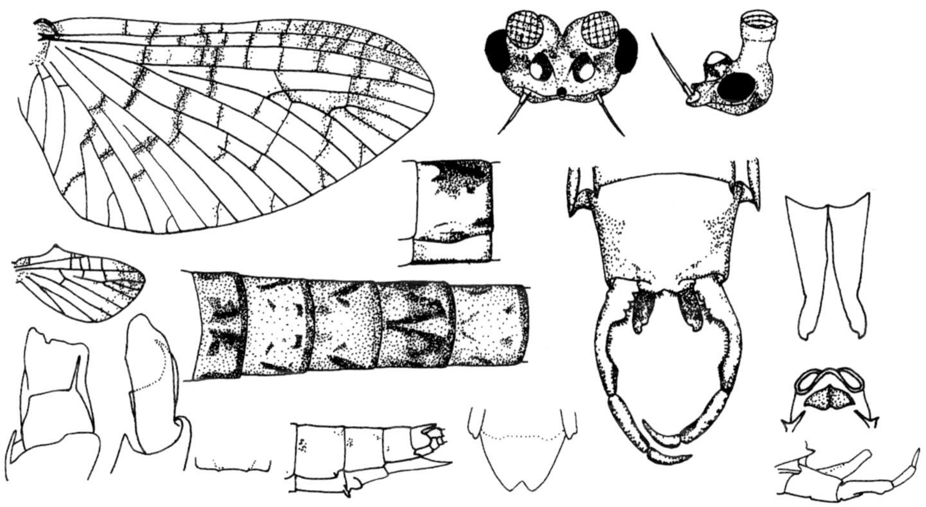

Fig. 2.413 *Miroculis fittkaui:* fore and hind wing (upper left), head of a male in dorsal (upper center) and lateral view (upper right), claw of an adult male and female (lower left, left and right, respectively), pattern on the third through seventh abdominal tergites (middle left center) and on the seventh segment in lateral view (above tergites), male genitalia in ventral view (lower right center) and in posterior and lateral view (lower right, above and below, respectively), an outline of the penis lobes (middle right), apex of the abdomen of a female in lateral view (lower left center), apex of its ninth sternite (lower center), and the posterior margin of its seventh sternite (right of claws).

8. The basal half of the fore-wing membrane is light brown or brown with a light blackish wash in the anterior cells, while the apical half is hyaline or tinged light brownish, making the wings appear distinctly of two colors. The hind wing

membrane is brown. There are 8 to 10 facets in the longest row across the upper portion of the eye. The male forceps narrows abruptly in the basal 1/8 (**Fig. 2.414**). Length of male: 7.1 to 8.6 mm. Fore-wing length: 6.7 to 7.5 mm. Coloration: light brown with darker brown or blackish markings. The first eight abdominal segments are almost hyaline. The female has not been described. ....*Miroculis (Miroculis) bicoloratus* Savage, 1987 (Venezuela).

- The entire membrane of the fore and hind wings of both males and known females is hyaline. The inner margin of the first segment of the male forceps narrows gradually in the basal half. The posteromedian margin of the styliger plate is gradually rounded without any indentation. There are brownish black marks in the median area of the sixth abdominal tergite, or the tergite has a brownish tone (**Fig. 2.415**). 9

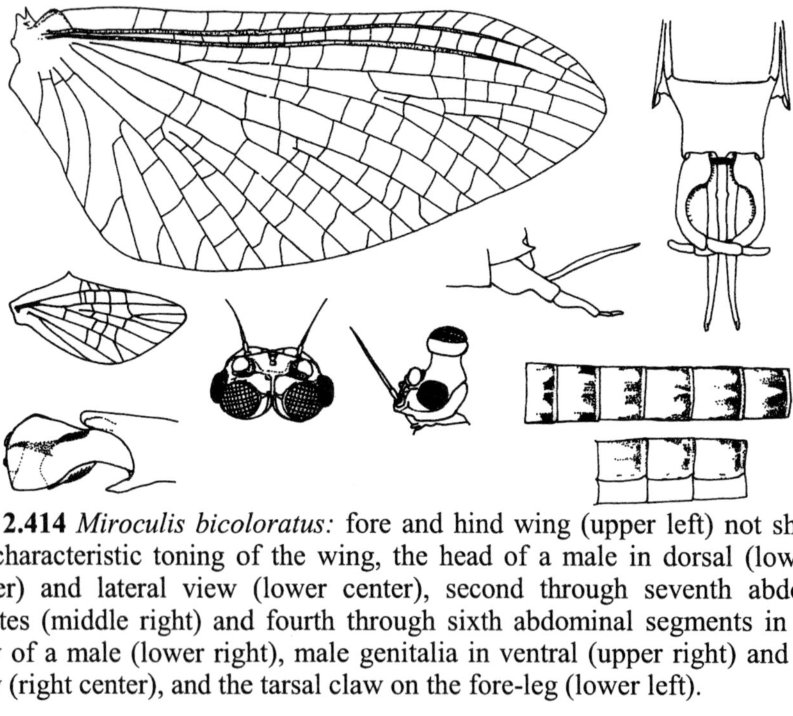

Fig. 2.414 *Miroculis bicoloratus:* fore and hind wing (upper left) not showing the characteristic toning of the wing, the head of a male in dorsal (lower left center) and lateral view (lower center), second through seventh abdominal tergites (middle right) and fourth through sixth abdominal segments in lateral view of a male (lower right), male genitalia in ventral (upper right) and lateral view (right center), and the tarsal claw on the fore-leg (lower left).

9. The longest row across the compound eye of the male consists of eight or nine facets. In the fore-wing of males and females, the base of vein MP_2 is free or is attached to vein IMP by a cross vein. The penis lobes are 0.8 to 1.1 times the length of the first segment of the forceps. Each penis is tubular and lacks spines (**Fig. 2.415**). Body length: 4.6 to 6.0 mm. Fore-wing length: 4.1 to 4.7 mm. Color: mainly brown with blackish markings. ....*Miroculis (Miroculis) marauiae* Savage and Peters, 1983 (Amazonas).

- The longest row across the compound eye of the male consists of six facets. In the fore-wing, the base of vein MP_2 is attached to veins MP_2 and CuA by cross veins. The penis lobes are 2.0 to 2.2 times the length of the first segment of the forceps. Each penis narrows apically and bears a row of about 15 stout spines on a ventral ridge near the base (**Fig. 2.306**). Body length: 4.7 mm. Fore-wing length: 5.0 mm. Color: the second through seventh abdominal segments are almost hyaline. The rest of the abdomen is yellowish brown, and the head and thorax are mainly brown with blackish markings. The female has not been described.
.... *Miroculis (Miroculis) rossi* Edmunds, 1963 (Peru).

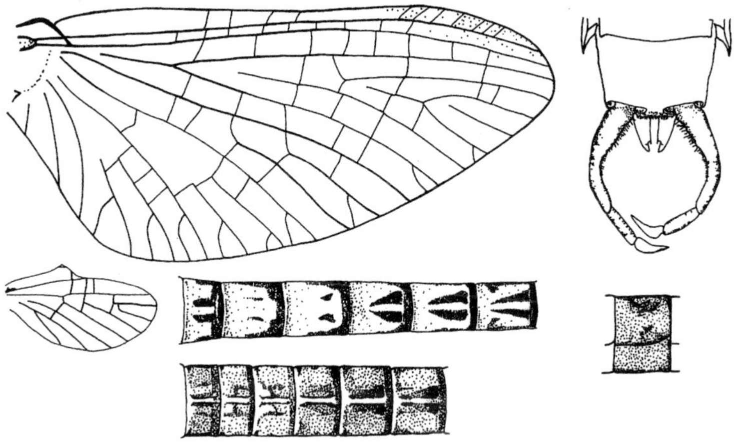

Fig. 2.415 *Miroculis (Miroculis) marauiae:* the fore and hind wing (left), the pattern on the third through eighth abdominal tergites of the male (center) and female (lower center), the sixth segment of the male abdomen in lateral view (lower right), and the male genitalia in ventral view (upper right).

Key to the species of the known *Miroculis* larvae in South America

Information for the key was taken from Savage and Peters (1983) and Savage (1983, 1987b). The larvae of *Miroculis (Miroculis) rossi* have not yet been described.

1. The color pattern on the each abdominal tergite from the second through the sixth is about the same, consisting of longitudinal, submedian, sublateral, and lateral brownish black bars. The anterolateral margins of the labrum are evenly rounded or somewhat flattened. The abdominal gills are small (**Fig. 2.331**). Length: 4 to 5 mm.
.............................*Miroculis (Ommaethus) mourei* Savage and Peters, 1983 (Paraná).

- The third, fourth, or fifth abdominal tergites or any two of them are lighter in color than some of the others, and some of the second through sixth tergites have color patterns different from the others. The anterolateral margins of the labrum are at least slightly flattened (**Fig. 2.416**). 2

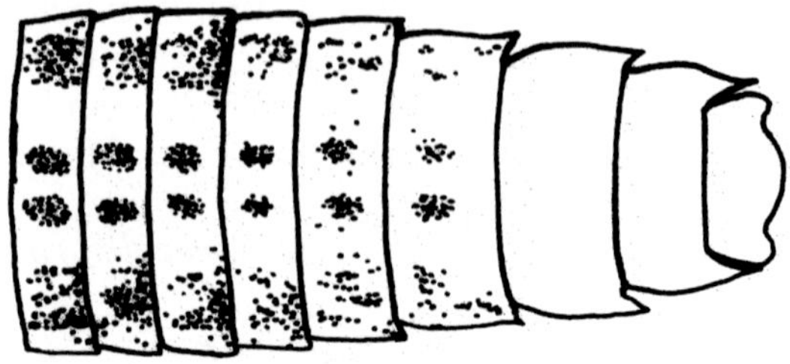

Fig. 2.416 The second through tenth tergites of an early larva instar presumed by Savage and Peters (1983) to belong to *Miroculis froehlichi.*

2. The eighth through tenth abdominal tergites are brownish yellow and lack a distinct pattern, while the second through seventh have a brownish black pattern of large lateral patches and small, round dots on either side of the midline, which become progressively lighter in color from the second to the seventh tergites (**Fig. 2.416**). The description is based on an early instar larva of a female, about 4 mm long, which was not identified with complete certainty. The last larval instar has not been described.

...............probably *Miroculis (Ommaethus) froehlichi* Savage and Peters, 1983 (São Paulo).

- At least the eighth abdominal tergite usually has a distinguishable dark pattern, and the spots on either side of the midline are usually not isolated and round (**Fig. 2.417**). 3

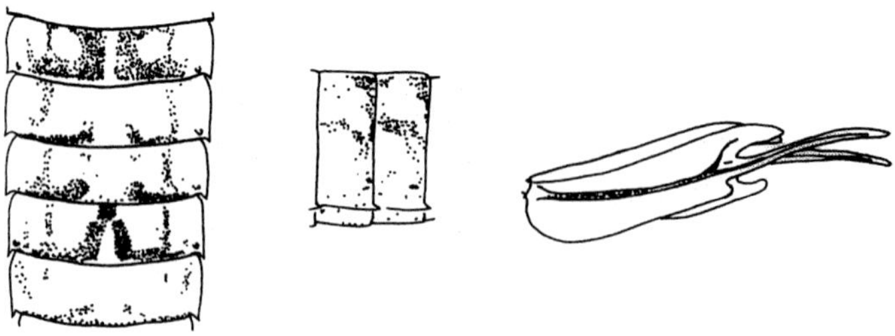

Fig. 2.417 *Miroculis brasiliaensis* larva (left to right): the fourth through seventh abdominal segments in dorsal view, the fourth and fifth abdominal segments in lateral view, and a gill from the fourth abdominal segment.

3. The fourth and fifth tergites are lighter than the second, third, and sixth. The sixth tergite has a V-shaped mark in the middle, or a pair of dark triangular marks flank the lighter midline on the sixth through eighth tergites (**Fig. 2.417**). 4

- The third, fourth, or both of these tergites are slightly lighter than the second and fifth. The fourth through eighth or fifth through eighth have the same color pattern, or each has a different pattern. The sixth tergite never has a V-shaped mark (**Fig. 2.418**). 6

Fig. 2.418 *Miroculis duckensis* larvae (left to right): the seventh through tenth abdominal tergites of a final instar larva, the color pattern on the abdomen of an earlier instar, and a gill from the fourth segment of the abdomen.

4. The sixth through eighth abdominal tergites have almost the same color pattern. The abdominal gills are narrow and small and lack prominent posterolateral projections (**Fig. 2.419**). Length: 4 to 5.5 mm. ....*Miroculis (Miroculis) marauiae* Savage and Peters, 1983 (Amazonas).

- There is a dark V-shaped mark on the sixth abdominal tergite with the open end of the V directed posteriad. The color patterns on the seventh and eighth tergites are relatively light, The abdominal gills bear prominent posterolateral projections (**Fig. 2.417**). 5

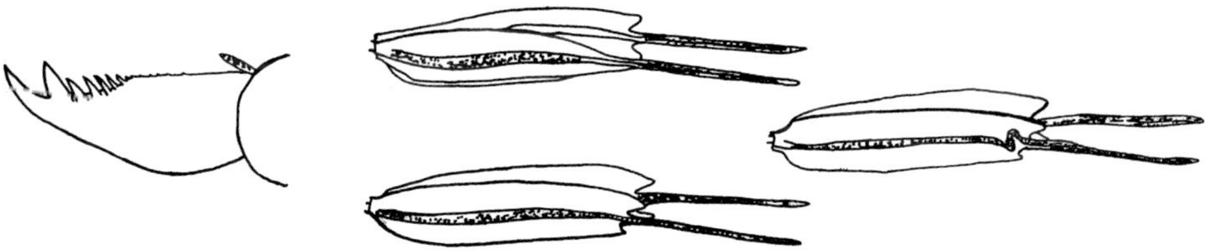

Fig. 2.419 *Miroculis (Miroculis) marauiae* larva (left to right): a tarsal claw and three gills from the fourth abdominal segments of three different specimens.

5. The abdominal gills are relatively large. There is a subapical row of 35 to 40 pectinate setae on the maxillae. The caudal filaments are brownish yellow with black bands on the fifth through ninth segments (**Fig. 2.417**). Length: 4.6 to 4.8 mm. ....*Miroculis (Miroculis) brasiliaensis* Savage and Peters, 1983 (Goiás).

- The abdominal gills are relatively small (**Fig. 2.420**). There is a subapical row of 44 to 49 pectinate setae on the maxillae. The caudal filaments are light brown without any black bands. Length: 4.3 to 6.2 mm. ....*Miroculis (Miroculis) fittkaui* Savage and Peters, 1983 (Venezuela, Surinam, Pará).

Fig. 3.420 Gill from the fourth abdominal segments of two specimens of *Miroculis fittkaui*.

6. The abdominal gills are relatively small and have small posterolateral projections. The abdomen is mainly brownish yellow with intersecting brownish black arcs (**Fig. 2.418**). Length: 4.0 to 5.5 mm. ....*Miroculis (Atroari) duckensis* Savage and Peters, 1983 (Amazonas).
- The abdominal gills are relatively large and have prominent posterolateral projections (**Fig. 2.421**). 7

Fig. 2.421 *Miroculis colombiensis* larva (left to right): the pattern on the fourth through tenth abdominal segments, a tarsal claw, and a gill from the fourth abdominal segment.

7. The fourth through eighth abdominal tergites become progressively darker toward the posterior end of the abdomen. They have circular submedian marks. The gill membrane is brownish yellow (**Fig. 2.421**). Length: 5.0 to 5.6 mm. ....*Miroculis (Atroari) colombiensis* Savage and Peters, 1983 (Colombia).

- The fourth through eighth abdominal tergites are equally dark, or the fourth tergite is slightly lighter than the others. They have submedian markings that are not circular. The gill membrane is grayish (**Fig. 2.422**). Length: 4 to 6 mm. ....*Miroculis (Atroari) amazonicus* Savage and Peters, 1983 (Amazonas).

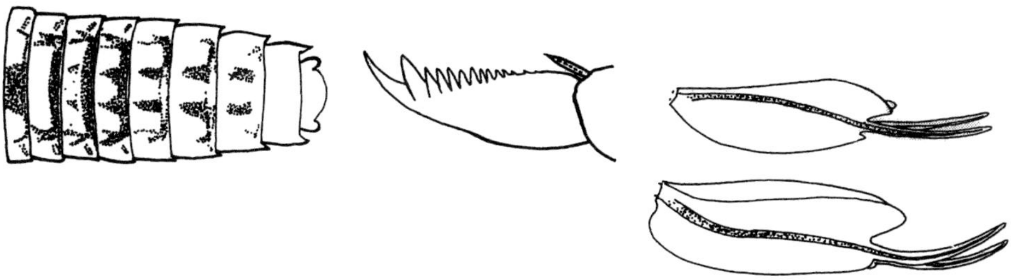

Fig. 2.422 *Miroculis amazonicus* larva (left to right): the pattern on the second through tenth abdominal segments, a tarsal claw, and a gill from the first (above) and one from the fourth abdominal segment (below).

Key to the species of adult *Hermanellopsis* in South America

Information for the key was taken from Savage and Peters (1983). A direct comparison of the two species is tentative, at best, because one is known only from an adult male and subadult female and the other is known only from a subadult male. One could venture the question of why Savage and Peters (1983) described the male subadult as a new species in light of the fact that they had ascertained a great variability in the morphological features on which the description was based, opening the possibility that their subadult male might turn out to be conspecific with the known species.

1. There are about 14 cross veins in the costal space of the fore-wing and about 15 in the subcostal space. The costal projection on the hind wing is acutely pointed. A cross vein joins MP_2 to MP_1 and CuA in the fore-wing (**Fig. 2.313**). Length of adult male: 3.5 to 5 mm; of subadult female: 3.1 mm. Fore-wing length of adult male: 4 mm; of subadult female: 3.2 to 3.3 mm. Color: generally brownish yellow with a blackish wash. Descriptions apply only to adult males and subadult females.
....*Hermanellopsis incertans* (Spieth, 1943) (Guyana, Surinam). Syn: *Hermanella incertans* Spieth, 1943.
- There are about three cross veins in the costal space of the fore-wing of the subadult and about six in the subcostal space. The apex of the costal projection on the hind wing rounded. MP_2 joins MP_1 in the fore-wing (**Fig. 2.423**). Length

of subadult male: 4.5 to 5 mm. Fore-wing length of subadult male: 4.2 to 4.9 mm. Color: generally brownish yellow with a brownish black markings. Descriptions apply only to subadult males.
....*Hermanellopsis arsia* Savage and Peters, 1983 (Amazonas).

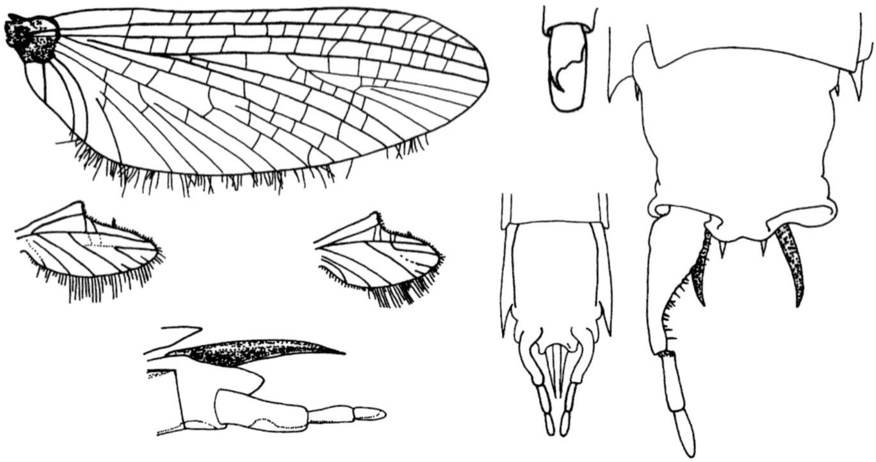

Fig. 2.423 *Hermanellopsis arsia* male subadult: fore-wing and the hind wings from two different specimens (upper left), the tarsal claw on the fore-leg (upper center), the apical segment of the abdomen in ventral view (lower right center), the male genitalia in ventral view with one limb of the forceps removed (right), and the male genitalia in lateral view (lower left).

Key to the subadult *Microphlebia* species in South America

Information for the key is taken from Savage and Peters (1983).

1. The longest row in the upper portion of the male eye consists of 23 to 30 facets. All femora are uniformly brownish yellow. The outline of the first segment of the forceps is crooked in dorsal or ventral view (**Fig. 2.324**). Length: 4.4 to 5.0; fore-wing length: 3.6 to 4.0 mm. Color: yellow, extensively washed with brownish black.
....*Microphlebia pallida* Savage and Peters, 1983 (Amazonas).

\- The longest row in the upper portion of the male eye consists of 17 to 24 facets. The hind femur is brownish yellow with a brownish black marking near the middle. At least the outer outline of the first segment of the forceps is evenly curved (**Fig. 2.315**). Length: 3.6 to 4.4 mm; fore-wing length: 3.4 to 3.9 mm. Color: brownish yellow, extensively washed with brownish black.
....................................*Microphlebia surinamensis* Savage and Peters, 1983 (Surinam, Venezuela).

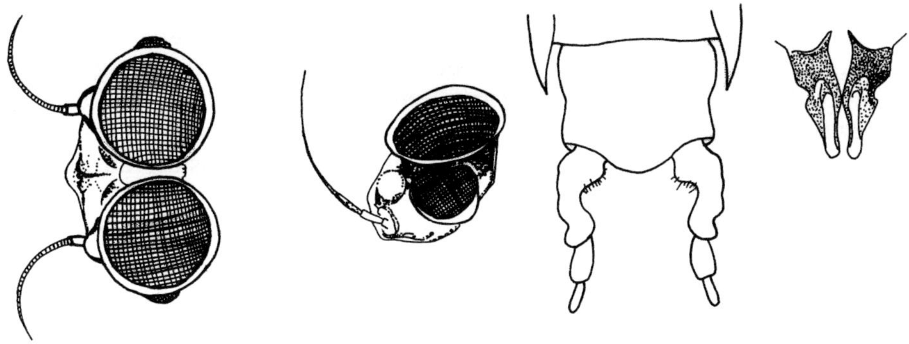

Fig. 2.424 *Microphlebia pallida* male subadult (left to right): the head in dorsal and lateral view, the genitalia in ventral view with the penis concealed beneath the apical sternite, and the penis in ventral view.

Key to the species of *Microphlebia* larvae in South America

Information for the key is taken from Savage and Peters (1983).

1. The femora, submentum and mentum of the labium are uniformly brownish yellow. The membrane of the gill is light gray. Length: 3.2 to 5.0 mm.
....*Microphlebia pallida* Savage and Peters, 1983 (Amazonas).
- The apical 1/5 to 1/3 of the femora and the submentum and mentum of the labium are brownish yellow washed with brownish black. The gill membrane is blackish gray (**Fig. 2.328**). Length: 3.2 to 4.4 mm.
................................*Microphlebia surinamensis* Savage and Peters, 1983 (Surinam, Venezuela).

Key to the species of adult *Meridialaris* in South America

The information for the key is taken from Ulmer (1904, 1938), Lestage (1931), Navás (1934a), Thew (1960), and Pescador and Peters (1987). The adults of *Meridialaris spina* Pescador and Peters, 1987, have not been described.

1. The length of the female is 10 mm, and the fore-wing length is 12 mm. The cross veins in the costal and subcostal areas of the fore-wing are darker than the other veins. There are 12 or 13 cross veins in the costal area distal to the bulla, the point along the anterior edge of the fore-wing near the midlength where the wing appears slightly bent. Proximal to the bulla, the veins in the costal area are very faint. The anterior margin of the hind wing is folded near the mid-length. The hind leg appears to have four tarsal segments because the two basal

segments are not completely separated (**Fig. 2.425**). The hind wing is about 3 mm long.

.... *Meridialaris patagonica* (Lestage, 1931) (Chile). Syn: *Atalophlebia chilensis* sensu Ulmer (1904); *Atalophlebia patagonica* Lestage, 1931; *Deleatidium patagonicum* (Lestage, 1931). This species is poorly defined, and Domínguez *et al.* (2001) consider the name a *nomen dubium*.

- The length is not more than 9 mm, and the fore-wing is not more than about 10.5 mm. 2

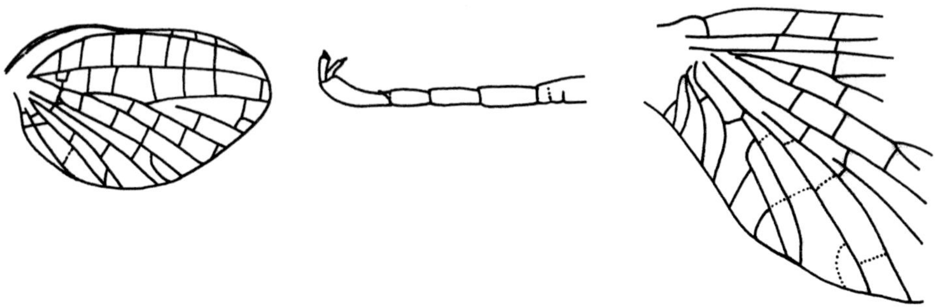

Fig. 2.425 *Meridialaris patagonica* female (left to right): hind wing, apex of the hind tibia and tarsal segments, base of the fore-wing.

2. The costal and subcostal areas of the fore-wings of the female are light reddish brown, and the rest of the wings are hyaline. The head between the lateral and median ocelli and the thorax have black stripes on a white or yellowish white background. Length: 8.1 mm; length of fore-wing: 10.3 mm. The male has not been described.

....*Meridialaris vittatum* (Thew, 1960) (Santa Catarina). Syn: *Deleatidium vittatum* Thew, 1960. Pescador and Peters (1987) stated that this species should be placed in a still undescribed genus. Until a name is given to this genus and a description provided, it must remain in *Meridialaris.*

- The costal and subcostal areas are hyaline like the rest of the wing except sometimes for a slight yellowing of the stigmal area. A distinct pattern of black stripes is not evident (**Fig. 2.426**). 3

3. The tergites or the male are translucent yellow with broad dark brown markings covering most of the dorsal surface, including almost all of the posterior half (**Fig. 2.426**). The tergites of the female are opaque yellow with similar dark brown markings. Veins ICu1 and CuA in the fore-wing are strongly divergent. Length: 6 to 7.5 mm; fore-wing length: 6 to 7.5 mm.

....*Meridialaris biobionica* (Ulmer, 1938) (Chile). Syn: *Deleatidium biobionicum* Ulmer, 1938.

- The light color on the abdominal tergites of the male is more extensive than the dark markings, and extensive areas on the posterior half of the tergites are light in color. If the abdominal tergites of the female are predominantly dark, then ICu1 is parallel or slightly divergent from CuA in the fore-wing (**Fig. 2.427**). 4

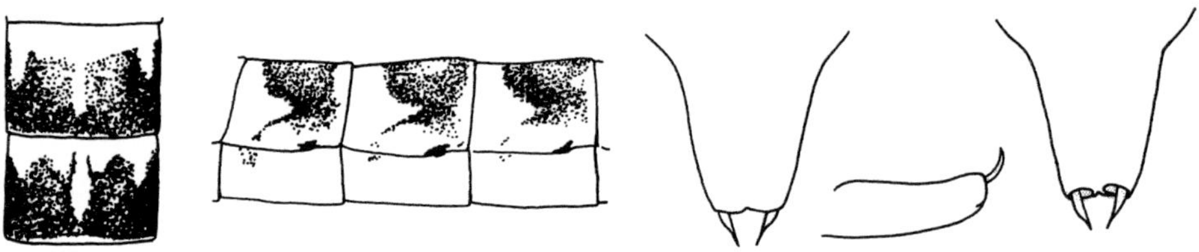

Fig. 2.426 *Meridialaris biobionica* (left to right): the fourth and fifth abdominal segments of a male in dorsal view, third through fifth segments in lateral view, and the penis in ventral, lateral, and dorsal view.

4. The central areas of the third and fourth tergites and the medial section of their posterior margins are almost entirely light in color, while the posterior margins of the second, fifth, and sixth segments are dark. Some of the dark dorsolateral markings on the third through fifth tergites of the male extend from the anterior to the posterior margin. The penis is bell-shaped with apical finger-like projections (**Fig. 2.316**). ICu1 is parallel or only slightly divergent from CuA in the fore-wing. Length: 8 mm; fore-wing length: 9 mm. The male subadult is similar to the male adult except for yellow or orange yellow coloration instead of brown on the head and parts of the thorax. Pescador and Peters (1987) stated that the female is unknown but included an illustration labeled as the fourth through sixth tergites of a female.
....*Meridialaris tintinnabula* Pescador and Peters, 1987 (Argentina).
- If the third and fourth tergites are much lighter in color than the others, then the penis is square or cylindrical (**Fig. 2.427**). 5
5. Some of the dark dorsolateral markings on the third through fifth tergites of the male extend from the anterior to the posterior margin. The femora do not have dark median bands. The penis is enlarged over the basal 60% of its length, and the apical projections are small (**Fig. 2.427**). The distal part of vein ICu_1 in the fore-wing diverges strongly from CuA. Length: 6 to 7 mm. Fore-wing length: 7 to 8 mm. Color, the head is mainly orange brown, and the thorax is dark brown.
....*Meridialaris inflata* Pescador and Peters, 1987 (Chile)
- None of the dark dorsolateral markings on the third through fifth tergites of the male extend from the anterior to the posterior margin (**Fig. 2.428**). 6

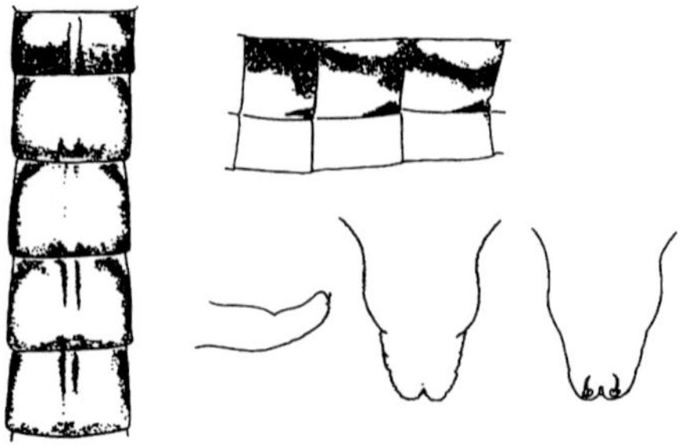

Fig. 2.427 *Meridialaris inflata:* the third through seventh abdominal segments of a male in dorsal view (left), the third through fifth in lateral view (upper right), the penis in lateral, ventral, and dorsal view (lower right, left to right).

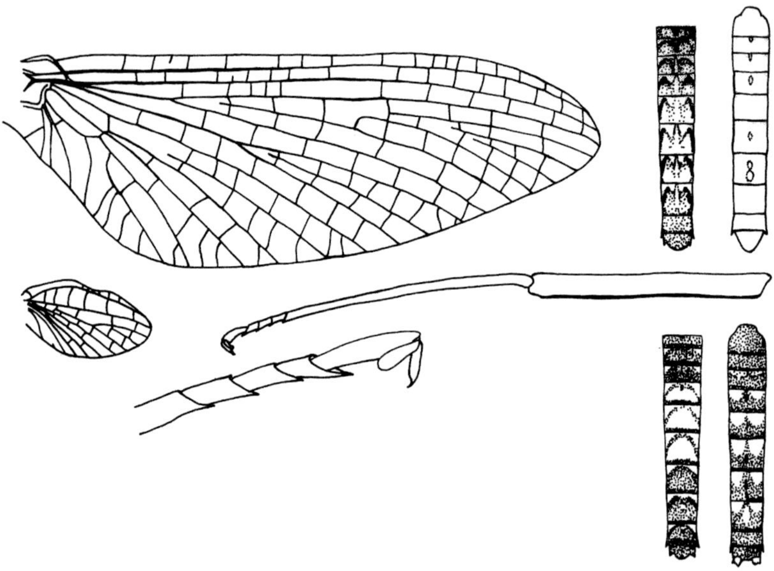

Fig. 2.428 *Meridialaris chiloeensis:* fore and hind wing of the male (upper left), middle leg (middle right); tarsi of the hind leg (middle left); patterns on the dorsal (left) and ventral (right) abdominal surfaces of the female (upper right) and male (lower right).

6. In the hind wing, there are more than 30 cross veins between the long veins. There is a blackish brownish undulating pattern on the reddish brown ground color of the second through seventh abdominal tergites of the female (**Fig. 2.428**). Length of the female: body, 9 mm; fore-wing, 10.5 mm; hind wing, 2.3 mm. Color: varying shades of brown on the head; compound eyes, blackish gray; thorax, shades of brown.
....*Meridialaris chiloeensis* (Demoulin, 1955) (Argentina, Chile). Syn: *Deleatidium penai* Demoulin, 1955, and *Deleatidium chiloeense* Demoulin, 1955.

- There are no more than 25 cross veins in the hind wing. There are no prominent undulating dark markings on the abdominal tergites. The abdominal tergites of the females do not have semicircular markings running from posterolateral margins to the middle of the anterior margin (**Fig. 2.429**). 7

7. There are no cross veins between the costal and subcostal veins on the basal half of the fore-wing. The longitudinal wing veins are brownish yellow. The second through tenth abdominal tergites of the female are blackish brown dorsally with large, roundish, pale brownish yellow markings on each side of the yellowish midline (**Fig. 2.429**). Length of the female: body, 8 mm; fore-wing, 10 mm, hind wing, 2.5 mm. Color: brown with lighter markings on head; compound eyes, grayish black; thorax, brownish yellow.
....*Meridialaris laminata* (Ulmer, 1920) (Argentina, Chile). Syn: *Deleatidium illapeli* Demoulin, 1955; *Deleatidium laminatum* Ulmer, 1920.

- There are several cross veins between the costal and subcostal vein on the basal half of the fore-wing. The longitudinal wing veins, except the costa and subcosta, are entirely hyaline. The dorsal surfaces of all abdominal tergites of the female are mainly brownish red or blackish brown without large lighter markings. The fourth through seventh tergites of the male abdomen are almost entirely translucent white in color, while the second, third, and eighth are translucent white with brownish red or blackish brown pigment along the posterior margins (**Fig. 2.430**). Length of male: body, 7 to 7.5 mm; fore-wing, 8 mm; hind wing, 2 mm. Color: head, light brown with blackish gray compound eyes; thorax, dark brown with lighter sutures.
....*Meridialaris diguillina* (Demoulin, 1955) (Argentina, Chile). Syn: *Deleatidium diguillinum* Demoulin, 1955.

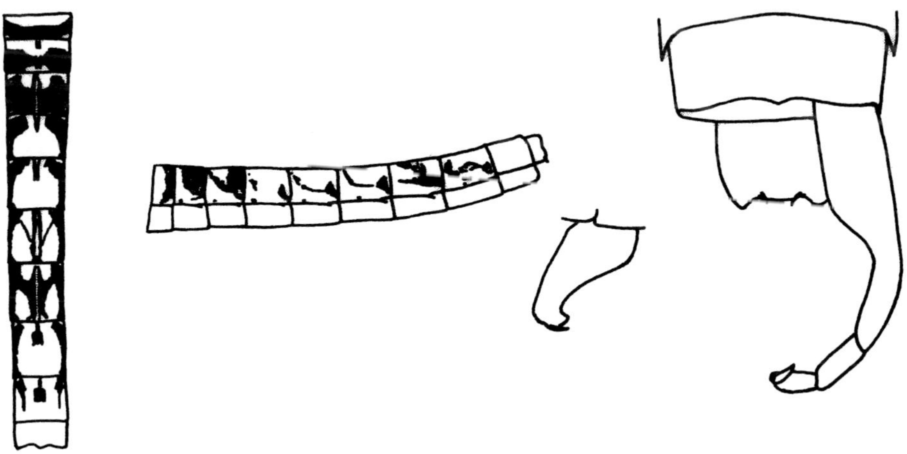

Fig. 2.429 *Meridialaris laminata* (left to right): the color pattern of the abdomen of both males and females in dorsal and lateral view, the penis in lateral view showing the small chitinous process, and the male genitalia in ventral view with the right forceps segments removed.

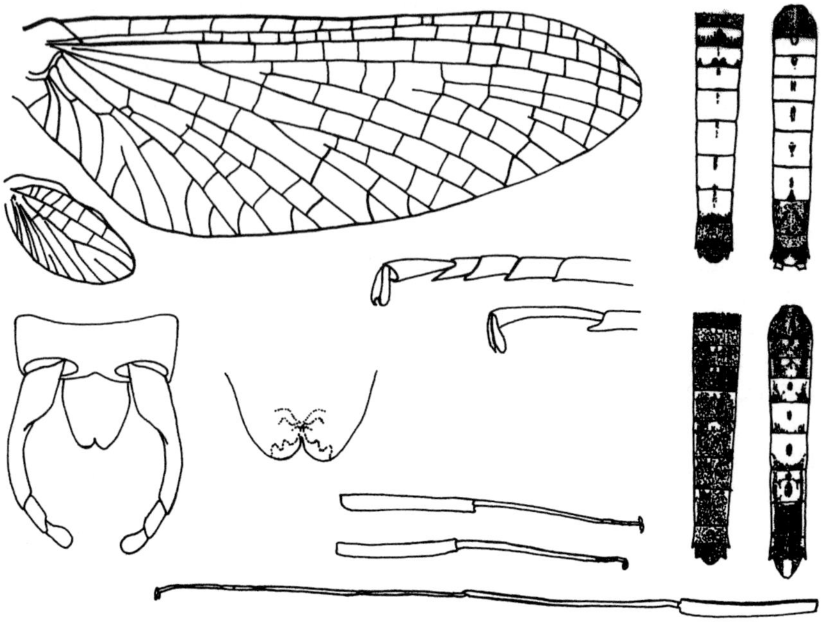

Fig. 2.430 *Meridialaris diguillina:* fore and hind wing of a male (upper left); male genitalia with the apex of the penis to the right (lower left); hind, middle, and fore-legs (below, top to bottom); enlargement of the middle tarsus and the apical tarsal segment of the fore-leg (middle right); patterns on the dorsal (left) and ventral (right) abdominal surfaces of the male (upper right) and female (lower right).

Key to the final larval stage of *Meridialaris* in South America

The information for the key was provided by Pescador and Peters (1987). The larvae of *Meridialaris biobionica, M. inflata, M. patagonica,* and *M. vittatum* have not been described.

1. The anteromedian emargination in the labrum is deep and V-shaped (**Fig. 2.431**). 2
- The anteromedian emargination in the labrum is shallow and U-shaped (**Fig. 2.344**). 4
2. There are long, needle-like spines on the tarsi. The lateral margin of the labrum is angular (**Fig. 2.431**). Length: 8 to 10 mm. ....*Meridialaris chiloeensis* (Demoulin, 1955) (Argentina, Chile). Syn: *Deleatidium penai* Demoulin, 1955, and *Deleatidium chiloeense* Demoulin, 1955.
- The tarsi lack long, needle-like spines. The lateral margins of the labrum are evenly curved (**Fig. 2.432**). 3

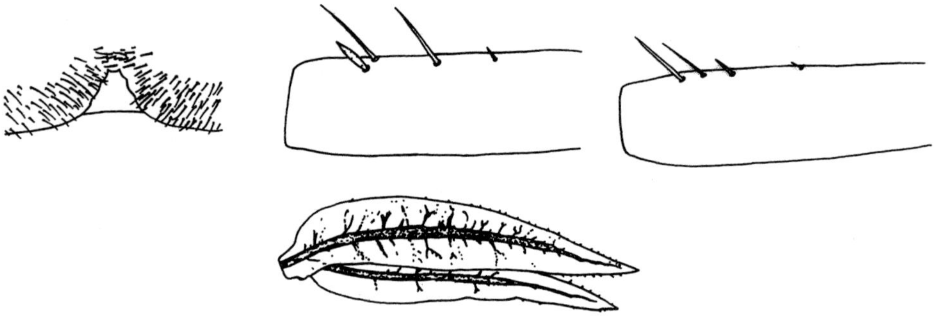

Fig. 2.431 *Meridialaris chiloeensis* larva (upper row, left to right): the emargination on the anterior border of the labrum and the spines on the metatarsi of two different specimens; and (below) the gill on the fourth abdominal segment.

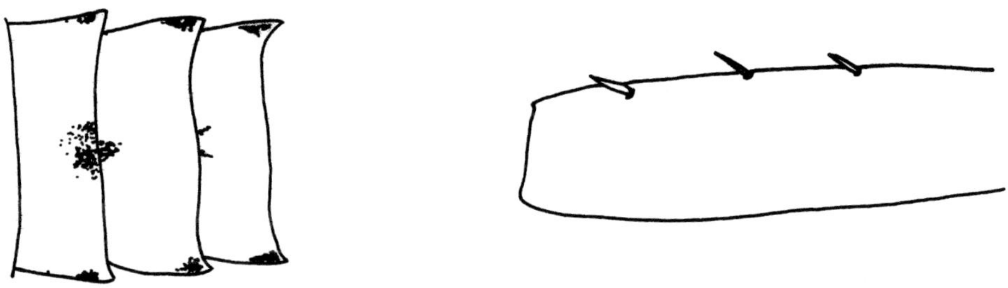

Fig. 2.432 *Meridialaris spina* larva: spots marking the abdominal tergites (left) and the spines on the tarsus (right).

3. The outer margin of the cardo bears fewer than 25 long spines. The segments of the antennal flagellum bear reddish brown annulations. There are two pale yellow spots on the femora. The tarsi are armed with about three short spines of about the same length (**Fig. 2.432**). Length: 8 to 10 mm.
....*Meridialaris spina* Pescador and Peters, 1987 (Chile, Argentina).

- The outer margin of the cardo bears more than 25 long spines. The segments of the antennal flagellum are not annulated. There are three pale yellow spots on the femora. (**Fig. 2.433**). Length: 7 to 9 mm.
...............................*Meridialaris tintinnabula* Pescador and Peters, 1987 (Argentina).

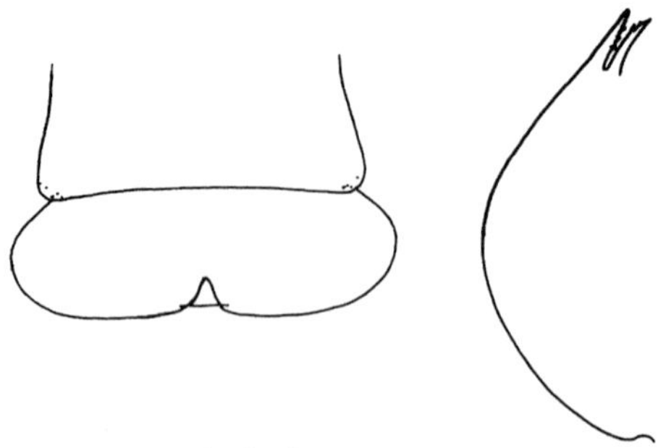

Fig. 2.433 *Meridialaris tintinnabula* larva: outlines of the labrum (left) and of the outer margin of a mandible (right).

4. The abdominal gills are rather broad, with several well-developed branches of the main tracheal trunk. The abdomen has dark lateral markings running almost from the anterior to the posterior margin (**Fig. 2.344**). Length: 7 to 9 mm. ………..……………………………………..*Meridialaris laminata* (Ulmer, 1920) (Argentina, Chile). Syn: *Deleatidium illapeli* Demoulin, 1955; *Deleatidium laminatum* Ulmer, 1920.

- The abdominal gills are slender with only a few weakly developed branches of the main tracheal trunk (**Fig. 2.434**). The abdomen is mainly light in color with dark lateral markings covering barely half of the distance from the anterior to the posterior margin. Length: 6 to 8.5 mm. ………..…………………………………...*Meridialaris diguillina* (Demoulin, 1955) (Argentina, Chile). Syn: *Deleatidium diguillinum* Demoulin, 1955.

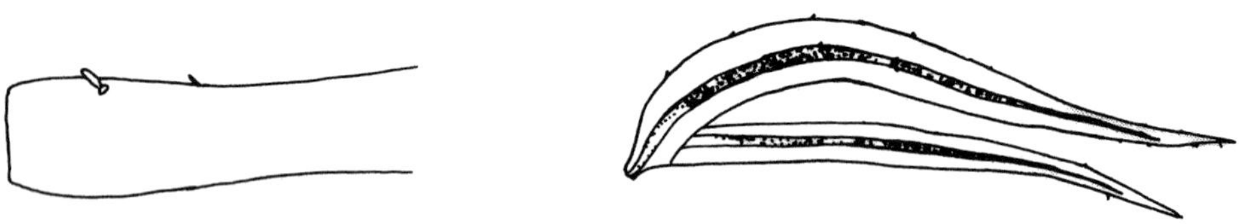

Fig. 2.434 *Meridialaris diguillina* larva: spines on the tarsus (left) and a gill from the fourth abdominal segment (right).

Key to the Neotropical species of adults in the genus *Ulmeritus*

The information for the key is taken from the descriptions of Traver (1956, 1959b), Thew (1960), Domínguez *et al.* (1992), and Domínguez (1991, 1995). In recent literature, the subgenera *Ulmeritus* Traver, 1956, and *Ulmeritoides* Traver, 1959, have often been treated as separate genera (Domínguez, 1991;

Domínguez *et al.* 1992). From the minimal morphological differences described by the authors, it seems scarcely justified to regard these even as valid subgenera, and there are no clear and consistent distinctions between all adults presently assigned the two genera. Although the known larvae can be consistently distinguished, the larvae of all species have not yet been described. Even the known larvae, however, show great similarity, as confirmed by the cladistic analysis of Flowers and Domínguez (1991). If, however, the two, and perhaps *Pseudulmeritus,* are to be treated as distinct genera, it is easiest to identify the species according to the key below and then refer it to the appropriate generic name, here designated as subgenus in parenthesis.

1. Both long and cross veins of the subadult are margined on all sides by brown. Each forceps limb has a rounded basal process on the inner side (**Fig. 2.435**). Length of body: 10.7 to 11 mm; of fore-wing: 10.9 to 11.4 mm. Cerci: 7 mm. Color: description of subimago - Pronotum yellow white with an oblique black-brown line on each side of the median line from the posterolateral corners to the median line. Lateral margins broadly blackish brown. The abdomen is light brown with yellow markings. The adult has not been described.
....*Ulmeritus (Ulmeritoides) patagiatus* Thew, 1960 (Santa Catarina).
- Only the cross veins might be margined by dark markings (**Fig. 2.436**). 2

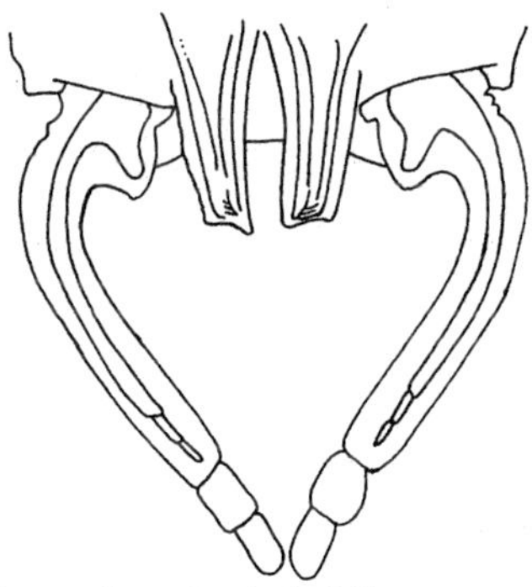

Fig. 2.435 The genitalia of a male subadult *Ulmeritus (Ulmeritoides) patagiatus.*

2. Vein Sc in the hind wing is about 0.9 times as long as the entire wing. The tip of each penis lobe is blunted rounded and bears ventral digitiform projection (**Fig. 2.436**).
....Subgenus *Ulmeritus*....................3
- Vein Sc in the hind wing is only about 0.6 times as long as the entire wing. The bases of both wings are clouded. The tip of each penis lobe is acutely pointed and bears one or more spines (**Fig. 2.437**). ....Subgenus *Ulmeritoides*....6
3. There is a small blackish brown spot on the wing membrane surrounding nearly every cross vein in the fore and hind wings (**Fig. 2.436**). 4
- The wings are not heavily spotted with numerous dark spots (**Fig. 2.438**). 5

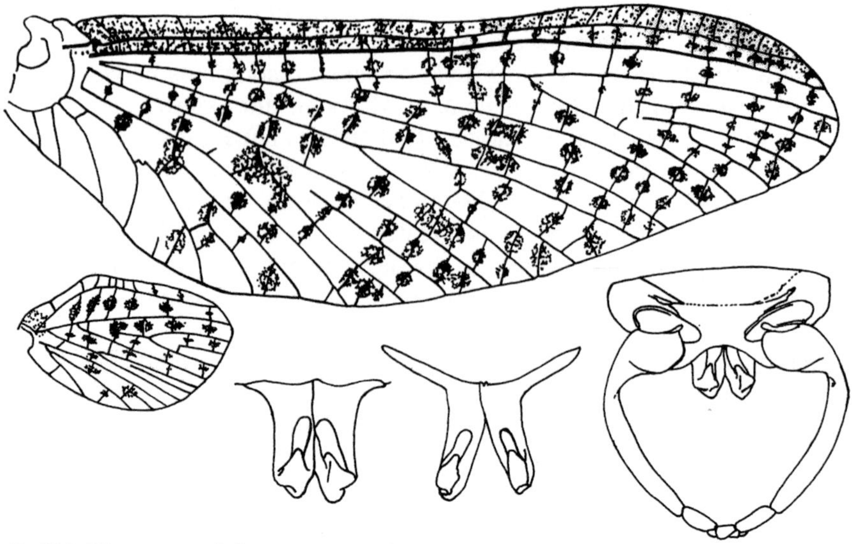

Fig. 2.436 *Ulmeritus (Ulmeritus) carbonelli:* the fore and hind wing (above and left), the male genitalia (lower right), and enlargements of the penis lobes of two different specimens (lower center).

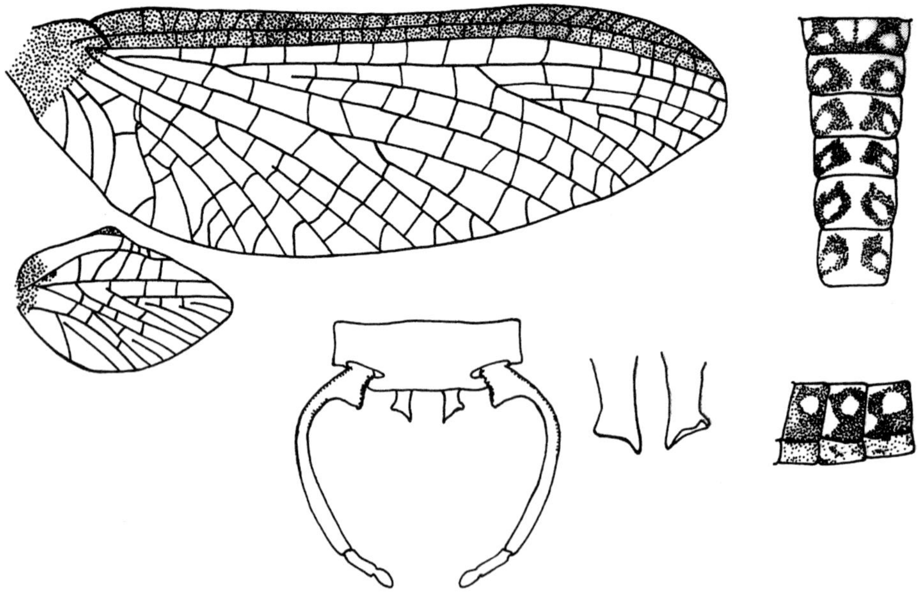

Fig. 2.437 *Ulmeritus (Ulmeritoides) misionensis* male sub-adult: fore and hind wings (left) with the hind wing enlarged slightly more than double normal size relative to the fore-wing, the pattern on the second through seventh abdominal segments in dorsal view (upper right), the fifth through seventh segments in lateral view (lower right), and the apex of the abdomen with genitalia in ventral view together with an enlargement of the penis lobes (below center).

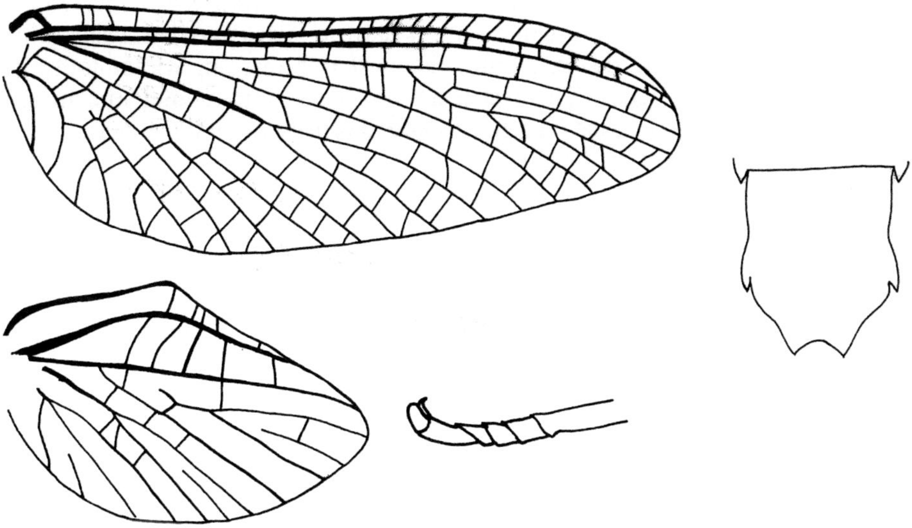

Fig. 2.438 *Ulmeritus haarupi* female: the fore and hind wings with the hind wing much larger than scale (left), the tarsal segments of the middle leg with the claw, and the ninth sternite.

4. The entire area of the hind wings is spotted (**Fig. 2.436**). Length of body: 8.5 mm; of fore-wing: 9 mm. Color: light reddish brown with a yellowish tinge to the thorax.
....*Ulmeritus (Ulmeritus) carbonelli* Traver, 1956 (Argentina, Uruguay). Most of the difference between this and the next species described by Traver (1956) are relative and rather slight, subjecting the validity of this species to doubt.
- The spots on the hind wing are confined to a single row around the cross veins in the subcostal space (**Fig. 2.317**). Length: 7 to 8.5 mm. Fore-wing length: 8 to 9.5 mm.
....*Ulmeritus (Ulmeritus) saopaulensis* (Traver, 1946) (São Paulo, Minas Gerais). Syn: *Atalophlebioides sao-paulense* Traver, 1946.
5. The body of the female is 8.5 mm long. The fore-wing length is 10.5 mm. The pronotum is yellowish with a reddish brown lateral tinge and five prominent black longitudinal streaks. There are two small lateral processes on the ninth sternite (**Fig. 2.438**). The male has not been described.
..................................*Ulmeritus (Ulmeritus) haarupi* (Esben-Petersen, 1912) (Uruguay, Argentina). Syn: *Thraulus haarupi* Esben-Petersen, 1912; *Deleatidium haarupi* (Esben-Petersen, 1912); *Atalophlebioides haarupi* (Esben-Petersen, 1912).

- The bodies of the male and female are about 10 mm long. The fore-wing is 10.5 mm. Pronotum yellow white with a narrow, black median stripe which forks like a Y a short distance from the posterior margin, which joins that borders the lateral edges. There is a narrow black longitudinal line in each lateral area.. There are spine-like posterolateral processes on the apical segment of the male abdomen (**Fig. 2.439**).
....*Ulmeritus (Ulmeritus) balteatus* Thew, 1960 (Uruguay, Santa Catarina).

6. The costal and subcostal areas of the fore-wing have a brownish tinge (**Fig. 2.437**). 7
- The fore-wings are hyaline in the costal and subcostal areas (**Fig. 2.440**). 9

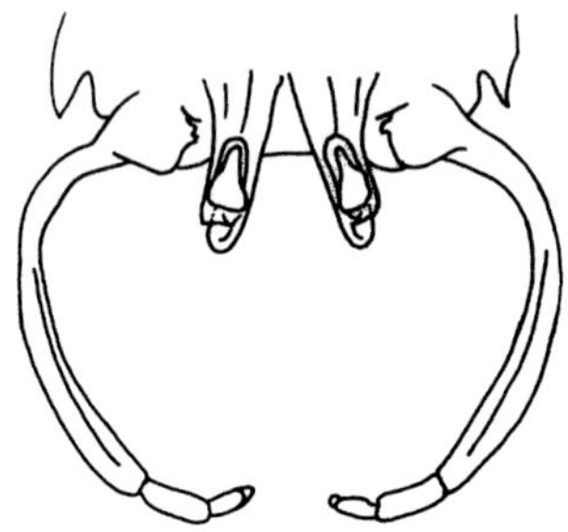

Fig. 2.439 The male genitalia of *Ulmeritus balteatus.*

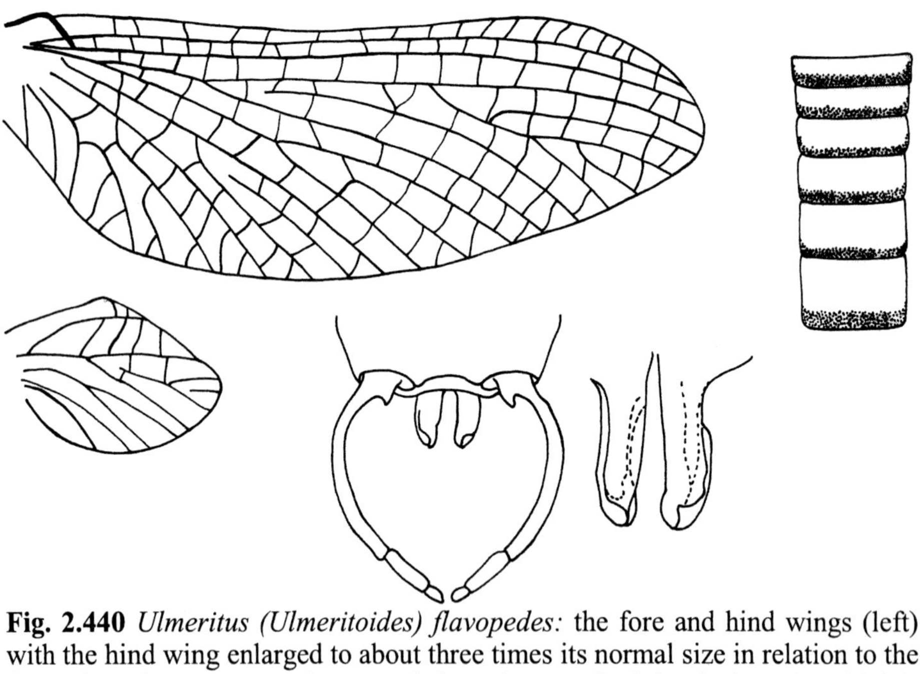

Fig. 2.440 *Ulmeritus (Ulmeritoides) flavopedes:* the fore and hind wings (left) with the hind wing enlarged to about three times its normal size in relation to the fore-wing, the pattern on the second through seventh abdominal tergites (right), the male genitalia in ventral view with an enlargement of the penis lobes (lower center).

7. The penis lobes have acute apical projections. The adult male and female sub-imago have yellowish brown tergites with black posterolateral markings, each with a light circular area near the center (**Fig. 2.437**). Body length of male: 7.9 to 8.2 mm, and of female sub-imago: 7.0 to 8.0 mm; fore-wing length of male: 9.2 to 9.6 mm, and of female sub-imago: 11.0 to 11.5 mm; hind wing length of male: 1.8 to 2.0 mm, and of female sub-imago: 2.0 to 2.2 mm.
............................*Ulmeritus (Ulmeritoides) misionensis* Domínguez, 1995 (Argentina).
- The apices of the penis lobes are rounded. The abdomen does not have black posterolateral markings with light circular areas near the center (**Fig. 2.441**). ...8

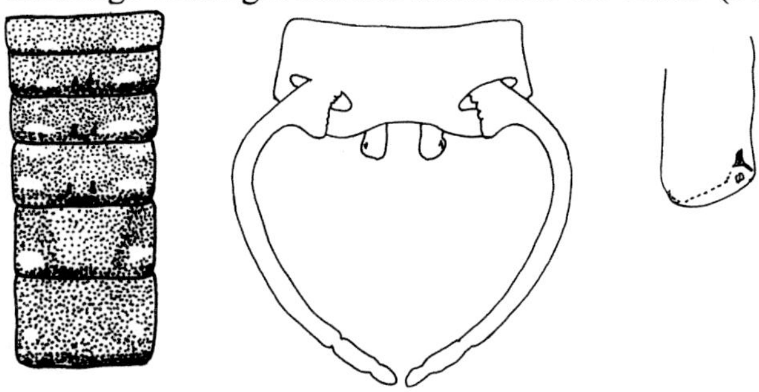

Fig. 2.441 *Ulmeritus fidalgoi* male: the pattern on the second through seventh abdominal tergites, the genitalia in ventral view (center), and an enlargement of the apex of one penis lobe (right).

8. The longitudinal veins of the fore-wing are brown, and the cross veins are blackish. There is a small, shallow prominence on the outer margin near the apex of the penis lobe. Body length of male: 6.8 to 7.2 mm, fore-wing length of male: 8.0 to 8.2 mm, hind wing length of male: 1.9 to 2.0 mm. The adult female has not been described.
..................................*Ulmeritus (Ulmeritoides) tifferae* Domínguez, 1995 (Known only from Costa Rica).
- Veins C, Sc, and R_1 in the fore-wing are brown. The other longitudinal veins are yellowish, and the cross veins are whitish. The penis lobes each have a small spine on at the outer corner of the ventral surface (**Fig. 2.441**). Body length of male: 6.5 to 7.2 mm, and of female: 7.3 to 7.8 mm; fore-wing length of male: 8.0 to 8.5 mm, and of female: 10.5 to 11.0 mm; hind wing length of male: 1.6 to 1.9 mm, and of female: 1.9 to 2.2 mm.
..................................*Ulmeritus (Ulmeritoides) fidalgoi* Domínguez, 1995 (Argentina).
9. The female has hyaline fore-wings but orange brown veins that are most strongly pigmented along the anterior margin. The abdomen is yellowish with a brownish tinge dorsally and ventrally, and the posterior margins of the segments are narrowly darker. The male imago has not been described, making its

comparison with the other species of this subgenus tentative. Length of body: 6.5 mm; of fore-wing: 8 mm.
......................................*Ulmeritus (Ulmeritoides) luteotinctus* Traver, 1959 (Argentina, Uruguay).
- The veins C, Sc, and R_1 in the fore-wing are yellow, except sometimes in the area of the stigma, where they are blackish. The tergites are orange brown dorsally, or they are brownish yellow with a distinct dark pattern.
....10
10. Penis lobes are rounded (**Fig. 2.440**). 11
- The penis lobes are not rounded (**Fig. 2.442**). 12

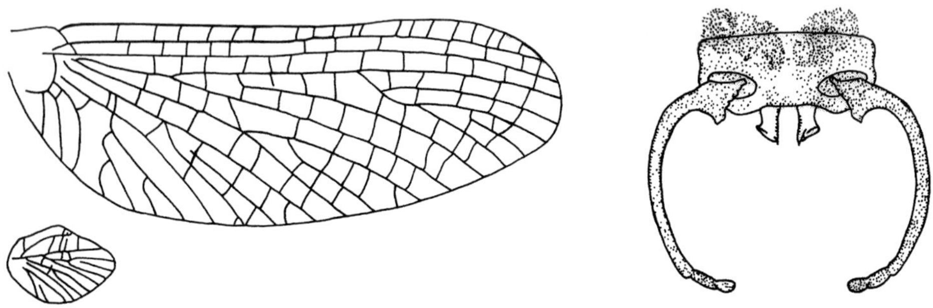

Fig. 2.442 *Ulmeritus (Ulmeritoides) uruguayensis:* the fore and hind wings (left) and the male genitalia (right).

11. Veins C, Sc, and R_1 in the fore-wing are entirely yellowish. Each penis lobe has a lateral groove and ends in thick, spine-like processes that curve perpendicular to the body axis (**Fig. 2.440**). Length of male body: 6.6 to 6.7 mm, fore-wing: 8.0 to 8.1 mm, hind wing: 1.5 to 1.6 mm. The female has not been described.
...................................*Ulmeritus (Ulmeritoides) flavopedes* (Spieth, 1943) (Surinam). Syn: *Thraulodes flavopedes* Spieth, 1943; *Atalophlebioides flavopedes* (Spieth, 1943); *Ulmeritus flavopedes* (Spieth, 1943).
- Veins C, Sc, and R_1 in the fore-wing are yellowish becoming blackish in the area of the stigma. There is a small prominence on the outer margin of each penis lobe. Length of male body: 6.6 to 6.7 mm, fore-wing: 8.0 to 8.1 mm, hind wing: 1.5 to 1.6 mm. The female has not been described.
.............................*Ulmeritus (Ulmeritoides) guanacaste* Domínguez, 1995 (Known only from Costa Rica).

12. The apical margin of each penis lobe ends in a small acute prominence (**Fig. 2.442**). Length: 5.4 to 8 mm. Fore-wing length: 5.9 to 8.5 mm.
.... *Ulmeritus (Ulmeritoides) uruguayensis* Traver, 1959 (Uruguay, Santa Catarina). Syn. *Ulmeritus (Ulmeritoides) adustus* Thew, 1960.
- The lobes of the penis bear several small spines on the apical margin (**Fig. 2.443**). Length of male body: 7.0 to 7.3 mm, fore-wing: 7.2 to 7.5 mm, hind wing: 1.4 to 1.5 mm. The female has not been described.
.... *Ulmeritus (Ulmeritoides) spinulipenis* Domínguez, 1995 (Argentina, Uruguay).

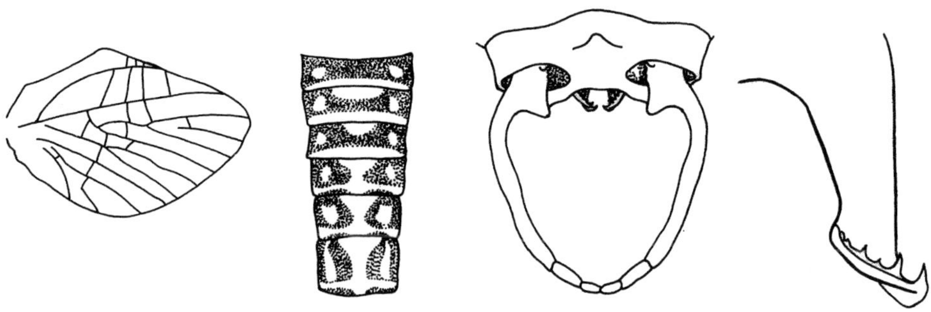

Fig. 2.443 *Ulmeritus (Ulmeritoides) spinulipenis* adult male (left to right): hind wing, second through seventh abdominal tergites, genitalia in ventral view, and an enlarged penis lobe in ventral view.

Key to the species of known larvae in the genus *Ulmeritus* in South America

Information for the key was provided by Domínguez (1991, 1995b). The larvae of *Ulmeritus (Ulmeritus) haarupi, U. (Ulmeritus) saopaulensis, U. (Ulmeritoides) patagiatus, U. (Ulmeritoides) fidalgoi, U. (Ulmeritoides) uruguayensis, U. (Ulmeritoides) luteotinctus, U. (Ulmeritoides) flavopedes*, and *U. (Ulmeritoides) spinulipenis* have not yet been described.

1. The row of long setae on the paraglossae are situated apically (**Fig. 2.339**). ..2
- The row of long setae on the paraglossae are situated medially (**Fig. 2.340**). ..3
2. The gills are grayish brown.
.... *Ulmeritus (Ulmeritus) balteatus* Thew, 1960 (Uruguay, Santa Catarina). This feature has not been confirmed because the identity of the larvae was determined only by the presence of many adults in the same stream (Domínguez, 1991).
- The gills are gray. The posterolateral projections on the ninth abdominal segment are sigmoid on their inner surfaces (**Fig. 2.339**).
.... *Ulmeritus (Ulmeritus) carbonelli* Traver, 1956 (Argentina, Uruguay).

3. There are two distinct black bands on the fore-tibia. On the anteromedian emargination of the labrum, the middle denticle is much larger than the other four (**Fig. 2.340**). Length in late stage: 7.0 to 8.8 mm.
.... *Ulmeritus (Ulmeritoides) misionensis* Domínguez, 1995 (Argentina).
- The fore-tibia lacks bands of dark pigment, but it may be almost entirely covered with blackish pigment, except at the base and apex. On the anteromedian emargination of the labrum, the middle denticle is not much larger than the others. 4
4. There are numerous short, pointed spines on the dorsal surfaces of the middle and hind femora and a black spot near the center of the middle femur. The fore-tibia is light with a blackish inner margin. The abdominal tergites are bright yellowish orange with distinct black markings. Length in late stage: 7.0 to 8.0 mm.
.... *Ulmeritus (Ulmeritoides) tifferae* Domínguez, 1995 (Known only from Costa Rica).
- There are a few short, blunt spines on the dorsal surfaces of the middle and hind femora and neither has a black spot near the center. The fore-tibia is almost entirely blackish, except at the base and apex. The abdominal tergites are orange brown with black pigment scattered on the background color, especially near the posterior margin, but without any distinct markings. Length in late stage: 5.1 to 5.2 mm.
.... *Ulmeritus (Ulmeritoides) guanacaste* Domínguez, 1995 (Known only from Costa Rica).

Key to the adult *Thraulodes* species in South America

The information for the key is taken from Navás (1934b, 1936b), Traver and Edmunds (1967), Domínguez (1986, 1987), Ferreira and Froehlich (1992), and Chacón *et al.* (1999). The characteristics reported are mainly those of the males. The larvae of the South American species have not been identified.

1. The general color of the thorax and abdomen is dark green, and the head is reddish yellow. The wings are hyaline but iridescent. The legs are light brown legs with femora darkened medially and apically. The general color of the thorax and abdomen is not dark green. The description of this species is very sketchy, and it may not be possible to identify it with certainty unless comparisons can be made with a sufficiently preserved type specimen.
.... *Thraulodes vitripennis* (Blanchard in Gay, 1851) (Chile). Syn: *Ephemera vitripennis* Blanchard in Gay, 1851; *Cloeon vitripennis* (Blanchard in Gay, 1851).
- The general color of the thorax and abdomen is not dark green. 2
2. There is no trace of a costal prominence along the anterior margin of the hind wing (**Fig. 2.444**). 3

- There is a well-developed costal prominence along the anterior margin of the hind wing (**Fig. 2.445**). 4

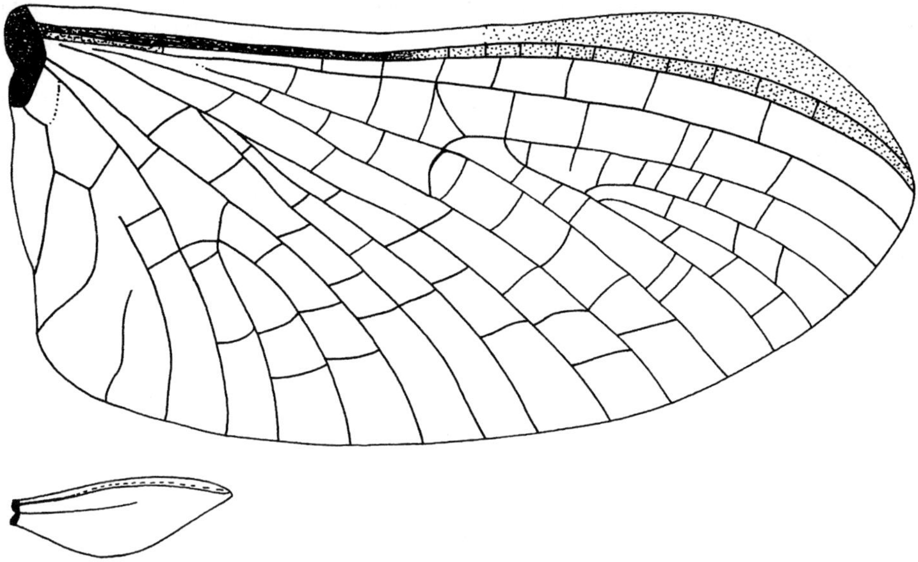

Fig. 2.444 The fore and hind wing of *Thraulodes vogleri.*

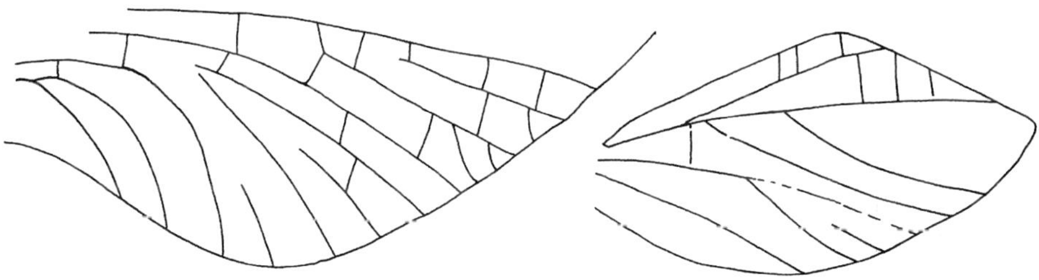

Fig. 2.445 *Thraulodes subfasciatus:* the veins of the inner posterior angle of the fore-wing (left) and the hind wing (right).

3. The fore-wing is entirely darkened and cloudy but lacks any spots. The head is brownish red, and the thorax is sepia. The abdomen is mainly dark gray. Length: 3 mm.

....*Thraulodes siewertii* (Weyenbergh, 1883) (Argentina). Syn: *Cloe siewertii* Weyenbergh, 1883; *Potamanthus siewertii* (Weyenbergh, 1883). That this species belongs to the genus *Thraulodes* is doubtful. It has appeared in recent lists as *Thraulus siewertii,* but *Thraulus* is now thought not to occur in South America. It is therefore usually treated as a *nomen dubium*.

- The apical half of the costal space and the entire subcostal space of the fore-wing are darkened, while the rest of the wing is clear. Vein Sc in the hind wing fades out beyond its short basal root (**Fig. 2.444**). Length: 3.5 mm. Color: head and thorax shiny black, and the abdomen of the male is dark. That of the female is somewhat lighter.
....*Thraulodes vogleri* (Weyenbergh, 1883). (Argentina). Syn: *Cloe vogleri* Weyenbergh, 1883; *Potamanthus vogleri* (Weyenbergh, 1883); *Thraulus vogleri* (Weyenbergh, 1883). That this species belongs to the genus *Thraulodes* is doubtful. It has appeared in recent lists as *Thraulus vogleri,* but *Thraulus* is now thought not to occur in South America. It is therefore usually treated as a *nomen dubium.*
4. The costal angulation of the hind wing is located in the apical half of the wing, about 3/7 to 1/3 of the wing length from the apex (**Fig. 2.445**). 5
- The costal angulation is located at about the mid-length of the hind wing (**Fig. 2.446**). In case of doubt, the size or color patterns differ from the descriptions in couplet 3. 6

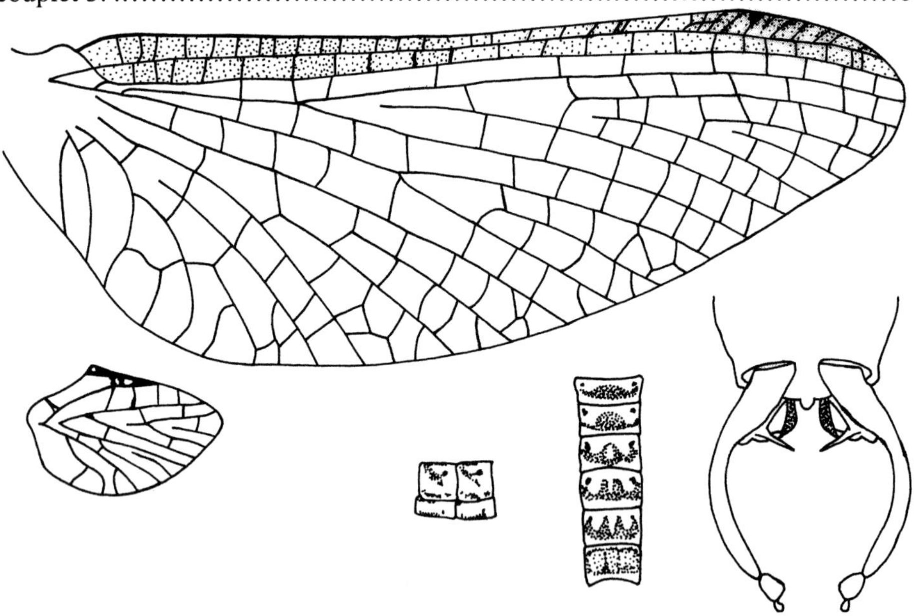

Fig. 2.446 *Thraulodes liminaris:* the fore and hind wings (left), the fifth and sixth abdominal segments in lateral view (lower left center), the second through seventh abdominal tergites in dorsal view (lower right center), and the male genitalia (lower right).

5. The costal angulation of the hind wing is located about 1/3 of the wing length from the apex. Only the female has been described, but the male is likely to share this feature. There are no bands on the femora. Only one or two faint cross veins cross the costal space in the proximal half of the fore-wing. The

abdominal tergites are light reddish yellow with the posterior margins blackened on the first eight. Fore-wing length of female: 15 mm. Color: mainly pale reddish brown on the thorax.
....*Thraulodes colombiae* (Walker, 1853) (Colombia). Syn: *Ephemera colombiae* Walker, 1853; *Leptophlebia colombiae* (Walker, 1853); *Adenophlebia colombiae* (Walker, 1853); *Thraulus colombiae* (Walker, 1853). Only one other known species of this genus, *Thraulodes ephippiatus* Traver and Edmunds, 1967, from Mexico, has a similarly located costal angulation, but the fore-wing of this species is only about half as long. This species has long been carried on species lists for South America, but Eaton (1883-88) listed the species for British Columbia.

- The costal angulation is located about 3/7 of the wing length from the apex (**Fig. 2.445**). The tergites are dark yellowish with dark spots in the basal third. Body length of male: 8.9 mm; female: 8.2 mm. Fore-wing length of male: 11.5 mm; female: 10.1 mm. Hind wing length of male and female: 2.2 to 2.3 mm. Color: head gray with black eyes; thorax light yellow with dark stripes; wings hyaline with somewhat darkened veins.
....*Thraulodes subfasciatus* Navás, 1934 (Rio de Janeiro).

6. The abdomen is a fairly uniform chestnut brown or brownish orange, sometimes with a faint pattern, and it does not have any whitish or translucent areas that contrast distinctly with the three apical segments (**Fig. 2.446**). 7

- The abdominal segments may be white, light yellow, or translucent with or without distinct spots or darker markings, or the segments are light and contrast distinctly with the three darker apical segments (**Fig. 2.447**). 9

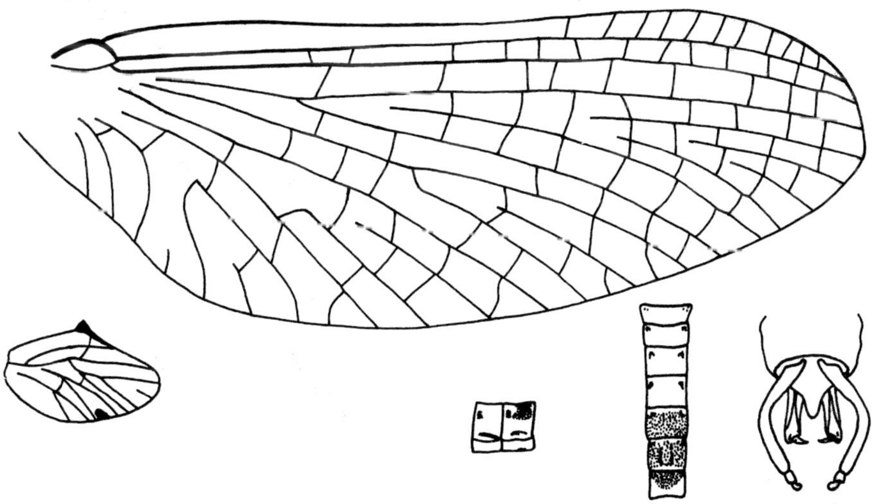

Fig. 2.447 *Thraulodes cochunaensis:* the fore and hind wings (left), the fifth and sixth abdominal segments in lateral view (lower left center), the second through seventh abdominal tergites in dorsal view (lower right center), and the male genitalia (lower right).

7. The wings are pale whitish or hyaline with yellow or colorless veins, except for the brownish bases of the Sc and R_1 and the base of the humeral vein in the fore-wing. Each femur has only one band. The joints are darkened only at the base of the caudal filaments. There is no dorsal extension on the tenth tergite (**Fig. 2.448**). Length: 6 to 6.5 mm; fore-wing length: 6 to 6.5 mm. Color: mainly dark brown on the head and thorax with lighter and darker markings. The abdomen is patterned mainly in pale and dark brown.
....*Thraulodes regulus* Traver and Edmunds, 1967 (Peru).
- There are brown markings or clouds in the costal and subcostal areas of the fore-wings (**Fig. 2.446**). 8

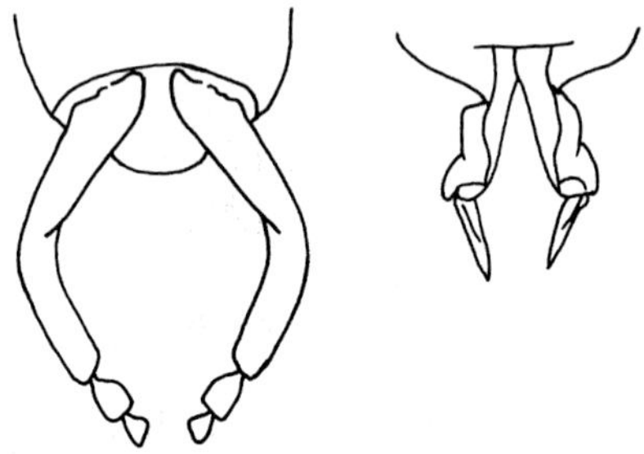

Fig. 2.448 The male genitalia of *Thraulodes regulus* with the forceps base and segments (left) separated from the penis (right).

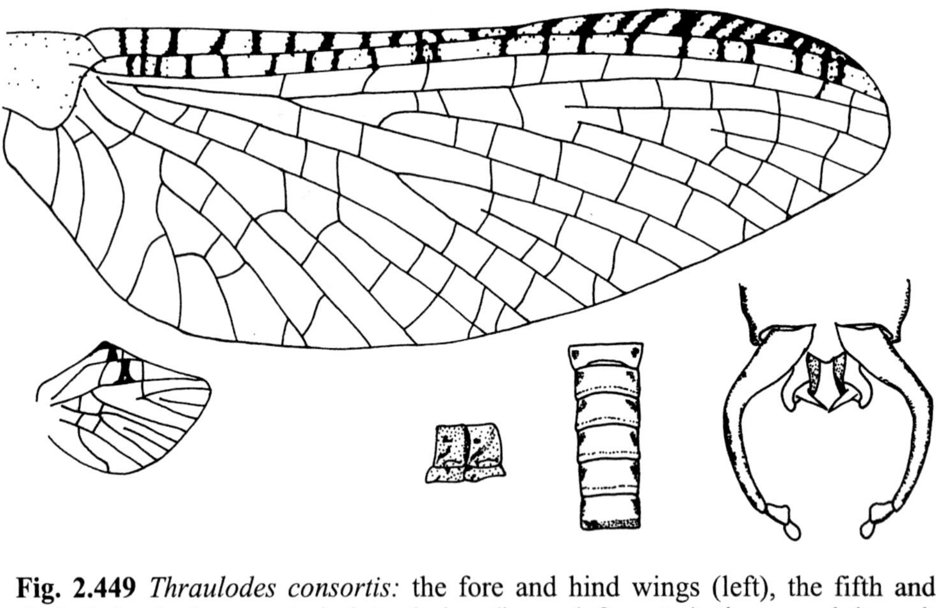

Fig. 2.449 *Thraulodes consortis:* the fore and hind wings (left), the fifth and sixth abdominal segments in lateral view (lower left center), the second through seventh abdominal tergites in dorsal view (lower right center), and the male genitalia (lower right).

8. The base of ICu_2 in the fore-wing arises from vein CuP. The spine on the penis is long and thin (**Fig. 2.446**). Length: 10.5 to 11.5 mm; fore-wing length: 12 to 13 mm; hind wing length: 2 to 2.5 mm.
....*Thraulodes liminaris* Domínguez, 1987 (Argentina).
- The base of ICu2 in the fore-wing does not arise from vein CuP. The spine on the penis is thick and strong (**Fig. 2.449**). Body length: 9.5 to 11 mm; fore-wing length: 10 to 12 mm; hind wing length: 1.9 to 2.2 mm.

....*Thraulodes consortis* Domínguez, 1987 (Argentina).
9. The abdominal segments are hazel brown on both the dorsal and ventral surfaces with the third through sixth segments translucent. The six distal abdominal tergites are lined with black along the posterior margins. There is a large black marking with a purplish tinge on each side of the seventh abdominal tergite. The second through sixth abdominal segments have very dark brown stigmatic spots, and the eighth through tenth are dark brown, somewhat lighter on the sternites. There are two bands on each femur. The wings are hyaline with brownish membranes in the costal and subcostal areas, light brown veins, blackish bases of C, Sc, and R in the fore-wing and a blackish humeral vein at the base of the hind wing. The penis lobes curve outward at the apex, and each bears a thick spine pointed (**Fig. 2.450**). Fore-wing length of male: 12 mm. The female has not been described.
....*Thraulodes venezuelana* Ulmer, 1943 (Venezuela).
- The third through fifth abdominal segments are yellowish or whitish, sometimes translucent with dark spots or other markings (**Fig. 2.447**). 10

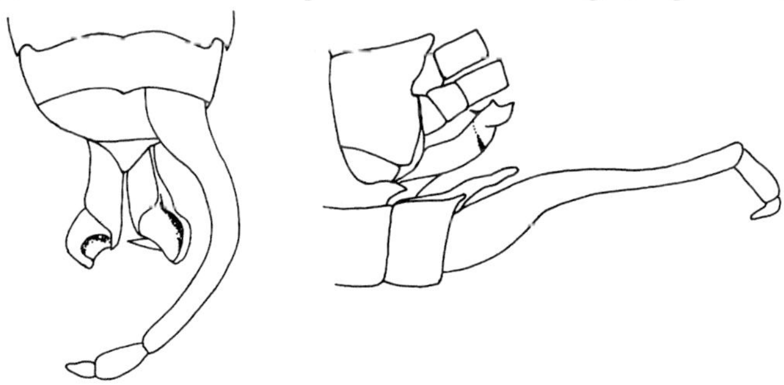

Fig. 2.450 The genitalia of a male *Thraulodes venezuelana* in ventral (left) and lateral view (right).

10. The second through fifth abdominal segments are translucent white with a reddish tinge in the central part. The sixth through tenth tergites are reddish and not translucent. Length: 4.9 to 5.6 mm. Fore-wing length: 5.1 to 6.0 mm. Hind

wing length: 1.0 to 1.2 mm. Color: mainly light brown with reddish markings on the abdomen. The female has not been described.
......................*Thraulodes marreroi* Chacón, Segnini, and Domínguez, 1999 (Venezuela).
- The second abdominal segment is not translucent, or there are no reddish markings on the dorsal surface of the abdomen. 11
11. The second through sixth abdominal segments are mainly translucent yellow without red markings on the dorsal side (**Fig. 2.451**). 12
- Either the second or the sixth abdominal segments or both are not translucent (**Fig. 2.452**). 14

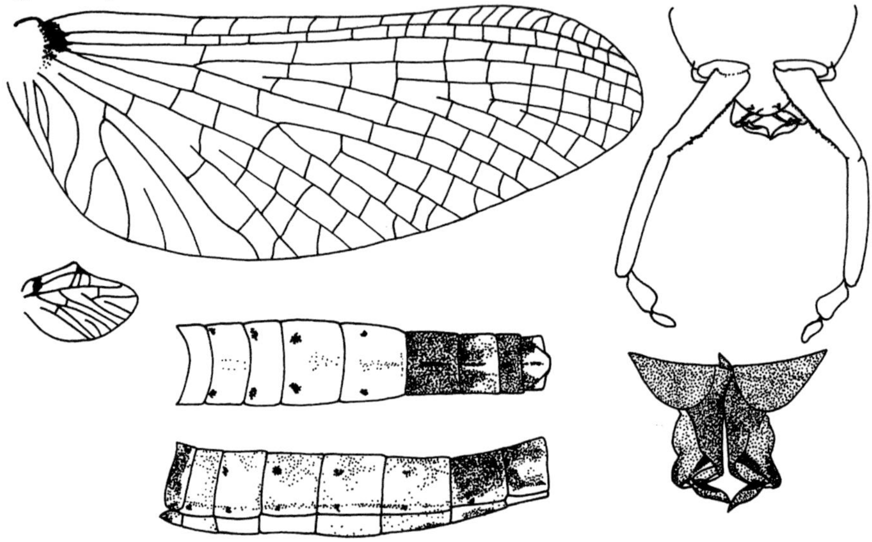

Fig. 2.451 *Thraulodes mucuy:* fore and hind wing (left), the male genitalia in ventral view (upper right) with and enlargement of the penis (lower right), the pattern on the second through tenth abdominal tergites (center), and the first through eighth abdominal segments in lateral view (lower center).

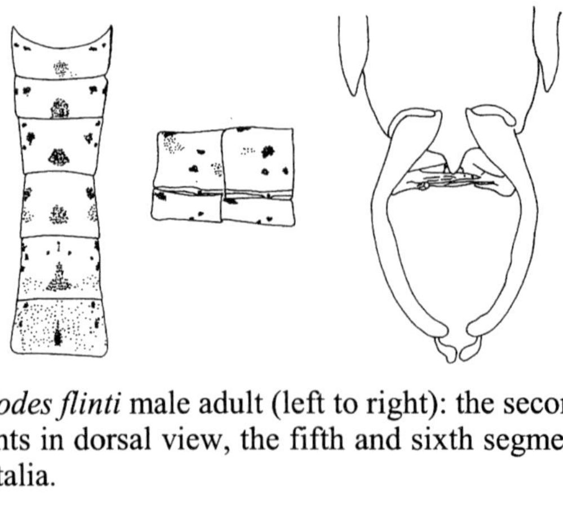

Fig. 2.452 *Thraulodes flinti* male adult (left to right): the second through seventh abdominal segments in dorsal view, the fifth and sixth segments in lateral view, and the male genitalia.

12. The second through sixth tergites are translucent yellowish with small dark median patches along the anterior margins, one dark spot on each side between the lateral margin and the middle about a third of the way from the anterior margin to the posterior border of the tergite, and another dark spot at the same level at the lateral margin (**Fig. 2.319**). Body length of male: 10.8 to 11 mm; female: 9.5 to 10 mm. Fore-wing length of male: 12.5 to 13 mm; female: 13.5 to 14 mm. Hind wing length of male and female: 2.5 to 2.6 mm.
.... *Thraulodes bolivianus* Domínguez, 1986 (Bolivia).

- There are no dark patches along the anterior margins of the second through sixth abdominal segments (**Fig. 2.451**). 13

13. The fore-wing has yellowish long veins that become darker toward the wing tip and brown cross veins (**Fig. 2.451**). Length: 10.3 to 10.8 mm. Fore-wing length: 12.4 to 12.8 mm. Hind wing length: 2.2 to 2.3 mm. Color: yellowish brown with orange brown and blackish markings. The female has not been described.
.... *Thraulodes mucuy* Chacón, Segnini, and Domínguez, 1999 (Venezuela).

- Both the long and cross veins in the fore-wing are yellowish white (**Fig. 2.453**). Length: 6.5 to 7 mm. Fore-wing length: 7 to 7.5 mm. Hind wing length of male: 1.5 to 1.8 mm; of female: 1.1 to 1.3 mm. Color: yellowish brown with darker brown, orange, or blackish markings.
.... *Thraulodes guanare* Chacón, Segnini, and Domínguez, 1999 (Venezuela).

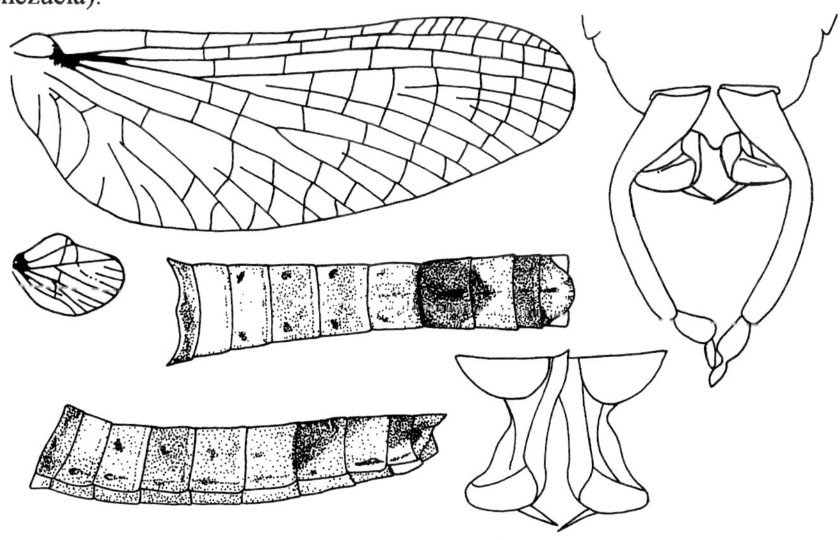

Fig. 2.453 *Thraulodes guanare:* fore and hind wing (upper left), the male genitalia in ventral view (right) with an enlargement of the penis (lower right center), the pattern on the first through tenth abdominal tergites (center), and the first through tenth abdominal segments in lateral view (lower center).

14. At least the third through sixth abdominal segments are pale and translucent with a small brown mark near the middle. On the abdominal sternites, there are two dark spots on each side of the midline and a larger spot in the posterolateral corner (**Fig. 2.452**). There are four fine cross veins through the costal space in the proximal half of the wing. There is a dark band and a dark spot on the fore-femur. Length: 8 to 8.2 mm. Fore-wing length: 8.5 to 8.7 mm. Hind wing length: 1.5 to 1.6 mm. Color: orange brown with a grayish head and dark markings. The female has not been described.
....*Thraulodes flinti* Domínguez, 1987 (Argentina).
- The sixth abdominal segment is not pale and translucent or the dark markings on the abdomen are not confined to a dark spot near the middle of each tergites and dark spots on the sternites that do not form a curved row (**Fig. 2.454**). 15
15. There is a curved row of small dark dots, usually three on each side, on the abdominal sternites. Midway, isolated spots and dorsal markings are present on the tergites. Segments 2 to 6 are transparent whitish or yellowish (**Fig. 2.454**).
....16
- The markings on the abdomen do not include a curved row of small dark dots on the sternites (**Figs. 2.447**). 18

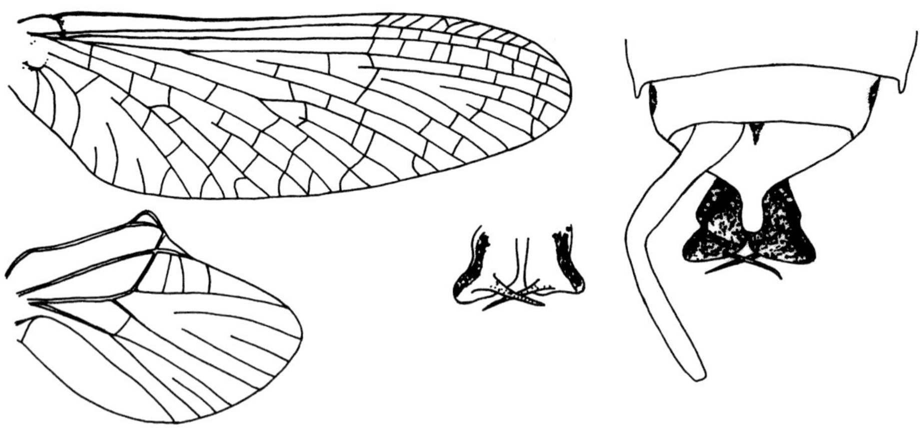

Fig. 2.454 *Thraulodes bomplandi:* fore and hind wings with the hind wing much larger than scale, the male genitalia in ventral view with one of the forceps limbs removed (right), and the penis is dorsal view.

16. The second through sixth abdominal segments are whitish. The second through seventh tergites have U-shaped, dark reddish-brown markings, which may be joined posteriorly, on each side of and parallel to the midline. Small midway spots are present. There are two small dark spots parallel to the pleural fold, and dark dash along the fold. The hind femur has two bands. Wing veins Sc, R_1, and R_2 have short thickened portions in the middle of the wing, and the

cross veins are very faint (**Fig. 2.454**). Length: 8.5 mm. Fore-wing length: 8 mm. Coloration: the thorax is light brown. The upper portion of the eyes is greenish blue, and the lower portion is black.

....*Thraulodes bomplandi* (Ebsen-Petersen, 1912) (Argentina). Syn: *Thraulus bomplandi* Ebsen-Petersen, 1912. Specimens identified as this species by Traver (1959b) actually belong to *Thraulodes daidaleus*.

- The markings on the tergites are different. The hind femur has only one band (**Fig. 2.455**). 17

17. Abdominal segments 2 to 6 pale yellowish. The second and fourth tergites have midway spots; and the fifth and sixth have a row of small dots near the anterior margin extending to dorsum. There is a dark dot below the midway points on the third through sixth tergites, and dark red dorsal triangles on the seventh; the eighth through tenth have rose-colored triangles and chalky white posterior angles. The costal prominence on the hind wing is broadly rounded (**Fig. 2.455**).

....*Thraulodes itatiajanus* Traver and Edmunds, 1967 (Rio de Janeiro).

- Abdominal segments 2 to 6 whitish. The second through sixth tergites each have a posterolateral median black spot, a midway black spot with a smaller black spot beside it; the seventh through tenth tergites are reddish with black median streak and lateral spot. The median process at the end of the tenth tergite is shorter than its width at its base (**Fig. 2.456**).

....*Thraulodes daidaleus* Thew, 1960 (Uruguay, Santa Catarina).

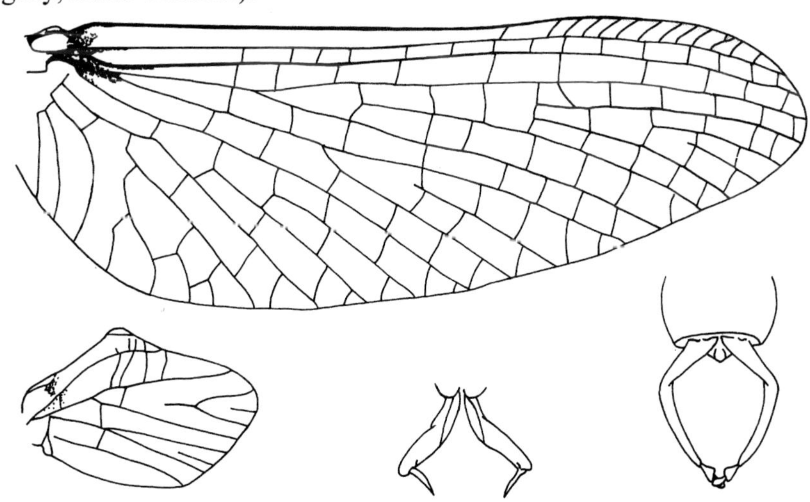

Fig. 2.455 *Thraulodes itatiajanus:* the fore and hind wings with the hind wing drawn larger than scale (left), the penis lobes of the male (lower center), and the apical segments of the male showing the forceps but without the penis lobes (lower right).

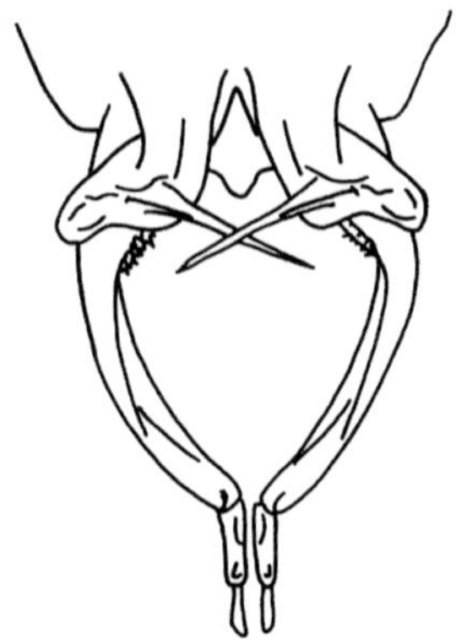

Fig. 2.456 The male genitalia of *Thraulodes daidaleus.*

18. There are dark dorsal or dorsolateral markings or both on at least some of the second through sixth abdominal tergites; if the spots are dorsolateral, then at least some of them join to form some dorsal markings (**Fig. 2.457**). 19
- There are no dark dorsal or dorsolateral markings or both on the second through sixth abdominal tergites, although small triangular spots that do not join to form dorsal spots may be present (**Fig. 2.447**). 22

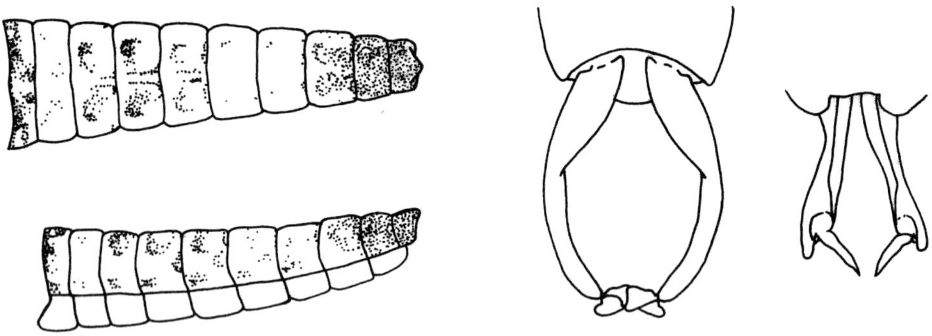

Fig. 2.457 *Thraulodes papilionis:* the abdomen of a male in dorsal (upper left) and lateral view (lower left) and the forceps base and segments of the male (center) separated from the penis (right).

19. Dorsal midway spots are not evident, and if they are faintly present, they are the same color as the dorsolateral markings. The second through sixth and the base of the seventh abdominal segments are transparent whitish with pale gray dorsolateral patches, which have a shape reminiscent of butterfly wings on the fourth through sixth tergites. There are dark stigmatic spots on the second through seventh abdominal segments. The cross wings in the fore-wing are pale

(**Fig. 2.457**). There is only one dark transverse band on each femur. Length: 6 mm. Fore-wing length: 6 mm. Color: head mainly blackish brown; thorax mainly pale yellow with black and reddish markings; abdomen mainly pale, darker at apex. Only the male adult has been described.
....*Thraulodes papilionis* Traver and Edmunds, 1967 (Colombia).

- Separate midway spots are evident, and if they are located within the dorsolateral spots, they are a different color (**Fig. 2.458**). 20

20. The cross veins in the fore-wing are dark. The only dark markings besides the dorsal and dorsolateral markings on the second through sixth tergites are purplish brown median triangles. There are two bands on each femur. The second segment of the forceps of the male is about three times as long as the third (**Fig. 2.458**). Fore-wing length: 14 mm. Color: thorax reddish brown with black markings. Legs yellow with black markings.
....*Thraulodes furficulus* Traver, 1946 (Guyana).

- The cross veins in the fore-wing are pale. There are dark spots scattered on the second through sixth abdominal tergites (**Fig. 2.459**). The fore-wings are about 8 mm long. 21

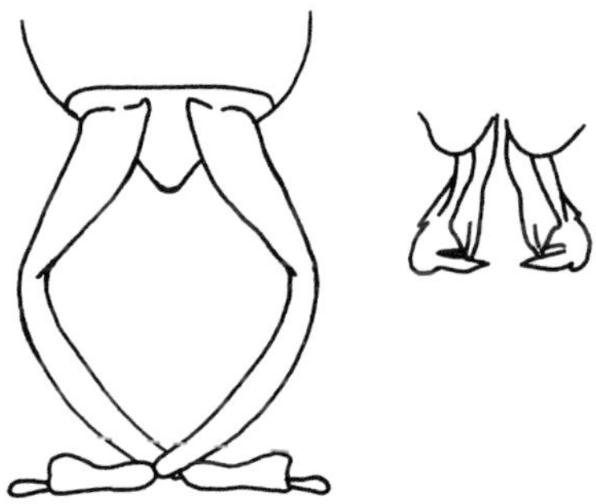

Fig. 2.458 The male genitalia of *Thraulodes furficulus* with the forceps base and segments (left) separated from the penis (right).

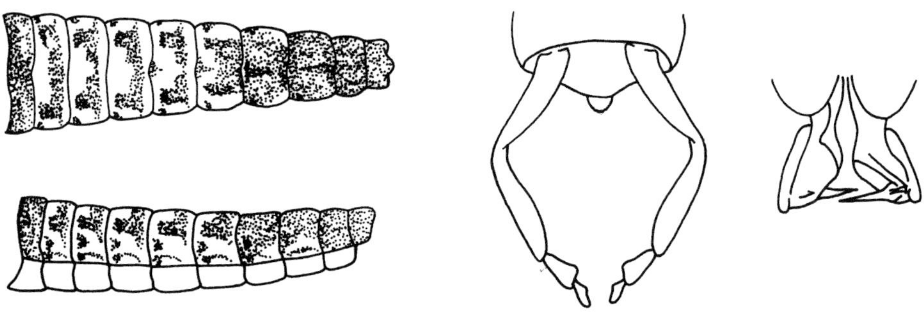

Fig. 2.459 *Thraulodes paysandensis:* the abdomen of a male in dorsal (upper left) and lateral view (lower left) and the forceps base and segments of the male (center) separated from the penis (right).

21. There are wide reddish brown bands along the posterior margins of the second through sixth abdominal tergites and reddish median triangles on the anterior margins of the fifth and sixth (**Fig. 2.459**). There is a curved dark line above the pleural fold.
....*Thraulodes paysandensis* Traver, 1964 (Argentina, Uruguay).
- The tergites have dorsal, submedian, midway, and isolated spots. Dark brown submedian spots are narrowly joined along the posterior margins on some of the second through sixth tergites, and no anterior triangles are present. There is no curved dark marking above the pleural fold. Each femur has two dark bands. The median process on the tenth tergite is considerably longer than its width at its base (**Fig. 2.460**).
....*Thraulodes traverae* Thew, 1960 (Uruguay, Santa Catarina).

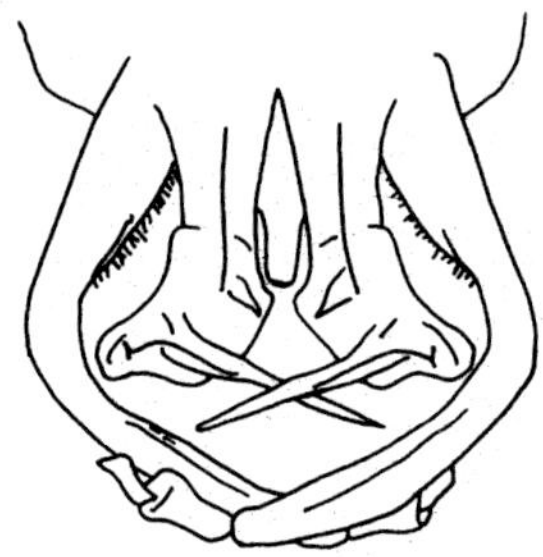

Fig. 2.460 The male genitalia of *Thraulodes traverae.*

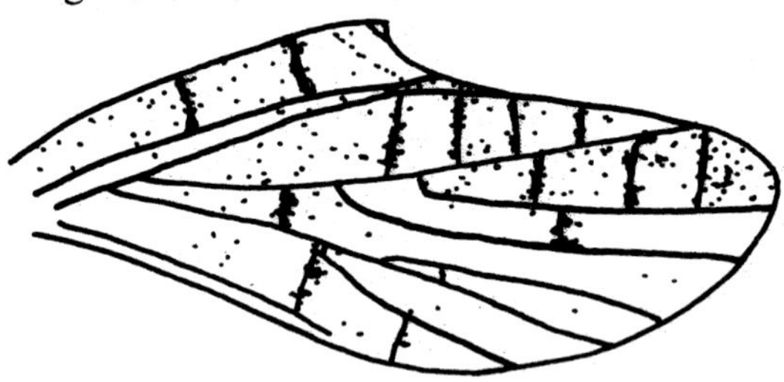

Fig. 2.461 The hind wing of *Thraulodes limbatus.*

22. The costal area of the hind wing has two prominent cross veins proximal to the costal prominence; the apical one is closer to the costal prominence than to the wing. The membrane of the fore and hind wings is darkened anteriorly and has some darkened reticulation (**Fig. 2.461**). The abdomen is yellowish with dark wash on the posterior half or each tergite. Length: 8 mm. Fore-wing length: 10.2 mm. Hind wing length: 2 mm. Color: yellow with dark markings.
....*Thraulodes limbatus* Navás, 1936 (Santa Catarina).

- If there are any cross veins in the costal zone proximal to the costal prominence, they are very faint. In case of doubt, the fore-wings are predominantly pale (**Fig. 2.462**). 23
23. In addition to the midway and stigmatic spots on the abdominal tergites, isolated dark spots are also present (**Fig. 2.462**). 24
- There are no isolated dark spots in addition to the midway and stigmatic spots on the abdominal tergites. There are no cross veins through the costal space in the basal half of the wings (**Fig. 2.447**). 25

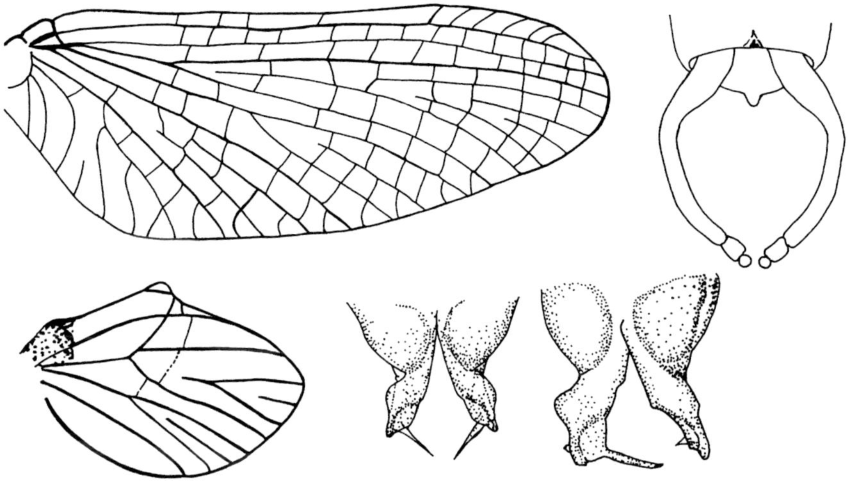

Fig. 2.462 *Thraulodes ulmeri:* fore and hind wing with the hind wing considerably enlarged (left), the foreceps of a male and its base in ventral view (upper right), and the penis in ventral and ventrolateral view (lower center).

24. The femora have only one band; they are fuscous on the proximal, dorsal and ventral edges and have a deep yellowish brown apical band. Tergites have midway spots, submedian patches, and stigmatic dots. The posterior margin of the sixth and the anterior margin of the seventh and eighth tergites are narrowly lined with purplish brown. The fore-wings are predominantly pale (**Fig. 2.462**). Length: 5.5 to 7 mm; fore-wing length: 6 to 8 mm; hind wing length: 1.2 to 1.3 mm. Color: mainly yellowish to yellowish brown with fuscous markings. ....*Thraulodes ulmeri* Edmunds, 1950 (Santa Catarina).
- Each femur has two bands. On each side of the second through sixth abdominal tergites, there are small triangular spots. The posterior margin of the eighth tergite is yellow, and the color forms a triangle that reaches the base of the seventh at its apex. Each penis lobe bears a subapical spine-like process directed toward the midline (**Fig. 2.463**). Fore-wing length: 8 mm. ....*Thraulodes laetus* (Eaton, 1883) (Argentina). Syn: *Calliarcys laetus* Eaton, 1883; *Thraulus laetus* (Eaton, 1883).

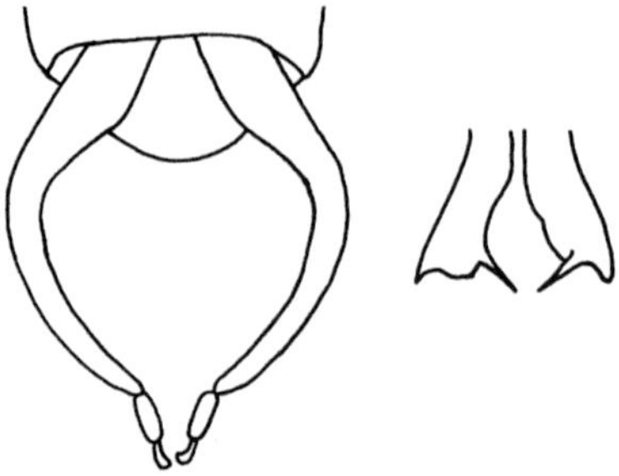

Fig. 2.463 The apex of the abdomen with the forceps (left) and the apical portion of the penis lobes (right) of a male *Thraulodes laetus.*

25. The apical portions of the penis lobes do not curve laterad or have laterally directed processes (**Fig. 2.447**). There is an incomplete dark band near the midlength of the fore-femur and a complete apical band. Length of male and female: 7 to 7.5 mm. Fore-wing length of male: 8.0 to 8.3 mm; of female: 8.9 to 9.3 mm. Hind wing length of male: 1.5 to 1.6 mm; of female: 1.2 to 1.4 mm. General color: brown with lighter and darker markings.
....*Thraulodes cochunaensis* Domínguez, 1987 (Argentina).
- The apical portions of the penis lobes curve laterad or bear laterally directed processes (**Fig. 2.464**). 26
26. The fore-wing of the male is at least 9.5 mm long, and that of the female is about 11 mm. Each femur has only one band. The second through sixth abdominal segments of the male are translucent without conspicuous dark markings dorsally but with dark spots or dashes laterally, while the seventh through tenth segments are rich chestnut brown bordered by white laterally. Beyond the yellowish bases, the cerci are irregularly banded with black. There is a broad medial band on the abdominal tergites of the female. The wings (**Fig. 2.464**) are hyaline with yellow veins and a curving brown spot running posteriad from the humeral cross vein. Body length: 8 to 10 mm; fore-wing length: 9.5 to 11.5 mm.
......................................*Thraulodes telegraphicus* Needham and Murphy, 1924 (Peru). Syn: *Thraulodes plicatus* Needham and Murphy, 1924.
- The fore-wing of the male is not more than about 8 mm long, and that of the female does not exceed 9 mm. 27
27. Each femur has two bands. The middle of the seventh abdominal tergite is dark brown, while its margins on all sides are pale. The eighth tergite has a dark median line and a basal band, while the ninth is darker and has a median stripe. The cross veins in the fore-wing are pale brown. The spears on the penis lobes are pointed at each other (**Fig. 2.465**). Fore-wing length: about 7.5 mm.
....*Thraulodes trijunctus* (Banks, 1918) (Peru). Syn: *Thraulus trijunctus* Banks, 1918.

- Each femur has only one band. The cross veins of the fore-wing are pale yellow or whitish (**Fig. 2.466**). 28

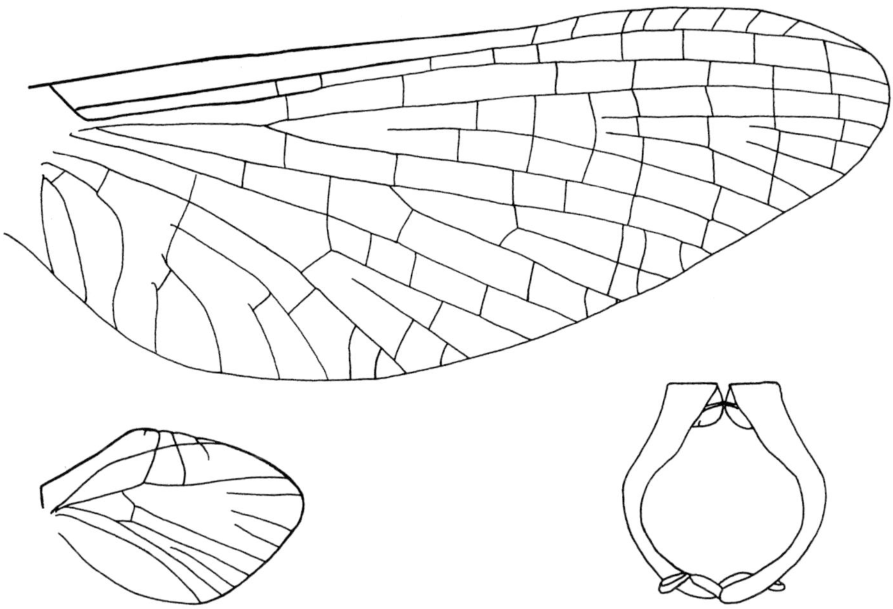

Fig. 2.464 *Thraulodes telegraphicus:* the veins of the fore and hind wings (left) with the dark marking in for fore-wing omitted and the male genitalia (lower right). The veins here depicted in black are actually yellow in living insects.

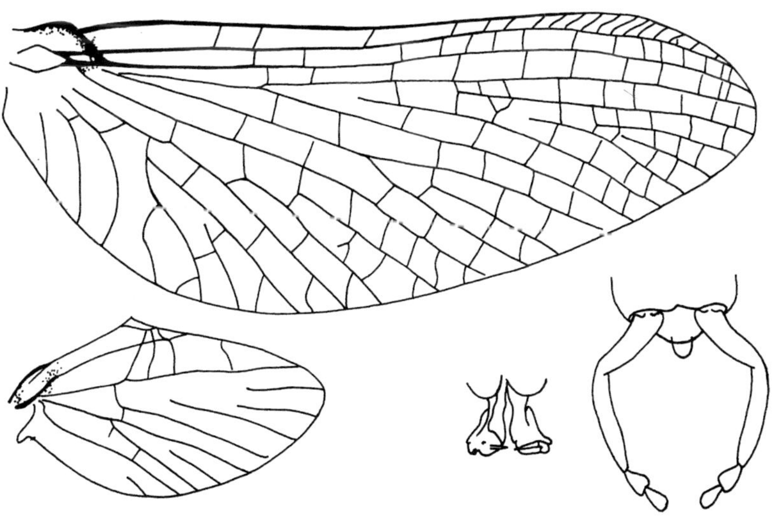

Fig. 2.465 *Thraulodes trijunctus:* the fore and hind wings with the hind wing drawn larger than scale (left) and the penis lobes (lower center) and apex of the male abdomen with the forceps but without the penis lobes (lower right).

28. The second through the fifth abdominal tergites and first, sixth, and seventh sternites are pale yellow and translucent with pale reddish to olive brown bands, while the sixth through tenth tergites are rose with the sixth brighter than the others. There are midway transverse dashes, stigmatic dots, and black longitudinal streaks on the pleurae of the eighth and ninth abdominal segments. The turbinate eyes of the male are very large and prominent. The penis lobes are strongly curved laterad so that the spears are directed almost posteriad (**Fig. 2.466**). Length of male: 6.5 to 7.5 mm; female: 7 to 8 mm. Fore-wing length of male: 6.5 to 7.3 mm; female: 8 to 9 mm.
....*Thraulodes schlingeri* Traver and Edmunds, 1967 (Peru).

- The second through sixth abdominal tergites have a pale reddish brown shade that is not more evident on the sixth segment than on the others. The second through fifth and first third of the sixth abdominal segment are translucent with transverse bands that are pale reddish brown. The seventh to tenth tergites are uniform in color except for narrow purplish brown anterior seams. The stigmatic spots are large and conspicuous. The apical segments of the male forceps are somewhat produced on the inner margins (**Fig. 2.467**). Length of male: 7 mm; of female: 6.5 mm. Fore-wing length of male: 7 mm; of female: 7.5 mm.
....*Thraulodes osiris* Traver and Edmunds, 1967 (Ecuador).

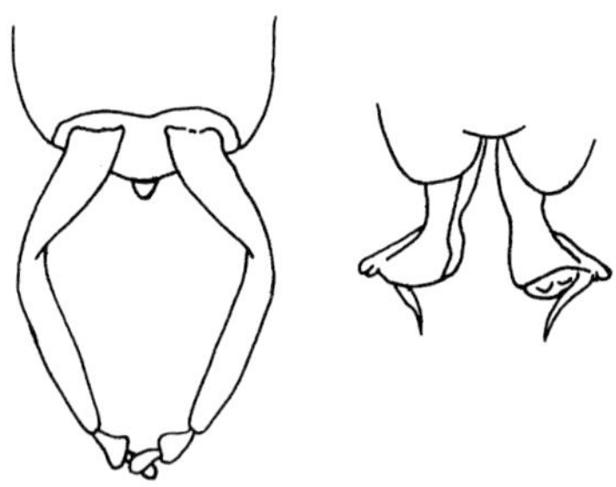

Fig. 2.466 The male genitalia of *Thraulodes schlingeri* with the forceps base and segments (left) separated from the penis (right).

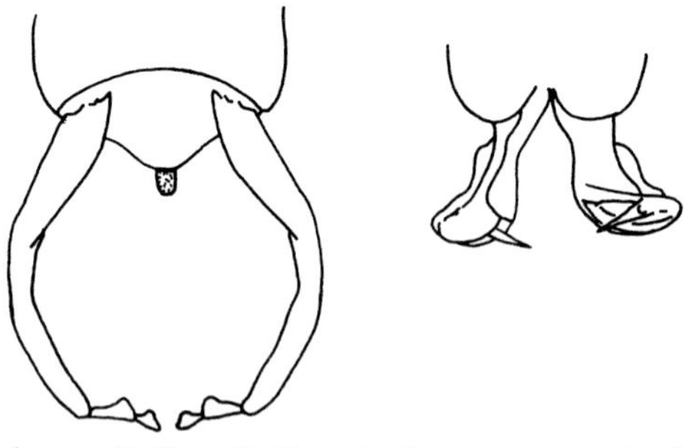

Fig. 2.467 The male genitalia of *Thraulodes osiris* with the forceps base and segments (left) separated from the penis (right).

Key to the species of subadult and adult *Homothraulus* in South America

Information for the key was provided by Needham and Murphy (1924), Navás (1926c), and Traver (1947b, 1960). The characteristics of the male are definitive. The larvae of the South American species have not been described.

1. The costal area of the subadult hind wing is blackened in its basal half. The base of the fore-wing has a conspicuously dark brownish gray crescent-shaped mark. The shape and venation of the hind wings is variable (**Fig. 2.468**). Body length of male subadult: 6 to 6.5 mm. Fore-wing length: 6 to 6.5 mm. Coloration: head reddish brown with black dashes near the mid-line. The thorax is light reddish brown with wavy blackish lines on the pronotum, a black outline around the mesonotum, and transverse black markings on the metanotum. The abdomen is deep reddish or blackish brown with conspicuous yellow markings. The adult has not been described.
....*Homothraulus lucretiae* Traver, 1960 (Uruguay).
- The basal half of the costal area of the adult hind wing is faint reddish brown (**Fig. 2.322**), with the coloration of the subadult presumed to be similar or somewhat lighter. 2

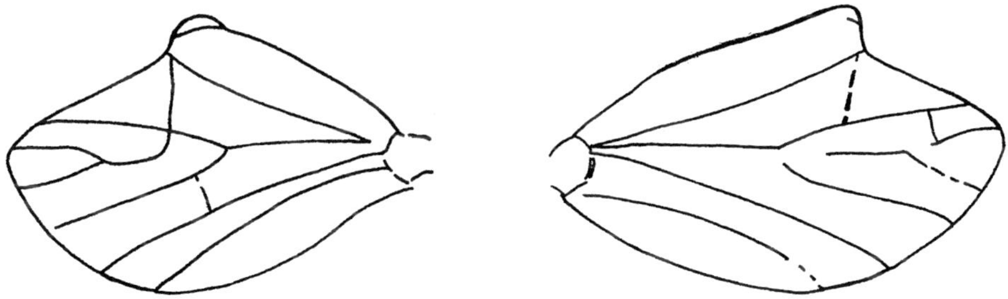

Fig. 2.468 Two differing hind wings from the same individual of *Homothraulus lucretiae.* The characteristic black portion of the costal space is not indicated so that the veins can be seen.

2. The length of the adult male is 5.5 mm. Fore-wing length: 7 mm. The long veins of the hind wing are somewhat thickened. There are about five cross veins in the hind wing (**Fig. 2.322**).
....*Homothraulus misionensis* (Esben-Petersen, 1912) (Argentina, Uruguay). Syn: *Thraulus misionensis* Esben-Petersen, 1912.
- The length of the adult male is 7.2 mm. Fore-wing length: 8.4 mm. There are more than ten cross veins in the hind wing (**Fig. 2.469**).
....*Homothraulus larensis* (Navás, 1926) (Argentina). The two species in the couplet may not be distinct, but due to sketchy descriptions and a loss of type specimens, this may require collection of new specimens to determine for sure.

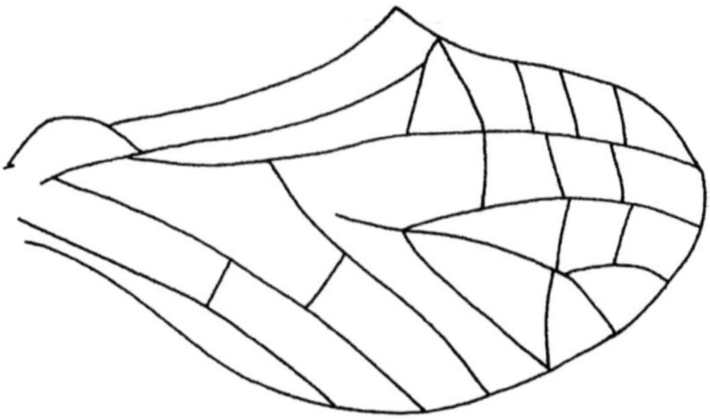

Fig. 2.469 The hind wing of *Homothraulus larensis.*

Key to the known species of adult *Hydrosmilodon*

Information for the key was taken from Flowers and Domínguez (1992).

1. The color is generally light orange-brown. The apical third of the forceps, the first segment and the underside of second are light yellow with the rest brown. .... *Hydrosmilodon primanus* (Eaton, 1892) (Mexico, Panama, not yet reported from South America).
 - The color is generally brownish black with a darker shade on the dorsal surface of the thorax. The forceps are entirely orange-brown (**Fig. 2.323**). Length of the male: 7.3 to 7.8 mm; female: 7.8 to 8.2 mm. Fore-wing length of male: 9.3 to 9.5 mm; female: 9.0 to 9.2 mm. ....*Hydrosmilodon saltensis* Flowers and Domínguez, 1992 (Argentina).

Key to the known species of *Hydrosmilodon* larvae

Information for the key was taken from Flowers and Domínguez (1992).

1. Exterior setae are present only at the base of the mandibles (**Fig. 2.351**). There are well-defined dark markings on the basal segment of the labial palp. Length of mature larva: 5.1 to 9.5 mm. General coloration: dark purplish brown with pale and black markings. ....*Hydrosmilodon primanus* (Eaton, 1892) (Mexico, Panama, not yet reported from South America).
 - The exterior setae on the mandibles extend from the base toward the exterior angle. The basal segment of the palps lack well defined markings. Length of mature larva: 5.2 to 8.5 mm. General coloration: dark orange brown with whitish, yellow, and black markings. ....*Hydrosmilodon saltensis* Flowers and Domínguez, 1992 (Argentina).

Caenidae

Key to the South American species of adult males

Information for the key was taken from Navás (1912, 1932), Puthz (1975), Soldán (1986), and Malzacher (1986). When he described the genus *Brasilocaenis*, Puthz (1975) was unable to find any consistent difference to distinguish the adults. Furthermore, because the descriptions of Navás's species were based exclusively on adults, Puthz (1975) suggested that some of these species may belong to *Brasilocaenis* rather than *Caenis*. It might be suggested, however, that the differences between the larvae of *Brasilocaenis* and those of *Caenis* are insufficient to merit the creation of a new genus. Although Malzacher (1986) was able to recognize characters to distinguish the two genera, all of them are based solely on the male genitalia and require careful examination to recognize. It could therefore be suggested that the differences are artificial, and the validity of Puthz's genus must be called into question. Another nominal genus, *Brachycercus*, was assigned to a species described by Curtis (1834) in such a way that Domínguez *et al.* (2001) regard the taxon as a *nomen dubium*. Moreover, *Brachycercus* is distinguished from a supposedly closely related genus, *Cercobrachys,* primarily on the basis of differences in the larvae. Worse still, the adults of neither species of *Cercobrachys* in South America have been described, making it impossible to know whether the key features used to distinguish the adults of these genera hold true for South American species. It may be premature to declare the groups recognized by Soldán (1986) to be valid genera. Considering the chaotic state of the taxonomy, major revisions can be expected in the future. The names in the key are those used in the most recent literature, and most of the species can be recognized, although the generic classification and synonymy will certainly be revised in the future. If female specimens must be identified, it is suggested that they be run through each of the generic keys to seek a description that fits. At the present time, there are no known features that can be used to identify the nominal genus of a female (Malzacher, 1986).

1. The length of the prosternum is two to three times the width resulting in the fore-coxae being close together **(Fig. 2.470)**. 2
- The length of the prosternum is only half its width, placing the fore-coxae far apart. Identifications of adults in this group remain tentative until those of the known South American species have been described. 3
2. The gonopods are nearly straight and extend stiffly away from the curvature of the sternal plate. They usually extend well beyond the apical margin of the sternal plate, or, if not, they diverge outward from the body axis (**Fig. 2.470**).
....*Caenis* Stephens, 1835..p. 371
- The gonopods curve around the lateral contour of the sternal plate, which they extend no more than slightly beyond (**Fig. 2.471**).
....*Brasilocaenis* Puthz, 1975..p. 382

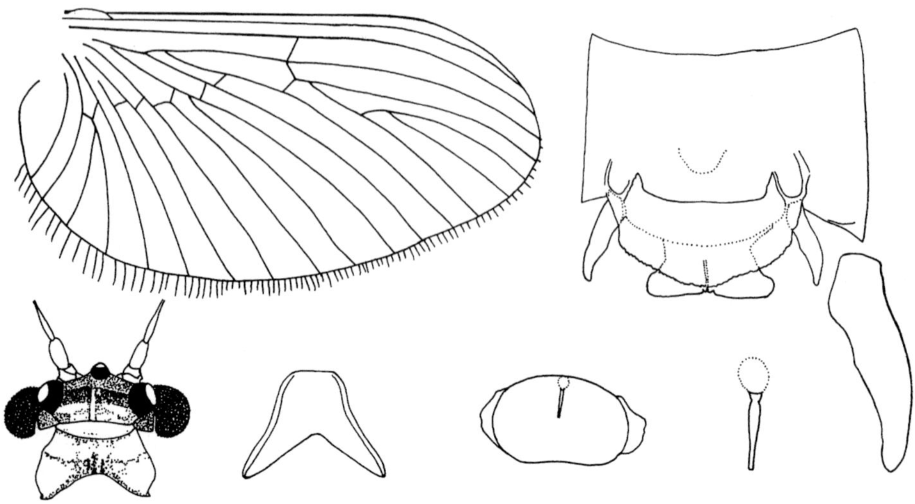

Fig. 2.470 *Caenis chamie:* The wing (upper left), head and pronotum with the basal part of the antennal flagellum of a male (lower left), male genitalia with one forceps limb, or "gonopod," enlarged (right), prosternum (lower left center), and egg with an enlargement of its micropyle to its right (lower right center).

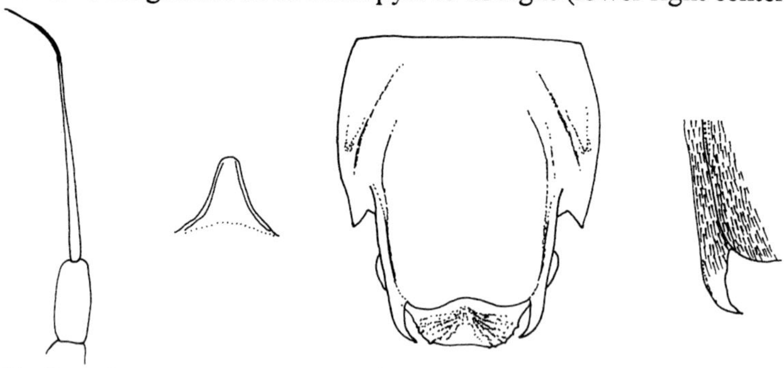

Fig. 2.471 *Brasilocaenis septentrionalis* adult male (left to right): antenna without an apical section of the flagellum, middle part of the prosternum, genitalia in ventral view, and apical part of a gonopod with the margin of the styliger plate.

3. The pedicel is at least 1.5 times as long as the scape. The posterolateral spines on the sixth segment of the abdomen curve only slightly dorsad. The oval eggs are twice to three times as long as wide, and their chorion has numerous rib-like structures. The pronotum lacks lateral, submarginal projections.
.... *Brachycercus* Curtis, 1834

The only South American species of this genus is *Brachycercus tenella* (Navás, 1912), also described as *Eurycaenis tenella* Navás, 1932, from

Argentina. Soldán (1986) stated that the types for this species could not be found in Navás's collection, and Thew (1960) suggested that the species does not belong in the genus *Brachycercus*. Because of these problems, Domínguez *et al.* (2001) regarded the name as a *nomen dubium*.
- The pedicel is as long as or only slightly longer than the scape. The posterolateral spines on the sixth segment of the abdomen curve strongly dorsad and around the dorsal side of the abdomen toward the mid-line. The spindle-shaped eggs are four to six times as long as wide, and their chorion has only four to six broad, rounded rib-like structures.
....*Cercobrachys* Soldán, 1986..p. 386

Key to the South American species of larvae in the family Caenidae

Information for the key was taken from Puthz (1975) and Soldán (1986).

1. The ocelli are set on processes with basal diameters equal to or larger than the diameter of the ocellus it bears. 2
- The ocelli are not set on prominent processes (**Fig. 2.29**). 3

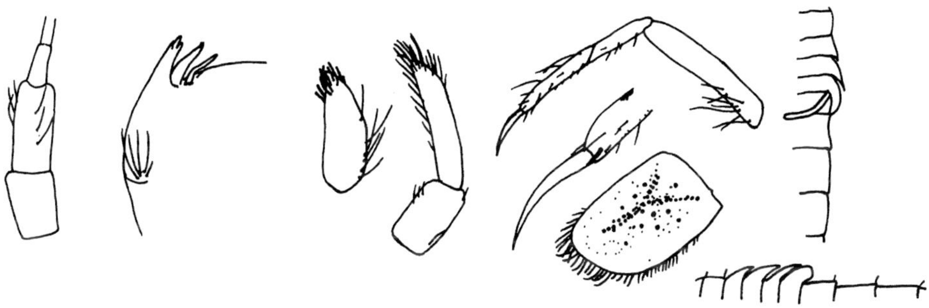

Fig. 2.472 *Cercobrachys colombianus* larva (left to right): the scape, pedicel, and first segment of the antennal flagellum; the outer margin of the mandible; labial and maxillary palps; middle leg above an enlargement of a claw and the gill cover; an outline of one side of the abdomen showing the posterolateral spines in dorsal (above) and lateral (below) view.

2. The pedicel of the antenna is as long as the scape or slightly longer. The anterior margin of the mesosternum is prominent and bears many long setae. There are long anterolateral processes on the sixth abdominal segment, which curve around toward the midline behind the gills and sometimes contact each other. There are two emarginations on the posterior margin of the ninth sternite. There are bristles on the legs as long as the tibiae (**Fig. 2.472**).
....*Cercobrachys* Soldán, 1986..p. 386

- The pedicel is 1.5 to 2 times as long as the scape. The anterior margin of the mesosternum is flat and lacks long setae. The sixth abdominal segment bears posterolateral processes that do not curve around toward each other. The bristles on the legs are all shorter than the tibiae.
.... *Brachycercus* Curtis, 1834

The only South American species of this genus is *Brachycercus tenella* (Navás, 1912), also described as *Eurycaenis tenella* Navás, 1932, from Argentina. Soldán (1986) stated that the types for this species could not be found in Navás's collection, and Thew (1960) suggested that the species does not belong in the genus *Brachycercus*. Because of these problems, Domínguez *et al.* (2001) regarded the name as a *nomen dubium*. Hence, the presence of this genus in South America remains unconfirmed.

3. The middle and hind coxae bear a thorn-like process. The anterolateral corners of the pronotum are extended considerably laterad (**Fig. 2.473**).
.... *Brasilocaenis* Puthz, 1975..p. 382

- The middle and hind coxae lack thorn-like processes. The anterolateral corners of the pronotum are broadly rounded (**Fig. 2.29**).
.... *Caenis* Stephens, 1835..p. 371

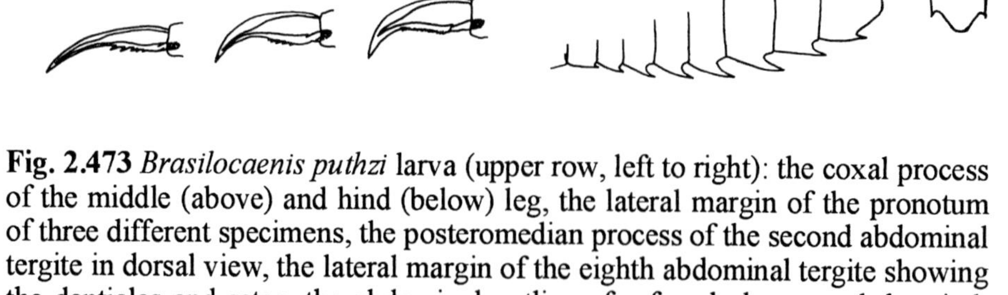

Fig. 2.473 *Brasilocaenis puthzi* larva (upper row, left to right): the coxal process of the middle (above) and hind (below) leg, the lateral margin of the pronotum of three different specimens, the posteromedian process of the second abdominal tergite in dorsal view, the lateral margin of the eighth abdominal tergite showing the denticles and setae, the abdominal outline of a female larva, and the ninth sternite of a male larvae; (lower row, left to right): the tarsal claws of the fore, middle, and hind legs, the abdominal outline of a second female specimen, and the ninth sternite of a second male larva.

Key to the species of adult *Caenis* known from South America

Information for the key was taken from Navás (1915, 1920a, 1922b, 1933b), Froehlich (1969), Malzacher (1986, 1990), Pereira and da Silva (1990), and Alba-Tercedor and Mosquera (1999). *Caenis chilensis* Navás is a *nomen nudum* without any taxonomic significance.

1. At least the distal parts of the femora and tibiae are darkly pigmented. 2
- The distal parts of the femora and tibiae are not both darkly pigmented, although limited dark subapical markings may be present on some of them. Caution: apparently some type specimens described had not been preserved well enough to fully determine the color pattern. In case of doubt, check both possibilities. The rest of the key is based mainly on characteristics of the male. 7
2. The legs are dark except for the white hind tibiae and tarsi. The wings are mostly hyaline with veins C, Sc, and R dark and the rest of the veins white. There is a slight darkening of the membrane in the basal part of the costal and almost the entire subcostal area. The color is mainly shiny black dorsally and dark brown ventrally. Length: 4.5 mm. Fore-wing length: 5.8 mm. Length of the cerci: 4.3 mm. The species has not been well described and will be difficult to identify with certainty because of the lack of a description of the male genitalia.
....*Caenis albata* (Navás, 1920) (Argentina). Syn: *Ordella albata* Navás, 1920. This is regarded as a *nomen dubium* Domínguez *et al*. (2001).
- Both the middle and hind femora are either entirely dark, or they are both light proximally and dark distally. 3

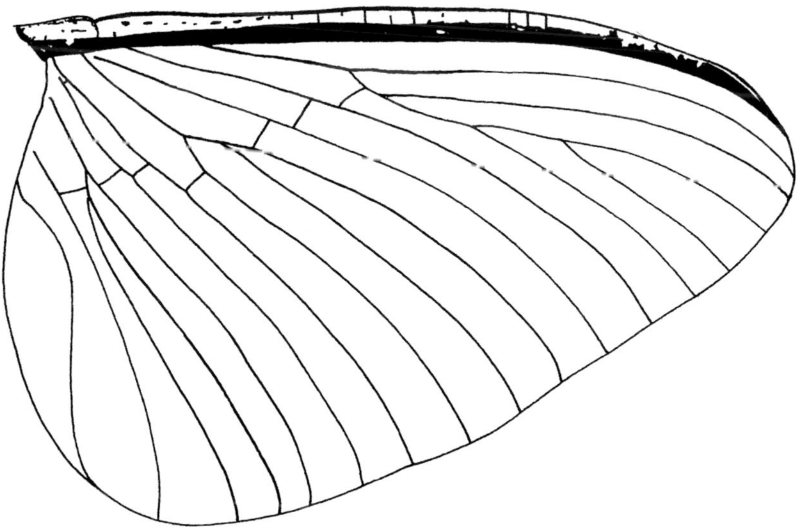

Fig. 2.474 The wing of *Caenis argentina*.

3. The middle and hind femora and tibiae are entirely dark. If the femora and tibiae of the fore-leg are somewhat lighter, then the abdomen is yellowish green with dark bands across the posterior margins of the tergites. The subcostal and at least part of the radial area of the wing are darkened (**Fig. 2.474**). 4
- The proximal parts of the femora or tibiae are lighter in color than the distal parts. The abdomen is not mainly yellowish green or blackish brown without any violet tinge. 6
4. The middle and hind legs are entirely dark, while the fore-leg is somewhat lighter. The abdomen is yellowish green with narrow dark bands across the posterior margins of the tergites. The thorax is black. The subcostal and radial areas of the wing are darkened. Length: 2.6 mm. Fore-wing length: 3.9 mm.
....*Caenis axillata* Navás, 1930 (Argentina). The description of this species is not precise enough to determine whether it really belongs to the genus *Caenis*. It is generally regarded as a *nomen dubium* (Domínguez *et al.*, 2001).
- All legs are blackish brown except for the tarsi, which are lighter in color. The head and thorax are shiny brownish black or gray. 5

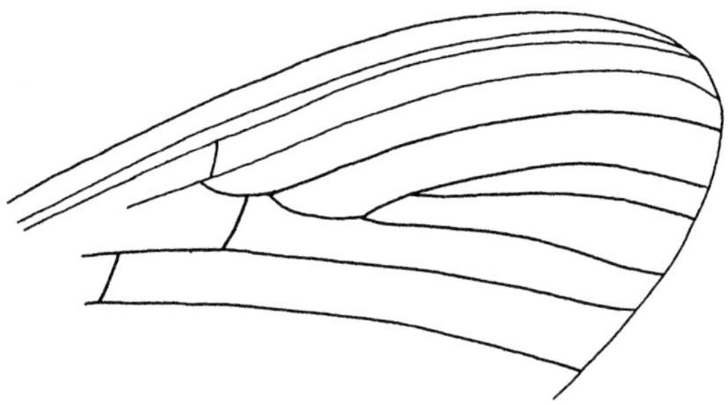

Fig. 2.475 The apical part of the wing of *Caenis nemoralis*.

5. The subcostal area of the wing is heavily darkened (**Fig. 2.474**). The head and thorax are shiny brownish black. The abdomen is blackish or dark reddish brown. Length: 4.5 to 5.2 mm. Fore-wing length: 4.5 to 6.5 mm.
....*Caenis argentina* Navás, 1915 (Argentina). Syn: *Ordella argentina* (Navás, 1915).
- The subcostal area of the wing is lightly colored with dark gray, and the rest of the membrane is somewhat reddish (**Fig. 2.475**). The head and thorax are gray. The abdomen is yellowish white with a whitish pubescence. Length of the female: about 3.8 mm. Fore-wing length of the female: about 4.4. Length of the cerci: about 3.3 mm. The male has not been described.
....*Caenis nemoralis* Navás, 1922 (Argentina). Syn: *Ordella nemoralis* Navás, 1932. *Nomen dubium*. The species is poorly known, and it may not be valid. The name is considered a *nomen dubium* by Domínguez *et al.* (2001).

6. A parthenogenetic species known only from females. The femur bears dark patches in the middle and at the apex, and the distal halves of the tibiae are darkened. The abdominal tergites are brownish, sometimes with a violet tinge on the first eight segments; their lateral margins are clear except for dark stigmatic spots. Length: 2.5 to 3.8 mm. Length of wing: 2.2 to 2.9 mm. The subadult has a similar color pattern.
....*Caenis cuniana* Froehlich, 1969 (São Paulo, Pará?).

- Males and females have both been found. The femora do not have large dark markings in the middle; but the femora and tibiae are conspicuously pigmented in the distal half. The gonopods, or limbs of the forceps, are curved and bluntly rounded at the apex, and they are covered with numerous fine setae, called trichomes (**Fig. 2.476**). Length of male: 2.4 to 2.8 mm; of female: 3.5 mm. Wing length of male: 1.8 to 2.0 mm; of female: 2.5 mm. Color: the mesothorax and metathorax are yellowish brown, and the rest of the body is mainly pale yellow with prominent cross bands on the first two abdominal tergites and lateral markings on other segments.
....*Caenis pflugfelderi* Malzacher, 1990 (Amazonas).

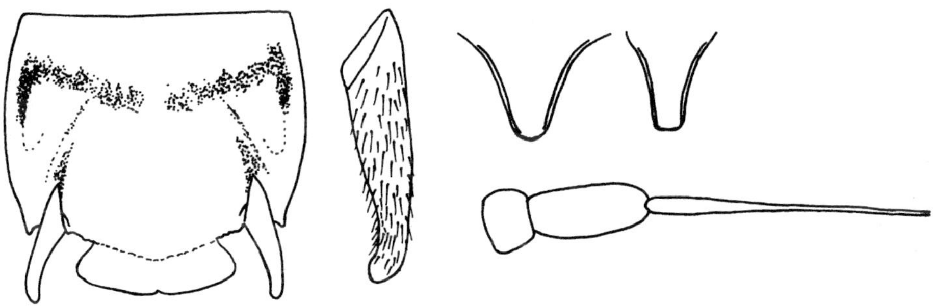

Fig. 2.476 *Caenis pflugfelderi:* the genitalia of a male in ventral view with an enlarged gonopod, or forceps limb, to its right (left); the middle portion of the prosternum of two different males, (upper right); the scape, pedicel, and basal section of the flagellum of the antenna of a male (lower right).

7. The wing length of the male is 3.4 to 3.7 mm, and its total length, without antennae, cerci, or other appendages is 3.4 to 4.0 mm. The head is heavily pigmented with a pale area in the middle of the vertex. The color on the mesonotum, pleurae, and metanotum is brown, while the ground color of the rest of the body is yellowish brown or white. Two black markings along the posterior margin of the pronotum are joined to a central marking, and there are narrow black lines around the margin and a narrow black line along the midline that runs the length of the pronotum. Plates on the underside of the head, the pleura, coxae, and underside of the thorax are also narrowly lined in black.

There are also black markings along the posterior margins of the abdominal tergites and on the apical sternites. Subapical spots are present on the femora. The antennal flagellum is not widened at the base. The apices of the forceps are acutely pointed and extend well beyond the penis lobes (**Fig. 2.477**). The female has not been described.
....*Caenis pseudamica* Malzacher, 1990 (Paraguay).
- The wing length of the male is less than 3 mm, or if it is 3.5 to about 4 mm, then the wing is almost 1.5 times as long as the body or longer. 8
8. The wing of the male, between 3.5 and about 4 mm long, is almost 1.5 times as long as the insect length or longer. The dorsal color is blackish, and at least the head and thorax are shiny. The cerci are white with narrow dark annulations. The wings are hyaline. Veins Sc and R are dark, but the rest of the veins are whitish. The femora are at least slightly darkened, and the tibiae are white. 9
- The wing length of the male is shorter than 3 mm, and it is shorter or only slightly longer than the length of the insect. 10
9. The length of the male is about 2.7 mm, and its wing length is about 4 mm. The dorsal surface of the body is shiny black, and the lower part is dark in color. The legs are white except for the slightly darkened femora. The stigmatic area of the wing is slightly darkened. The female has not been described.
....*Caenis ludicra* Navás, 1920 (Argentina). Syn: *Ordella ludiera* (Navás, 1920); *Caenis ludiera* Navás, 1920. Navás (1924) corrected his original spelling to *Ordella ludicra*. In the absence of a decision by the International Commission on Zoological Nomenclature, the name in use should remain valid. In recent taxonomic literature, *Caenis ludicra* has been used.
- The length of the male is about 2.5 mm, and its wing length is about 3.6 mm. The dorsal surface of the head and thorax are shiny black, and the abdomen is black. The underside of the thorax is pale. The legs are dark except for the whitish tibiae. The female has not been described.
....*Caenis nigella* Navás, 1933 (Chile). Both species in this couplet seem very similar, and due to the sketchy descriptions, their identities are in doubt. The name is considered a *nomen dubium* by Domínguez *et al.* (2001).
10. There are three intercalary veins in the R_2 cell, two of which join vein R_3 at a distance from the wing margin equal to about 1/5 of R_3. The genital forceps are short and rounded apically, and the penis lobes are completely fused and have a mushroom shape (**Fig. 2.18**). Length of male: 2.6 to 2.7 mm; female: 3.2 to 4.8 mm. Wing length of male: 2.6 to 2.9 mm; female: 3.2 to 3.6 mm. Color: generally brownish; darker on the thorax.
....*Caenis cigana* Pereira and da Silva, 1990 (Rio de Janeiro). The description of this species is so poor that it could scarcely be recognized from its other features.

- Three intercalary veins are not present in the R_2 cell, or two of them do not join vein R_3 at a distance from the wing margin equal to about 1/5 of R_3. The genital forceps are not short, are not rounded apically, or the penis lobes are not completely fused and mushroom shaped (**Fig. 2.478**). 11

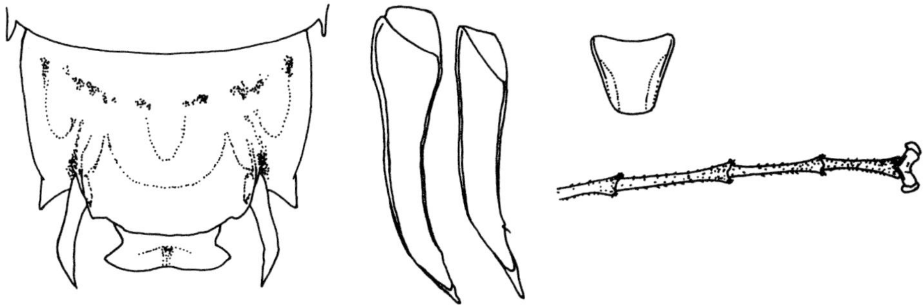

Fig. 2.477 *Caenis pseudamica* (left to right): the male genitalia, two forceps limbs enlarged, and the middle portion of the prosternum above the tarsal segments and claws.

11. The antenna flagellum of the male is expanded over a length greater than that of the pedicel (**Fig. 2.478**). 12
- The antenna flagellum of the male tapers to a narrow filament in the basal half over a length equal to or shorter than that of the pedicel (**Fig. 2.479**). 13

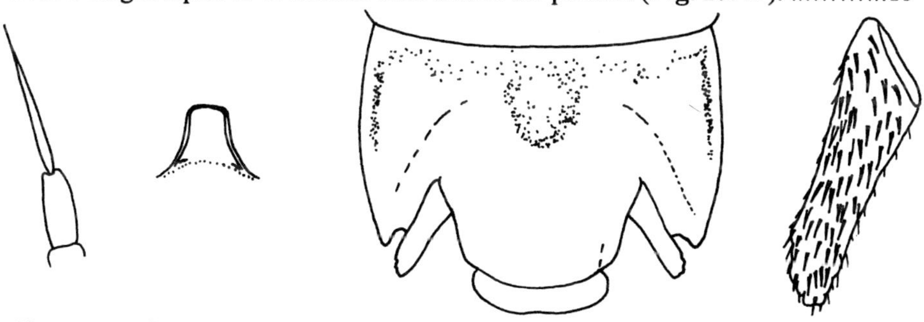

Fig. 2.478 *Caenis quatipuruica* male (left to right): antenna base, prosternum, genitalia, and a forceps limb enlarged.

12. The gonopod of the male is broadly rounded at the apex. The anterior corner of the prosternum is broad and truncate (**Fig. 2.478**). Length: 2.3 mm; fore-wing length: 1.9 mm. The female has not been described.
....*Caenis quatipuruica* Malzacher, 1986 (Pará).

- The gonopods of the male end in narrow, acutely pointed processes (**Fig. 2.480**). Length: 2.6 to 2.8 mm; fore-wing length: 2.0 mm.
....*Caenis candelata* Malzacher, 1986
(Probably Amazonas, the location reported by Malzacher could not be identified with certainty).

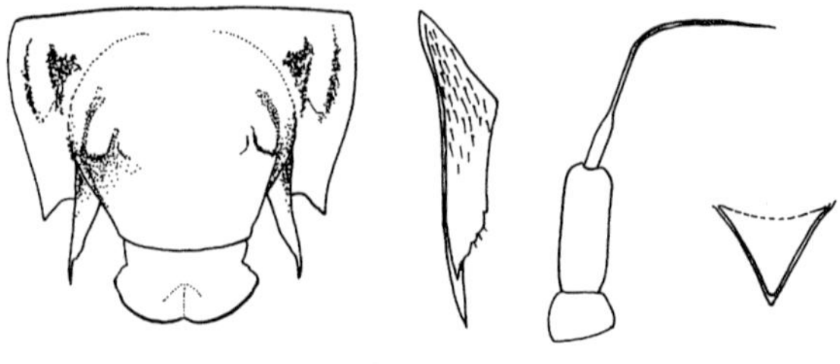

Fig. 2.479 *Caenis burmeisteri* (left to right): the male genitalia; one of the forceps limbs enlarged; the scape, pedicel, and base of the flagellum of the antenna; the middle portion of the prosternum.

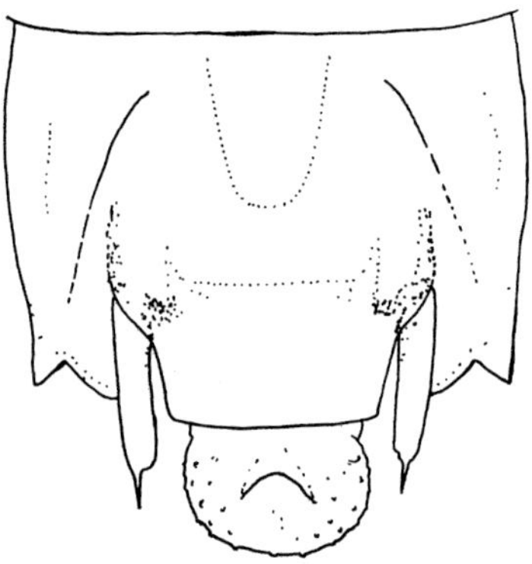

Fig. 2.480 The male genitalia of *Caenis candelata.*

13. The basal portion of the antennal flagellum of the male is enlarged over a length roughly equal to that of the pedicel. The penis lobes are shaped roughly like and anvil with a narrow incision separating them apically (**Fig. 2.470**). Length: 2.9 to 3.7 mm. Wing length of male: 2.3 to 2.9 mm; of female: 3.6 to 3.8 mm. Color: there is a light brown pattern on the pronotum, while the rest of the thorax is yellowish with darkened sutures. The apical 1/3 of the first two abdominal tergites and the apical 2/3 of the other abdomen tergites of the male are grayish brown. The female has a similar pattern on a yellowish brown ground color. The subcostal area of the wing and veins C, Sc, and R_1 have a light purplish gray tint. There are dark longitudinal stripes on the brownish fore-femora, while the tibiae and tarsi are white.
.... *Caenis chamie* Alba-Tercedor and Mosquera, 1999
(Colombia).

- The basal portion of the antennal flagellum of the male is enlarged over a length much shorter than that of the pedicel (**Fig. 2.479**). 14
14. The gonopods are long, and each ends in a narrow acutely pointed process (**Fig. 2.479**). The length is at least about 2.8 mm. 15
- The gonopods are broadly rounded at the apex. The abdominal segments lack distinct black lateral markings (**Fig. 2.481**). The length is not greater than 2.4 mm. 16

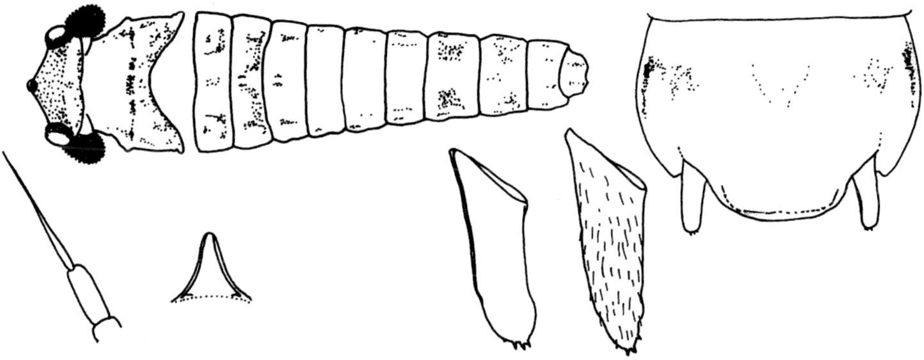

Fig. 2.481 *Caenis sigillata* male: the pigment pattern on the head and body in dorsal view (upper left); the genitalia (right) with enlarged forceps limbs from two individuals, one showing the spines (lower right center); the base of the antenna (lower left) and the prosternum (lower left center).

15. The black pigment on the abdominal tergites is diffuse and concentrated in the posterolateral corners of all tergites and the middle of the first three. The forceps arms taper from the base to the apex for most of their length (**Fig. 2.479**). Length: 3.3 to 3.5 mm; fore-wing length: 2.5 to 2.6 mm. Color: yellowish brown or paler with dark brown or black seams between the plates of the thorax. The female has not been described.
....*Caenis burmeisteri* Malzacher, 1990 (Paraguay).
- There are distinct black lateral marks on the abdominal segments. The forceps arms are parallel sided for much more than half of their length (**Fig. 2.482**). Length: 2.8 to 3.7 mm; fore-wing length: 2.0 to 2.7 mm.
....*Caenis fittkaui* Malzacher, 1986 (Pará).

16. The anterior corner of the prosternum is narrowly rounded. The color pattern of the male includes dark lateral marks and cross bands on the very light ground color of the abdominal tergites and sternites of the as well as dark markings on the head and thorax (**Fig. 2.481**). Length of male: 1.7 to 2.3 mm; fore-wing length: 1.5 to 1.7 mm. The female has not been described. .... *Caenis sigillata* Malzacher, 1986 (Pará).

- The anterior corner of the prosternum is broadly rounded, almost truncate (**Fig. 2.483**). The color pattern of the male and female usually includes dark cross bands and two lateral marks only on the first two abdominal sternites. Length: 2.0 to 2.4 mm; fore-wing length: 1.7 to 2.0 mm. .... *Caenis reissi* Malzacher, 1986 (Pará).

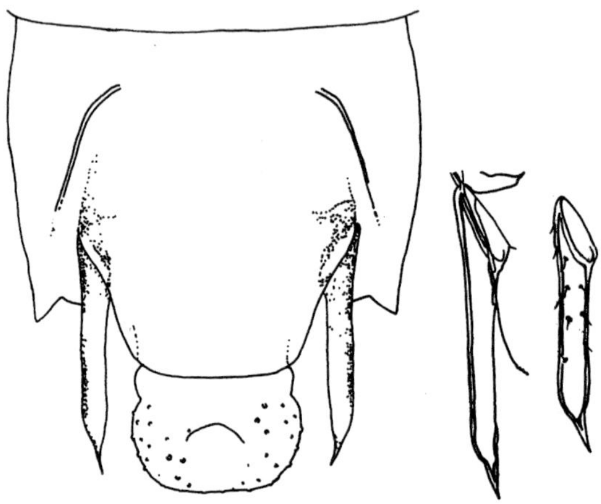

Fig. 2.482 *Caenis fittkaui:* the male genitalia (left), and enlarged forceps limbs from two different individuals (right).

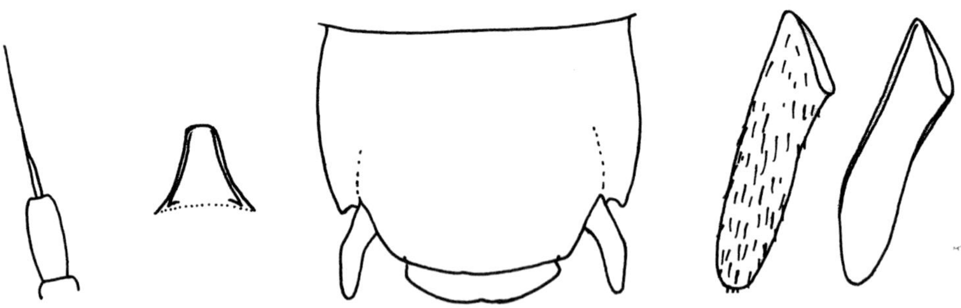

Fig. 2.483 *Caenis reissi* male (left to right): the base of the antenna, prosternum, genitalia, and enlarged gonopods from two individuals, one showing the spines.

Key to the species of *Caenis* larvae known from South America

Information for the key was taken from Froehlich (1969) and Malzacher (1986). The larvae of *Caenis albata, Caenis argentina, Caenis axillata, Caenis burmeisteri, Caenis candelata, Caenis cigana, Caenis ludrica, Caenis nemoralis, Caenis nigella, Caenis pseudamica, Caenis quatipuruica, Caenis reissi,* and *Caenis sigillata* have not been described.

1. The apex of the abdomen is clearly truncate or concave in ventral view. The posterolateral process on the second abdominal tergite is acutely pointed in dorsal view (**Fig. 2.484**). The pronotum does not bear two black dots on either side of the midline. 2

- The apex of the abdomen is convexly rounded (**Fig. 2.485**). 3

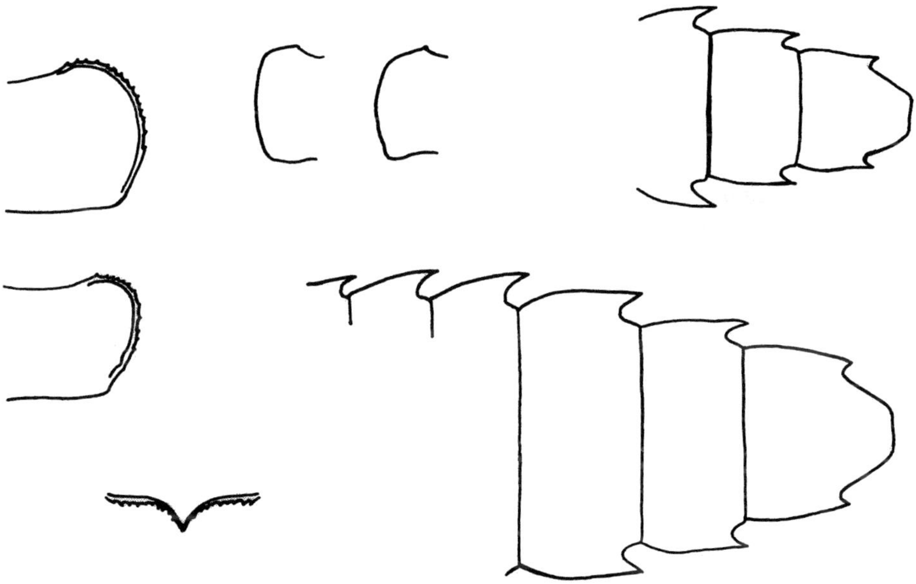

Fig. 2.484 *Caenis fittkaui* larva: coxal process on the middle (upper left) and hind (lower left) leg, the lateral margins of the pronotum of two different specimens (upper center), the outline of the apex of an abdomen of a male (upper right) and female larva (lower right), and the posteromedian process of the second abdominal tergite in dorsal view (lower left center).

2. The apex of the abdomen is truncate in ventral view (**Fig. 2.484**). The frons and vertex of the head display dark transverse markings. The pronotum does not bear two black dots on either side of the midline but has short dark lines along its lateral and posterior margins. Length: 2.0 to 3.7 mm. ....*Caenis fittkaui* Malzacher, 1986 (Pará).

- The apex of the abdomen is concave in ventral view (**Fig. 2.486**). There is a dark band on the head between the eyes. The vertex of the head, pronotum, and metanotum display a reticulation. Length of male larva: 3.0 to 3.5 mm; female larva: 4.3 to 5.3 mm.
.... *Caenis chamie* Alba-Tercedor and Mosquera, 1999 (Colombia).

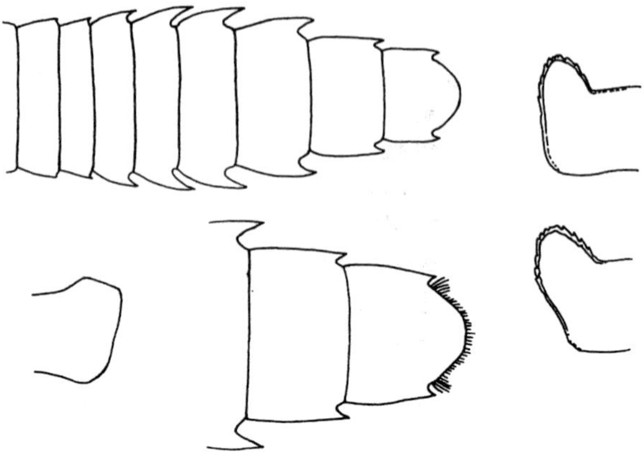

Fig. 2.485 *Caenis pflugfelderi* larva: lateral margin of the pronotum (lower left), abdomen of a male larva in ventral view (upper left), apex of the abdomen of a female larva (lower center), coxal process on the middle (upper right center) and hind (lower right) leg, and two modified setae on the fore-femur (upper right).

3. The coxal processes on the middle and hind legs are prominent and bear large denticles along the edges (**Fig. 2.485**). There are prominent posterolateral processes on the fourth through ninth sternites, with only acute corners on the third and sometimes second sternites. There is a narrow, acute posteromedian process on the second tergite, which has prominent teeth along its posterior margin. Length: 2.8 to 4.0 mm.
....*Caenis pflugfelderi* Malzacher, 1990 (Amazonas).
- The coxal processes on the middle and hind legs are hardly produced at all and are not notably denticulate (**Fig. 2.29**). The posterolateral process on the second abdominal tergite is bluntly rounded in dorsal view. The frons and vertex of the head lack distinct dark transverse bands. The pronotum bears two black dots on either side of the midline. Length: 3.8 to 4.5 mm.
....*Caenis cuniana* Froehlich, 1969 (São Paulo, Pará?). Unfortunately, Malzacher (1986) based his description of this species on specimens identified as *Caenis cuniana* with a question mark added to the name. Assuming that this indicates doubt concerning the identity of the illustrations and descriptions on which its distinction from *C. fittkaui* are based, and considering the minor differences he sets forth, the validity of *C. fittkaui* as a distinct species has to be questioned.

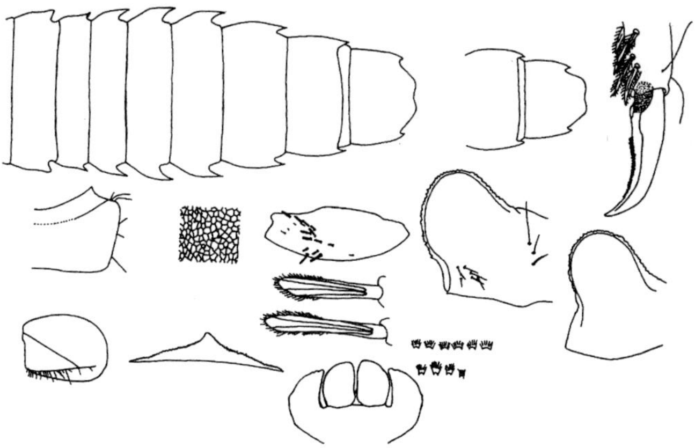

Fig. 2.486 *Caenis chamie* larva: the abdomen of a female in ventral view (upper left), apex of the abdomen of a male in ventral view (upper right center), right half of the pronotum to the left of an enlargement of the fine reticulated pattern on its surface (middle left), operculate second gill (lower left) and distal and middle rows of minute scales on its surface (lower right center), apophyses of the middle (right center) and hind coxae (lower right), posterior view of the second abdominal tergite showing the mid-dorsal elevation (lower left center), apex of the hind leg with the tarsal claw (upper right), labium (lower center), and femur of the fore-leg with two of its spines enlarged below it (center).

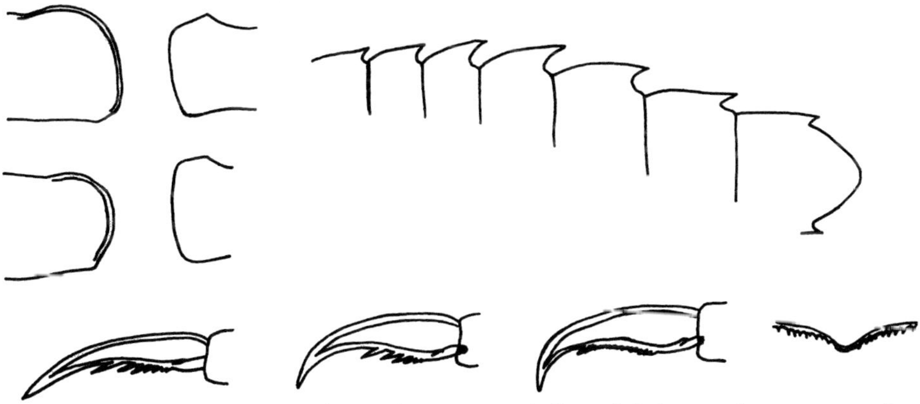

Fig. 2.487 *Caenis cuniana* larva (upper row, left to right): coxal process on the middle (above) and hind legs (below), the lateral margins of the pronotum of two different specimens, the outline of the right side of the abdomen of a female larva; (lower row, left to right): the tarsal claws of the fore, middle, and hind legs; the posteromedian process of the second abdominal tergite in dorsal view.

Key to the species of adult male *Brasilocaenis* known from South America

Information for the key was taken from Malzacher (1986, 1990, 1998). Differences among the known females are insignificant. The adult and subadult of *Brasilocaenis intermedia* have not yet been described.

1. The sternal sclerite ventral to the male genitalia is broadly truncate or concave at the apex (**Fig. 2.488**). 2
- The sternal sclerite is evenly rounded (**Fig. 2.489**). 4
2. The sternal sclerite ventral to the male genitalia is deeply emarginated and strongly concave (**Fig. 2.488**). Length of male: 2.6 to 3.3 mm. Wing length of male: 2.3 to 2.5 mm. Color: the pattern is blackish brown, and the background color is pale. The female has not been described.
....*Brasilocaenis mendesi* Malzacher, 1998 (Mato Grosso).
- The sternal sclerite ventral to the male genitalia is no more than shallowly emarginate and lightly concave (**Fig. 2.490**). 3

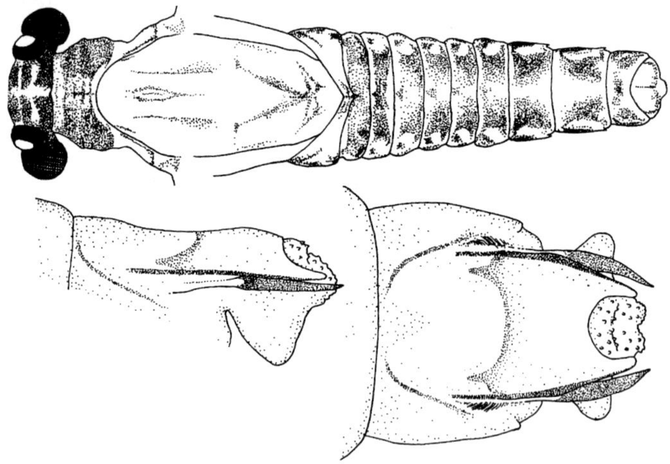

Fig. 2.488 *Brasilocaenis mendesi:* the pattern on the dorsal surface of the body (above), a lateral view of the genitalia with the ventral side up and a portion of the dorsal tergites omitted (lower left), and the apex of the abdomen in ventral view (lower right).

3. The gonopods follow the contour of the apical sternal sclerite without curving notably at the apex; both the gonopods and sclerite extend beyond the apex of the penis, concealing it in ventral view (**Fig. 2.490**). Male body length: 2.8 mm; fore-wing length: 2.0 mm. Length of female: 3.7 to 4.0 mm. Wing length: 2.3 to 2.5 mm. Color: light yellowish or reddish brown.
....*Brasilocaenis renata* Malzacher, 1986 (Amazonas).

- The gonopods extend beyond the apex of the apical sternal sclerite and curve strongly at the apex. The penis extends well beyond the apex of the sclerite, making it easily visible in ventral view (**Fig. 2.471**). Length of male: 3 mm. Wing length: 2 mm.
.....................................*Brasilocaenis septentrionalis* Malzacher, 1990 (Pará).

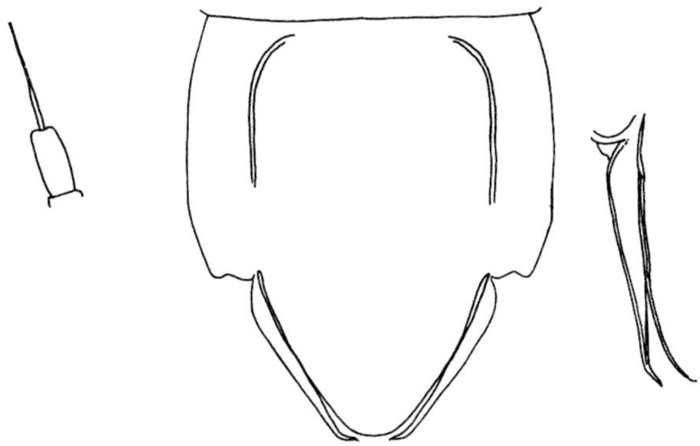

Fig. 2.489 *Brasilocaenis irmleri* male adult (left to right): the scape, pedicel, and base of the flagellum of the antenna; the genitalia; and a gonopod enlarged.

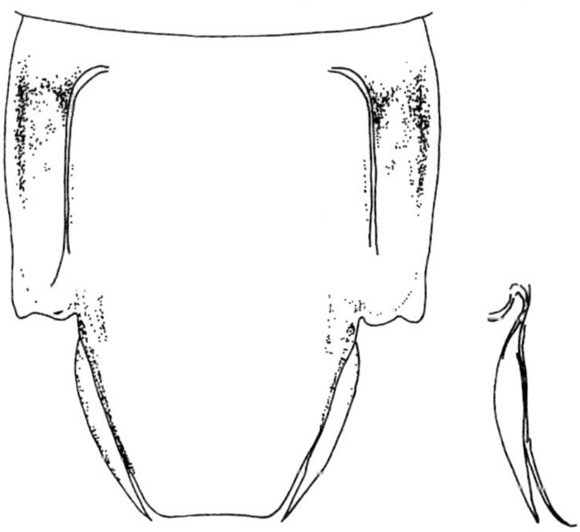

Fig. 2.490 The genitalia and enlarged gonopod of an adult male *Brasilocaenis renata.*

4. The gonopods curve evenly around the apex of the sternal sclerite. The posterolateral corners of the ninth sternite are not drawn out into distinct processes (**Fig. 2.489**). Male body length: 3.1 to 3.5 mm; fore-wing length: 2.2 to 2.5 mm. Female body length: 3.5 to 4.6 mm; fore-wing length: 2.3 to 2.8 mm. Color: light yellowish or reddish brown.
....*Brasilocaenis irmleri* Puthz, 1975 (Venezuela, Amazonas, Pará, Mato Grosso).

- The gonopods are recurved at the apex and extend beyond the apex of the sternal sclerite. The posterolateral corners of the ninth sternite form short processes (**Fig. 2.491**). Male body length: 2.5 to 3.0 mm; fore-wing length: 2.0 to 2.1 mm. Female body length: 2.7 to 3.5 mm; fore-wing length: 2.0 to 2.2 mm. Color: pale yellowish or reddish brown.
....*Brasilocaenis puthzi* Malzacher, 1975 (Venezuela, Amazonas, Pará, Mato Grosso).

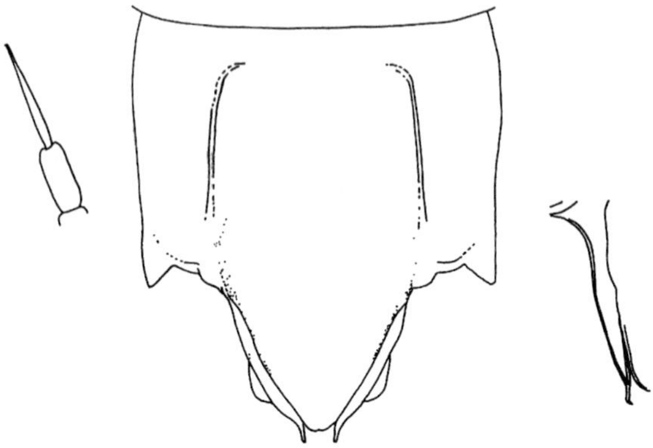

Fig. 2.491 *Brasilocaenis puthzi* male adult (left to right): the scape, pedicel, and base of the flagellum of the antenna; the genitalia; a gonopod enlarged.

Key to the species of *Brasilocaenis* larvae known from South America

Information for the key was taken from Malzacher (1986, 1990, 1998). The larva of *Brasilocaenis septentrionalis* has not been described.

1. The posterior margin of the ninth sternite is broadly truncate or slightly emarginated (**Fig. 2.492**). This characteristic is known only from a male larval exuvium. The female larva has not been described.
....*Brasilocaenis renata* Malzacher, 1986 (Amazonas). The larvae are reportedly very similar to those of the widespread *B. irmleri*, casting some doubt on the validity of this species.
- The posterior of margin of the ninth sternite is convex to somewhat truncate at the apex (**Fig. 2.493**). 2

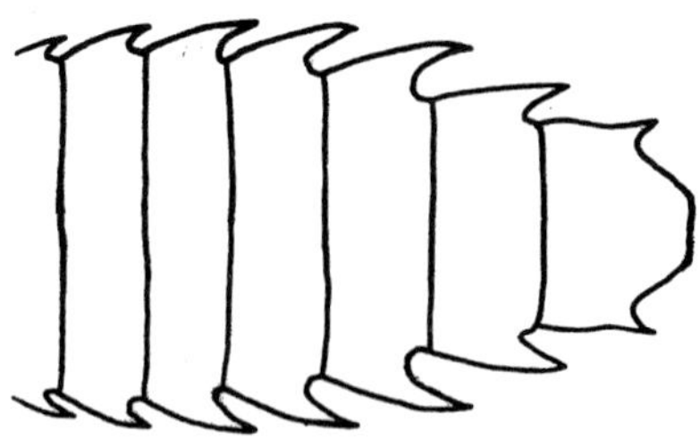

Fig. 2.492 Outline of the abdomen of a larval *Brasilocaenis renata.*

Fig. 2.493 *Brasilocaenis intermedia* larva (left to right): the coxal process of the middle (above) and hind (below) leg, the lateral margin of the pronotum of two different specimens, the posteromedian process of the second abdominal tergite in lateral view, the abdominal outline of a female larva, and the ninth sternites of two different male larvae.

2. The middle and hind coxal processes are broadly rounded apically (**Fig. 2.493**). The larvae were described only from exuvia. ....*Brasilocaenis intermedia* Malzacher, 1986 (Pará). Recently, Malzacher (1998) expressed doubt about the identity of this species.
- The middle and hind coxal processes are narrowly rounded to acute (**Fig. 2.494**). 3

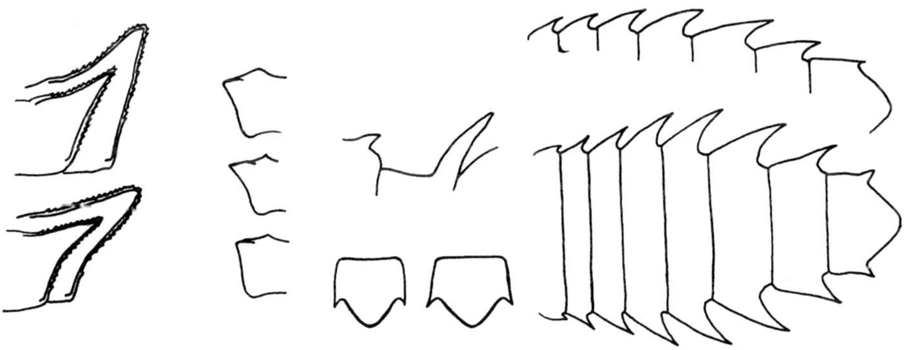

Fig. 2.494 *Brasilocaenis irmleri* larva (left to right): coxal processes of the middle (above) and hind (below) legs of two specimens superimposed, the lateral margin of the pronotum of three different individuals, the posteromedian process of the second abdominal tergite in lateral view (above), the ninth sternites of two different larvae, and the abdominal outline of a female larva.

3. The apex of the ninth sternite is clearly convex and evenly rounded. The middle and hind coxal processes are acutely pointed to narrowly rounded (**Fig. 2.494**). Length of the male larva: 2.9 to 3.2 mm; female larva: 3.6 to 4.2 mm.
....*Brasilocaenis irmleri* Puthz, 1975 (Amazonas, Pará).
- The apex of the ninth sternite is usually truncate. The middle and hind coxal process are almost acute but clearly rounded at the apex (**Fig. 2.473**). Length of the male larva: 2.6 to 3.0 mm; female larva: 3.3 to 3.8 mm.
....*Brasilocaenis puthzi* Malzacher, 1986 (Amazonas, Pará). The two species in this couplet are very similar, making the validity of *B. puthzi* doubtful.

Key to the species of *Cercobrachys* larvae known from South America

Information for the key was taken from Soldán (1986).

1. There is a submarginal ridge on the pronotum. The bases of the lateral ocellar tubercles are narrower than the ocellus. The second segment of the maxillary palp is half as broad as the first segment and more than twice as long. The gill cover is at least 1.6 times longer than wide. The posterolateral spines on the sixth abdominal segment are three times longer than wide in dorsal view (**Fig. 2.495**). Length: 4.5 to 4.7 mm. Length of cerci 1.3 to 1.4 mm. Color: head pale yellow, whitish between the ocelli, and yellowish brown on the frons, vertex, and occiput. Thorax and abdomen pale yellowish with some browning stippling and darker shading.
....*Cercobrachys peruanicus* Soldán, 1986 (Colombia).

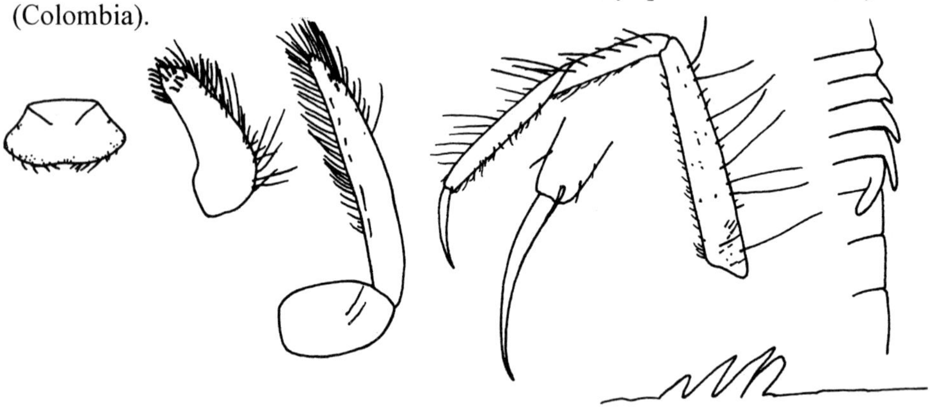

Fig. 2.495 *Cercobrachys peruanicus* larva (left to right): the labrum, labial and maxillary palps, middle leg above an enlargement of a claw, an outline of one side of the abdomen showing the posterolateral spines in dorsal (above) and lateral (below) view. The spine curving inward is on the sixth abdominal segment.

- The pronotum lacks a submarginal ridge. The bases of the lateral ocellar tubercles are the same width as the ocellus. The second segment of the maxillary palp is nearly as broad the first segment and 1.7 to 1.9 times as long. The gill cover is no more than 1.4 times longer than wide. The posterolateral spines on the sixth abdominal segment are five to six times longer than wide in dorsal view (**Fig. 2.472**). Length: 4 mm. Length of the cerci: 1.8 mm. Color: head and thorax dark yellow or yellowish brown with darker markings that are inconspicuous. The abdomen is dark yellowish with a dark median band on the first, second, seventh, eighth, and ninth tergites.

……..*Cercobrachys colombianus* Soldán, 1986 (Colombia, Amazonas).

Oniscigastridae

The only genus in this family known from South America is *Siphlonella* Needham and Murphy, 1924. These relatively large mayflies inhabit the cool streams of southern Chile.

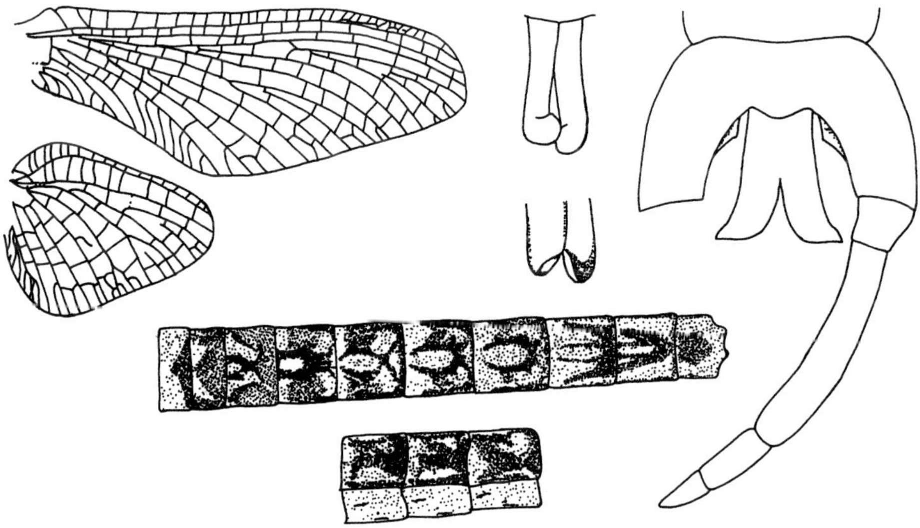

Fig. 2.496 *Siphlonella ventilans:* fore and hind wing (upper left), the male genitalia in ventral view with one of the forceps arms removed (upper right), the apex of the penis from a sub-adult (upper center) and an adult (center) in dorsolateral view, the dorsal color pattern on the abdomen above the fifth through seventh abdominal segments in lateral view (lower center).

Key to the species of adults in the genus *Siphlonella*

Information was provided by Pictet (1843-45), Needham and Murphy (1924), and Navás (1930e). The larva of *Siphlonella guttata* has not been described, so a key to the larvae cannot yet be produced, while that of *Siphlonella ventilans* is shown in **Figure 2.28**.

1. The abdomen is dark dorsally with yellow markings on the anterior parts of the middle tergites and sternites that are entirely pale except for single black dots at the spiracles (**Fig. 2.496**). The dark rings on the cerci are very narrow. Length: 10.5 to 11 mm. Fore-wing length: 10.5 to 11 mm. Hind wing length: 5 to 5.1 mm.
......................................*Siphlonella ventilans* Needham and Murphy, 1924 (Chile). Syn: *Siphlonella fazi* Navás, 1930; *Siphlonurus fazi* (Navás, 1930).
- The abdomen is yellow dorsally with narrowly black posterior borders of the segments and black oblique markings on each side of the mid-line (**Fig. 2.20**). There are also lateral rows of dashes and streaks visible in dorsal and ventral view. Broad and narrow dark rings alternate on the cerci.
.... *Siphlonella guttata* (Pictet, 1843) (Chile). Syn: *Baetis guttata* Pictet, 1843; *Ecdyonurus guttatus* (Pictet, 1843); *Heptagenia guttata* (Pictet, 1843); *Ecdyurus guttata* (Pictet, 1843).

Literature

Alba-Tercedor, J., and S. Mosquera, 1999: *Caenis chamie,* a new species from Colombia (Ephemeroptera: Caenidae). – Pan-Pacific Entomologist 75(2):61-67.

Allen, R. K., 1966: *Haplohyphes*, a new genus of Leptohyphinae (Ephemeroptera: Tricorythidae). Journal of the Kansas Entomological Society 39:565-568.

Allen, R. K., 1967: New species of New World Leptohyphinae (Ephemeroptera: Tricorythidae). - Canadian Entomologist 99:350-375.

Allen, R. K., 1973: New species of *Leptohyphes* Eaton (Ephemeroptera: Tricorythidae). - Pan-Pacific Entomologist 49:363-372.

Allen, R. K., and C. D. Murvosh, 1987: Mayflies (Ephemeroptera: Tricorythidae) of southwestern United States and northern Mexico. – Annals of the Entomological Society of America 80:35-40.

Allen, R. K., and S. S. Roback, 1969: New species and records of New World Leptohyphinae (Ephemeroptera: Tricorythidae). – Journal of the Kansas Entomological Society 42:372-379.

Banks, N., 1913: The Stanford Expedition to Brazil, 1911. Neuropteriod insects from Brazil. – Psyche 20:83-89 + 1 pl.

Banks, N., 1914: New neuropteroid insects, native and exotic. – Proceedings of the Academy of Natural Sciences of Philadelphia 66:608-632 + pl. 28.

Banks, N., 1918: New neuropteroid insects. – Bulletin of the Museum of Comparative Zoölogy at Harvard College 62(1):1-22 + 2 pl.

Berner, L., 1978: The status of *Asthenopus curtus* (Hagen) (Ephemeroptera: Polymitarcyidae). – Acta Amazonica 8(1):103-105.

Berner, L. and T. B. Thew, 1961: Comments on the mayfly genus *Campylocia* with a description of a new species (Euthyplociidae: Euthyplociinae). - American Midland Naturalist 66:329-336.

Burmeister, H., 1839: Ephemerina. Handbuch der Entomologie 2(2):788-804, 1016. – Berlin.

Chacón, M. M., S. Segnini, and E. Domínguez, 1999: Three new species of *Thraulodes* (Ephemeroptera: Leptophlebiidae: Atalophlebiinae) from Venezuela. – Aquatic Insects 21:249-257.

Cockerell, T. D. A., 1926: A new mayfly from Peru. – Bulletin of the Brooklyn Entomological Society 21:189-190.

Curtis, J., 1834: Descriptions of some nondescript British species of mayflies of anglers. – Philosophical Magazine Journal of Science 4:1-11.

Da Silva, E. R., 1993: Efemerópteros da Serra dos órgãos, Estado do Rio de Janeiro. II. Descrição de uma nova espécie de *Leptohyphes* Eaton, 1882 (Ephemeroptera, Tricorythidae). – Revista brasileira Ent. 37(2):313-316.

Da Silva, E. R., and S. M. Pereira, 1993: Efemerópteros da Serra dos Órgãos, Estado do Rio de Janeiro. III. Descrição de uma nova espécie de *Lachlania* Hagen, 1868 (Ephemeroptera: Oligoneuriidae). – Anais da Academia Brasileira de Ciências 65(3):295-301.

Day, W. C., 1955: New genera of mayflies from California. – Pan-Pacific Entomologist 31:121-137.

Day, W. C., 1963: Ephemeroptera. In: R. L. Usinger (Ed.), Aquatic Insects of California. - Univ. Calif. Press, Berkeley and Los Angeles. pp. 79-105.

Demoulin, G., 1952a: Contribution a l'étude des Ephoronidae Euthyplociinae (Insectes Ėphéméroptères). – Bulletin de la Institut royal des Sciences naturelles de Belgique 28(45):1-22.

Demoulin, G., 1952b: Un faux Baëtidae, *Bruchella* Navas (Insectes Ephéméroptères). - Bulletin et Annales de la Sociète Entomologique de Belgique 88:281-285.

Demoulin, G., 1952c: Essai de nouvelle cle, pour la determination des Oligoneuriidae (Insectes Ėphéméroptères). - Bulletin de la Institut royal des Sciences naturelles de Belgique 28(43):1-4.

Demoulin, G., 1952d: À propos des *Ameletopsis* Phillips, Ėphéméroptères paléantarctiques. – Bulletin et Annales de la Sociète Entomologique de Belgique 88:170-172.

Demoulin, G., 1955a: Une mission biologique Belge au Bresil. Ėphéméroptères. - Bulletin de la Institut royal des Sciences naturelles de Belgique 31(20):1.32.

Demoulin, G., 1955b: Ėphéméroptères nouveaux ou rares du Chili. - Bulletin de la Institut royal des Sciences naturelles de Belgique 31(22):1.15.

Demoulin, G., 1955c: Ėphéméroptères nouveaux ou rares du Chili. II. - Bulletin de la Institut royal des Sciences naturelles de Belgique 31(58):1.16.

Demoulin, G., 1955d: Ėphéméroptères nouveaux ou rares du Chili. III. - Bulletin de la Institut royal des Sciences naturelles de Belgique 31(73):1.30.

Demoulin, G., 1955e: Brèves remarques sur le genre *Metamonius* Eaton (Ėphéméroptères Siphlonuridae). - Bulletin de la Institut royal des Sciences naturelles de Belgique 31(24):1-3.

Demoulin, G., 1955f: Ėphéméroptères neotropicaux. Une synonymie probable: *Atalophlebia axillata* Navas = *Massartella brieni* (Lestage). – Bulletin et Annales de la Sociète Entomologique de Belgique 91:59-60.

Demoulin, G., 1966a: Contribution a l'étude des Ėphéméroptères du Surinam. - Bulletin de la Institut royal des Sciences naturelles de Belgique 42(37):1-22.

Demoulin, G., 1966b: Contribution a l'étude des Euthyplociidae. III. (Insectes Ėphéméroptères). – Zoologische Mededelingen 41(7):137-141.

Domínguez, E., 1982: Una especie nueva de Tricorythidae de la Argentina (Insecta: Ephemeroptera). - Revista de la Sociedad Entomológica Argentina 41:331-335.

Domínguez, E., 1984: Dos especies nuevas del genero *Haplohyphes* Allen (Ephemeroptera: Tricorythidae) de la Argentina. – Revista de la Sociedad Entomológica Argentina 43:103-112.

Domínguez, E., 1986: *Thraulodes bolivianus*, una nueva especie de la familia Leptophlebiidae, Insecta: Ephemeroptera) de Bolivia. - Acta Zoologica Lilloana 38(2):149-153.

Domínguez, E., 1987: El genero *Thraulodes* (Ephemeroptera: Leptophlebiidae) en la Republica Argentina. – Acta Zoologica Lilloana 39:47-65.

Domínguez, E., 1988a: *Asthenopus gilliesi* sp. n. y su importancia en la taxonomia de la subfamilia Asthenopodinae (Ephemeroptera: Polymitarcyidae). – Anales Museo de Historia Natural Valparaiso 19:21-26.

Domínguez, E., 1988b: *Ecuaphlebia*: a new genus of Atalophlebiinae (Ephemeroptera: Leptophlebiidae) from Ecuador. – Aquatic Insects 10(4):227-235.

Domínguez, E., 1991: The status of the genus *Ulmeritus* (Ephemeroptera: Leptophlebiidae: Atalophlebiinae) and related taxa. In: J. Alba-Tercedor and A. Sanchez-Ortega (Eds.), Overviews and strategies of Ephemeroptera and Plecoptera. – Sandhill Crane Press, Gainesville, Florida. pp. 157-168.

Domínguez, E., 1995a: Description of a new subgenus and a new species of *Traverella* Edmunds, with comments on its phylogeny and biology. In: L. D. Corkum and J. J. H.

Ciborowski (Eds.), Current Directions in Research on Ephemeroptera. – Canadian Scholars' Press, Toronto. pp. 423-431.

Domínguez, E., 1995b: Cladistic analysis of the *Ulmeritus-Ulmeritoides* group (Ephemeroptera, Leptophlebiidae), with descriptions of five new species of *Ulmeritoides*. – Journal of the New York Entomological Society 103(1):15-38.

Domínguez, E., 1999: Systematics, cladistics and biogeography of the American genus *Farrodes* (Ephemeroptera: Leptophlebiidae: Atalophlebiinae). Zoological Journal of the Linnaean Society 126:155-189.

Domínguez, E., M. J. Ferreira, and C. Nieto, 2001: Redescription and phylogenetic relationships of *Leentvaaria* Demoulin (Ephemeroptera: Leptophlebiidae). In: E. Domínguez (Ed.), Trends in Research in Ephemeroptera and Plecoptera. Kluwer, Dordrecht. pp. 313-320.

Domínguez, E., and R. W. Flowers, 1989: A revision of *Hermanella* and related genera (Ephemeroptera: Leptophlebiidae; Alalophlebiinae) from subtropical South America. – Annals of the Entomological Society of America 82(5):555-573.

Domínguez, E., M. D. Hubbard, and W. L. Peters, 1992: Clave para ninfas y adultos de las familias y géneros de Ephemeroptera (Insecta) sudamericanos. - Biologia Acuatica 16:1-32 + 8 pl.

Domínguez, E., M. D. Hubbard, and W. L. Peters, 1995: Insecta Ephemeroptera. In: E. C. Lopretto and G. Tell (Eds.), Ecosistemas de Aguas Continentales. Ed. Sur, La Plata. pp. 1069-1089.

Domínguez, E., M. D. Hubbard, and W. L. Peters, 2001: Checklist of the Ephemeroptera of South America. – Ephemeroptera Galactica, Internet, http://www.famau org/mayfly/sacat.html.

Domínguez, E., and C. Molineri, 1996: A new species of *Atopophlebia* (Ephemeroptera: Leptophlebiidae: Atalophlebiinae) from Bolivia and Argentina. – Aquatic Insects 18(1):55-59.

Domínguez, E., and M. L. Pescador, 1983: A new species of *Penaphlebia* (Ephemeroptera: Leptophlebiidae) from Argentina. – Entomological News 94(1):21-24.

Domínguez, E., W. Peters, J. G. Peters, and H. M. Savage, 1997: The imago of *Simothraulopsis* Demoulin with a redescription of the nymph (Ephemeroptera: Leptophlebiidae: Atalophlebiinae). – Aquatic Insects 19:141-150.

Domínguez, E., and H. M. Savage, 1987: Two new species of *Farrodes* from continental South America with comments on the distribution of faunal components in Argentina

(Ephemeroptera: Leptophlebiidae). – Studies on Neotropical Fauna and Environment 22(1):43-52.

Dominique, Y., A. Thomas, K. Orth, and C. Dauta, 2000: Les Ephémères de la Guyane Française. 2. *Camelobaetidius billi* et *C. janae* n. spp. (Ephemeroptera, Baetidae). – Ephemera 2(1):39-48.

Eaton, A. E., 1868: An outline and re-arrangement of the genera of Ephemeridae. Entomological Monthly Magazine 5:82-91.

Eaton, A. E., 1871: A monograph on the Ephemeridae. – Transactions of the Royal Entomological Society of London 1871:1-164 + pl. 1-6.

Eaton, A. E., 1881: An announcement of new genera of the Ephemeridae. – Entomologist's Monthly Magazine 18:21-27.

Eaton, A. E., 1882: An announcement of new genera of the Ephemeridae. – Entomologist's Monthly Magazine 18:207-208.

Eaton, A. E., 1883-1888: A revisional monograph of recent Ephemeridae or mayflies. – Transactions of the Linnean Society of London, Second Series, Zoology 3:1-352 + 65 pl.

Edmunds, G. F., Jr., 1959: Ephemeroptera. In: W. T. Edmondson (Ed.), Fresh-Water Biology. 2nd ed. University of Washington, Seattle. pp. 908-916.

Edmunds, G. F., Jr., 1963: A new genus and species of mayfly from Peru. - Pacific Entomologist 39:34-36.

Edmunds, G. F., Jr., 1984: 10. Ephemeroptera. In: R. W. Merritt and K. W. Cummins (Eds.), An Introduction to the Aquatic Insects of North America. 2nd ed. - Kendall, Dubuque. pp. 94-125.

Edmunds, G. F., Jr., R. K. Allen, and W. L. Peters, 1963: An annotated key to the nymphs of the families and subfamilies of mayflies (Ephemeroptera). – University of Utah Biological Series 13(1):1-55.

Edmunds, G. F., S. L. Jensen, and L. Berner, 1976: The Mayflies of North and Central America. University of Minnesota, Minneapolis. 330 pp.

Elliott, J. M., and U. H. Humpesch, 1983: A key to the adults of the British Ephemeroptera. – Freshwater Biological Association, Scientific Publication 47. Titus Wilson, Kendal. 101 pp.

Esben Petersen, T., 1909: New Ephemeridae from Denmark, Arctic Norway and the Argentine Republic. Deutsche Entomologische Zeitschrift 1909:551-556.

Esben Petersen, T., 1912: New and little-known species of Ephemerida from Argentine (Neuropt.). Deutsche Entomologische Zeitschrift 1912:333-342.

Ferreira, M. J. N., and E. Domínguez, 1992: A new species of *Hermanella* (Ephemeroptera: Leptophlebiidae: Atalophlebiinae) from Southeastern Brazil. – Aquatic Insects 14(3):179-182.

Ferreira, M. J. N., and C. G. Froehlich, 1992: Estudo da fauna de Ephemeroptera (Insecta) do córrego do Pedregulho (Pedregulho, SP, Brasil) com aspectos da biologia de *Thraulodes schlingeri* Traver & Edmunds, 1967. – Revista brasileira Ent. 36(3):541-548.

Flowers, R. W., 1980: *Atopophlebia fortunensis*, a new genus and species from Panama (Leptophlebidae: Ephemeroptera). - Florida Entomologist 63:163-165.

Flowers, R. W., 1985: *Guajirolus*, a new genus of neotropical Baetidae (Ephemeroptera). – Studies on Neotropical Fauna and Environment 20(1):27-31.

Flowers, R. W., 1987: New species and life stages of *Atopophlebia* (Ephemeroptera: Leptophlebiidae: Atalophlebiinae). - Aquatic Insects 9:203-209.

Flowers, R. W., and E. Domínguez, 1991: Preliminary cladistics of the *Hermanella* complex (Ephemeroptera: Leptophlebiidae, Atalophlebiinae). In: J. Alba-Tercedor and A. Sanchez-Ortega (Eds.), Overview of Strategies of Ephemeroptera and Plecoptera. – Sandhill Crane Press, Gainesville. pp. 49-62.

Flowers, R. W., and E. Domínguez, 1992: New genus of Leptophlebiidae (Ephemeroptera) from Central and South America. – Annals of the Entomological Society of America 85:655-661.

Froehlich, C. G., 1969: *Caenis cuniana* sp. n., a parthenogenetic mayfly. Beiträge zur neotropischen Fauna 6:103-108.

Gillies, M. T., 1990: A revision of the Argentine species of *Callibaetis* Eaton (Baetidae: Ephemeroptera). – Revista de la Sociedad Entomológica Argentina 48(1-4):15-39.

Gillies, M. T., 2001: A new species of *Aturbina* (Ephemeroptera, Baetidae) Lugo-Ortiz and McCafferty from Uruguay. In: E. Domínguez (Ed.), Trends in Research in Ephemeroptera and Plecoptera. Kluwer, Dordrecht. pp. 331-335.

Hagen, H., 1855: Die Ephemeren-Gattung *Oligoneuria*. Stettiner entomologische Zeitung 16:262-270 + 1 pl.

Hagen, H., 1861: Synopsis of the Neuroptera of North America with a list of the South American species. – Smithsonian Miscellaneous Collection. XX + 347 pp.

Heckman, C. W., 1998a: The seasonal succession of biotic communities in wetlands of the tropical wet-and-dry climatic zone: V. Climatological influences on the life cycles of the invertebrates in the Pantanal of Mato Grosso, Brazil. – Internationale Revue der gesamten Hydrobiologie 83:31-63.

Heckman, C. W., 1998b: The Pantanal of Poconé. – Kluwer, Dordrecht. 622 pp.

Hofmann, C., M. Sartori, and A. Thomas, 1999: Les Ephéméroptères (Ephemeroptera) de la Guadalupe (petites Antilles Françaises). Mémoires de la Société Vaudoise des Sciences Naturelles 20(1):1-96.

Hubbard, M. D., 1974: Some specific homonyms in mayflies (Ephemeroptera). – Journal of the Kansas Entomological Society 47:357-359.

Hubbard, M. D., 1979: Genera and subgenera of recent Ephemeroptera. Eatonia, Supplement 2:1-8.

Hubbard, M. D., 1982: Catálogo abreviado de Ephemeroptera da América do Sul. – Papéis Avulsos de Zoologia 34(24):257-282.

Hubbard, M. D., and E. Domínguez, 1988: Synonymy of the Neotropical mayfly genera *Asthenopus* and *Asthenopodes* (Ephemeroptera: Polymitarcyidae: Asthenopodinae). - Florida Entomologist 71:207-210.

Keffermüller, M., 1979: Un changement dans la taxonomie de la familie Baetidae. In: J. F. Flannagan and K. E. Marshall (Eds.), Advances in Ephemeroptera Biology. - Plenum, New York. pp. 115-121.

Kluge, N. J., D. Studemann, P. Landolt, and T. Gonser, 1995: A reclassification of Siphlonuroidea (Ephemeroptera). – Mitteilungen der Schweizerischen Entomologischen Gesellschaft 68:103-132.

Lestage, J. A., 1924: *Atalophlebia Brieni*, Ėphéméroptère nouveau du Brésil. – Annales de la Sociète Entomologique de Belgique 64:21-24.

Lestage, J. A., 1930: Notes sur le genre *Massartella* nov. gen. de la famille Leptophelbiidae (Ephemeroptera) et la génotype *Massartella brieni* Lest. – Mission Biologique Belge au Brésil 2:249-258.

Lestage, J. A., 1931: Contribution à l'étude des Ėphéméroptères. VIII. Les Ėphéméroptères du Chili. - Bulletin et Annales de la Sociète Entomologique de Belgique 71:41-60.

Lestage, J. A., 1935: Contribution à l'étude des Ėphéméroptères. IX. Le groupe Siphlonuridien. - Bulletin et Annales de la Sociète Entomologique de Belgique 75:77-139.

Lugo-Ortiz, C. R., and W. P. McCafferty, 1994: The mayfly genus *Acerpenna* (Insecta, Ephemeroptera, Baetidae) in Latin America. – Studies on Neotropical Fauna and Environment 29:69-74.

Lugo-Ortiz, C. R., and W. P. McCafferty, 1995: Three distinctive new genera of Baetidae (Insecta, Ephemeroptera) from South America. – Annales de Limnologie 31(4):233-243.

Lugo-Ortiz, C. R., and W. P. McCafferty, 1996a: Taxonomy of the Neotropical genus *Americabaetis*, new status (Insecta: Ephemeroptera: Baetidae). Studies on Neotropical Fauna and Environment 31:156-169.

Lugo-Ortiz, C. R., and W. P. McCafferty, 1996b: *Aturbina georgei* gen. et sp. n.: a small minnow mayfly (Ephemeroptera: Baetidae) without turbinate eyes. – Aquatic insects 18:175-183.

Lugo-Ortiz, C. R., and W. P. McCafferty, 1996c: The genus *Paracloeodes* (Insecta: Ephemeroptera: Baetidae) and its presence in South America. – Annales de Limnologie 32(3):161-169.

Lugo-Ortiz, C. R., and W. P. McCafferty, 1996d: Phylogeny and classification of the *Baetodes* complex (Ephemeroptera: Baetidae), with description of a new genus. – Journal of the North American Benthological Society 15(3):367-380.

Lugo-Ortiz, C. R., and W. P. McCafferty, 1997: First report and new species of the genus Apobaetis (Ephemeroptera: Baetidae) from South America. – Aquatic insects 19(4):243-246.

Lugo-Ortiz, C. R., and W. P. McCafferty, 1998: Five new genera of Baetidae (Insecta: Ephemeroptera) from South America. – Annales de Limnologie 34(1):57-73.

Lugo-Ortiz, C. R., and W. P. McCafferty, 1999a: Three new genera of small minnow mayflies (Insecta: Ephemeroptera: Baetidae) from the Andes and Patagonia. – Studies on Neotropical Fauna and Environment 34:88-104.

Lugo-Ortiz, C. R., and W. P. McCafferty, 1999b: An unusual new species of *Camelobaetidius* (Ephemeroptera: Baetidae) from Paraguay. – Entomological News 110(4):221-224.

Lugo-Ortiz, C. R., W. P. McCafferty, and R. D. Waltz, 1994: Contribution to the taxonomy of the Panamerican genus *Fallceon* (Ephemeroptera: Baetidae). – Journal of the New York Entomological Society 102(4):460-475.

Malzacher, P., 1986: Caenidae aus dem Amazonasgebiet (Insecta, Ephemeroptera). – Spixiana 9(1):83-103.

Malzacher, P., 1990: Neue Arten der Eintagsfliegen-Familie Caenidae (Insecta, Ephemeroptera) aus Südamerika. – Studies on Neotropical Fauna and Environment 25(1):31-39.

Malzacher, P., 1998: Remarks on the genus *Brasilocaenis* (Ephemeroptera: Caenidae), with the descriptions of a new species: *Brasilocaenis mendesi*. – Stuttgarter Beiträge zur Naturkunde, Serie A (Biologie) 580:1-6.

Mayo, V. K., 1968a: Some new mayflies of the subfamily Leptohyphinae. – Pan-Pacific Entomologist 44(4):301-308.

Mayo, V. K., 1968b: Two new species of the genus *Baetodes* from Ecuador. - Pan-Pacific Entomologist 44:251-257.

Mayo, V. K., 1972: New species of the genus *Baetodes* (Ephemeroptera: Baetidae). - Pan-Pacific Entomologist 48:226-241.

Mayo, V. K., 1973a: Four new species of the genus *Baetodes* (Ephemeroptera: Baetidae). - Pan-Pacific Entomologist 49:308-314.

Mayo, V. K., 1973b: A new species of *Baetis* from Ecuador (Ephemeroptera: Baetidae). - Pan-Pacific Entomologist 49:285-288.

McCafferty, W. P., 1968: The mayfly genus *Hexagenia* in Mexico. – Proceedings of the Entomological Society of Washington 70:358-359.

McCafferty, W. P., 1970: Neotropical nymphs of the genus *Hexagenia* (Ephemeroptera: Ephemeridae). Journal of the Georgia Entomological Society 5:224-228.

McCafferty, W. P., 1991: Toward a phylogenetic classification of the Ephemeroptera (Insecta): a commentary on systematics. – Annals of the Entomological Society of America 84:343-360.

McCafferty, W. P., 2000: Notations on South American Baetidae (Ephemeroptera). – Entomological News 111(5):375-379.

McCafferty, W. P., and C. R. Lugo-Ortiz, 1995: *Cloeodes hydation*, n. sp. (Ephemeroptera: Baetidae): an extraordinary, drought tolerant mayfly from Brazil. – Entomological News 106(1):29-35.

Meglitsch, P. A., 1967: Invertebrate Zoology. Oxford University Press, Oxford. 901 pp.

Mol, A. W. M., 1986: *Harpagobaetis gulosus* gen. nov., spec. nov., a new mayfly from Suriname (Ephemeroptera: Baetidae). – Zoologische Mededelingen 60(4):63-70.

Molineri, C., 1999a: Revision of the genus *Tricorythopsis* (Ephemeroptera: Leptohyphidae) with the Description of four new species. - Aquatic Insects 21:285-300.

Molineri, C., 1999b: Descripción de una nueva especie de *Haplohyphes* Allen (Ephemeroptera: Leptohyphidae) y clave para separar los imagos machos del género. Acta Zoologica Lilloana 45(1):29-32.

Molineri, C., 2001a: A new genus of Leptohyphidae (Insecta: Ephemeroptera). In: E. Domínguez (Ed.), Trends in Research in Ephemeroptera and Plecoptera. Kluwer, Dordrecht. pp. 337-345.

Molineri, C., 2001b: *Traverhyphes*: a new genus of Leptohyphidae for *Leptohyphes* indicator and related species (Insecta, Ephemeroptera). – Spixiana 24(2):129-140.

Molineri, C., 2002a (2001): El género *Tricorythopsis* (Ephemeroptera: Leptohyphidae): nuevas combinaciones y descripción de neuvas especies y estadios. – Revista de la Sociedad Entomologica Argentina 60:217-238.

Molineri, C., 2002b (2001): Una especie nueva de *Tricorythodes* (Ephemeroptera: Leptohyphidae) de la Argentina con notas sobre su biología. – Revista de la Sociedad Entomologica Argentina 60:61-66.

Molineri, C., and R. W. Flowers, 2001: Primera cita de *Allenhyphes* Hofmann y Sartori, 1999 (Ephemeroptera: Leptohyphidae) para Amárica continental. – Entomotropica 16(1):61-62.

Morihara, D. K., and G. F. Edmunds, Jr., 1980: Notobaetis: a new genus of Baetidae (Ephemeroptera) from South America. – Internationale Revue der gesamten Hydrobiologie 65:605-610.

Navás, R. P. L., 1912: Neurópteros nuevos de América. - Broteria 10:194-202.

Navás, R. P. L., 1915: Neuropteros Sudamericanos. – Broteria (Serie Zoologica) 13:5-13.

Navás, R. P. L., 1916: Neuroptera nova Americana. – Memorie della Pontificia Accademia Romana Dei Nuovi Lincei. 2nd Series 2:59-80.

Navás, R. P. L., 1918: Insectos Chilenos. – Boletin Sociedad Aragonesa de Ciencias Naturales 17:212-230.

Navás, R. P. L., 1919 (1918): Algunos insectos neurópteros de la Argentina. Serie 2ª (1). – Physis 4:80-89.

Navás, R. P. L., 1920a: Algunos insectos del Brasil. 3ª Serie. Revista do Museu Paulista 12:413-417.

Navás, R. P. L., 1920b: Insectos Sudamericanos. 3ª Serie. - Anales de la Sociedad Cientifica Argentina 90:52-72.

Navás, R. P. L., 1920c (1919): Insecta nova. VI Series. – Memorie della Pontificia Accademia Romana Dei Nuovi Lincei. 2nd Series 5:11-20.

Navás, R. P. L., 1920d: Algunos insectos de Santa Fe (Republica Argentina) recogidos por el P. Juan C. Muhn, S.J. – Estudios (Buenos Aires) 18:131-135.

Navás, R. P. L., 1920e: Insectos Sudamericanos. 1ª Serie. - Anales de la Sociedad Cientifica Argentina 90:33-43.

Navás, R. P. L., 1922a: Efemerópteros nuevos o poco conocidos. – Boletin de la Sociedad Entomologica de España 5:54-63.

Navás, R. P. L., 1922b: Insectos nuevos o poco conocidos. – Memorias de la Real Academia de Ciencias y Artes de Barcelona. Tercera Época – 17(15):383-400.

Navás, R. P. L., 1923a (1921): Algunos insectos de Chile. – Revista Chilena de Historia Natural 25:443-445.

Navás, R. P. L., 1923b (1922): Estudos sobre neuròpters (Insectes). – Arxius de l'Institut de Ciències (Barcelona) 7:179-203.

Navás, R. P. L., 1924: Insectos de la Argentina y Chile. – Estudios 22:358-368.

Navás, R. P. L., 1925: Insectos Neotrópicos. – 1.a Serie. - Revista Chilena de Historia Natural 29:305-313.

Navás, R. P. L., 1926a: Insectos de la Argentina y Chile. Secunda Serie – Estudios 31:103-111.

Navás, R. P. L., 1926b: Algunos insectos del Museo de Paris. 3ª Serie. Broteria (Série Zoológica) 23:95-115.

Navás, R. P. L., 1926c: Efemerópteros nuevos de la República Argentina. Revista de la Sociedad Entomológica Argentina 1(2):33-35.

Navás, R. P. L., 1927: Insectos nuevos de la República Argentina. Revista de la Sociedad Entomológica Argentina 1(3):27-29 + pl. 3.

Navás, R. P. L., 1929: Insectos de la Argentina. Quinta Serie. Revista de la Sociedad Entomológica Argentina 2(10):219-225 + pl. 3.

Navás, R. P. L., 1930a: Insectos de la Argentina. Sexta Serie. Revista de la Sociedad Entomológica Argentina 3(13):125-132.

Navás, R. P. L., 1930b (1929): Algunos insectos de Chile. 3.ª Serie. – Revista Chilena de Historia Natural 33:326-334.

Navás, R. P. L., 1930c: Insectos Neotropicos. 6.ª Serie (1). - Revista Chilena de Historia Natural 34:62-75.

Navás, R. P. L., 1930d: Algunos insectos de Guayaquil (Ecuador). – Revista Chilena de Historia Natural 34:18-19.

Navás, R. P. L., 1930e (1929): Algunos insectos de Chile. 4.ª Serie (1). – Revista Chilena de Historia Natural 34:350-366.

Navás, R. P. L., 1931a (1929): Insectos del Brasil. 4.ª Serie. Revista do Museu Paulista 17:455-458.

Navás, R. P. L., 1931b: Insectos de la Argentina. Séptima Serie. Revista de la Sociedad Entomológica Argentina 3(17):317-324.

Navás, R. P. L., 1932: Insectos de la Argentina. Revista de la Academia de Ciencias exactas, fisicoquimicas y naturales de Zaragosa 16:87-120.

Navás, R. P. L., 1933a: Algunos insectos de Chile. 4. Serie. – Revista Chilena de Historia Natural 37:230-234.

Navás, R. P. L., 1933b (1932): Insectos de la Argentina y Chile. Revista de la Sociedad Entomológica Argentina 22(5):79-86.

Navás, R. P. L., 1934a: Insectos Suramericanos. Octava Serie. - Revista de la Academia Cientifica de Madrid 31(11):9-28.

Navás, R. P. L., 1934b: Insectos Suramericanos. Novena Serie. - Revista de la Academia Cientifica de Madrid 31:155-184.

Navás, R. P. L., 1936a (1935): Algunos insectos de Chile. 7ª Serie.- Revista Chilena de Historia Natural 39:138-140.

Navás, R. P. L., 1936b (1931): Insectos del Brasil. 5ª Serie. Revista do Museu Paulista 20:721-734.

Needham, J. G., 1899: Directions for collecting and rearing dragon flies, stone flies, and May flies. – Bulletin of the United States National. Museum No. 39, part. 9 pp.

Needham, J. G., and H. E. Murphy, 1924: Neotropical Mayflies. - Bulletin of the Lloyd Library of Botany, Pharmacy and Materia Medica 24 (Entomological Series 4):1-79.

Nolte, U., R. S. Tietboehl, and W. P. McCafferty, 1996: A mayfly from tropical Brazil capable of tolerating short-term dehydration. Journal of the North American Benthological Society 15: 87-94.

Orth, K., A. Thomas, C. Dauta, V. Horeau, S. Brosse, and C. Ademmer, 2000: Les Ephémères de la Guyane Française. 1. Primier inventaire générique, à but de biosurveillance (Ephemeroptera). – Ephemera 2(1):25-38.

Pennak, R. W., 1978: Fresh-water Invertebrates of the United States. 2nd ed. – Wiley Interscience, New York. 803 pp.

Percheron, A., 1838: Vol. 6. In: Guérin-Meneville and Percheron (1835-38): Genera des Insectes ou exposition detaillée de tous le caractères propres à chacun des genres de cette classe d'animaux. – Méquiguen-Marvis, Paris.

Pereira, S. M., 1989: Descrição de uma nova espécie de *Lachlania* Hagen, 1868, imago macho, da Argentina (Ephemeroptera, Oligoneuridae). – Boletim do Museu Nacional, Rio de Janeiro, Nova Série Zoologia 326:1-4.

Pereira, S. M., and E. R. da Silva, 1990: Nova espécie de *Caenis* Stephens, 1835 do sudeste do Brasil (Ephemeroptera, Caenidae). – Boletim do Museu Nacional, Nova Série, No. 341:1-8.

Pereira, S. M., and E. R. da Silva, 1991: Descrição de uma nova espécie de *Campsurus* Eaton, 1868 do sudeste do Brasil, com notas biológicas (Ephemeroptera: Polymitarcyidae: Campsurinae). – Revista Brasileira Biol. 51(2):321-326.

Pescador, M. L., 1997: *Gonserellus*: a new genus of Leptophlebiidae (Ephemeroptera) from southern South America. – Aquatic Insects 19(4):237-242.

Pescador, M. L., and G. F. Edmunds, Jr., 1994: A new genus of Oligoneuriidae (Ephemeroptera) from South America. – Annals of the Entomological Society of America 87:263-269.

Pescador, M. L., and T. Gonser, 2001: The nymph of *Secochela illiesi* (Ephemeroptera: Leptophlebiidae: Atalophlebiinae) from South America. In: E. Domínguez (Ed.), Trends in Research in Ephemeroptera and Plecoptera. Kluwer, Dordrecht. pp. 347-352.

Pescador, M. L., and W. L. Peters, 1979: Phylogenetic relationships and zoogeography of cool-adapted Leptophlebiidae (Ephemeroptera) in Southern South America. In: J. F. Flannagan and K. E. Marshall (Eds.), Advances in Ephemeroptera Biology. – Plenum, New York. pp. 43-56.

Pescador, M. L., and W. L. Peters, 1980: A revision of the genus *Homoeoneuria* (Ephemeroptera: Oligoneuriidae). – Transactions of the American Entomological Society 106:357-393.

Pescador, M. L., and W. L. Peters, 1982: Four new genera of Leptophlebiidae (Ephemeroptera: Atalophlebiinae) from southern South America. - Aquatic Insects 4:1-19.

Pescador, M. L., and W. L. Peters, 1985: Biosystematics of the genus *Nousia* from southern South America (Ephemeroptera: Leptophlebiidae: Atalophlebiinae). – Journal of the Kansas Entomological Society 58(1):91-123.

Pescador, M. L., and W. L. Peters, 1987: Revision of the genera *Meridialaris* and *Massartellopsis* (Ephemeroptera: Leptophlebiidae: Atalophlebiinae) from South America. – Transactions of the American Entomological Society 112:147-189.

Pescador, M. L., and W. L. Peters, 1990: Biosystematics of the genus *Massartella* Lestage (Ephemeroptera: Leptophlebiidae: Atalophlebiinae) from South America. - Aquatic Insects 12:145-160.

Pescador, M. L., and W. L. Peters, 1991: Biosystematics of the genus *Penaphlebia* (Ephemeroptera: Leptophlebiidae: Atalophlebiinae) from South America. – Transactions of the American Entomological Society 117:1-38.

Peters, W. L., 1969: *Askola froehlichi* a new genus and species from southern Brazil (Leptophlebiidae: Ephemeroptera). - Florida Entomologist 52:253-258.

Peters, W. L., 1981: *Coryphorus aquilus*, a new genus and species of Tricorythidae from the Amazon Basin (Ephemeroptera). - Aquatic Insects 3:209-217.

Peters, W. L., and E. Domínguez, 2001: The identity of *Hagenulopsis minuta* Spieth (Leptophlebiidae: Atalophlebiinae). – In: In: E. Domínguez (Ed.), Trends in Research in Ephemeroptera and Plecoptera. Kluwer, Dordrecht. pp. 353-358.

Peters, W. L., and G. F. Edmunds, 1972: A revision of the generic classification of certain Leptophlebiidae from southern South America (Ephemeroptera). – Annals of the Entomological Society of America 65(6):1398-1414.

Pictet, F. J., 1843-45: Histoire naturelle générale et particulière des insects néuroptères. Famille des éphémérines. – J. Kessmann et Ab. Cherbuliz, Geneva. XIX + 300 pp. + 47 pl.

Puthz, V., 1973: Eintagsfliegen (Ephemeroptera) aus Südamerika. – Opuscula Zool. Budapest 12:91-97.

Puthz, V., 1975: Eine neue Caenidengattung aus dem Amazonasgebiet (Insecta, Ephemeroptera: Caenidae). - Amazoniana 5:411-415.

Sattler, W., 1967: Über die Lebensweise, insbesondere das Bauverhalten, neotropischer Eintagsfliegen-Larven (Ephemeroptera, Polymitarcidae). – Beiträge zur neotropischen Fauna 5(2):89-110.

Savage, H. M., 1982: A curious new genus and species of Atalophlebiinae (Ephemeroptera: Leptophlebiidae) from the southern coastal mountains of Brazil. – Studies on Neotropical Fauna and Environment 17:209-217.

Savage, H. M., 1983: *Perissophlebiodes*, a replacement name for *Perissophlebia* Savage nec Tillyard (Ephemeroptera: Leptophlebiidae). – Entomological News 94(5):204.

Savage, H. M., 1983: Wing evolution within *Miroculis* and related genera (Ephemeroptera: Leptophlebiidae) from northern South America. – Zeitschrift für zoologische Systematik und Evolutionsforschung 21:124-142.

Savage, H. M., 1986: Systematics of the *Terpides* lineage from the Neotropics: definition of the *Terpides* lineage, methods, and revision of *Fittkaulus* Savage and Peters (Insecta, Ephemeroptera, Leptophlebiidae). – Spixiana 9(3):255-270.

Savage, H. M., 1987a: Biogeographic classification of Neotropical Leptophlebiidae (Ephemeroptera) based upon geological centers of ancestral origin and ecology. – Studies on Neotropical Fauna and Environment 22(4):199-222.

Savage, H. M., 1987b: Two new species of *Miroculis* from Cerro de la Neblina, Venezuela with new distribution records for *Miroculis fittkaui* and *Microphlebia surinamensis* (Ephemeroptera: Leptophlibiidae). – Aquatic Insects 9(2):97-108.

Savage, H. M., and E. Domínguez, 1992: A new genus of Atalophlebiinae (Ephemeroptera: Leptophlebiidae) from northern South America. - Aquatic Insects 14:243-248.

Savage, H. M., and W. L. Peters, 1978: *Fittkaulus maculatus*, a new genus and species from northern Brazil (Leptophlebiidae: Ephemeroptera). - Acta Amazonica 8:293-298.

Savage, H. M., and W. L. Peters, 1983: Systematics of *Miroculis* and related genera from northern South America (Ephemeroptera: Leptophlebiidae). – Transactions of the American Entomological Society 108:491-600.

Soldán, T., 1986: A revision of the Caenidae with ocellar tubercles in the nymphal stage (Ephemeroptera). – Acta Universitatis Carolinae – Biologica 1982-1984:289-362.

Spieth, H. T., 1941: Taxonomic studies on the Ephemeroptera. II. The genus *Hexagenia*. - American Midland Naturalist 26:233-280.

Spieth, H. T., 1943: Taxonomic studies on the Ephemeroptera. III. Some interesting ephemerids from Surinam and other Neotropical localities. - American Museum Novitates 1244:1-13.

Thew, T. B., 1960: Taxonomic studies on some Neotropical leptophlebiid mayflies (Ephemeroptera: Leptophlebiidae). Pan-Pacific Entomologist 36:119-132.

Thomas, A., K. Orth, V. Horeal, and Y. Dominique, 2000: Les Ephémères de la Guyane Française. 3. *Bessierus doloris* n. gen., n. sp. (Ephemeroptera, Leptophlebiidae). – Ephemera 2(1):49-57.

Traver, J. R., 1938: Mayflies of Puerto Rico. – The Journal of Agriculture of the University of Puerto Rico 22(1):5-42 + 3 pl.

Traver, J. R., 1943: New Venezuelan Mayflies. - Bolletin de Entomologia Venezolana 2:79-98.

Traver, J. R., 1944: Notes on Brazilian mayflies. – Boletim do Museu Nacional, Rio de Janeiro, Zoologia 22:1-53.

Traver, J. R., 1946: Notes on Neotropical mayflies. Part I. Family Baetidae, subfamily Leptophlebiinae. - Revista de Entomologia 17:416-436.

Traver, J. R., 1947a: Notes on Neotropical mayflies. Part III. Family Ephemeridae. - Revista de Entomologia 18:370-395.

Traver, J. R., 1947b: Notes on Neotropical mayflies. Part II. Family Baetidae, subfamily Leptophlebiinae. – Revista de Entomologia 18:149-160.

Traver, J. R., 1950: Notes on Neotropical mayflies. Part IV. Family Ephemeridae (continued). – Revista de Entomologia 21:593-614.

Traver, J. R., 1956: The genus *Asthenopodes* (Ephemeroptera). - Comunicaciones Zoologicas del Museo de Historia Natural de Montevideo 75(4):1-10.

Traver, J. R., 1958: The subfamily Leptohyphinae (Ephemeroptera: Tricorythidae), Part I. – Annals of the Entomological Society of America 51:491-503.

Traver, J. R., 1959a: The subfamily Leptohyphinae. Part II: five new species of *Tricorythodes*. - . Proceedings of the Entomological Society of Washington 61:121-131.

Traver, J. R., 1959b: Uruguayan mayflies. – Revista de la Sociedad Uruguaya de Entomologia 3:1-13 + 3 pl.

Traver, J. R., 1960: Uruguayan mayflies. Family Leptophlebiidae. Part III. - Revista de la Sociedad Uruguaya de Entomologia 4:73-132.

Traver, J. R., 1963: Uruguayan mayflies. Family Leptophlebiidae. Part IV. - Revista de la Sociedad Uruguaya de Entomologia 5:25-31.

Traver, J. R., 1964: A new species of *Thraulodes* from Uruguay (Ephemeroptera: Family Leptophlebiidae). Revista de la Sociedad Uruguaya de Entomologia 6:33-37.

Traver, J. R., 1971: Four new species of Neotropical *Baetis* (Ephemeroptera: Baetidae). – Proceedings of the Entomological Society of Washington 73:58-63.

Traver, J. R., and G. F. Edmunds, Jr., 1967: A revision of the genus *Thraulodes* (Ephemeroptera: Leptophlebiidae). – Miscellaneous Publications of the Entomological Society of America 5:349-395.

Traver, J. R., and G. F. Edmunds, Jr., 1968: A revision of the Baetidae with spatulate-clawed nymphs (Ephemeroptera). Pacific Insects 10:629-677.

Ulmer, G., 1904: Ephemeriden. – Hamburger Magalhaensische Sammelreise. 1904:1-8 + 1 pl.

Ulmer, G., 1920a (1919): Neue Ephemeropteren. Archiv für Naturgeschichte 85A(2):1-80.

Ulmer, G., 1920b: Übersicht über die Gattungen der Ephemeropteren, nebst Bemerkungen über einzelne Arten. - Stettiner entomologische Zeitung 81:97-144.

Ulmer, G., 1920c: Über die Nymphen einiger exotischer Ephemeropteren. – Festschrift für Zschokke 25:1-25.

Ulmer, G., 1921: Über einige Ephemeropteren-Typen älterer Autoren. - Archiv für Naturgeschichte 87A(6):229-267.

Ulmer, G., 1924: Einige alte und neue Ephemeropteren. – Konowia 3:23-37.

Ulmer, G., 1930: Key to the genera of Ephemerida. – Bulletin of the Department of Biology, Yenching University 1(3):1-18.

Ulmer, G., 1932: Bemerkungen über die seit 1920 neu aufgestellten Gattungen der Ephemeropteren. - Stettiner entomologische Zeitung 93:204-219.

Ulmer, G., 1938: Chilenische Ephemeropteren, hauptsächlich aus dem Deutschen Entomologischen Institut, Berlin-Dahlem. – Arbeiten über morphologische und taxonomische Entomologie aus Berlin-Dahlem 5(2):85-108.

Ulmer, G., 1942: Alte und neue Eintagsfliegen (Ephemeropteren) aus Süd- und Mittelamerika. Part 1. - Stettiner entomologische Zeitung 103:98-128.

Ulmer, G., 1943: Alte und neue Eintagsfliegen (Ephemeropteren) aus Süd- und Mittelamerika. Part 2. - Stettiner entomologische Zeitung 104:14-46.

Walker, F., 1853: Ephemerinae. List of the specimens of neuropterous insects in the collection of the British Museum 3:533-585.

Walker, F., 1860: Characters of undescribed Neuroptera in the collection of W. W. Saunders, Esq., F. R. S. – Transactions of the Entomological Society of London 5:176-199.

Waltz, R. D., and W. P. McCafferty, 1985: *Moribaetis*: a new genus of Neotropical Baetidae (Ephemeroptera). – Proceedings of the Entomological Society of Washington 87(1):239-251.

Waltz, R. D., and W. P. McCafferty, 1987a: Revision of the genus *Cloeodes* Traver (Ephemeroptera: Baetidae). – Annals of the Entomological Society of America 80(2):191-207.

Waltz, R. D., and W. P. McCafferty, 1987b: Generic revision of *Cloeodes* and description of two new genera (Ephemeroptera: Baetidae). – Proceedings of the Entomological Society of Washington 89(1):177-184.

Waltz, R. D., and W. P. McCafferty, 1987c: New genera of Baetidae for some Nearctic species previously included in *Baetis* Leach (Ephemeroptera). – Annals of the Entomological Society of America 80:667-670.

Waltz, R. D., and W. P. McCafferty, 1999: Additions to the taxonomy of *Americabaetis* (Ephemeroptera: Baetidae): *A lugoi* n. sp., adult of *A. robacki*, and key to larvae. – Entomological News 110(1):39-44.

Wang, T. Q., R. W. Sites, and W. P. McCafferty, 1998: Two new species of *Leptohyphes* (Ephemeroptera: Leptohyphidae) from Ecuador. – Florida Entomologist 81(1):68-75.

Weyenbergh, H., 1883: Bijdrage tot de kennis der Suid-Amerikaansche Ephemeriden. – Tijdschrift Entomologie 26:159-174.

Wiersema, N. A., and W. P. McCafferty, 2000: Generic revision of the North and Central American Leptohyphidae (Ephemeroptera: Pannota). – Transactions of the American Entomological Society 126(3-4):337-371.

Index